Lecture Notes in Computer Science 12342

More information about this series at http://www.springer.com/series/7409

Zhisheng Huang · Wouter Beek ·
Hua Wang · Rui Zhou ·
Yanchun Zhang (Eds.)

Web Information Systems Engineering – WISE 2020

21st International Conference
Amsterdam, The Netherlands, October 20–24, 2020
Proceedings, Part I

 Springer

Editors
Zhisheng Huang
VU Amsterdam
Amsterdam, The Netherlands

Wouter Beek (iD)
VU Amsterdam
Amsterdam, The Netherlands

Hua Wang (iD)
Victoria University
Melbourne, VIC, Australia

Rui Zhou (iD)
Swinburne University of Technology
Hawthorn, VIC, Australia

Yanchun Zhang (iD)
Victoria University
Melbourne, VIC, Australia

ISSN 0302-9743   ISSN 1611-3349   (electronic)
Lecture Notes in Computer Science
ISBN 978-3-030-62004-2   ISBN 978-3-030-62005-9   (eBook)
https://doi.org/10.1007/978-3-030-62005-9

LNCS Sublibrary: SL3 – Information Systems and Applications, incl. Internet/Web, and HCI

This Springer imprint is published by the registered company Springer Nature Switzerland AG
The registered company address is: Gewerbestrasse 11, 6330 Cham, Switzerland

Preface

Welcome to the proceedings of the 21st International Conference on Web Information Systems Engineering (WISE 2020), held in Amsterdam and Leiden, The Netherlands, during October 20–24, 2020. The series of WISE conferences aims to provide an international forum for researchers, professionals, and industrial practitioners to share their knowledge in the rapidly growing area of web technologies, methodologies, and applications. The first WISE event took place in Hong Kong, China (2000). Then the trip continued to Kyoto, Japan (2001); Singapore (2002); Rome, Italy (2003); Brisbane, Australia (2004); New York, USA (2005); Wuhan, China (2006); Nancy, France (2007); Auckland, New Zealand (2008); Poznan, Poland (2009); Hong Kong, China (2010); Sydney, Australia (2011); Paphos, Cyprus (2012); Nanjing, China (2013); Thessaloniki, Greece (2014); Miami, USA (2015); Shanghai, China (2016); Pushchino, Russia (2017); Dubai, UAE (2018); Hong Kong, China (2019); and this year, WISE 2020 was held in Amsterdam and Leiden, The Netherlands, supported by Vrije Universiteit Amsterdam.

A total of 190 research papers were submitted to the conference for consideration, and each paper was reviewed by at least three reviewers. Finally, 37 submissions were selected as regular papers (with an acceptance rate of 20% approximately), plus 44 as short papers. The research papers cover the areas of network embedding, graph neural network, social network, graph query, knowledge graph and entity linkage, spatial temporal data analysis, service computing, cloud computing, information extraction, text mining, security and privacy, recommender systems, database system, workflow, and data mining and applications.

In addition, special thanks are due to the members of the International Program Committee and the external reviewers for a rigorous and robust reviewing process. We are also grateful to Vrije Universiteit Amsterdam, Springer Nature, Atlantis Press, Ztone, Triply, and the International WISE Society for supporting this conference. The WISE Organizing Committee is also grateful to the workshop organizers for their great efforts to help promote web information system research to broader domains.

We expect that the ideas that have emerged in WISE 2020 will result in the development of further innovations for the benefit of scientific, industrial, and social communities.

October 2020

Zhisheng Huang
Wouter Beek
Hua Wang
Rui Zhou
Yanchun Zhang

Organization

General Co-chairs

Yanchun Zhang — Victoria University, Australia
Frank van Harmelen — Vrije Universiteit Amsterdam, The Netherlands
Marek Rusinkiewicz — New Jersey Institute of Technology, USA

Program Co-chairs

Zhisheng Huang — Vrije Universiteit Amsterdam, The Netherlands
Wouter Beek — Triply, The Netherlands
Hua Wang — Victoria University, Australia

Workshop Co-chairs

Rui Zhou — Swinburne University of Technology, Australia
Haiyuan Wang — Ztone Beijing, China

Tutorial and Panel Co-chairs

Kamal Karlapalem — International Institute of Information Technology, India
Yunjun Gao — Zhejiang University, China

Industry Chair

Wouter Beek — Triply, The Netherlands

Demo Chair

Yi Cai — South China University of Technology, China

Sponsor Chair

Ran Dang — Atlantis Press, France

Finance Chair

Qing Hu — Ztone International BV, The Netherlands

Local Arrangement Co-chairs

Ting Liu   Vrije Universiteit Amsterdam, The Netherlands
Xu Wang   Vrije Universiteit Amsterdam, The Netherlands
Wei Wei   Holiday Inn Leiden, The Netherlands

Publication Chair

Rui Zhou   Swinburne University of Technology, Australia

Publicity Co-chairs

Panagiotis Bouros   Johannes Gutenberg University of Mainz, Germany
An Liu   Soochow University, China
Wen Hua   The University of Queensland, Australia

Website Co-chairs

Haiyuan Wang   Ztone Beijing, China
Di Wang   Ztone Beijing, China

WISE Steering Committee Representative

Qing Li   The Hong Kong Polytechnic University, Hong Kong

Program Committee

Karl Aberer   EPFL, Switzerland
Marco Aiello   University of Stuttgart, Germany
Bernd Amann   LIP6, Sorbonne Université, France
Chutiporn Anutariya   Asian Institute of Technology, Thailand
Nikolaos Armenatzoglou   Amazon, USA
Wouter Beek   Vrije Universiteit Amsterdam, The Netherlands
Devis Bianchini   University of Brescia, Italy
Xin Cao   University of New South Wales, Australia
Jinli Cao   La Trobe University, Australia
Tsz Nam Chan   The Hong Kong Polytechnic University, Hong Kong
Jiefeng Cheng   The Chinese University of Hong Kong, Hong Kong
Dickson K. W. Chiu   The University of Hong Kong, Hong Kong
Dario Colazzo   LAMSADE, Université Paris-Dauphine, France
Alexandra Cristea   Durham University, UK
Valeria De Antonellis   University of Brescia, Italy
Anton Dignös   Free University of Bozen-Bolzano, Italy
Lei Duan   Sichuan University, China
Yixiang Fang   University of New South Wales, Australia
Yunjun Gao   Zhejiang University, China

Daniela Grigori	LAMSADE, Université Paris-Dauphine, France
Tobias Grubenmann	University of Bonn, Germany
Hakim Hacid	Zayed University, UAE
Jiafeng Hu	Google, USA
Haibo Hu	The Hong Kong Polytechnic University, Hong Kong
Zhisheng Huang	Vrije Universiteit Amsterdam, The Netherlands
Xin Huang	Hong Kong Baptist University, Hong Kong
Jyun-Yu Jiang	University of California, Los Angeles, USA
Panos Kalnis	King Abdullah University of Science and Technology, Saudi Arabia
Verena Kantere	University of Ottawa, Canada
Georgia Kapitsaki	University of Cyprus, Cyprus
Panagiotis Karras	Aarhus University, Denmark
Kyoung-Sook Kim	National Institute of Advanced Industrial Science and Technology (AIST), Japan
Hong Va Leong	The Hong Kong Polytechnic University, Hong Kong
Jianxin Li	Deakin University, Australia
Hui Li	Xiamen University, China
Kewen Liao	Australian Catholic University, Australia
An Liu	Soochow University, China
Guanfeng Liu	Macquarie University, Australia
Siqiang Luo	Harvard University, USA
Fenglong Ma	Penn State University, USA
Hui Ma	Victoria University of Wellington, New Zealand
Jiangang Ma	Federation University, Australia
Yun Ma	City University of Hong Kong, Hong Kong
Abyayananda Maiti	Indian Institute of Technology Patna, India
Silviu Maniu	Université Paris-Sud, France
Yannis Manolopoulos	Open University of Cyprus, Cyprus
George Papastefanatos	Information Management Systems Institute, Athena Research Center, Greece
Kostas Patroumpas	Information Management Systems Institute, Athena Research Center, Greece
Dimitris Plexousakis	Institute of Computer Science, FORTH, Greece
Nicoleta Preda	Université Paris-Saclay, France
Dimitris Sacharidis	Vienna University of Technology, Austria
Heiko Schuldt	University of Basel, Switzerland
Caihua Shan	The University of Hong Kong, Hong Kong
Jieming Shi	National University of Singapore, Singapore
Kostas Stefanidis	University of Tampere, Finland
Stefan Tai	TU Berlin, Germany
Bo Tang	Southern University of Science and Technology, China
Chaogang Tang	China University of Mining and Technology, China
Xiaohui Tao	University of Southern Queensland, Australia
Dimitri Theodoratos	New Jersey Institute of Technology, USA
Hua Wang	Victoria University, Australia

Jin Wang	University of California, Los Angeles, USA
Lizhen Wang	Yunnan University, China
Haiyuan Wang	Beijing University of Technology, China
Hongzhi Wang	Harbin Institute of Technology, China
Xin Wang	Tianjin University, China
Shiting Wen	Zhejiang University, China
Mingjun Xiao	University of Science and Technology of China, China
Carl Yang	University of Illinois at Urbana-Champaign, USA
Zhenguo Yang	City University of Hong Kong, Hong Kong
Hongzhi Yin	University of Queensland, Australia
Sira Yongchareon	Auckland University of Technology, New Zealand
Demetrios Zeinalipour-Yazti	University of Cyprus, Cyprus
Yanchun Zhang	Victoria University, Australia
Detian Zhang	Jiangnan University, China
Jilian Zhang	Jinan University, China
Yudian Zheng	Twitter, USA
Rui Zhou	Swinburne University of Technology, Australia
Lihua Zhou	Yunnan University, China
Feida Zhu	Singapore Management University, Singapore
Yi Zhuang	Zhejiang Gongshang University, China

Additional Reviewers

Olayinka Adeleye	Haridimos Kondylakis	Xiangchen Song
Ebaa Alnazer	Kyriakos Kritikos	Wenya Sun
Yixin Bao	Christos Laoudias	Xiangguo Sun
Dong Chen	Rémi Lebret	Pei-Wei Tsai
Hongxu Chen	Bing Li	Fucheng Wang
Tong Chen	Veronica Liesaputra	Jun Wang
Xuefeng Chen	Ting Liu	Qinyong Wang
Constantinos Costa	Tong Liu	Xu Wang
Alexia Dini Kounoudes	Xin Liu	Xin Xia
Thang Duong	Steven Lynden	Yuxin Xiao
Vasilis Efthymiou	Stavros Maroulis	Yanchao Tan
Negar Foroutan	Grace Ngai	Costas Zarifis
Eugene Fu	Rebekah Overdorf	Chrysostomos Zeginis
Ilche Georgievski	Xueli Pan	Shijie Zhang
Yu Hao	Jérémie Rappaz	Zichen Zhu
Wenjie Hu	Mohammadhadi Rouhani	Nikolaos Zygouras
Xiaojiao Hu	Brian Setz	
Firas Kassawat	Panayiotis Smeros	

Contents – Part I

Social Network

Graph Query

Knowledge Graph and Entity Linkage

Service Computing and Cloud Computing

Contents – Part II

Security and Privacy

Recommender Systems

Database System and Workflow

Data Mining and Applications

Network Embedding

Higher-Order Graph Convolutional Embedding for Temporal Networks

Xian Mo[1], Jun Pang[2], and Zhiming Liu[1($\boxtimes$)]

[1] College of Computer and Information Science,
Southwest University, Chongqing 400715, China
zhimingliu88@swu.edu.cn
[2] Faculty of Science, Technology and Medicine,
Interdisciplinary Centre for Security, Reliability and Trust,
University of Luxembourg, 4364 Esch-sur-Alzette, Luxembourg

Abstract. Temporal networks are networks that edges evolve over time. Network embedding is an important approach that aims at learning low-dimension latent representations of nodes while preserving the spatial-temporal features for temporal network analysis. In this paper, we propose a spatial-temporal higher-order graph convolutional network framework (ST-HN) for temporal network embedding. To capture spatial-temporal features, we develop a truncated hierarchical random walk sampling algorithm (THRW), which randomly samples the nodes from the current snapshot to the previous one. To capture hierarchical attributes, we improve upon the state-of-the-art approach, higher-order graph convolutional architectures, to be able to aggregate spatial features of different hops and temporal features of different timestamps with weight, which can learn mixed spatial-temporal feature representations of neighbors at various hops and snapshots and can well preserve the evolving behavior hierarchically. Extensive experiments on link prediction demonstrate the effectiveness of our model.

Keywords: Temporal networks · Representation learning · Higher-order graph convolutional network · Spatial-temporal features · Link prediction

1 Introduction

In recent years, network science has become very popular for modeling complex systems and is applied to many disciplines, such as biological networks [4], traffic networks [14], and social networks [17]. Networks can be represented graphically: $G = \langle V, E \rangle$, where $V = \{v_1, \ldots, v_n\}$ represents a set of nodes, and n is the number of nodes in the network, and $E \subseteq \{V \times V\}$ represents a set of links (edges). However, in the real world, most networks are not static but evolve with time, and such networks are called temporal networks [2]. A temporal network can be defined as $G_t = \langle V, E_t \rangle$, which represents a network $G = \langle V, E \rangle$ evolving over

© Springer Nature Switzerland AG 2020
Z. Huang et al. (Eds.): WISE 2020, LNCS 12342, pp. 3–15, 2020.
https://doi.org/10.1007/978-3-030-62005-9_1

time and generates a sequence of snapshots $\{G_1, \ldots, G_T\}$, where $t \in \{1, \ldots, T\}$ represents the timestamps. The key point for temporal network analysis is how to learn useful temporal and spatial features from network snapshots at different time points [5]. One of the most effective temporal network analysis approaches is temporal network embedding, which aims to map each nodes of each snapshot of the network into a low-dimensional space. Such a temporal network embedding method is proved to be very effective in link prediction, classification and network visualisation [12]. However, one of the biggest challenges in temporal network analysis is to reveal the spatial structure at each timestamp and the temporal property over time [5].

Many network embedding methods have been proposed in the past few years [4]. DeepWalk [10] adopt neural network to learn the representation of nodes. GraphSAGE [7] leverages node feature information to efficiently generate node embeddings for previously unseen data. But both methods focus on static networks. In order to obtain temporal network embedding, the following methods have been proposed in the literature. BCGD [17] only captures the spatial features, and LIST [16] and STEP [2] capture the spatial-temporal features by matrix decomposition. However, they cannot represent the highly nonlinear features [13] due to the fact that they are based on matrix decomposition. At present, the emergence of deep learning techniques brings new insights into this field. tNodeEmbed [12] learns the evolution of a temporal network's nodes and edges over time, while DCRNN [8] proposes a diffusion convolutional recurrent neural network to captures the spatio-temporal dependencies. To achieve effective traffic prediction, STGCN [14] replaces regular convolutional and recurrent units that integrating graph convolution and gated temporal convolution. The flexible deep embedding approach (NetWalk) [15] utilises an improved random walk to extract the spatial and temporal features of the network. More recently, DySAT [11] computes node representations through joint self-attention along with the two dimensions of the structural neighborhood and temporal dynamics, and dyngraph2vec [6] learns the temporal transitions in the network using a deep architecture composed of dense and recurrent layers. However, these methods are not considered learning mixed spatial-temporal feature representations of neighbours at various hops and snapshots. Therefore, the representation ability of temporal networks is still insufficient.

To tackle the aforementioned problems, we propose in the current paper a spatial-temporal Higher-Order Graph Convolutional Network framework (ST-HN) for temporal network embedding hierarchically (the overview of ST-HN is described in Fig. 1). Some work [10] shows that extracting the spatial relation of each node can be used as a valid feature representation for each node. Moreover, the current snapshot topological structure of temporal networks is derived from the previous snapshot topology, it is necessary to combine the previous snapshot to extract the spatial-temporal features for the current snapshot. Inspired by these ideas, we develop a truncated hierarchical random walk sampling algorithm (THRW) to extract both spatial and temporal features of the network, which randomly samples the nodes from the current snapshot to the previous one and it

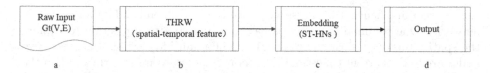

a   b   c   d

Fig. 1. Overview of the framework ST-HN. (a) Model input which is a temporal network G_t. (b) A spatial-temporal feature extraction layer which extracts $\Gamma(v, t_{Sta}, t_{End})$ of each node v of each network snapshot. (c) An embedding layer (ST-HNs) which maps each node of each snapshot to its D-dimensional representation. (d) Model output: Y_t with $t \in \{1, \ldots, T\}$, where each $Y_t \in R^{N \times k}$ (N is the number of nodes and K is the dimension) is the representation of G_t;

can well extract networks' spatial-temporal features. Because snapshots closer to the current snapshot contributes more to the current snapshot for temporal features. The THRW also incorporates a decaying exponential to assign longer walk length to more recent snapshots, which can better preserve the evolving behavior of temporal networks. Inspired by social networks, a user's friends contain some information, and friends of their friends have different information, all of which are useful, so we should consider these features together. Then, we improve upon the state-of-the-art approach, higher-order graph convolutional architectures, to embed the nodes, which can aggregate spatial-temporal features hierarchically and further reinforces the time-dependence for each snapshot. Besides, it can learn mixed spatial-temporal feature representations of neighbors at various hops and snapshots. Finally, we test the embedded vector's performance on link prediction task to verify its performance.

Our major contributions in this work can be summarised as follows. (1) We propose a model ST-HN to perform temporal network embedding. The model improves upon a Higher-Order Graph Convolutional Architecture (MixHop) [1] to hierarchically aggregate temporal and spatial features, which can better learn mixed spatial-temporal feature representations of neighbours at various hops and snapshots and can further reinforces the time-dependence for each network snapshot. (2) We also propose the THRW method for both spatial and temporal feature extraction. It adopts a random walk to sample neighbours for the current node v from the current snapshot to the previous snapshots, which can well extract networks' spatial-temporal features. It also incorporates a decaying exponential to assign longer walk length to more recent snapshots, which can better preserve the evolving behavior of temporal networks. (3) Extensive experiments on link prediction demonstrate that our ST-HN consistently outperforms a few state-of-the-art baseline models.

2   Related Work

In this section, we briefly summarise related work for temporal network embedding. Network embedding for static networks has been extensively studied (e.g., see a recent literature survey [4]). Since networks are continually evolving with time in real-life, hence it is necessary to study network embedding for temporal networks. One common approach is based on matrix decomposition to explore the spatial topology of the networks [17]. The main idea is that the closer two nodes are in the current timestamp, the more likely they are to form a link in the future timestamp. However, real-life networks are often evolving, methods considering only spatial information may have poor performance. There exist a few other methods focusing on both spatial and temporal evolution features such as STEP [2] and LIST [16]. STEP constructs a sequence of higher-order proximity matrices to capture the implicit relationships among nodes, while LIST defines the network dynamics as a function of time, which integrates the spatial topology and the temporal evolution. However, they have a limited ability to extract the correlation of high dimensional features since they are based on matrix decomposition. In recent years, neural network-based embedding methods have gained great achievements in link prediction and node classification [11]. tNodeEmbed [12] presents a joint loss function to learns the evolution of a temporal network's nodes and edges over time. DCRNN [8] adopts the encoder-decoder architecture, which uses a bidirectional graph random walk to model spatial dependency and recurrent neural network to capture the temporal dependencies. STGCN [14] integrates graph convolution and gated temporal convolution through Spatio-temporal convolutional blocks to capture spatio-temporal features. The DySAT model [11] stacks temporal attention layers to learn node representations, which computes node representations through joint self-attention along with the two dimensions of the structural neighborhood and temporal dynamics, while dyngraph2vec [6] learns the temporal transitions in the network using a deep architecture composed of dense and recurrent layers, which learns the structure of evolution in dynamic graphs and can predict unseen links. NetWalk [15] is an flexible deep embedding approach, and it uses an improved random walk to extract the spatial and temporal features. However, these methods are not considered to learn mixed spatial-temporal feature representations of neighbours at various hops and snapshots. Therefore, the representation ability of these methods for temporal networks is still insufficient.

3   Problem Formulation

We introduce some definitions and formally describe our research problem.

Definition 1 (Node W-walks neighbours). *Let $G = \langle V, E \rangle$ be a network. For a given node v, its W-walks neighbours are defined as the multi-set $N(v, W)$ containing all the nodes with W steps using a random walk algorithm on the network G.*

Definition 2. *Let $G_t = \langle V, E_t \rangle$ be a temporal network. For a node v, its all W-walks neighbours from time t_{Sta} up to time t_{End} (i.e., $t_{Sta} \leq t_{End}$) are defined as $\Gamma(v, t_{Sta}, t_{End}) = \bigcup_{t=t_{Sta}}^{t_{End}} \{N^t(v, W^t)\}$, where W^t represents the number of steps of random walk at timestamp t and $W^{t-1} = a * W^t$, where a represents the decaying exponential between 0 and 1, and $N^t(v, W^t)$ is a multi-set of W-walks neighbours of v in a network snapshot G_t where $t \in \{t_{Sta}, \ldots, t_{End}\}$.*

Algorithm 1: Truncated hierarchical random walk sampling (THRW)

Input : $G_t\langle V, E_t\rangle$: a temporal network;
$\qquad\qquad W^t$: the number of steps at snapshot t;
$\qquad\qquad N$: sampling temporal window size;
$\qquad\qquad a$: decaying exponential;
Output: $X[i]$ with $i \in \{1, \ldots, T\}$, where each $X[i]$ consists of the neighbour
$\qquad\qquad$ sets within the network snapshots of window size N for every node v;

1 **for** $i \in \{1, \ldots, T\}$ **do**
2 $\quad$ **if** $i - N \leq 0$ **then**
3 $\quad\quad$ **for** $v \in V$ **do**
4 $\quad\quad\quad$ $X[i].\text{add}(\Gamma(v, 1, i))$
5 $\quad\quad$ **end**
6 $\quad$ **else**
7 $\quad\quad$ **for** $v \in V$ **do**
8 $\quad\quad\quad$ $X[i].\text{add}(\Gamma(v, i - N + 1, i))$
9 $\quad\quad$ **end**
10 $\quad$ **end**
11 **end**

Temporal Network Embedding: For a temporal network $G_t = \langle V, E_t \rangle$, we can divide it evenly into a sequence of snapshots $\{G_1, \ldots, G_t\}$ by timestamp t. For each snapshot G_t, we aim to learn a mapping function $f^t : v_i \to R^k$, where $v_i \in V$ and k represents dimensions and $k \ll |V|$. The purpose of the function f^t is to preserve the similarity between v_i and v_j on the network structure and evolution patterns of a given network at timestamp t.

4 Our Method

In this section, we introduce our model ST-HN. We firstly propose a truncated hierarchical random walk sampling algorithm (THRW) to extract both spatial and temporal features for each node (Sect. 4.1). Then, we use the proposed ST-HNs to embed the node for each snapshot (Sect. 4.2).

4.1 Spatial-Temporal Feature Extraction

We propose the THRW algorithm (Algorithm 1) to sampling $\Gamma(v, t_{Sta}, t_{End})$ for each node v. Algorithm 1 has three parameters: W^t the number of steps in random walks of a given node v at snapshot t, N a sampling window size defining how many previous network snapshot are taken into account when sampling v's nodes, and a is a decaying exponential defining the current snapshot to have more steps than the previous snapshot (see Definitions 1 and 2 for more details of these parameters). In Algorithm 1, each $X[i]$ represents the sets of nodes for all nodes in the network at time i, i.e, $\Gamma(v, i - N + 1, i)$, thus X contains all such sets, i.e., $X[i]$ with $i \in \{1, \ldots, T\}$, for the complete network snapshot

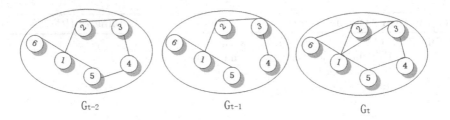

Fig. 2. The THRW algorithm: an illustrative example

sequence. The first layer loop is used to select snapshot at time i from the temporal network $G_t(V, E)$, and the purpose is to traverse all the snapshots. The algorithm extracts spatial and temporal features from N snapshots to better simulate the evolutionary behavior of the temporal network. If the number of the previous snapshots is greater than N, sampling neighbour nodes is performed between $i - N + 1$ and i snapshots. Otherwise, it is only sampled from the very first snapshot to the current snapshot i. The second loop is to sample the sets of neighbours for every node in the network. As shown in Fig. 2, there are 6 nodes in the temporal network, and we extract features with the previous 2 snapshots for the node 1 in the current snapshot G_t. If we set $W^t = 8$ and $a = 0.5$, then for the snapshot G_t the walk length is 8 and the multi-set of sampled nodes is $\{1,2,3,1,5,4,3,1\}$. For the snapshot G_{t-1}, the walk length is 4 and the multi-set of sampled nodes is $\{1,2,3,2\}$, and for the snapshot G_{t-2} the walk length is 2 and the multi-set of sampled nodes is $\{1,2\}$. Then we combine the previous 2 snapshots sampled nodes as the final features for the node 1 of the snapshot G_t: $\{1,2,3,1,5,4,3,1,1,2,3,2,1,2\}$.

4.2   ST-HNs

Our proposed model refers to Higher-Order Graph Convolutional Architectures via Sparsified Neighborhood Mixing (MixHop) [1], which mixes feature representations of neighbors at various distances to learn neighborhood mixing relationships. More precisely, it combine 1-hop, 2-hop, ..., neighbours in distinct feature spaces such that it can effectively aggregate features of different hops in the network. Each layer of MixHop is formally defined as follows:

$$H^{(i+1)} = \|_{j=0}^{P} \sigma(L^j H^{(i)} W_j^{(i)}) \tag{1}$$

where $H^{(i)} \in R^{N \times d_i}$ and $H^{(i+1)} \in R^{N \times d_{i+1}}$ are input and output of layer i. N represents the number of network nodes. $W_j^{(i)} \in R^{d_i \times d_{i+1}}$ is a weight matrix. σ is a nonlinear activation function. L^j is a symmetrically normalized Laplace matrix and can be constructed by $L^j = D^{-\frac{1}{2}} A D^{\frac{1}{2}}$. A is an adjacency matrix with self-connections: $A = A + I_N$, where I_N is a self-connections matrix. D is a diagonal degree matrix with $D_{mm} = \sum_n A_{mn}$, where m represents a row and n represents a column, and j is the powers of L, ranging from 0 to P. L^j

denotes the matrix L multiplied by itself j times, and $\|$ denotes column-wise concatenation. For example, L^2 represents 2-hop neighbours in feature spaces. The MixHop model can learn features in difference feature space between immediate and further neighbours, hence it has a better representational capability. However, for temporal networks, MixHop cannot capture their temporal features. Our model, ST-HNs, improves upon MixHop and utilizes an aggregator to learn mixed spatial-temporal feature representations of neighbours at various hops and snapshots.

To solve the problem above, we propose Spatial-Temporal Higher-Order Graph Convolutional Temporal Network (ST-HNs), where each node of each snapshot aggregate spatial and temporal features from their 1-hop neighbours to P-hop neighbours through the previous A snapshot. In this way, it can learn mixed spatial-temporal feature representations of neighbours at various hops and snapshots. Our model is formally defined as follows:

$$
H^i = \begin{cases} \| \{\oplus(Y_{t-A}, \ldots, Y_{t-1}), X_t\} & \text{if } i = 0 \\ \|_{j=0}^{P} \sigma(L^j H^{(i-1)} W_j^{(i-1)}) & \text{if } i \in [1, \ldots, M] \end{cases} \tag{2}
$$

where P is the number of powers defining aggregation features of different hops in the network, M is the number of layers, $\|$ represents concatenation operate, A is the temporal window size defining the number of the snapshots used to aggregate spatial-temporal features, and the operator $\oplus$ represents an aggregator. We adopt the GRU aggregator to aggregate the temporal features of the previous A snapshots, which can further reinforce the time-dependence for each snapshot of temporal networks. The GRU aggregator is based on an GRU architecture [3]. The input is a number of network snapshots $\{Y_{t-A}, \ldots, Y_{t-2}\}$ and the ground truth is the most recent snapshot Y_{t-1}, where the $Y_{t-1} \in R^{N \times k}$ is the representation of G_{t-1}, where N represents the number of nodes, k represents dimensions and $k \ll N$. Y_1 can be obtained from the MixHop (Eq. 1) with M layers. If the number of the previous snapshots is greater than A, aggregation is performed between $t - A$ and $t - 1$ snapshots. Otherwise, it is only aggregate from the very first snapshot to the snapshot $t - 1$. We use the input and ground truth to train the GRU model and update parameter. After training, we shift the window one step forwards to obtain the representation of temporal features Y. The Y and X_t are concatenate to get the aggregated spatial-temporal feature H^0, where $X_t \in R^{N \times d_0}$ is a feature matrix of each node of the snapshot at time t and it is obtained by the THRW algorithm (see Algorithm 1). The model input is the aggregated spatial-temporal feature H^0. Through M times ST-HNs layers, the model output is $H^M = Z \in R^{N \times d_M}$, where Z is a learned feature matrix. The Z contains mixed spatial-temporal feature representations of neighbours at various hops and snapshots and it can better preserve the temporal and spatial features of temporal networks. For each snapshot, to learn representations, $z_u, u \in \mathbf{V}$, in a fully unsupervised setting, we apply a graph-based loss function to tune the weight matrices $w_j^i, j \in (1, \ldots, P), i \in (1, \ldots, M)$. The loss function

Table 1. The statistics of four temporal networks.

Network	#Nodes	#Links	Clustering coefficient	Format
Hep-Ph	28,093	4,596,803	28.0%	Undirected
Digg	30,398	87,627	0.56%	Directed
Facebook wall posts	46,952	876,993	8.51%	Directed
Enron	87,273	1,148,072	7.16%	Directed

encourages nearby nodes to have similar representations and enforces that the representations of disparate nodes are highly distinct:

$$Loss = \frac{1}{N} \sum_{u=1}^{N} (z_u - Mean(Adj(z_u)))^2 \qquad (3)$$

where N is the number of nodes, $Adj(z_u)$ obtains neighborhood node representation of u. $Mean$ means average processing.

5   Experiments

We describe the datasets and baseline models, and present the experimental results to show ST-HN's effectiveness for link prediction in temporal networks.

5.1   Datasets and Baseline Models

We select four temporal networks from different domains in the KONECT project.[1] All networks have different sizes and attributes. Their statistic properties present in Table 1. Hep-Ph dataset is the collaboration network of authors of scientific papers. Nodes represent authors and edges represent common publications. For our experiment, we select 5 years (1995–1999) and denote them as H_1 to H_5. Digg dataset is the reply network of the social news website. Nodes are users of the website, and edges denote that a user replied to another user. We evenly merged it into five snapshots by day and denote it as D_1 to D_5. Facebook wall posts dataset is the directed network of posts to other user's wall on Facebook. The nodes represent Facebook users, and each directed edge represents one post. For our experiment, we combined 2004 and 2005 data into one network snapshot and defined it as W_1. The rest of the data is defined as W_2 to W_5 by year and each snapshot contains a one-year network structure. Enron dataset is an email network that sent between employees of Enron between 1999 and 2003. Nodes represent employees and edges are individual emails. We select five snapshots in every half year during 2000-01 to 2002-06 and denote them as E_1 to E_5. For our experiments, the last snapshot was used as ground-truth of network inference and the other snapshots was used to train the model.

---

[1] http://konect.uni-koblenz.de/.

Table 2. Prediction results (AUC values).

Model	Hep-Ph	Digg	Facebook wall posts	Enron
BCGD	0.60	0.68	0.74	0.64
LIST	0.63	0.73	0.72	0.67
STEP	0.61	0.74	0.76	0.71
NetWalk	0.69	0.71	0.74	0.72
ST-HN	**0.74**	**0.81**	**0.83**	**0.79**

Baselines: We compare ST-HNs with the following baseline models: LIST [16] describes the network dynamics as a function of time, which integrates the spatial and temporal consistency in temporal networks; BCGD [17] proposes a temporal latent space model for link prediction, which assumes two nodes are more likely to form a link if they are close to each other in their latent space; STEP [2] utilises a joint matrix factorisation algorithm to simultaneously learn the spatial and temporal constraints to model network evolution; NetWalk [15] model updates the network representation dynamically as the network evolves by clique embedding, and it focuses on anomaly detection, and we adopt its representation vector to predict the link.

Parameter Settings: In our experiments, we randomly generate the number of non-linked edges smaller than twice linked edges to ensure data balance [17]. For the Hep-Ph and Digg, we set the embedding dimensions as 256. For the Facebook wall posts dataset and Enron, we set the embedding dimensions as 512. For different datasets, the parameters for baselines are tuned to be optimal. Other settings include: the learning rate of the model is set as 0.0001; the number of powers P is set as 3; the number of steps W is set as 200; the temporal window A is set as 3; the sampling temporal window N is set as 3; the M for layers of the ST-HNs is set as 5; the decaying exponential a is set as 0.6. For the result of our experiments, we adopt the area under the receiver operating curve (AUC) to evaluate predictive power for future network behaviors of different methods and we carry out five times independently and reported the average AUC values for each dataset.

5.2   Experimental Results

We compare the performance of our proposed model on the four datasets with four baselines for link prediction. We use ST-HNs to embed each node into a vector at each snapshot. Then we use the GRU to predict the vector representation of each node for the very last snapshot. In the end, we use the obtained representations to predict the network structure similar to [11]. Table 2 summarises the AUCs of applying different embedding methods for link prediction over the four datasets. Compared with other models, our model, ST-HN, achieves the best performance. Essentially, we use the THRW algorithm to sampling $\Gamma(v, t_{Sta}, t_{End})$

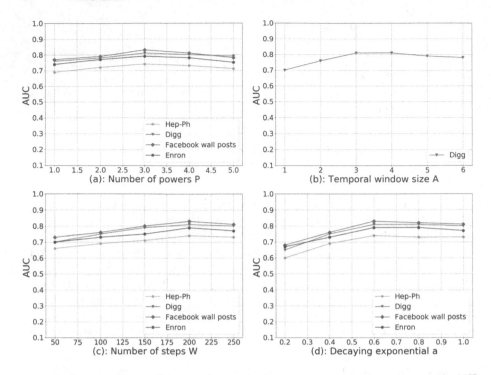

Fig. 3. Results of parameter sensitivity analysis.

for each node v, which can better capture both spatial and temporal features for each node. It also incorporates a decaying exponential to assign longer walk length to more recent snapshots, which can better preserve the evolving behavior of temporal networks. Then we apply ST-HNs to embed spatial-temporal features of nodes, which can aggregate spatial-temporal features hierarchically and learn mixed spatial-temporal feature representations of neighbours at various hops and snapshots and can further reinforcing the time-dependence for each snapshot. In this way, our embedding method can well preserve the evolving behaviour of the networks.

5.3   Parameter Sensitivity Analysis

We further perform parameter sensitivity analysis in this section, and the results are summarised in Fig. 3. Specifically, we estimate how different the number of powers P and the temporal window A and the number of steps W and the decaying exponential a can affect the link prediction results.

The Number of Powers P: We vary the power from 1 to 5, to prove the effect of varying this parameter. As P increases from 1 to 3, performance continues to increase due to the mixing of neighbour information with more hops in the current node. The best result is at $P = 3$ (the detail is described in Fig. 3a).

After which the performance decreases slightly or remains unchanged, while P continues to increase. The reason might be that the further away from the current node, the less information there is about the current node.

Temporal Window Size A: Due to the fact that the Digg dataset contains sixteen-days records, we split it by day and it will generates 16 snapshots. Since the Digg dataset has more snapshots than others, we select 7 snapshots to analyze the parameter A. We vary the window size from 1 to 6 to check the effect of varying this parameter. The results show that the best results are obtained when $A = 3$. The reason might be that the closer snapshot is to the current snapshot, the more information can be captured about the current snapshot. But the accuracy no longer increases, when A continuously increases (see Fig. 3b).

The Number of Steps W and the Decaying Exponential a: Since W and a jointly determine the sampling size of the current node, we analyse these two parameters together. When analysing W, a is set to 0.6. The experimental results show that the performance is the best when $W = 200$. As W increases from 50 to 200, performance continues to increase. The reason might be that the more nodes sampled, the more features of the current node are included. But the accuracy no longer increases, when W continuously increases. When analyzing a, W is set to 200. The experimental results show that the performance is the best when $a = 0.6$. As a increases from 0.2 to 0.6, performance continues to increase. The reason might be that the previous snapshot contains useful features for the current snapshot, as the number of sampled features increases. After which the performance decreases slightly or remains unchanged, while a continue to increase (the detail is described in Fig. 3c and 3d).

6 Conclusion

We have proposed a new and effective framework, ST-HN, for temporal network embedding. Through intensive experiments we demonstrated its effectiveness in link prediction. In particular, we proposed the THRW algorithm to extract spatial-temporal features in each snapshot to model network evolution. Moreover, we proposed the ST-HNs framework to embed nodes in the network, which can learn mixed spatial-temporal feature representations of neighbours at various hops and snapshots and can well preserve the evolving behavior of the network hierarchically and can further reinforcing the time-dependence for each snapshot. Our future work will study performance improvements from different aggregation methods and aggregate other information [9] in temporal networks.

Acknowledgements. This work has been supported by the Chongqing Graduate Research and Innovation Project (CYB19096), the Fundamental Research Funds for the Central Universities (XDJK2020D021), the Capacity Development Grant of Southwest University (SWU116007), the China Scholarship Council (202006990041), and the National Natural Science Foundation of China (61672435, 61732019, 61811530327).

References

1. Abu-El-Haija, S., et al.: MixHop: higher-order graph convolution architectures via sparsified neighborhood mixing. In: Proceedings of the 36th International Conference on Machine Learning, pp. 21–29 (2019)
2. Chen, H., Li, J.: Exploiting structural and temporal evolution in dynamic link prediction. In: Proceedings of the 27th ACM International Conference on Information and Knowledge Management, pp. 427–436. ACM (2018)
3. Cho, K., Van Merriënboer, B., Bahdanau, D., Bengio, Y.: On the properties of neural machine translation: encoder-decoder approaches. In: Proceedings of the 8th Workshop on Syntax, Semantics and Structure in Statistical Translation, pp. 103–111 (2014)
4. Cui, P., Wang, X., Pei, J., Zhu, W.: A survey on network embedding. IEEE Trans. Knowl. Data Eng. **31**(5), 833–852 (2018)
5. Cui, W., et al.: Let it flow: a static method for exploring dynamic graphs. In: Proceedings of the 2014 IEEE Pacific Visualization Symposium, pp. 121–128. IEEE (2014)
6. Goyal, P., Chhetri, S.R., Canedo, A.: dyngraph2vec: capturing network dynamics using dynamic graph representation learning. Knowl.-Based Syst. **187**, 104816 (2020)
7. Hamilton, W., Ying, Z., Leskovec, J.: Inductive representation learning on large graphs. In: Proceedings of the Advances in Neural Information Processing Systems, pp. 1024–1034 (2017)
8. Li, Y., Yu, R., Shahabi, C., Liu, Y.: Diffusion convolutional recurrent neural network: data-driven traffic forecasting. In: Proceedings of the 2018 International Conference on Learning Representations, pp. 1–16. IEEE (2018)
9. Pang, J., Zhang, Y.: DeepCity: a feature learning framework for mining location check-ins. In: Proceedings of the 11th International AAAI Conference on Web and Social Media, pp. 652–655 (2017)
10. Perozzi, B., Al-Rfou, R., Skiena, S.: DeepWalk: online learning of social representations. In: Proceedings of the 20th ACM SIGKDD International Conference on Knowledge Discovery and Data Mining, pp. 701–710. ACM (2014)
11. Sankar, A., Wu, Y., Gou, L., Zhang, W., Yang, H.: DySAT: deep neural representation learning on dynamic graphs via self-attention networks. In: Proceedings of the 13th International Conference on Web Search and Data Mining, pp. 519–527. ACM (2020)
12. Singer, U., Guy, I., Radinsky, K.: Node embedding over temporal graphs. In: Proceedings of the 2019 International Joint Conference on Artificial Intelligence, pp. 4605–4612. Morgan Kaufmann (2019)
13. Tenenbaum, J.B., De Silva, V., Langford, J.C.: A global geometric framework for nonlinear dimensionality reduction. Science **290**(5500), 2319–2323 (2000)
14. Yu, B., Yin, H., Zhu, Z.: Spatio-temporal graph convolutional networks: a deep learning framework for traffic forecasting. In: Proceedings of the 27th International Joint Conference on Artificial Intelligence, pp. 3634–3640. Morgan Kaufmann (2018)
15. Yu, W., Cheng, W., Aggarwal, C.C., Zhang, K., Chen, H., Wang, W.: NetWalk: a flexible deep embedding approach for anomaly detection in dynamic networks. In: Proceedings of the 24th ACM SIGKDD International Conference on Knowledge Discovery and Data Mining, pp. 2672–2681. ACM (2018)

16. Yu, W., Wei, C., Aggarwal, C.C., Chen, H., Wei, W.: Link prediction with spatial and temporal consistency in dynamic networks. In: Proceedings of the 26th International Joint Conference on Artificial Intelligence, pp. 3343–3349 (2017)
17. Zhu, L., Dong, G., Yin, J., Steeg, G.V., Galstyan, A.: Scalable temporal latent space inference for link prediction in dynamic social networks. IEEE Trans. Knowl. Data Eng. **28**(10), 2765–2777 (2016)

RolNE: Improving the Quality of Network Embedding with Structural Role Proximity

Qi Liang[1,2(✉)], Dan Luo[1,2], Lu Ma[1,2], Peng Zhang[1(✉)], MeiLin Zhou[1,2], YongQuan He[1,2], and Bin Wang[3]

[1] Institute of Information Engineering,
Chinese Academy of Sciences, Beijing, China
{liangqi,luodan,malu,pengzhang,zhoumeilin,heyongquan}@iie.ac.cn
[2] School of Cyber Security,
University of Chinese Academy of Sciences, Beijing, China
[3] AI Lab, Xiaomi Incy, Beijing, China
wangbin11@xiaomi.com

Abstract. The structural role of a node is an essential network structure information, which provides a better perspective to understand the network structure. In recent years, network embedding learning methods have been widely used in node classification, link prediction, and visualization tasks. Most network embedding learning algorithms attempt to preserve the neighborhood information of nodes. However, these methods are hard to recognize the structural role proximity of nodes. We propose a novel method, RolNE, which learns structural role proximity of nodes through clustering the degree vector of nodes and uses an aggregation function to learn node embedding that contains both neighborhood information and structural role proximity. Experiments on multiple datasets show that our algorithm outperforms other state-of-the-art baselines on downstream tasks.

Keywords: Network embedding · Social network analysis · Graph neural network

1 Introduction

Network describes complex information in daily life. For example, mail communication constitutes a social network between people [1], and transportation traffic between cities constitutes a transportation network. How to efficiently perform analysis tasks on large-scale networks, such as node classification and clustering [2], has been the research foundation and focus of this field. The current mainstream method is to use network representation learning to learn the features of the nodes in the network. The network embedding algorithm converts the network information into a low-dimensional dense real vector and used as the input of downstream machine learning algorithm [2,3].

© Springer Nature Switzerland AG 2020
Z. Huang et al. (Eds.): WISE 2020, LNCS 12342, pp. 16–28, 2020.
https://doi.org/10.1007/978-3-030-62005-9_2

At present, the work of network embedding learning algorithms mainly focuses on preserving the microstructure of the network, such as first-order proximity, second-order proximity, and high-order proximity between nodes. Deep-Walk [7] introduced word2vec [13] to the network embedding learning task, and more work [8,9,14,15,28] expanded the definition of the neighborhood from different ranges and captured the neighborhood information to improve DeepWalk. GraphSAGE [16] is an important algorithm for network embedding learning in recent years. This algorithm mainly learns to summarize the feature information from the local neighborhood of the node and then learns the node embedding. However, the structural information in the network not only has microscopic structures [3], but also contains mesoscopic structures, such as structural role proximity.

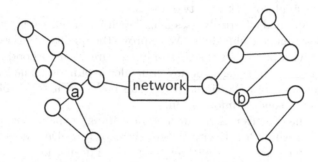

Fig. 1. An illustrative example of structural role proximity between the two nodes that are far apart in the network.

The structural role proximity describes the similarity between nodes that have similar roles in the network, such as the edge of a chain, the center of a star, and the bridge between two communities [3]. Especially in e-mail networks and transportation networks, node structure role is significant for characterizing nodes. For example, the secretaries of the two departments in the social network formed by e-mail connections, who are far apart in the network, indeed have similar structural roles. Unlike first-order proximity, second-order proximity, and high-order proximity, which capture the neighborhood information of nodes, the structural role proximity attempts to capture the similarity between nodes with the same structural role. As shown in Fig. 1, nodes a and b are far apart, but they have similar structural roles. Intuitively, we consider nodes with similar roles should have similar embeddings, but if two nodes are far apart in the network without common neighbors, embedding learning based on neighborhood information cannot capture the structural role similarity of nodes. If the classification of nodes in a network is more dependent on the structural role proximity, then the embedding that captures the proximity of structural roles will perform better.

Existing network embedding learning algorithms that preserve structural role similarity usually need to manually label topological characteristics [17], perform complex feature engineering, and then calculate the similarity score between nodes to obtain structural similarity between nodes [10]. When these algorithms define the structural roles of nodes that correspond to different topologies in the neighborhood of the local network, such as nodes on a chain, the center of a star in a discrete space, they need define such discrete roles in advance, which require domain expertise and manual inspection of the network structure.

We have designed a novel algorithm RolNE which uses a simple heuristic method to discover node structure role, and then jointly learns the characteristics of structural role proximity and neighborhood information of nodes through an aggregation function to enhance the quality of node embedding. We conclude two main contributions of our research.

First, we use a heuristic method to find nodes with similar structural roles that do not require manual labeling of the topological features in the network or calculate the similarity between two nodes.

Second, our method effectively uses neighbor information and structural role proximity to learn node embedding. We improve the quality of network embedding compared with previous ways that only capture neighborhood characteristics of nodes or just preserving structure equivalence of nodes. The learned node embeddings contain both the similarity of the nodes' structural roles and the node connection's sequence information.

We verified the effectiveness of our method through experiments on multiple networks. We compared RolNE with the classic algorithms DeepWalk, Node2vec, GraphSAGE, which are capturing neighborhood information, and struct2vec that learned structural identify information. Experimental results show that our algorithm has a better effect.

The arrangement of the remaining chapters of this article is as follows. The second part introduces the related work of network embedding learning in recent years. The third part presents the RolNE framework in detail. The fourth part introduces the experimental dataset, comparison algorithm, and experimental results. Finally, the fifth part summarizes and discusses this article.

2   Related Work

In recent years, network embedding learning technology has developed vigorously. Early work was mainly to obtain node embedding through matrix decomposition. LLE (Locally Linear Embedding) [4] assumes that the embedding of a node and its neighbors are located in a locally linear region of the manifold. LE (Laplacian Eigenmaps) [5] obtains the embedding of each node by solving the eigenvalues of the Laplacian matrix corresponding to the network. GraRep [6] decomposes a specific relationship matrix by SVD to perform feature dimensionality reduction to obtain node embedding. DeepWalk [7] makes the node embedding learning algorithm based on random walk develop rapidly. The algorithm uses a random walk to capture the structural relationship between vertices. DeepWalk algorithm first uses a random walk to generate sequences, and then uses skip-gram model to learn node embedding. The Node2vec [8] algorithm proposes a biased random walk algorithm to generate sequences based on DeepWalk and explores the neighbor structure more effectively. HARP [9] folded related nodes in the network to form super nodes and then used Deep-Walk and Node2vec to generate node embedding. LINE [11] is an edge modeling algorithm that uses the direct connection between nodes to describe the first-level similarity relationship, and the common neighbors of two nodes that are not directly connected to describe the second-level similarity relationship between the nodes. With the continuous expansion of the application range of deep learning, many network embedding learning methods using deep learning have appeared in recent years. DNGR [18] uses a random surfing model to capture network structure information directly and applies deep learning methods to the PPMI matrix. To enhance the robustness of the model, DNGR uses a stacked denoising autoencoder to learn multiple layers of embedding. SDNE [12] proposes a semi-supervised depth model with multiple layers of nonlinear functions to capture highly nonlinear network structures. GraphSAGE [16] uses graph neural networks to aggregate node neighborhood information to generate node embeddings. GCN [19] improves the idea of neighborhood aggregation and normalizes the information of each neighbor. This normalization is not a simple average value but introduces the differences between each neighbor, and reduce the weight of high degree neighbors. [26] use anonymous walking and Graph Anchor LDA to build topic models on graphics to reduce complexity and effectively generate structural topics.

The above method is mainly to preserve the neighborhood information of the node, but the structural role of the node is also essential information in Network. RolX [17] is a method that uses a network structure to identify the role of a node, this method requires enumerating the local structural characteristics of the node, such as degree, triangle count, and PageRank [21] score. SNS [20] uses orva [22] to help calculate the Graphlet Degree Vector of each node and then calculates the Euclidean distance between the vectors. Struct2vec encodes nodes with similar structural roles into a multi-layer graph and captures structural information by random walk. GraphWave [23] uses a thermal wavelet diffusion model to learn the structure information of nodes. GraphCSC [27] uses linkage information and centrality information to learn low-dimensional vector representations of network vertices. The embeddings learned by GraphCSC can preserve different central information of the node. The methods described above basically only learn part of the structural information of the node, such as neighborhood information or structural role information. The embedded information of the node is not comprehensive.

3   RolNE

The key idea of our algorithm is to aggregate feature information from the local neighborhood and the set of nodes with similar roles of the node. This method should have two characteristics: a) The distance of node embedding should be in strong correlation with the proximity of node structural roles, that is, nodes with similar structural roles should have similar embedding, and nodes with different structural roles should be far away. b) The embedding distance of nodes should be in strong correlation with the neighborhood information of the nodes, that is, the embedding distance of directly connected nodes or nodes with common neighbors should be relatively close.

We propose RolNE, which mainly has two steps. (1) Find the nodes with similar structural roles by clustering the node's degree vectors. As shown in Fig. 1, nodes a and b have the same degree, and the sum of the degrees of their neighbors is also roughly the same, and the two nodes have similar structural roles. Intuitively, we can use these two-dimensional vectors to determine whether nodes have similar structural roles. (2) We use the aggregation function to learn node embedding containing both neighborhood information and structural role information. The training process includes multiple iterations. In each iteration, the node will aggregate information from nodes in the local neighborhood and the nodes with the same structural role. As the number of iterations increases, the node will aggregate more information from a more in-depth and broader range.

Algorithm 1: RolNE Model Training

Require: $G=\{V,E\}$:a Graph with nodes and links;K: depth

   for all nodes $u \in V$ **do**

      $x(u) = \text{concat} (\ \text{degree}\ (u),\quad \sum \text{degree}(w, w \in N(u))$

   end for

   $R \leftarrow$ clustering $(X)x \in X$

   for each *iterator* $\in [1, 2, 3, ..., K]$ **do**

      for all nodes $u \in V$ **do**

         $h_{N(u)}^{k} \leftarrow \text{MEAN} \left(\{ h_v^{k-1}, \forall v \in N(u) \} \right)$

         $h_{R(u)}^{k} \leftarrow \text{MEAN} \left(\{ h_w^{k-1}, \forall w \in R(w) \} \right)$

         $h_u^{k} \leftarrow \sigma \left(W^k \cdot \text{CONCAT} \left(h_u^{k-1}, h_{N(u)}^{k}, h_{R(u)}^{k} \right) \right).$

      end for

      $h_u^k \leftarrow h_u^k / \left\| h_u^k \right\|_2, \forall u \in V$

   end for

   $z_u \leftarrow h_u^k \cdot \forall u \in V$

Algorithm 1 gives the main process of RolNE. As shown in Eq. 1, to discover the similar structural role nodes, for each node u in the graph, we calculate the degree vector $x(u)$ of the node. This degree vector contains two dimensions, the first dimension is the $degree(u)$ of the node; the second dimension is the sum of the degrees of the neighbors of the node $\sum \text{degree}(w, w \in N(u))$. After generating the degree vectors of all nodes, we use the clustering algorithm Kmeans [24] to cluster the nodes and then obtain clusters R of nodes with similar structural roles.

$$x(u) = \text{concat} \left(\text{degree}\ (u),\quad \sum \text{degree}(w, w \in N(u)) \right) \tag{1}$$

K represents the number of iterations of the aggregation operation, k represents the current step, and h_u^k represents the embedding of node u at step k.

For each node $u \in V$, we calculate the mean of the embedded sum of its directly connected neighbors $h_{N(u)}^k$.

$$h_{N(u)}^k \leftarrow \text{MEAN}\left(\{h_v^{k-1}, \forall v \in N(u)\}\right) \tag{2}$$

Then we calculate the mean of the embedding sum of its nodes in the same structural role cluster $h_{R(u)}^k$.

$$h_{R(u)}^k \leftarrow \text{MEAN}\left(\{h_w^{k-1}, \forall w \in R(w)\}\right) \tag{3}$$

We connect these two embeddings with the node's current embedding h_u^{k-1}, and input the connected vector to a fully connected layer. The fully connected layer uses a nonlinear activation function, and the output is h_u^k, which will be used as the next step iterative input.

$$h_u^k \leftarrow \sigma\left(W^k \cdot \text{CONCAT}\left(h_u^{k-1}, h_{N(u)}^k, h_{R(u)}^k\right)\right). \tag{4}$$

After completing all the iterations, the output h_u^k will be used as the final embedding of each node. W^k is the weight matrix of the k layer. We take the same loss function as GraphSAGE, a graph-based loss function applied to Z_u, and all parameters are trained by stochastic gradient descent. The entire training process is unsupervised, and the loss function is shown below. Q is the number of negative samples, v is the node near u that can be reached by a random walk of fixed length and the node of the same cluster as u.

$$J_{\mathcal{G}}(\mathbf{z}_u) = -\log\left(\sigma\left(\mathbf{z}_u^\top \mathbf{z}_v\right)\right) - Q \cdot \mathbb{E}_{v_n \sim P_n(v)} \log\left(\sigma\left(-\mathbf{z}_u^\top \mathbf{z}_{v_n}\right)\right) \tag{5}$$

4   Experimental Estimate

In this section, we evaluate the effect of RolNE on multiple datasets.

4.1   Barbell Graph

Datasets and Settings. We consider a barbell graph, which consists of a long chain connecting two densely connected node clusters. There are a large number of nodes with similar structural roles in the barbell graph as shown in Fig. 2(a). Nodes with the same structural role have the same color. We hope that not only nodes with the same neighbors have similar embeddings, but also nodes with similar structural roles should have similar embeddings.

Result. Figure 2 shows the embeddings learned by DeepWalk, Node2vec, Struct2vec, GraphSAGE, and RolNE for barbell graphs. We use T-SNE tools to visualize the embeddings learned. After the visualization, the distance between the nodes of the same color projected in the graph should be relatively close. We can see that RolNE has correctly learned that nodes with similar structural roles have similar embeddings, and project all node vectors with the same color onto space closer to each other. DeepWalk, Node2vec, and GraphSAGE have captured the neighborhood information of the node; that is, the closely connected nodes have a closer distance on the network. However, they have not captured the characteristics of the nodes with similar structural roles. Although struct2vec has learned the structural characteristics of the node, it cannot recover more accurate structural equivalence and has not learned the neighborhood information of the node. The nodes with green, pink, purple colors are further from the red nodes than the black and blue nodes. On a given barbell graph, the blue and black nodes are farther from the red nodes than the green, pink, and purple nodes. Our model learns nodes with similar structures together, the distance between the same color nodes is relatively close, and the projected node order is also consistent with the original graph. The nodes closest to red are orange, green, pink, purple, black, and blue in order. RolNE effectively captured the structure information and neighborhood information of the original graph.

4.2   Mirror Karate Club

Datasets and Settings. The Karate Club Network is a classic real social network [25]. Each node in the network represents a member of the club. The side indicates that there is an interaction between members, that is, they are friends. The dataset contains 34 nodes and 78 edges. We use the same experimental settings as struct2vec, by copying the original karate club network and adding an edge between node 1 and node 37 in the original network to generate a mirrored karate club network. The purpose of the experiment is to identify mirror nodes by capturing the structural similarity of the nodes. Fig. 3 (a) is a schematic diagram of a mirrored karate club network. Nodes of the same color are nodes and their mirror nodes, and they have the same structural role.

Result. We use DeepWalk, Node2vec, SDNE, GraphSAGE, and struct2vec as the comparison algorithms for RolNE. Figure 3 shows the node embeddings learned by these algorithms. We visualize the node embeddings through T-SNE. Deep-Walk, Node2vec, and GraphSAGE are network embedding learning algorithms that mainly maintain the neighborhood information of the node. They have learned the neighborhood information of the node well. The dataset is divided into two closely connected groups of nodes. However, for the classification of labels that have a strong correlation with the characteristics of node structural roles, these methods cannot capture the structural similarity characteristics of nodes. For the visualization results of RolNE and Struct2vec, in order to facilitate observation, we use ellipses to circle the nodes that are closer in the network. RolNE, like Struct2vec, can correctly capture the structural similarity characteristics of the nodes. The mirrored nodes have relatively similar spatial distances after mapping.

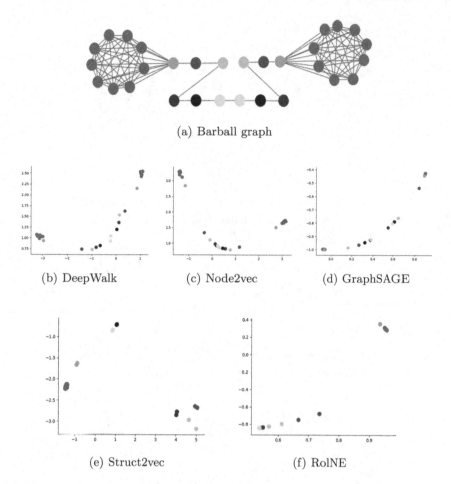

Fig. 2. (a) Visualization of Barball network. Network representations learned by (b) DeepWalk, (c) Node2vec, (d) GraphSAGE, (e) Struct2vec, (f) RolNE (Color figure online)

4.3   Air Traffic Network

A typical application of network embedding learning is node classification. We use the air traffic network to perform node classification tasks and use accuracy to evaluate the effectiveness of the algorithm.

Datasets and Settings. The air traffic network refers to a network consisted of airports and their air flights. The dataset contains the flight data of the National Civil Aviation Agency (ANAC), the Bureau of Transportation Statistics, the Statistical Office of the European Union (Eurostat) [10]. The node refers to the airport, the edge refers to the existence of a flight between the two airports, and the node label corresponds to the activity of the airport. The dataset contains

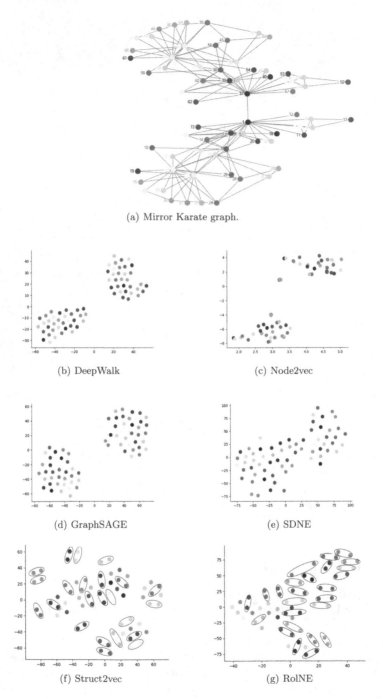

(a) Mirror Karate graph.

(b) DeepWalk

(c) Node2vec

(d) GraphSAGE

(e) SDNE

(f) Struct2vec

(g) RolNE

Fig. 3. (a) Mirror karate graph. Graph representations learned by (b) DeepWalk, (c) Node2vec, (d) GraphSAGE, (e) SDNE, (f) Struct2vec, (g) RolNE on the mirror karate network. (Color figure online)

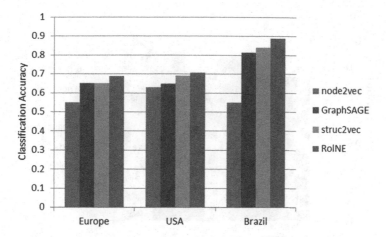

Fig. 4. In Brazil, the United States and Europe air traffic networks, the average accuracy of different algorithms in multi-class node classification task.

three subsets. The first dataset is the Brazilian air-traffic network: the network contains 131 nodes and 1038 edges. The data span is from January to December 2016. Airport activity is a mapping of the number of airports taking off and landing in that year. The second subset is the American air-traffic network: the network contains a total of 1190 nodes and 13599 edges. The airport's activity depends on the number of people entering and leaving the airport. The third subset is the European air-traffic network: the network contains 399 nodes and 995 edges. Landings plus takeoffs are used to measure the airport's activity. For each subset, the dataset is evenly divided into four parts according to the Airport activity, and each piece of data is labeled. Tag 0 indicates the least active airport, and tag 3 indicates the more active airport. After learning the node embedding without using labels, the node embedding is used as a feature and input to the one-vs-rest logistic regression with L2 regularization classifier. We use accuracy to evaluate the classification effect. For each algorithm, 80% of the data is used for training, and 20% of the data is used for testing. After the experiment is repeated ten times, we will calculate the average performance.

Result. Figure 4 shows the effect of all algorithms on three datasets. It can be seen that RolNE is significantly better than other algorithms. Compared with struct2vec, RollGE maintains both the neighborhood characteristics of nodes and the structural equivalence characteristics of nodes.

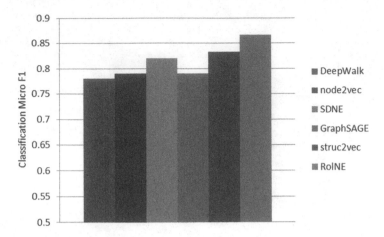

Fig. 5. In enron email network, the average micro F1 scores of different algorithms in multi-class node classification task.

4.4   Enron Email Network

We consider a typical social network, Enron email Network, which contains email data for Enron's 150 users, most of whom are Enron's senior management.

Datasets and Settings. The Enron mail network consists of 151 nodes and 2235 edges and each node in the network represents one employee in Enron, and the edge represents mail communication between the two users. These employees have one of eight functions in the company, such as CEO, Employee, In House Lawyer, Manager, Director, President, Trader, Vice President. These functions provide ground-truth information about the roles of the corresponding nodes in the network. We want to perform node classification tasks on the Enron mail network, and the eight functions above correspond to the nodes' labels. We used DeepWalk, node2vec, SDNE, GraphSAGE, struct2vec, and RolNE to study the embedding of each node. We then use these embeddings as features into the logical regression classifier, which all used the same parameter settings, with a training set ratio of 80%. We calculate average Micro F1 scores after ten experiments.

Result. In Fig. 5, We find that RolNE and stru2vec are superior to the methods based on neighborhood information, and show that RolNE and stru2vec can effectively capture node structure role information when the label of nodes in the network depends more on the structural equivalence of the nodes. In Enron mail networks, the labels of nodes respond to neighborhood information and structural information, such as the relationship between the employees is more closely, the relationship between the presidents is looser, but their structure in the network is more similar. RolNE works better than stru2vec because it captures both the neighborhood characteristics of the node and the structural equivalence characteristics.

5   Conclusion

We propose RolNE, a novel method that captures both the neighborhood characteristics and the structural role of the node. We proved the effectiveness of the algorithm on multiple datasets through compared with state-of-the-art techniques such as DeepWalk, Node2vec, GraphSAGE, struc2vec. Different algorithms have captured different structural features, and RolNE has captured the more comprehensive structural information that generates a rich embedding of nodes.

Acknowledgment. The research work is supported by the National Key R&D Program with No. 2016QY03D0503, Strategic Priority Research Program of Chinese Academy of Sciences, Grant No. XDC02040400, National Natural Science Foundation of China (No. 61602474, No. 61602467, No. 61702552).

References

1. Crawford, W.: Successful Social Networking in Public Libraries. American Library Association, Chicago (2014)
2. Hamilton, W.L., Tang, J.: Graph representation learning. In: AAAI 2019 (2019)
3. Zhang, D., Yin, J., Zhu, X., et al.: Network representation learning: a survey. IEEE Trans. Big Data **6**, 3–28 (2018)
4. Roweis, S.T., Saul, L.K.: Nonlinear dimensionality reduction by locally linear embedding. Science **290**(5500), 2323–2326 (2000)
5. Belkin, M., Niyogi, P.: Laplacian eigenmaps and spectral techniques for embedding and clustering. In: Advances in Neural Information Processing Systems, pp. 585–591 (2002)
6. Cao, S., Lu, W., Xu, Q.: GraRep: learning graph representations with global structural information. In: Proceedings of the 24th ACM International Conference on Information and Knowledge Management, pp. 891–900 (2015)
7. Perozzi, B., Al-Rfou, R., Skiena, S.: DeepWalk: online learning of social representations. In: Proceedings of the 20th ACM SIGKDD International Conference on Knowledge Discovery and Data Mining, pp. 701–710 (2014)
8. Grover, A., Leskovec, J.: Node2vec: scalable feature learning for networks. In: Proceedings of the 22nd ACM SIGKDD International Conference on Knowledge Discovery and Data Mining, pp. 855–864 (2016)
9. Chen, H., Perozzi, B., Hu, Y., Skiena, S.: HARP: hierarchical representation learning for networks. In: Proceedings of the 32nd AAAI Conference on Artificial Intelligence, pp. 2127–2134 (2018)
10. Ribeiro, L.F., Saverese, P.H., Figueiredo, D.R.: struc2vec: learning node representations from structural identity. In: Proceedings of the 23rd ACM SIGKDD International Conference on Knowledge Discovery and Data Mining, pp. 385–394 (2017)
11. Tang, J., Qu, M., Wang, M., Zhang, M., Yan, J., Mei, Q.: LINE: large-scale information network embedding. In: Proceedings of the 24th International Conference on World Wide Web, pp. 1067–1077 (2015)
12. Wang, D., Cui, P., Zhu, W.: Structural deep network embedding. In: Proceedings of the 22nd ACM SIGKDD International Conference on Knowledge Discovery and Data Mining, pp. 1225–1234 (2016)

13. Rehurek, R., Sojka, P.: Software framework for topic modelling with large Corpora. In LREC (2010)
14. Chen, J., Zhang, Q., Huang, X.: Incorporate group information to enhance network embedding. In: Proceedings of the 25th ACM International Conference on Information and Knowledge Management, pp. 1901–1904 (2016)
15. Li, J., Zhu, J., Zhang, B.: Discriminative deep random walk for network classification. In: Proceedings of the 54th Annual Meeting of the Association for Computational Linguistics, vol. 1, pp. 1004–1013 (2016)
16. Hamilton, W., Ying, Z., Leskovec, J.: Inductive representation learning on large graphs. In: Advances in Neural Information Processing Systems, pp. 1024–1034 (2017)
17. Henderson, K., et al.: RolX: structural role extraction & mining in large graphs. In: KDD, pp. 1231–1239 (2012)
18. Cao, S., Lu, W., Xu, Q.: Deep neural networks for learning graph representations. In: Proceedings of the 30th AAAI Conference on Artificial Intelligence, pp. 1145–1152 (2016)
19. Kipf, T.N., Welling, M.: Variational graph auto-encoders. In: NIPS Workshop on Bayesian Deep Learning (2016)
20. Lyu, T., Zhang, Y., Zhang, Y.: Enhancing the network embedding quality with structural similarity. In: Proceedings of the. ACM on Conference on Information and Knowledge Management, vol. 2017, pp. 147–156 (2017)
21. Xing, W., Ghorbani, A.: Weighted pagerank algorithm. In: Proceedings of the Second Annual Conference on Communication Networks and Services Research, pp. 305–314. IEEE (2004)
22. Hočevar, T., Demšar, J.: A combinatorial approach to graphlet counting. Bioinformatics **30**(4), 559–565 (2014)
23. Donnat, C., Zitnik, M., Hallac, D., et al.: Learning structural node embeddings via diffusion wavelets. In: Proceedings of the 24th ACM SIGKDD International Conference on Knowledge Discovery & Data Mining, pp. 1320–1329 (2018)
24. Krishna, K., Murty, M.N.: Genetic k-means algorithm. IEEE Trans. Syst. Man Cybern. Part B (Cybern.) **29**(3), 433–439 (1999)
25. Zachary, W.W.: An information flow model for conflict and fission in small groups. J. Anthropol. Res. **33**(4), 452–473 (1977)
26. Long, Q., Jin, Y., Song, G., Li, Y., Lin, W.: Graph Structural-Topic Neural Network. arXiv preprint arXiv:2006.14278 (2020)
27. Chen, H., Yin, H., Chen, T., Nguyen, Q.V.H., Peng, W.C., Li, X.: Exploiting centrality information with graph convolutions for network representation learning. In: 2019 IEEE 35th International Conference on Data Engineering (ICDE), pp. 590–601. IEEE, April 2019
28. Chen, H., Yin, H., Wang, W., Wang, H., Nguyen, Q.V.H., Li, X.: PME: projected metric embedding on heterogeneous networks for link prediction. In: Proceedings of the 24th ACM SIGKDD International Conference on Knowledge Discovery & Data Mining, pp. 1177–1186, July 2018

Weighted Meta-Path Embedding Learning for Heterogeneous Information Networks

Yongjun Zhang$^{(\boxtimes)}$, Xiaoping Yang, and Liang Wang

School of Information, Renmin University of China, Beijing 100872, China
{zyjun,yang,wangliang}@ruc.edu.cn

Abstract. A low-dimensional embedding can be easily applied in the downstream tasks for network mining and analysis. In the meantime, the popular models of random walk-based network embedding are viewed as the form of matrix factorization, whose computational cost is very expensive. Moreover, mapping different types of nodes into one metric space may result in incompatibility. To cope with the two challenges above, a weighted meta-path embedding framework (WMPE) is proposed in this paper. On one hand, a nearly-linear approximate embedding approach is leveraged to reduce the computational cost. On the other hand, the meta-path and its weight are learned to integrate the incompatible semantics in the form of weighted combination. Experiment results show that WMPE is effective and outperforms the state-of-the-art baselines on two real-world datasets.

Keywords: Heterogeneous information networks · Network embedding · Representation learning · Meta-path learning

1 Introduction

In the real world, a variety of entities and relationships can be abstracted and represented as heterogeneous information networks (HINs), on which data mining and analysis have raised increasing attention in the past decade due to the complex structures and rich semantic information. In the meantime, network embedding, mapping a high-dimensional and sparse network into a low-dimensional and dense space to learn the latent representation of nodes and edges while preserving the multiplex semantics, is extremely convenient for network analysis, e.g., classification [1], clustering [2], link prediction [3], and recommendation system [4].

Matrix factorization is widely utilized for network embedding. Transform network into the form of matrix and decompose it, so that each node can be represented as a distribution on the latent semantics. However, it is hardly suitable for large-scale networks due to the expensive computational cost. Inspired by Word2Vec algorithm, a series of random walk-based models [6–8] introduce neural network into the task of graph embedding, and achieve significant successes. Nevertheless, a recent study [9] shows that methods aforementioned can be viewed as asymptotically and implicitly process of matrix factoring in essence.

© Springer Nature Switzerland AG 2020
Z. Huang et al. (Eds.): WISE 2020, LNCS 12342, pp. 29–40, 2020.
https://doi.org/10.1007/978-3-030-62005-9_3

Heterogeneous information network brings not only the rich semantic information, but also the conflicts among variety of relationships, which is more complex and difficult to deal with. For example, a node is associated with two different types of nodes, which are irrelevant each other. It is hard to represent the semantic that a node is simultaneously close to two distant nodes in one metric space. Therefore, embedding different types of nodes and edges into one feature space will lead to semantic incompatibility, which brings the special challenge for HINs embedding.

With the intention to solve the problems aforementioned, we propose a weighted meta-path embedding framework, referred to as WMPE, to learn the nodes representation for HINs. Firstly, bypassing the eigen-decomposition, an approximate commute embedding approach is utilized to embed the heterogeneous network into a low-dimensional metric space, which reduce the computational cost in nearly-linear time. Aiming at the existence of semantic incompatibility in HINs, a set of meta-paths are automatically generated. Furthermore, the weight of each meta-path is learned by optimizing the loss function with a small number of labeled nodes. At last, relatedness between nodes are obtained in the form of weighted combination of meta-paths atop node embeddings. In this way, we can preserve the different semantics specified by the generated meta-paths even in the presence of incompatibility.

2 Related Work

2.1 Meta-Path of HIN

Meta-path describes the rich semantic via a sequence of relations amongst objects in HINs and using meta-paths to characterize the heterogeneity is distinct from that of homogeneous networks. The major line of work focusing on meta-path-based similarity or meta-path-based neighbors has been footstone for broad applications of heterogeneous information networks. PathSim [10] leverages the normalized number of reachable paths complying with the schema of meta-path to compute the similarity between a pair of nodes, and PathSelClus [11] integrates a suite of meta-paths to extract network structure for objects clustering, where the paths are defined to be symmetrical and the relatedness is only measured between objects with the same type. In consideration of the asymmetric meta-paths, HeteSim [12] improves the method and focuses on relevance measure for objects with the same type or different types. Recently, some models [20–22] extract the meta-paths-level structural property and integrate other methods for HINs mining and analysis. However, the meta-paths are prepared in advance and treated equally, ignoring the fact that importance of the semantic information represented by corresponding meta-path is different from each other.

2.2 Network Embedding

Network embedding can be traced back to the usage of matrix factorization. Initially, matrix factorization is used to the dimension reduction, such as Multi-dimension Scaling (MDS) [13], Spectral Clustering [5], and Laplacian eigenmaps [14]. These methods transform the network into a matrix, and then a low-dimensional representation of the network can be obtained by matrix eigen-decomposition on each eigenvector. GraRep [15] decomposes k-step transition probability matrixes and integrates the k relationships to represent global characteristics of weighted network. HOPE [16] and M-NMF [17] capture network properties by depicting the n-th order proximity to construct relation matrixes for directed edges and macroscopic community structure respectively, where the optimizing process of matrix decomposition is quadratic time complexity at least. Therefore, the matrix factorization-based embedding is difficult to apply to large-scale networks.

Inspired by word embedding, the random walk-based models introduce neural network into network representation learning. DeepWalk [6] samples the network by depth-first random walk to generate nodes sequence, which can be served as "word-context" sentences for Word2Vec model to learn the nodes vectors. As an extension of Deep-Walk, node2vec [8] designs a biased walk to explore the neighborhood of nodes in combination with BFS and DFS styles. LINE [7] defines the similarity of first-order proximity and second-order proximity to optimize objective function. However, reference [9] expounds the theoretical connections among these models of random walk-based network embedding, which can be induced to the form of matrix factorization.

In consideration of the diverse semantics amongst multiple nodes in heterogeneous information network, metapath2vec [18] generates the context sequences guided by meta-paths, and improves the Skip-Gram model to adapt to the negative sampling. In addition, HIN2Vec [19] takes nodes and their relationships specified in forms of meta-paths together as the input data for training, and also some efficient strategies are developed for data preparation. Recently, motivated by the great success of deep learning, some graph neural network models [20–22] are designed to extract "latent" semantics directly from a continuous distribution for HIN embedding. Reference [23] points out that, compared with homogenous networks, those methods above map multiple types of nodes and edges into one metric space, which will inevitably lead to incompatibility in heterogeneous information networks.

3 Preliminaries

In this section, we introduce some related concepts and formalize the notations.

DEFINITION 1. **Heterogeneous Information Network**. An information network is a directed graph $G = (V, E)$ with a node type mapping function $\phi : V \rightarrow A$ and a link type mapping function $\varphi : E \rightarrow R$, where each node $v \in V$ is mapped to one particular node type in A, i.e., $\phi(v) \in A$, and each edge $e \in E$ is mapped to one particular link type in R, i.e., $\varphi(e) \in R$. When $|A| > 1$ or $|R| > 1$, the network is called a heterogeneous information network.

DEFINITION 2. **Network Schema**. A network schema is denoted as $T_G = (A, R)$ to abstract the meta-information of the given HIN $G = (V, E)$ with node type mapping function $\phi : V \rightarrow A$ and a link type mapping function $\varphi : E \rightarrow R$.

DEFINITION 3. **Meta-path**. A meta-path is defined as a path schema in the form of $A_1 \xrightarrow{R_1} A_2 \xrightarrow{R_2} \cdots \xrightarrow{R_l} A_{l+1}$, which describes a composite relation $R = R_1 \circ R_2 \circ \cdots \circ R_l$ between type A_1 and A_{l+1}, where $\circ$ denotes the composition operator on relations.

DEFINITION 4. **Path Weight Matrix**. Given a meta-path $p : A_1 \rightarrow A_2 \rightarrow \cdots \rightarrow A_{l+1}$, the path weight matrix is defined as $M = W_{A_1A_2} \times W_{A_2A_3} \times \cdots \times W_{A_lA_{l+1}}$, where $W_{A_iA_j}$ is the adjacency matrix between the nodes of A_i and A_j. M_{ij} denotes the number of instances complying with the meta-path p from node $i \in A_1$ to node $j \in A_j$.

4   Framework of Proposed WMPE

To cope with the expensive cost of matrix factorization in embedding process and incompatibility among heterogeneous semantics, we propose a weighted meta-path embedding framework, referred to as WMPE, to learn the low-dimension embedding representation for heterogeneous information networks. The framework of WMPE includes two phases. In the first phase, all the nodes in HIN are pretrained to learn the vectorized representation by a nearly-linear approximate embedding approach. In the second phase, a set of meta-paths are automatically generated, and the weight of each meta-path is learned by optimizing the loss function with a small number of labeled nodes. Afterward, the proposed WMPE framework is described in detail.

4.1   Approximate Commute Embedding

Instead of the eigen-decomposition, an approach of approximate commute embedding is applied to project the network into a low-dimension metric space for embedding learning, which can greatly reduce the computational complexity into nearly-linear time.

A HIN can be denoted by a directed graph G with n nodes and s edges. Given signed edge-vertex incidence matrix $B_{s \times n}$, defined as

$$B(e, v) = \begin{cases} 1, & \text{if } v \text{ is the head of } e, \\ -1, & \text{if } v \text{ is the tail of } e, \\ 0, & \text{otherwise,} \end{cases}$$

and diagonal matrix of edges weights $W_{s \times s}$, the Laplacian $L = B^T WB$ [24].

Therefore, the commute distance between node i and j can be written as:

$$
\begin{aligned}
c_{ij} &= V_G (e_i - e_j)^T L^+ (e_i - e_j) \\
&= V_G (e_i - e_j)^T L^+ B^T WBL^+ (e_i - e_j) \\
&= \left[\sqrt{V_G} W^{1/2} BL^+ (e_i - e_j) \right]^T \cdot \left[\sqrt{V_G} W^{1/2} BL^+ (e_i - e_j) \right]
\end{aligned}
$$

where $V_G = \sum w_{ij}$ and e_i the unit vector with a 1 at location i and zero elsewhere. That is, $\theta = \sqrt{V_G} W^{1/2} BL^+ \in \mathbb{R}^{s \times n}$ is a commute embedding for HIN G where c_{ij} is the squared Euclidean distance between the i-th and j-th column vectors in space θ. Since it takes $O(n^3)$ for the pseudo-inversion L^+ of L in θ, an approximate commute embedding method is adopted more efficiently.

LEMMA 1 [24]. Given vectors $v_1, \ldots, v_n \in R^{n \times s}$ and $\varepsilon > 0$, let $Q_{k_r \times s}$ be a random matrix where $Q(i, j) = \pm 1/\sqrt{k_r}$ with equal probability $k_r = O(\log n/\varepsilon^2)$. For any pair

v_i, v_j, exist

$$(1 - \varepsilon)\|v_i - v_j\|^2 \le \|Qv_i - Qv_j\|^2 \le (1 + \varepsilon)\|v_i - v_j\|^2$$

with probability at least $1 - 1/n$.

Therefore, construct a matrix $Z_{k_r \times s} = \sqrt{V_G}QW^{1/2}BL^+$ and we have:

$$(1 - \varepsilon)c_{ij} \le \|Z(e_i - e_j)\|^2 \le (1 + \varepsilon)c_{ij}$$

for $\forall v_i, v_j \in G$, with probability at least $1 - 1/n$ from Lemma 1. That is, $c_{ij} \approx \|Z(e_i - e_j)\|^2$ with an error ε. Due to the expensive computational cost of L^+, a nearly-linear time method ST-solver [25] is used instead. Let $Y = \sqrt{V_G}QW^{1/2}B$, and $Z = YL^+$ which is equivalent to $ZL = Y$. Then each z_i (the i-th row of Z) is computed by solving the equation $z_iL = y_i$ where y_i is the i-th row of Y, and the solution is denoted as $\tilde{Z}_i$. Since $\|z_i - \tilde{z}_i\|_L \le \varepsilon\|z_i\|_L$, we have:

$$(1 - \varepsilon)^2 c_{ij} \le \|\tilde{Z}(e_i - e_j)\|^2 \le (1 + \varepsilon)^2 c_{ij} \tag{1}$$

where $\tilde{Z}$ is the matrix consisting of $\tilde{Z}_i$. Equation (1) indicates that $c_{ij} \approx \|\tilde{Z}(e_i - e_j)\|^2$ with the error ε^2 and $\tilde{Z}$ is the approximate commute embedding of G.

4.2 Meta-Path Generation and Weight Learning

In order to measure the semantic-level relatedness of nodes, a set of meta-paths are generated automatically by a random walk strategy.

Given the start node type A_s and end node type A_e, the target is to find all the meta-paths in the form of $A_s - * - A_e$ within the maximum length l. On the network schema, start from node A_s, and randomly jump to the neighbor of current node for each step within l. Each time the node A_e is reached, the sequential passed nodes compose a meta-path. Repeating this process for the number of iterations, a set of candidate meta-paths are generated automatically. Note that, the start and end nodes can be of arbitrary types, not requiring the same type, and the path does not have to be symmetric. Figure 1 illustrates the process of meta-paths generation, the form is defined as *author* − * − *author*, with the maximum length 4. After the exploration on DBLP schema, a set of meta-paths are obtained including: APA, APPA, APPPA, APAPA, APCPA, and APTPA.

The contributions of different semantic relationships specified by the meta-paths are not equivalent to a certain target. In fact, the conflicting, irrelevant semantics will result in low performances in the following tasks. Also, more meta-paths will increase the computation cost. A proper way is to assign higher weights to the meta-paths that can promote the downstream tasks, and discard the irrelevant and conflicting meta-paths.

Therefore, we design an objective function to transform the weight learning process of meta-paths into an optimization problem.

For a meta-path p_k, we build the meta-path relatedness of node pair (u, v) atop their approximate commute embeddings in the first phase as

$$sim_k(u, v) = \frac{\exp(M_{uv}^k \cdot g(u, v))}{\sum_{\phi(\tilde{v})=\phi(v)} \exp(M_{uv}^k \cdot g(u, \tilde{v})) + \sum_{\phi(\tilde{u})=\phi(u)} \exp(M_{uv}^k \cdot g(\tilde{u}, v))},$$

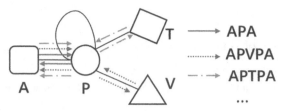

Fig. 1. An illustration of meta-paths generation. The walk paths are represented by lines with arrows in the form of *author–*–author* on DBLP schema.

where M_{uv}^k denotes the number of paths between u and v following meta-path p_k, and $g(u, v)$ is the nodes closeness from u to v. Considering the existence of both undirected and directed relationships in heterogeneous networks, $g(u, v) \neq g(v, u)$, if the path between the two nodes contains a directed relation. Especially, we decompose the pre-trained node embedding into two sections $x = \begin{bmatrix} x^O \\ x^I \end{bmatrix}$, where x^O and x^I are two column vectors of the same dimension, and define nodes closeness on approximate commute embedding as

$$g(u, v) = \begin{cases} 2x_u^o \cdot x_v^I, & \text{directed from } u \text{ to } v \\ x_u^o \cdot x_v^I + x_v^o \cdot x_u^I, & \text{undirected} \end{cases},$$

where $\cdot$ denotes the inner-product.

On the contrary to the assumption that nodes of different types are independent each other, we hold the view that closely related nodes in a network are not only structurally but also semantically consistent. Therefore, taking all meta-paths into account, the lose function is designed to learn the weight of each meta-path as the follows:

$$L(\Theta) = \left\| 1 - sign(u, v) \sum_{k=1}^{K} \theta_k sim_k(u, v) \right\|^2 + \alpha \|\Theta\|^2,$$

where $\Theta = \{\theta_1, \theta_2, \cdots, \theta_K\}$, θ_k is the importance of meta-path p_k, and α is the regularization parameter. The $sign()$ is an indicator function defined as

$$sign(u, v) = \begin{cases} 1, & u \text{ and } v \text{ have the same label} \\ -1, & \text{otherwise} \end{cases}.$$

The loss function is to maximize the importance of meta-path connecting nodes with same label and minimize that connecting nodes with different labels.

In order to minimize $L(\Theta)$, for each $k = 1, 2, \cdots, K$, equate the partial derivative of loss function with respect to θ_k to zero. Iterating this process until the error converges, the importance of each meta-path is obtained. The meta-paths with negative or too small value will be discarded, which represent conflict and low correlative semantic meanings to the following tasks. Weights are computed by normalizing the importance values of reduced meta-paths. In this way, the weighted combination of independent semantics specified by meta-paths is to measure the similarities among nodes in heterogeneous information network, which can be easy-to-use for the downstream tasks, e.g., classification.

4.3 Complexity Analysis

As mentioned before, the framework of the WMPE contains two major parts:

Network Embedding. Let $|E|$ and $|V|$ be the average sizes of edges and nodes in the HIN, and computing $Y = \sqrt{V_G}QW^{1/2}B$ costs $O(2k_r|E| + |E|)$ since there are $2|E|$ non-zero elements in B and $|E|$ in diagonal matrix W. Then each $\tilde{z}_i$ of $\tilde{z}$ is obtained by solving the equation $z_iL = y_i$ in $\tilde{O}(k_r|E|)$ time. Therefore, learning approximate commute embeddings of the nodes is in $\tilde{O}(3k_r|E| + |E|) = \tilde{O}(k_r|E|)$ time.

Meta-path Generation and Weight Learning. Set the maximum length as l and the iterations as t_1, it takes $O(t_1 l)$ to generate a group of meta-paths. As a very small number of labeled nodes are randomly extracted to learn the weights of meta-paths, the $|E|$ is reduced to $|E'|$, $|E'| \ll |E|$. Constructing the path weight matrix M costs $O(K \cdot \sum_{i=1}^{l} |E'_i|) \approx O(|E'|)$, where E'_i denotes the labeled edges with type R_i and K is the number of meta-paths. Optimizing the loss function to obtain the $\Theta = \{\theta_1, \theta_2, \cdots, \theta_K\}$ over t_2 iterations, it takes $O(t_2 KN)$, where N is the number of instances for all the K meta-paths.

Therefore, the overall complexity of WMPE is $\tilde{O}(k_r|E| + t_1 l + |E'| + t_2 KN)$.

5 Experiments

5.1 Datasets

The proposed WMPE is evaluated on two real-world datasets: DBLP and IMDB.

DBLP is a bibliographical network in computer science, which includes four types of nodes: paper (P), author (A), venue (V), and term (T). We construct two subsets of DBLP. DBLP_1 contains 240 venues (30 venues for each area), 189,378 authors, 217,764 papers and 263,332 terms in eight research areas. Because of the bias that a few authors publish most of the papers, DBLP_2 extracts 20 venues, 9,737 authors (publish at least five papers), 12,098 papers, and 9,936 terms in four research areas. To evaluate the methods, the research area is taken as ground truth.

IMDB is a movie network consisting of five types of nodes: movie (M), actor (A), director (D), year (Y) and genre (G). There are 5000 movies, 2342 actors, 879 directors and 67 genres, where we use the movie genre as ground truth.

5.2 Baselines

We compare the proposed WMPE framework with the state-of-the-art baseline models: DeepWalk [6], LINE [7], HIN2Vec [19] and metapath2vec [18]. DeepWalk and LINE are designed for homogeneous network embedding. HIN2Vec and metapath2vec aim at heterogeneous information network, but all nodes and relationships are mapped into one feature space. Due to the random initialization, all the experimental results reported in the following sections are the average over 10 trials, and the parameters are set to the optimal.

5.3 Meta-Path Filtration

A set of meta-paths are generated and the corresponding importance values are also learned respectively on DBLP and IMDB. The results are shown in Table 1, and the preserved meta-paths are indicated in bold.

Table 1. Results of meta-path generation and the importance.

	Meta-path	Length	Importance
DBLP	**author-paper-author (APA)**	2	0.13
	author-paper-paper-author (APPA)	3	0.09
	author-paper-paper-paper-author (APPPA)	4	0.05
	author-paper-author-paper-author (APAPA)	4	0.11
	author-paper-term-paper-author (APTPA)	4	0.31
	author-paper-venue-paper-author (APVPA)	4	0.78
IMDB	**movie-actor-movie (MAM)**	2	0.17
	movie-director-movie (MDM)	2	0.44
	movie-year-movie (MYM)	2	−0.30
	movie-actor- movie-director-movie (MAMDM)	4	0.03
	movie-actor-movie-year-movie (MAMYM)	4	−0.46
	movie-director-movie -year-movie (MDMYM)	4	−0.42

For DBLP dataset, we set the maximum length as 4, and the meta-paths are generated in the form of *author − * − author*, in which both the start node type and end node type are authors. The order of importance for each meta-path in DBLP is as follows: APVPA > APTPA > APA > APAPA > APPA > APPPA. It turns out that, compared with other relations, the authors publishing papers at the same conference are more likely to be in the same field.

For IMDB dataset, we set the maximum length as 4, and the meta-paths are generated in the form of *movie − * − movie*. Since the meta-paths are undirected, reciprocal paths are filtrated away. We note that all the importance values of the meta-path including the type of year (Y) are negative, which means there is no connection between the year a film is made and its genre. Therefore, those meta-paths will disturb the following tasks.

5.4 Results of Classification

We conduct a classification experiment using KNN classifier to verify the effectiveness of the algorithm. The embeddings of nodes in networks are represented by 128-dimension vectors for all the methods. Randomly select 2%, 4% and 6% of the labelled nodes for weight learning respectively. The average performances are reported in Table 2 with the metrics of Accuracy (Acc) and macro F1-score (F1), in which a larger value implies a better effect.

Table 2. Performance comparison of classification. The best results are indicated in bold.

Method	Metric	DBLP_1			DBLP_2			IMDB		
		2%	4%	6%	2%	4%	6%	2%	4%	6%
DeepWalk	Acc	65.38	65.38	65.38	74.32	74.32	74.32	53.20	53.20	53.20
	F1	62.71	62.71	62.71	70.10	70.10	70.10	52.02	52.02	52.02
LINE	Acc	81.74	81.74	81.74	85.23	85.23	85.23	68.27	68.27	68.27
	F1	80.43	80.43	80.43	81.82	81.82	81.82	65.54	65.54	65.54
HIN2Vec	Acc	83.61	84.67	84.73	90.09	91.87	93.04	72.45	72.68	73.26
	F1	84.27	85.40	85.62	88.34	89.34	90.73	71.78	71.88	72.09
metapath2vec	Acc	86.93	86.21	87.64	**93.73**	**95.49**	95.55	72.56	72.56	73.42
	F1	83.57	84.07	86.32	91.65	91.64	91.89	70.30	70.30	71.35
WMPE	Acc	**90.76**	**91.73**	**91.89**	93.57	94.29	**95.58**	**74.24**	**75.09**	**78.22**
	F1	**89.01**	**90.38**	**91.23**	**92.10**	**92.66**	**93.21**	**74.93**	**76.32**	**77.06**

As can be seen, on the whole, WMPE outperforms the other baselines on three datasets for two metrics, which validate the effectiveness of our proposed model. In terms of datasets, all methods on DBLP_2 achieve best results. We conclude that DBLP_2, compared with DBLP_1, discards large numbers of authors who publish only a few papers to void the sparsity of network, which is instrumental for classification. The worst overall performances are on IMDB. One explanation could be that an actor/actress or director does not exactly associated with a certain movie genre, which can be confirmed from weights of the meta-paths including the type actor. Compared to the best performances of baselines on DBLP_1, a large-scale and biased network, WMPE achieves the improvements of 5.55% Acc, and 6.31% F1 over metapath2vec. For baselines, the observations in Table 3 show that the results of heterogeneous network embedding models (HIN2Vec and metpath2vec) are generally better than that of the homogeneous (DeepWalk and LINE). DeepWalk and LINE cannot use the meta-paths and the results will not change, although the labelled nodes are increasing. Since meta-paths are introduced to represent the semantic information among different types of nodes. Besides, because of discarding the conflicting and irrelevant meta-paths and assigning higher weights to the effective meta-paths, our method WMPE is superior to all the methods above.

5.5 Impact of Different Meta-Paths

We also conduct an experiment to analyze the effect of different meta-paths and the results are given in Fig. 2.

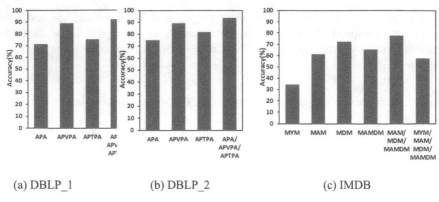

(a) DBLP_1   (b) DBLP_2   (c) IMDB

Fig. 2. Impact of meta-paths on different datasets

As shown in Fig. 2(a) and (b), the effect of the experiment is consistent with the importance of the meta-paths that the meta-path with the highest weight gets the best results. This suggests that the relation of publishing papers in common venues is more sensitive than the relations of sharing common terms and co-author to research areas.

For IMDB dataset, shown in Fig. 2(c), we add a meta-path MYM to compare with the preserved ones. The accuracy of MDM is better than that of MAM. That is, directors, compared with actors/actresses, are more likely to have their inherent genres. We note that, increasing meta-paths can improve the accuracy in general, but it does not mean the more, the better. A possible reason is that some meta-paths may contain noisy or conflicting information, e.g., MYM.

5.6   Analysis of Network Embedding Time

In this subsection, we discuss the computation time in the process of network embedding and Table 3 reports the results on three datasets. DeepWalk takes the longest running time, because it uses Skip-Gram with hierarchical softmax, whose time complexity is proportional to the value of $|V|\log|V|$, where $|V|$ is the number of nodes. Since LINE, HIN2Vec and metapath2vec leverage negative sampling to reduce time complexity, the running time is almost linear to the number of edges $|E|$. However, their computational cost is still affected by the vector dimension, walk length, neighborhood size and number of negative samples, etc. The time complexity of network embedding for WMPE is $\tilde{O}(k_r|E|)$, and k_r is a very small number in practice (parameter k_r is discussed in the following subsection). Hence, WMPE achieve the best performance in terms of network embedding time.

5.7   Parameter k_r

The time complexity of approximate commuting embedding is $\tilde{O}(k_r|E|)$, and parameter k_r is also related to the error of the learned representation. A proper value of k_r can not only ensure the effectiveness, but also reduce the learning time of network embedding. We conducted an experiment with different k_r in each dataset. Figure 3 illustrates the

Table 3. Comparison of running time (s) for network embedding

Method	DBLP_1	DBLP_2	IMDB
DeepWalk	9791.76	783.12	96.25
LINE	1542.47	108.04	21.73
HIN2Vec	2085.30	184.72	32.89
metapath2vec	1738.91	129.46	24.64
WMPE	960.19	62.33	15.08

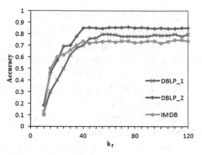

Fig. 3. Classification results with different k_r

influence of k_r on classification accuracy. With the increase of k_r, the accuracy curve is rising, but tend to be smooth. According to the curve, k_r is set to 60, 50 and 40 on dataset DBLP_1, DBLP_2 and IMDB respectively, which can keep the balance between accuracy and speed.

6   Conclusion

This paper presents a weighted meta-path embedding learning for heterogeneous information networks, called WMPE, which can easily integrate incompatible semantics by a weighted combination of the effective meta-paths. Also, a nearly-linear approximate approach reduces the time complexity in embedding process. Experimental results on two real-world datasets prove that the proposed WMPE is effective and feasible.

References

1. Lee, G., Kang, S., Whang, J.: Hyperlink classification via structured graph embedding. In: Proceedings of the 42nd International ACM SIGIR Conference on Research and Development in Information Retrieval, pp. 1017–1020 (2019)
2. Fan, S., et al.: One2Multi graph autoencoder for multi-view graph clustering. In: WWW'20: Proceedings of The Web Conference 2020, pp. 3070–3076 (2020)
3. Rosso, P., Yang, D., Philippe, C.M.: Beyond triplets: hyper-relational knowledge graph embedding for link prediction. In: WWW'20: Proceedings of The Web Conference 2020, pp 1885–1896 (2020)
4. Wen, Y., et al.: Network embedding based recommendation method in social networks. In: Companion of the Web Conference 2018, pp. 11–12 (2018)
5. Ng, A.Y., Jordan, M.I., Weiss, Y.: On spectral clustering: analysis and an algorithm. In: Advances in Neural Information Processing Systems, pp. 849–856 (2002)
6. Perozzi, B., Al-Rfou, R., Skiena, S.: DeepWalk: online learning of social representations. In: Proceedings of the 20th ACM SIGKDD International Conference on Knowledge Discovery and Data Mining, pp. 701–710 (2014)
7. Tang, J., et al.: LINE: large-scale information network embedding. In: Proceedings of 24th International Conference on World Wide Web, pp. 1067–1077 (2015)
8. Grover, A., Leskovec, J.: Node2vec: scalable feature learning for networks. In: Proceedings of the 22nd ACM SIGKDD International Conference on Knowledge Discovery and Data Mining, pp. 855–864 (2016)

9. Qiu, J., et al.: Network embedding as matrix factorization: unifying DeepWalk, LINE, PTE, and node2vec. In: Proceedings of the 11th ACM International Conference on Web Search and Data Mining, pp. 459–467 (2018)
10. Sun, Y., et al.: Pathsim: meta path-based top-k similarity search in heterogeneous information networks. Proc. VLDB Endow, pp. 992–1003 (2011)
11. Sun, Y., et al.: Integrating meta-path selection with user-guided object clustering in heterogeneous information networks. In: Proceedings of 18th ACM SIGKDD International Conference on Knowledge Discovery and Data Mining, pp. 723–724 (2012)
12. Shi, C., et al.: HeteSim: a general framework for relevance measure in heterogeneous networks. IEEE Trans. Knowl. Data Eng. **26**, 2479–2492 (2014)
13. Agarwal, A., Phillips, J.M., Venkatasubramanian, S.: Universal multi-dimensional scaling. In: Proceedings of the 16th ACM SIGKDD International Conference on Knowledge Discovery and Data Mining, pp. 1149–1158 (2010)
14. Belkin, M., Niyogi, P.: Laplacian eigenmaps and spectral techniques for embedding and clustering. In: Advances in Neural Information Processing Systems, pp. 585–591 (2002)
15. Cao, S., Lu, W., Xu, Q.: GraRep: learning graph representations with global structural information. In: Proceedings of the 24th ACM International Conference on Information and Knowledge Management, pp. 891–900 (2015)
16. Ou, M., et al.: Asymmetric transitivity preserving graph embedding. In: Proceedings of the 22nd ACM SIGKDD International Conference on Knowledge Discovery and Data Mining, pp. 1105–1114 (2016)
17. Wang, X., et al.: Community preserving network embedding. In: 31st AAAI Conference on Artificial Intelligence, pp. 203–209 (2017)
18. Dong, Y., Chawla, N.V., Swami, A.: metapath2vec: Scalable representation learning for heterogeneous networks. In: Proceedings of the 23rd ACM SIGKDD International Conference on Knowledge Discovery and Data Mining, pp. 135–144 (2017)
19. Fu, T., Lee, W., Lei, Z.: HIN2Vec: explore meta-paths in heterogeneous information networks for representation learning. In: Proceedings of the 26th ACM on Conference on Information and Knowledge Management, pp. 1797–1806 (2017)
20. Hu, Z., et al.: Heterogeneous graph transformer. In: Proceedings of the World Wide Web Conference (2020)
21. Wang, X., et al.: Heterogeneous graph attention network. In: Proceedings of the World Wide Web Conference, pp. 2022–2032 (2019)
22. Hu, B., Fang, Y., Shi, C.: Adversarial learning on heterogeneous information networks. In: Proceedings of the 25th ACM SIGKDD International Conference on Knowledge Discovery and Data Mining, pp. 120–129 (2019)
23. Shi, Y., et al.: Easing embedding learning by comprehensive transcription of heterogeneous information networks. In: Proceedings of the 24th ACM SIGKDD International Conference on Knowledge Discovery and Data Mining, pp. 2190–2199 (2018)
24. Khoa, N.L.D., Chawla, S.: Large scale spectral clustering using resistance distance and Spielman-Teng solvers. In: International Conference on Discovery Science, pp. 7–21 (2012)
25. Spielman, D.A., Srivastava, N.: Graph sparsification by effective resistances. In: Proceedings of the 40th Annual ACM Symposium on Theory of Computing, pp. 563–568 (2008)

A Graph Embedding Based Real-Time Social Event Matching Model for EBSNs Recommendation

Gang Wu[1,2]([✉])(iD), Xueyu Li[1], Kaiqian Cui[1], Zhiyong Chen[1], Baiyou Qiao[1],
Donghong Han[1], and Li Xia[1]

[1] Northeastern University, Shenyang, China
wugang@mail.neu.edu.cn
[2] State Key Laboratory for Novel Software Technology,
Nanjing University, Nanjing, China

Abstract. Event-based social networks (EBSNs), are platforms that provide users with event scheduling and publishing. In recent years, the number of users and events on such platforms has increased dramatically, and interactions have become more complicated, which has made modeling heterogeneous networks more difficult. Moreover, the requirement of real-time matching between users and events becomes urgent because of the significant dynamics brought by the widespread use of mobile devices on such platforms. Therefore, we proposed a graph embedding based real-time social event matching model called GERM. We first model heterogeneous EBSNs into heterogeneous graphs, and use graph embedding technology to represent the nodes and their relationships in the graph which can more effectively reflect the hidden features of nodes and mine user preferences. Then a real-time social event matching algorithm is proposed, which matches users and events on the premise of fully considering user preferences and spatio-temporal characteristics, and recommends suitable events to users in real-time efficiently. We conducted experiments on the Meetup dataset to verify the effectiveness of our method by comparison with the mainstream algorithms. The results show that the proposed algorithm has a good improvement on the matching success rate, user satisfaction, and user waiting time.

Keywords: EBSNs · Event publishing · Graph embedding · Real-time matching

1 Introduction

With the popularization of mobile Internet and smart devices, the way people organize and publish events is gradually networked. Event-based social networks

Supported by the National Key R&D Program of China (Grant No. 2019YFB1405302), the NSFC (Grant No. 61872072 and No. 61672144), and the State Key Laboratory of Computer Software New Technology Open Project Fund (Grant No. KFKT2018B05).

© Springer Nature Switzerland AG 2020
Z. Huang et al. (Eds.): WISE 2020, LNCS 12342, pp. 41–55, 2020.
https://doi.org/10.1007/978-3-030-62005-9_4

(EBSNs) platforms, such as Meetup[1] and Plancast[2], provide a new type of social network that connects people through events. On these platforms, users can create or join different social groups. Group users can communicate online or hold events offline, which significantly enriches people's social ways and brings huge business value through a large number of active users. For example, Meetup currently has 16 million users with more than 300,000 monthly events [9]. Obviously, in order to improve user satisfaction, effective social event recommendation methods are needed to help users select more appropriate events.

However, existing recommendation methods face two challenges in social event recommendation scenarios. On the one hand, as the increase of user and event types and complexity of interactions, traditional graph-based event recommendation algorithms cannot make good use of the rich hidden information in EBSNs. On the other hand, existing event recommendation algorithms are usually designed for those highly planned events rather than scenarios where impromptu events may be initiated and responded immediately. In the paper, recommending impromptu events to users is called "real-time events matching".

To address the above problems, we proposed a graph embedding based real-time social event matching model for EBSNs recommendation called GERM. It considers not only the rich potential preferences of users, but also the spatio-temporal dynamics of both users and events. For the first challenge, in order to discover more hidden information in the EBSNs, we constructed a heterogeneous information network using various types of EBSNs entities (i.e., user, event, group, tag, venue, and session) and their relations. Then, a meta-path node sampling [16] is performed on the heterogeneous information network to obtain node sequences that retain both the structure information and the interpretable semantics. Hence negative sampling based skip-gram model [11] is used for training the vector representation of different types of nodes in the same feature space by taking previous sampled node sequences as sentences and nodes as words. For the second challenge, we defined the interest similarity matrix, the cost matrix between the user and the event, and a set of matching constraints. Finally, a greedy heuristic algorithm is designed to perform the real-time matching between users and events based on the above definitions. We also designed and implemented a real-time social event publishing prototype system for the experimental comparisons.

In summary, the main contributions of this paper are as follows:

(1) Analyzed the characteristics of EBSNs. Constructed heterogeneous information network by considering the rich types and relations of entity nodes in the EBSNs. First applied the graph embedding method to event recommendation systems based on EBSNs.
(2) Proposed a real-time social event matching algorithm for improving users' satisfaction. According to the real-time user location, time, travel expense, and venue capacity constraints, it dynamically and efficiently matches events for users to maximize the overall matching degree.

---

[1] https://www.meetup.com/.
[2] https://www.plancast.com/.

(3) We verified the proposed algorithm on the real dataset from Meetup and compared it with the existing mainstream algorithms. The experimental results confirm the effectiveness of the proposed algorithm.

2   Related Work

2.1   Recommendation Algorithms for EBSNs

There are two main categories of recommendation algorithms for EBSNs, i.e., algorithms based on history data and algorithms based on graph model.

Algorithms Based on History Data: These algorithms use the historical data in the system to learn the model parameters, which is used to estimate the user's selection preferences and make recommendations. Heterogeneous online+offline social relationships, geographical information of events, and implicit rating data from users are widely used in the event recommendation problems, e.g., [10,14]. Lu Chen et al. considered the constraints of users' spatial information to find highly correlated user groups in social networks more efficiently [2]. Li Gao et al. further considered the influence of social groups to improve the event recommendation performance [6]. I^2Rec [5] is an Iterative and Interactive Recommendation System, which couples both online user activities and offline social events together for providing more accurate recommendation.

Algorithms Based on Graph Model: This kind of algorithms model the system as a graph, and take the event recommendation problem as a proximity node query calculation problem. Since the graph embedding technology can represent the structural and semantic information of the nodes in a graph, it has been widely used in these algorithms.

Tuan-Anh NP et al. [13] proposed a general graph-based model, called HeteRS, which considers the recommendation problem as a query-dependent node proximity problem. After that, more recommendation algorithms based on heterogeneous graphs were proposed. Then Yijun Mo et al. proposed a scale control algorithm based on RRWR-S to coordinate the arrangement of users and activities [12]. Lu Chen et al. [1] proposed a new parameter-free contextual community model, which improves the efficiency of community search and the matching effect. The MKASG [3] model considers multiple constraints, which effectively narrows the search space, and then finds the best matching result.

For the recommendation of events, most existing research work focus on exploiting historical data. Literature [7,8,13], and [12] used heterogeneous information networks while they did not use graph embedding techniques to learn the feature vectors of nodes. Considering that the graph embedding technology has been widely used in the recommendation system and achieved great success, we propose to employ it to calculate more meaningful feature vectors of nodes and further construct nodes similarity matrix for better matching results.

2.2 Event Planning

Event planning refers to developing a plan for users to choose events for participate. Yongxin Tong [19], Jieying She [15], Yurong Cheng [4] and others proved
that such kind of event scheduling is a NP-hard problem. They respectively proposed greedy-based heuristic method, two-stage approximation algorithm, and
other approximate solutions to solve the problem. These methods do not take
into account the needs for users to participate in events dynamically in real-time.
We focus on designing real-time social event matching algorithm in this paper.

3 Graph Embedding Based Real-Time Social Event Matching Model

3.1 Heterogeneous Information Network of EBSNs

We follow the definition of heterogeneous information network introduced by
[13]. It identifies a group of basic EBSNs entity types including *User*, *Event*,
interest Groups, *Tag*, and *Venues*. Further, a composite entity, namely *Session*,
is defined to model the temporal periodicity of a user participating certain event.
In this way, it is natural to connect various types of entities through appropriate
relations to construct a heterogeneous information network for EBSNs.

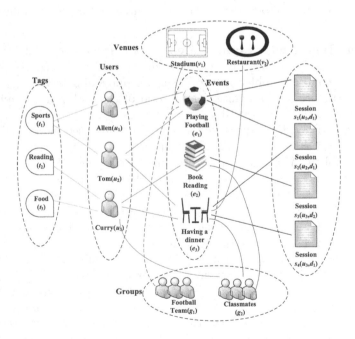

Fig. 1. The heterogeneous information network of EBSNs

Definition 1 (EBSNs Heterogeneous Information Network [13]). *Let* $U = \{ u_1, u_2, \cdots, u_{|U|} \}$, $E = \{e_1, e_2, \cdots, e_{|E|}\}$, $G = \{g_1, g_2, \cdots, g_{|G|}\}$, $T = \{t_1, t_2, \cdots, t_{|T|}\}$, $V = \{v_1, v_2, \cdots, v_{|V|}\}$, *and* $S = \{s_1, s_2, \cdots, s_{|S|}\}$ *be the entity types for describing sets of users, events, groups, tags, venues, and sessions respectively. Let* $C = \{U, E, G, T, V, S\}$ *be the set of entity types. Let* $R = \{\langle U, E \rangle, \langle E, G \rangle, \langle E, V \rangle, \langle U, G \rangle, \langle U, T \rangle, \langle G, T \rangle, \langle E, S \rangle\}$ *be the set of relations. EBSNs heterogeneous information network is defined to be a directed graph* $\mathcal{G} = (\mathcal{V}, \mathcal{E})$. *Here,* $\mathcal{V} = U \cup E \cup G \cup T \cup V \cup S$ *is the node set. And* $\mathcal{E} = \{\langle v_1, v_2 \rangle \mid v_1 \in C_1 \wedge v_2 \in C_2 \wedge \{\langle C_1, C_2 \rangle, \langle C_2, C_1 \rangle\} \cap R \neq \emptyset\}$ *is the edge set where the entity type* $C_1 \in C$ *and* $C_2 \in C$.

The heterogeneous network can clearly express the user's interest preference and spatio-temporal preference for events. The detailed description of a sample heterogeneous network is as follows: As shown in Fig. 1, users u_1 and u_2 together participated in the event e_1 held by the group g_1 in the football field v_1. Both u_1 and u_2 have tag t_1, indicating that they both like sports. We can also find that u_2 and u_3 together participated in the dinner event e_3 held by group g_2 at venue v_2 on the same date d_1. Furthermore, session nodes are created for associating date information with events and users. Suppose that events e_1 and e_3 occurred on the same date d_1, and users u_1 and u_2 together participated in e_1. Then the session node $s_1(u_1, d_1)$ and $s_2(u_2, d_1)$ are created. At the same time, u_2 participated in event e_3 on d_1 as well. Then the event e_3 is connected to s_2.

3.2   Feature Vector Representation Method

In order to make full use of the rich information embedded in the heterogeneous information network for event matching, an effective node feature vector representation method is needed.

Meta-Path Sampling. The first step is to sample the heterogeneous information network. Since both the structure of the network and the semantic relationship between nodes are useful, the meta-path sampling method [16] is employed here. The meta-path is defined as "the sequence of relationships between different types of nodes" [16]. Therefore, meta-path based sampling is more interpretable compared with the traditional random walk sampling method, and has been proved to be helpful for mining heterogeneous information networks [4,17].

Definition 2 (Meta-path Pattern). *Given a heterogeneous information network* $\mathcal{G} = (\mathcal{V}, \mathcal{E})$, *the meta-path pattern* $\mathcal{P}$ *is defined as* $C_1 \xrightarrow{R_1} C_2 \xrightarrow{R_2} \cdots C_t \xrightarrow{R_t} C_{t+1} \cdots \xrightarrow{R_{l-1}} C_l$, *where* $\mathcal{R} = R_1 \circ R_2 \circ \cdots R_{l-1}$ *is a composite relation defined between the node types* C_1 *and* C_l. *And* $C_t \in C$ *and* $R_t \in R$ *are the entity type and the relation type under Definition 1.*

For example, the relation "UEU" indicates that two users have participated in the event together.

Given a meta-path pattern $\mathcal{P}$ and a heterogeneous graph $\mathcal{G} = (\mathcal{V}, \mathcal{E})$, then the transition probability of the meta-path-based sampling is defined as Eq. 1:

$$p\left(v^{i+1}|v_t^i, \mathcal{P}\right) = \begin{cases} \dfrac{1}{\left|N_{t+1}\left(v_t^i\right)\right|} & \langle v^{i+1}, v_t^{i+1}\rangle \in \mathcal{E}, v^{i+1} \in C_{t+1} \\ 0 & \langle v^{i+1}, v_t^{i+1}\rangle \in \mathcal{E}, v^{i+1} \notin C_{t+1} \\ 0 & \langle v^{i+1}, v_t^{i+1}\rangle \notin \mathcal{E} \end{cases} \tag{1}$$

where $v_t^i \in C_t$, and $N_{t+1}\left(v_t^i\right)$ represents the set of neighbor nodes whose type is C_{t+1}, i.e., $v^{i+1} \in C_{t+1}$. Then the next node selected at the current node v_t^i depends on the previously defined meta-path pattern $\mathcal{P}$.

For the case of social event matching for EBSNs recommendation, we defined a meta-path pattern "$UTUGUESEVESEUGUT$" for sampling. In the pattern, the relation $\langle U, T\rangle$ represents the interest tag chosen by the user. The relation $\langle U, G\rangle$ represents the group to which the user belongs. The relation $\langle U, E\rangle$ represents those events that the user participated in. The relation $\langle E, V\rangle$ represents the venue of the event. And the relation $\langle E, S\rangle$ represents when the user participates in the event.

Therefore, given a heterogeneous information network $\mathcal{G}$, by applying meta-path sampling with pattern "$UTUGUESEVESEUGUT$" according to Eq. 1, a set of sampled paths will be generated as a corpus for node feature vector representation learning in the next step.

Representation Learning. As we know, graph embedding is a common technique for node feature vector representation. Given a heterogeneous network $\mathcal{G} = (\mathcal{V}, \mathcal{E})$, the graph embedding based node feature representation for $\mathcal{G}$ is $\mathbf{X} \in \mathbb{R}^{|\mathcal{V}| \times d}$ where d is the dimension of the learnt feature vector and $d \ll |\mathcal{V}|$. We expect the dense matrix $\mathbf{X}$ to effectively capture both the structural and semantic information of nodes and relations.

Here, the Skip-gram model based on negative sampling [11] is used to obtain the feature vector that contains the hidden information of the node context. Let the obtained sampled path set by inputting $\mathcal{G}$ and the meta path pattern $\mathcal{P}$ to the meta-path sampling algorithm be θ. In order to apply the negative sampling Skip-gram model, θ is used as a corpus for subsequent representation learning. First, initialize the node feature matrix $\mathbf{X}$, the iteration termination threshold ϵ, the learning rate η, and the maximum number of iterations Max_iter. Second, train each path in the set θ with respect to the skip-gram objective function in Eq. 2.

$$\arg\max_\theta \prod_{j=1}^{2c} P\left(D = 1|v, t_j, \theta\right) + \prod_{m=1}^{M} P\left(D = 0|v, u_m, \theta\right) \tag{2}$$

where t_j is the j-th positive sample, u_m is the m-th negative sampling sample, θ is the corpus, D indicates whether the current sample is a positive sample. Each node v in the path is taken as a central word whose context with window size c forms a positive sample node set T. And the set of nodes U that do not belong to the context of v are obtained through negative sampling, which is recorded as a negative sample of the central word v. We maximize the probability of $D = 1$ when taking positive samples and the probability of $D = 0$ when taking negative samples.

Then, we can derive Eq. 3 from Eq. 2:

$$\arg\min_{\theta} - \log \sigma \left(X_{t_j} \cdot X_v\right) - \sum_{m=1}^{M} E_{u_m \sim p(u)} \left[\log \sigma \left(-X_{u_m} \cdot X_v\right)\right] \tag{3}$$

Where $\sigma(x) = \frac{1}{1+e^{-x}}$, X_v is the v-th row in X, referring to the feature vector of node v, $p(u)$ is the probability of sampling negative samples. The method of gradient descent is used to optimize the objective function, and the iteration is stopped when it is less than the termination threshold ϵ. The above is the method of learning the hidden feature representation of nodes.

Measuring Proximity. In this paper, "proximity" is employed to represent the similarity between nodes in the heterogeneous information network of EBSNs. The proximity is measured by the cosine similarity between two node feature vectors. Therefore, a similarity matrix $\mathbf{S}$ can be constructed for recording the proximity between any two nodes in a network $\mathcal{G} = (\mathcal{V}, \mathcal{E})$, where element S_{ij} is the proximity between node $v_i \in \mathcal{V}$ and node $v_j \in \mathcal{V}$. The greater the value of S_{ij}, the more similarities between the two nodes, and vice versa.

Suppose $\mathbf{X} \in \mathbb{R}^{|\mathcal{V}| \times d}$ is the learnt node feature matrix representation of $\mathcal{G} = (\mathcal{V}, \mathcal{E})$. And $X_i \left(x_i^{(1)}, x_i^{(2)}, \cdots, x_i^{(d)}\right) (i = 1, 2, \cdots, |\mathcal{V}|)$ is the i-th d-dimensional feature vector of $\mathbf{X}$ corresponding to node v_i. Then the calculation of similarity between node v_i and v_j is as follows:

$$S_{ij} = \frac{X_i \cdot X_j}{\|X_i\| \cdot \|X_j\|} = \frac{\sum_{k=1}^{d} X_i^{(k)} \times X_j^{(k)}}{\sqrt{\sum_{k=1}^{d} \left(X_i^{(k)}\right)^2} \times \sqrt{\sum_{k=1}^{d} \left(X_j^{(k)}\right)^2}} \tag{4}$$

3.3 Real-Time Social Event Matching

Though the above EBSNs representation learning approach facilitates the evaluation of the similarity between users and events, to make the matching achieve real-time EBSNs status response, the dynamics of constraints on users and events should be effectively expressed.

User Profile. Firstly, as a location-based platform, each EBSNs event can set its own unique location information, while users have their respective location at the time when initiating event participation request. Secondly, in order to avoid time conflicts, users may clearly define their available time intervals for participating in events. Moreover, the travel expense is another important factor affecting event participation. Therefore, a user is defined to be a quintuple $u(s, t, l, r, b)$, where $u.s$ and $u.t$ respectively represent the earliest start time (lower limit) and the latest terminate time (upper limit) respectively that the user can participate in an event, $u.l$ represents the current location, $u.r$ is the maximum radius that the user can travel, and $u.b$ is the travel budget.

Event Profile. Events usually have limits on the number of participants. Some sports are typical examples, such as basketball usually involves at least 3 to 5 people. Furthermore, due to the limitation of the venue size, the upper limit on the number of participants should be controlled. In addition, it is also necessary to specify the start and end time of each event. In summary, an event is defined to be a quintuple $e(s, t, l, min, max)$, where $e.s$ and $e.t$ represent the start and terminate time of the event respectively, $e.l$ is the location of the event venue, $e.min$ and $e.max$ are the minimum and maximum number of participants allowed by the event.

Matching Constraints. According to the definitions of user profile and event profile, the following constraints are essential to be applied in our real-time social event matching.

(1) **Budget.** The travel expense of user u_i participating event e_j should not exceed the user's budget, i.e., $\delta_{ij} \leq u_i.b$, where δ_{ij} is the expense that u_i move from his location $u_i.l$ to the event venue $e_j.l$. In Eq. 5, δ_{ij} is calculated based on the Manhattan distance between the two geographic locations where $cost()$ is a given cost function, and (lon_i, lat_i) and $(long_j, lat_j)$ are the coordinate representations of $u_i.l$ and $e_j.l$ respectively.

$$\delta_{ij} = cost\left(|lon_i - lon_j| + |lat_i - lat_j|\right) \tag{5}$$

(2) **Time.** The duration of the event must be within the user's predefined available time intervals so as to guarantee the successful participation, i.e., $u.s < e.s < e.t \leq u.t$ should hold for any matched user u and event e.

(3) **Capacity.** The number of people participating in an event must meet the constraint on the number of people in the event. Let $\sigma_e = |\{u|\forall(u, e) \in \mathcal{E}\}|$ be the total number of users participating in the event e, then there must be $e.min \leq \sigma_e \leq e.max$.

Problem Definition. Once we have determined the similarity measurement and clarified the necessary constraints, the problem of real-time social event matching can be defined as follow.

Definition 3 (Real-Time Social Event Matching). *Let U be the current available user set and E be the current available event set. The real-time social event matching is to assign each $e \in E$ a group of users $U_e = \{u_e^1, u_e^2, \dots, u_e^i, \dots\}$ under constraints (1)–(3) so that the overall matching degree of all current events* *can be maximized. In other words, it is to recommend each $u \in U$ a set of events $E_u = \{e_u^1, e_u^2, \ldots, e_u^j, \ldots\}$ under constraints (1)–(3) for choosing from so that the overall matching degree of all current events can be maximized.*

Here, A_{ij} is used to denote the matching degree between user u_i and event e_j, which considers the effects from both the proximity S_{ij} (Eq. 4) and the travel expense δ_{ij} (Eq. 5). Therefore, we have Eq. 6.

$$A_{ij} = \alpha * S_{ij} - \beta * F\left(\delta_{ij}\right) \tag{6}$$

where weights $0 \leq \alpha \leq 1$, $0 \leq \beta \leq 1$, $\alpha + \beta = 1$, and $F(x)$ is a mapping function with value range $(0, 1)$. Hence, the problem of real-time social event matching is a optimization problem as shown in Eq. 7.

$$\max \sum_{e_j \in E} \sum_{u_i \in U_{e_j}} A_{ij}$$

$$s.t. \qquad \delta_{ij} \leq u_i.b \tag{7}$$
$$u.s < e.s < e.t \leq u.t$$
$$e.min \leq \sigma_e = |\{u | \forall(u, e) \in \mathcal{E}\}| \leq e.max$$

Matching Algorithm. According to the above problem definition of real-time social event matching, we propose a greedy-based heuristic real-time algorithm. The workflow is shown in Fig. 2.

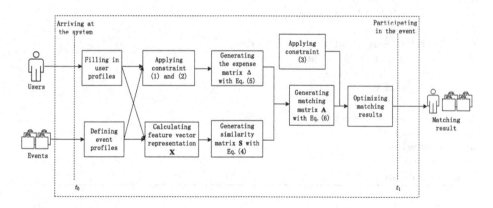

Fig. 2. The workflow of real-time social event matching algorithm

Firstly, using the method described above to calculate the matching degree matrix **A** between users and events, and then use the greedy-based heuristic real-time matching algorithm to calculate the final matching results for recommendation as shown in Algorithm 1.

In line 2, a queue Q is initialized to store the currently active events in the system, and a dictionary R is initialized to store the matching results with users as keys and corresponding sorted candidate events as values. In line 3, events are sorted in ascending order according to their start time, and then stored in Q, which is based on the assumption that events with earlier start time are preferentially matched to users for real-time matching purpose. From line 4 to 15, events are taken from the head of the queue to match with the users in the corresponding column of the matching degree matrix according to the constraints (3), where matched events are stored in R as candidates for future recommendations. Line 6 sorts the columns corresponding to the event e_j in descending order according to the value of the matching degree to ensure that users who meet the constraints (3) are those with a higher matching degree. From line 8 to 14, for an event $e_j \in E$, if the number of users who meet the constraints is greater than $e_j.max$, to maximize the overall matching degree, any user $u_i \in U$ with lower matching degree value A_{ij} should be ignored, so that $e_j.min \leq \sigma_e = |\{u|\forall(u,e) \in \mathcal{E}\}| \leq e_j.max$ is satisfied. Finally, R is returned in line 16.

Algorithm 1. Greedy-based heuristic real-time matching algorithm

---

Input: user set U, event set E, matching matrix $\mathbf{A}$
Output: recommended matching for each user $\{u_1 : E_{u_1}, \cdots\}$
1: **function** GREEDYMATCH(U,E,$\mathbf{A}$)
2:     Initialize a queue Q, a dictionary R
3:     $Q \leftarrow$ sorted(E) by $e.s$ asc
4:     **while** Q is not empty **do**
5:         $e_j \leftarrow Q$.poll()
6:         sorted($\mathbf{A}_{\cdot j}$) by A_{ij} desc
7:         sum $\leftarrow 0$
8:         **for** $u_i \in \mathbf{A}_{\cdot j}$ and sum $\leq e.max$ **do**
9:             **if** u_i **not in** R **then**
10:               $R[u_i] \leftarrow \emptyset$
11:             **end if**
12:             $R[u_i] \leftarrow R[u_i] \cup e_j$
13:             sum $\leftarrow$ sum$+1$
14:         **end for**
15:     **end while**
16:     **return** R
17: **end function**

After the matching algorithm obtains the final matching result, the matching event lists are recommended to users. Then the user can choose to participate in one of the events. Since all the recommended matching events are to be held within the user's expected participation interval, the proposed algorithm is able to meet the user's real-time event participation requirements, and maximize the overall matching degree of the system as well, i.e., maximizing users' interest preferences and minimizing travel expenses.

4 Experiments and Evaluation

In this section, we show the experiments and results, including the data set description, parameter settings and model comparisons.

4.1 Dataset Description

In order to make the evaluation meaningful, we use a real data set that is from Meetup records of user participation event in California (CA) from January 1 to December 31, 2012 [13]. After preprocessing, the data set consists of users who have participated in at least 5 events, events that have at least 5 participants, and groups that have initiated more than 20 events. See Table 1 for detailed.

Table 1. The details of the dataset

Entity type	Events	Users	Groups	Tags	Venues
Quantity	15588	59989	631	21228	4507

The experiment uses 20,000 users and 2,000 events as the test data set. Due to the lack of user budget, event capacity, location and time information in the profile of user and event in the original data set, the above attribute values were randomly set for each user and event. The start and terminate time attributes were integers generated randomly between 10 and 24. The user's budget was randomly generated between 0 and 2000. The location coordinate is a pair of numbers between 0 and 2000, which means the locations of users and events appear in a rectangle of 2000 × 2000. The minimal number of participants min was randomly generated between 5 and 10, while the upper limit max was randomly generated between $min + 5$ and 20.

A heterogeneous information network was constructed according to the method in Sect. 3.1. The experiment was conducted in batches of 10 days, and 2,000 users and 400 events were randomly selected from the test data set every day. User nodes were connected to the network according to their tags relations. Event nodes were connected to the network according to their group relations. Every day after applying the matching algorithm, users and events in the matched result were appended to the network. The above steps repeated for the next day.

4.2 Evaluation Criteria

Real-time social event matching technology not only needs to satisfy users' interest preferences, but also ensures that users can participate in events in a timely manner. For this purpose, the performance of matching algorithms is measured from the following aspects.

1. **User matching success rate**. This value represents the ratio of the every-day number of users who successfully participate in the event to the total number of users. The higher the user matching success rate, the better the algorithm performance.
2. **Event matching success rate**. This value represents the ratio of the every-day number of events successfully held to the total number of events. Similarly, the higher the value, the better the performance of the matching algorithm.
3. **The average matching degree of users**, which represents the ratio of the sum of the matching degrees of users who successfully participated in the events to the total number of successful participants. The higher the value, the better the performance of the matching algorithm. Since the matching degree defined in Eq. 6 involves both the preference similarity and the travel expense satisfaction, this measurement somewhat reflects the ability of the algorithm to reveal implicit feature representations and save expenses.
4. **The average user waiting time**, which represents the average value of the differences between the time t_0 of all users arriving to the system who participating in the event and t_1 of the start time of the corresponding event. Apparently, it is an important indicator for measuring real-time performance of the algorithm. The smaller the average waiting time, the better the performance of the algorithm.

4.3   Performance Comparisons

For the convenience of description, we named our model as *GERM*. The random matching algorithm *Random* and *AGMR* [13] are the baselines for comparing with *GERM*.

Figure 3 shows the experimental results. Doubtless, the Random algorithm has a stable but relative low performance in all aspects.

For GERM and AGMR, the success rate of user matching is even worse at the very beginning of the experiment because of the lack of data in the network as shown in Fig. 3(a). And hence, since the total number of events remains unchanged (i.e., 2000), the success rate of event matching is not better than that of Random as shown in Fig. 3(b). However, as the experiment progressed, the heterogeneous information network became more complicated, which provides richer information to GERM and AGMR for improving the overall accuracy of matching. Thus, the matching success rate of user and event, and the average matching degree gradually increased until they became stable as shown in Fig. 3(a),(b),(c). Moreover, the result plots of GERM was generally above those of AGMR. Because GERM uses the graph embedding method to obtain node feature vectors and calculate the similarity of nodes, as the data gets richer, it can better mine feature information. Secondly, AGMR recommends static event lists for users, but GERM can dynamically recommend events for users according to location and time. Figure 3(d) shows the comparisons of the average user waiting time. It can be seen that the GERM result curve is always below those

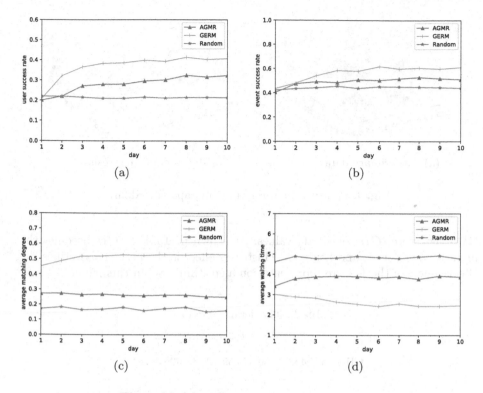

Fig. 3. Model performance comparison result

of AGMR and Random. The average user waiting time is the smallest, which can meet the user's real-time requirements.

In summary, the model GERM proposed in this paper is superior to the AGMR and Random.

4.4   Discussion on Graph Embedding

We also performed clustering analysis to verify the effectiveness of the graph embedding based feature representation method used in this paper.

The cluster analysis was conducted on the feature vectors of the five types of entities shown in Table 1. Because the original data set is too large to visualize, only 100 samples were randomly selected from each type of data in the experiment. The KMeans algorithm in Sklearn toolkit [18] was used to cluster the extracted samples. The visualization result after clustering is shown in Fig. 4. We can see from Fig. 4(b) that after embedding into feature vectors, entities of the same type will still be clustered together, which shows that the feature vectors generated based on the graph embedding method in this paper can effectively reflect both the structural information and semantic relationship of the original data.

Furthermore, since the experimental data contains five types of entities, we tried to set different number of clusters to calculate corresponding Calinski-

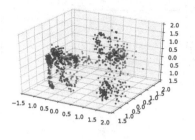
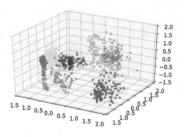

(a) The primary data   (b) The clustering result

Fig. 4. Performance verification of graph embedding

Harabasz score (CH for short) values. As shown in Table 2, CH becomes the highest when k is 5, and hence the clustering effect is the best. This verifies the effectiveness of the feature representation algorithm used in this paper.

Table 2. The details of data set

k	3	4	5	6	7
CH-score	48.23	50.47	59.72	42.69	39.13

5 Conclusion and Future Work

In this paper, we have proposed a real-time social event matching model based on heterogeneous graph embedding, named GERM. Compared with existing models, our model uses graph embedding method to calculate the feature vector of nodes, which can effectively mine deeper preferences of users' interests and events. Considering the user's preferences and the spatio-temporal characteristics of users and events, to achieve more accurate real-time matching of users and events. Experiment results on Meetup dataset have shown that the proposed algorithm has improvements on the matching success rate, average matching degree, user satisfaction, and user waiting time.

In this study, there are still many aspects worth continuing to improve. As the scale of heterogeneous information network increases, more time will be consumed in the sampling process of node sequences. Therefore, it will be very interesting to explore efficient sampling algorithms. In addition, GERM lacks the analysis of social relationships between users. It will be meaningful to take social relationships into consideration to help users discover more interesting events.

References

1. Chen, L., Liu, C., Liao, K., Li, J., Zhou, R.: Contextual community search over large social networks. In: 2019 IEEE 35th International Conference on Data Engineering (ICDE) (2019)
2. Chen, L., Liu, C., Zhou, R., Li, J., Wang, B.: Maximum co-located community search in large scale social networks. Proc. VLDB Endow. **11**, 1233–1246 (2018)
3. Chen, L., Liu, C., Zhou, R., Xu, J., Li, J.: Finding effective geo-social group for impromptu activity with multiple demands (2019)
4. Cheng, Y., Yuan, Y., Chen, L., Giraudcarrier, C., Wang, G.: Complex event-participant planning and its incremental variant, pp. 859–870 (2017)
5. Dong, C., Shen, Y., Zhou, B., Jin, H.: I^2Rec: an iterative and interactive recommendation system for event-based social networks, pp. 250–261 (2016)
6. Gao, L., Wu, J., Qiao, Z., Zhou, C., Yang, H., Hu, Y.: Collaborative social group influence for event recommendation, pp. 1941–1944 (2016)
7. Li, B., Bang, W., Mo, Y., Yang, L.T.: A novel random walk and scale control method for event recommendation. In: 2016 International IEEE Conferences on Ubiquitous Intelligence & Computing, Advanced and Trusted Computing, Scalable Computing and Communications, Cloud and Big Data Computing, Internet of People, and Smart World Congress (UIC/ATC/ScalCom/CBDCom/IoP/SmartWorld) (2016)
8. Liu, S., Bang, W., Xu, M.: Event recommendation based on graph random walking and history preference reranking. In: International ACM SIGIR Conference (2017)
9. Liu, X., He, Q., Tian, Y., Lee, W., Mcpherson, J., Han, J.: Event-based social networks: linking the online and offline social worlds, pp. 1032–1040 (2012)
10. Macedo, A.Q.D., Marinho, L.B.: Event recommendation in event-based social networks. In: 1st International Workshop on Social Personalisation (SP 2014) (2014)
11. Mikolov, T., Chen, K., Corrado, G.S., Dean, J.: Efficient estimation of word representations in vector space (2013)
12. Mo, Y., Li, B., Bang, W., Yang, L.T., Xu, M.: Event recommendation in social networks based on reverse random walk and participant scale control. Future Gener. Comput. Syst. **79**(PT.1), 383–395 (2018)
13. Pham, T.A.N., Li, X., Gao, C., Zhang, Z.: A general graph-based model for recommendation in event-based social networks. In: 2015 IEEE 31st International Conference on Data Engineering (2015)
14. Qiao, Z., Zhang, P., Cao, Y., Zhou, C., Guo, L., Fang, B.: Combining heterogenous social and geographical information for event recommendation, pp. 145–151 (2014)
15. She, J., Tong, Y., Chen, L.: Utility-aware social event-participant planning, pp. 1629–1643 (2015)
16. Sun, Y., Han, J., Yan, X., Yu, P.S., Wu, T.: Pathsim: meta path-based top-k similarity search in heterogeneous information networks. Proc. VLDB Endow. **4**(11), 992–1003 (2011)
17. Sun, Y., Norick, B., Han, J., Yan, X., Yu, P.S., Yu, X.: Pathselclus: integrating meta-path selection with user-guided object clustering in heterogeneous information networks. ACM Trans. Knowl. Discov. Data **7**(3), 11 (2013)
18. Swami, A., Jain, R.: Scikit-learn: machine learning in python. J. Mach. Learn. Res. **12**(10), 2825–2830 (2012)
19. Tong, Y., Meng, R., She, J.: On bottleneck-aware arrangement for event-based social networks, pp. 216–223 (2015)

Competitor Mining from Web Encyclopedia: A Graph Embedding Approach

Xin Hong[1,2], Peiquan Jin[1,2(✉)], Lin Mu[1,2], Jie Zhao[3], and Shouhong Wan[1,2]

[1] University of Science and Technology of China, Hefei 230027, Anhui, China
jpq@ustc.edu.cn
[2] Key Laboratory of Electromagnetic Space Information, China Academy of Science, Hefei 230027, Anhui, China
[3] School of Business, Anhui University, Hefei 230601, Anhui, China

Abstract. Mining competitors from the web has been a valuable and emerging topic in big data and business analytics. While normal web pages may include incredible information like fake news, in this paper, we aim to extract competitors from web encyclopedia like Wikipedia and DBpedia, which provide more credible information. We notice that the entities in web encyclopedia can form graph structures. Motivated by this observation, we propose to extract competitors by employing a graph embedding approach. We first present a general framework for mining competitors from web encyclopedia. Then, we propose to mine competitors based on the similarity among graph nodes and further present a similarity computation method combing graph-node similarity and textual relevance. We implement the graph-embedding-based algorithm and compare the proposed method with four existing algorithms on the real data sets crawled from Wikipedia and DBpedia. The results in terms of precision, recall, and F1-measure suggest the effectiveness of our proposal.

Keywords: Web encyclopedia · Competitor mining · Similarity measurement · Heterogeneous graph

1 Introduction

Competitor mining [1] has been an important issue in the big data era. With the advance of the Internet, competition is becoming more serious in almost every business area. Especially in Internet-based businesses, new companies appear frequently, which makes it difficult to detect competition situations in the market. For instance, if one new company plans to start research and development of Internet-based intelligent tutoring, it is much hard for the company to know the competence situation because of the rapid development of the area. Therefore, an automatic intelligent competitor mining tool is needed to provide the newest and complete information about concerned competitors. Meanwhile, the web has become an encyclopedia especially due to the appearance of some web knowledge bases like Wikipedia (www.wikipedia.org) and DBpedia (https://wiki.dbpedia.org/). As a result, mining competitors from web encyclopedia has been a valuable and emerging topic in big data and business analytics.

© Springer Nature Switzerland AG 2020
Z. Huang et al. (Eds.): WISE 2020, LNCS 12342, pp. 56–68, 2020.
https://doi.org/10.1007/978-3-030-62005-9_5

Some existing works are focusing on competitor mining from the Web [1–4]. However, previous studies in competitor mining focused on web pages or entities and neglected the graph-structure information among competitors. In addition. the incredible information like fake news in web pages will make the extracted information untrusted. On the other hand, we found that the rich information in the current web encyclopedia like Wikipedia and DBpedia is helpful to construct a graph structure among competitors, products, business areas, and other information.

In this work, we concentrate on mining competitors from web encyclopedia. Specially, we focus on Wikipedia and DBpedia, as they are among the biggest web encyclopedia in the world. Differing from previous rule-based competitor mining approaches, we consider the graph structure about competitors and propose a new graph embedding approach to mine competitors from web encyclopedia. Briefly, the differences of this study from existing works and the unique contributions of the paper can be summarized as follows:

(1) We propose to mine competitors from web encyclopedia rather than from web pages or search engines. Further, we notice that the entities in web encyclopedia can form intrinsic graph structures. Thus, we propose to mine competitors by combining graph embedding and text mining. To the best of our knowledge, this is the first study that integrates graph embedding and text mining to extract competitors from web encyclopedia.
(2) We develop a framework for mining competitors from web encyclopedia. Particularly, we use the similarity among graph nodes to measure the competitive relationships and propose a similarity computation method based on graph-node similarity and textual relevance. In addition, we give the detailed implementation of the graph embedding approach.
(3) We evaluate the proposed method on data sets crawled from Wikipedia and DBpedia, and compare our algorithm with four existing algorithms. The experimental results suggest the effectiveness of our proposal.

2 Related Work

In this section, we review the related works of this study, including competitor mining, Wikipedia-related studies, and graph embedding.

Competitor Mining. There have been a few similar works studying competitors mining problems based on text mining techniques [1–7]. Bao et al. [1] proposed a novel algorithm called CoMiner, which conducted web-scale mining in a domain-independent method. Bondarenko et al. [2] studied the comparative web search problem in which the user inputted a set of entities and the system returned relevant information about the entities from the web. Chen et al. [5] presented a framework user-oriented text mining. And they discovered knowledge by the form of association rules. Li et al. [6] developed a weakly-supervised bootstrapping method for competitive question identification and comparable entity extraction by adopting a mass online question archive. In addition, Ruan et al. [7] leveraged seeds of competition between companies and competition between products to distantly supervise the learning process to find text patterns in free texts, i.e., Yahoo! Finance and Wikipedia.

Wikipedia Data Analysis. There are also some previous works focusing on Wikipedia data analysis. Lange et al. [8] presented a system that automatically populates the infoboxes of Wikipedia articles by extracting attribute values from the article's text. This system achieved an average extraction precision of 91% for 1,727 distinct infobox template attributes. Lara Haidar-Ahmad et al. [9] leveraged RDF to build a sentence representation and SPARQL to model patterns and their transformations so that easing the querying of syntactic structures and the reusability of the extracted patterns.

Graph Embedding. Our work is also related to graph mining, which is an important problem in network analysis. Perozzi et al. [10] used local information obtained from truncated random walks to learn latent representation by treating walks as the equivalent of sentences. Grover et al. [11] proposed an algorithm to generalize prior work which was based on rigid notions of network neighborhoods. It defined a flexible notion of a node's network neighborhood and designed a biased random walk procedure. Dong et al. [12] formalized meta-path-based random walks to construct the heterogeneous neighborhoods of a node and then leveraged a heterogeneous skip-gram model to perform node embeddings.

Differing from all previous studies, in this study we propose to integrate graph embedding with textual relevance to improve the effectiveness of competitor mining from web encyclopedia. To the best of our knowledge, this is the first work that considers graph embedding and textual relevance for the competitor-mining problem toward web encyclopedia.

3   Framework for Competitor Mining from Web Encyclopedia

Figure 1 shows the framework of mining competitors from Wikipedia and DBpedia. It consists of several modules, including data crawling, entity extraction, similarity computation, and results ranking.

(1) *Data Crawling.* We craw raw data from Wikipedia and DBpedia. Wikipedia is an online encyclopedia that has a lot of knowledge about entities in the real world. Wikipedia not only contains mass unstructured data but also contains some semi-structured data. Similarly, DBpedia contains much structured information that is made available on the world wide web. Finally, we crawl raw data from Wikipedia and DBpedia using the pattern matching technology [13].

(2) *Entity Extraction.* We mainly extract entities using the infobox information of Wikipedia and some valuable knowledge from DBpedia which contains many kinds of structured knowledge. We utilize the redirection [14] of Wikipedia pages to solve the ambiguity problem between entities. What's more, we also use stem extracting technology to extract the stemming of an industry entity.

(3) *Graph Generating.* We use different node types to create a company knowledge graph. Based on the business background of competitor mining and analysis on some example webpages in Wikipedia, we define four node types, i.e., Industry,

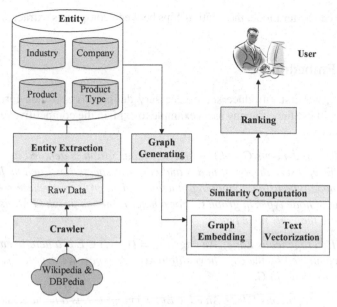

Fig. 1. The framework of competitor mining

Company, Product, Product Type. Industry nodes are regarded as inter-linking nodes between companies. We can also link two companies according to the product information from the companies. There are also some other relationships between entity types, e.g., a company owns an industry and an industry is owned by a company. Consequently, the constructed graph is stored in a Neo4j database.

(4) *Similarity Computation.* The competitive relationships between two companies are mainly determined by their similarity in the real world. Previous works mainly focused on document similarity, but in this paper, we consider graph-structure-based similarity in addition to document similarity. Specially, we first compute graph-node similarity by generating entity vectors according to the graph embedding approach. Further, we employ a heterogeneous graph embedding approach based on a random walk [15] and SkipGram [16] to generate entity vectors. In addition, we compute textual relevance by transforming each entity into a vector using Doc2vec [17]. These two kinds of vectors are finally concatenated and we use the cosine similarity between two concatenated vectors to measure the similarity between two companies.

(5) *Ranking.* The ranking module returns a sorted competitor list for a given company based on the cosine similarity computed by the similarity computation model.

The key point of the proposed framework for competitor mining is that we introduce the graph embedding approach in addition to the textual relevance which has been widely used in text mining. As graphs can describe rich relationships between entities, we can extract competitor relationships by constructing a company knowledge graph derived from web encyclopedia. Based on graph-node similarity and traditional textual relevance, we can better model the relationships between companies extracted from web encyclopedia.

4 Graph Embedding

In this section, we first introduce some necessary definitions and then formulate the problem. We will use the company as an example to explain the competitive relationship mining problem.

Definition 1 (Heterogeneous Graph). A heterogeneous graph is denoted as $G = (V, E)$, where V and E represent the set of nodes and edges, respectively. Each node $v \in V$ is mapped to a node type using a mapping function ϕ, i.e., $\phi(v) = r$, where $r \in R$ and R denotes the set of node types in graph G. For a heterogeneous graph G, $|R|$ is over 1 as R contains more than one node types. ∎

Definition 2 (Heterogeneous Edge). An edge $E_{he} = (v_s, v_t) \in E$ is a heterogeneous edge if two nodes connected by the edge have different node types. We denote E_{he} as the set of heterogeneous edges in G. ∎

Definition 3 (Homogeneous Edge). An edge $E_{ho} = (v_s, v_t) \in E$ is a homogeneous edge if the two nodes connected by the edge are from the same domain, i.e., $\phi(v_s) = \phi(v_t)$. We use E_{ho} to denote the set of homogeneous edges in G. ∎

Definition 4 (Company Heterogeneous Graph). We create a directed label graph of companies $G = (V, E, \phi, \varphi)$ where:

(1) *V is a set of entities (nodes) in the graph.*
(2) *E is a set of directed edges in the graph representing relationships between entities.*
(3) *$\phi: V \to P(\tau)$ is a mapping function that assigns each entity v $V \to$ to an entity type. Here, $\phi(v) \subseteq \tau$.*
(4) *$\varphi: E \to R$ is a mapping function that assigns each edge $e \in E$ to a relation type. Here, $\varphi(E) \in R$.* ∎

Based on the above definitions, we can describe the basic idea of the graph-embedding-based approach for competitor mining as follows:

- First, we build a company heterogeneous graph $G = (V, E, \phi, \varphi)$.
- Second, for a given company, we employ a random walk technique to traverse the graph to generate the candidate nodes for the company. These candidate nodes are regarded as competitors candidates.
- Third, we perform a graph-node embedding learning process to learn a vector of company entity embedding from the company heterogeneous graph. We use the learned vector to compute the graph-node similarity between the given company and all the candidate competitors.
- Fourth, we compute the textual relevance between the given company and all the candidate competitors.
- Finally, we combine the graph-node similarity and the textual relevance to rank all candidates. The top-k competitors are returned as the result.

4.1 Random Walk in the Company Heterogeneous Graph

We use random walk technology to traverse the graph to generate candidate nodes, which are considered to contain the relationship between competitors. When we perform a random walk on a homogeneous graph, we select a node by sampling a node from the current node uniformly. In contrast, for a heterogeneous company graph, there are two strategies to choose the next node.

(1) *Skip-Based Random Walk.* This refers to uniformly sampling a node from the nodes connected to v_i through heterogeneous edges. The candidate node set for skipping from v_i to the target node type t is represented by Eq. (1).

$$V_{he}(v_i) = \{v | (v_i, v) \in E_{he} \cup (v, v_i) \in E_{he}\} \tag{1}$$

(2) *Keep-Based Random Walk.* This policy is to uniformly sample one node from those linked to v_i via homogeneous edges. The corresponding candidate set of nodes is denoted by Eq. (2).

$$V_{ho}(v_i) = \{v | (v_i, v) \in E_{ho} \cup (v, v_i) \in E_{ho}\} \tag{2}$$

Algorithm 1. Random-walk-based graph construction

Input:
      (1) A heterogeneous graph $G = (V,E)$
      (2) the initial stay probability α
      (3) the number of memorized domains m
      (4) the number of random walks per node p
      (5) the maximum walk length L_{max}
Output: W, the node-set of random walks.
```
1:    W = ∅;
2:    for i = 1 to p do
3:        for each v ∈ V do
4:            Initialize a random walk by adding v;
5:            while |W| < L_max do
7:                Make a Skip or Keep decision according to Eq.(3);
8:                if Keep then
9:                    | Continue W by keeping;
10:               else if Jump then
11:                   Sample a target node type r according to Eq.(4);
12:                   Continue W by Skipping to node type r;
13:                   Update Q by skipping only last m domains;
14:               end if
15:           end while
16:           Add w to W
17:       end for
18:   end for
19:   return W
```

In the graph traversal process based on random walks, we aim to uniformly select nodes from a candidate set that mainly includes three aspects. For example, the neighborhood of v_i has only homogeneous edges or only heterogeneous edges. It is also possible to have both homogeneous and heterogeneous edges. Based on the above two choices, we manage the random walk process probabilistically. Then, we calculate the probability by Eq. (3) to decide whether to skip. Here, the parameter of ℓ means that when the node has the same node type as the current node v_i, they will be continuously accessed. If no edge is connected to v_i then we can only link to another node with a different node type. If the neighbor nodes of v_i have no heterogeneous edges, the random walk will be restricted to the same node type. In addition, if there are both heterogeneous and homogeneous edges linked to the current node, we choose the probability $e^{-\alpha\ell}$ to stay in a node with the same node type as the current node. Here, we use an exponential decay function to avoid the long-stay in nodes with the same node type during the random walk process. The next node to jump to is determined by Eq. (4).

$$P(v_i) = \begin{cases} 0 & \text{if } V_{he}(v_t) = \emptyset \\ 1, & \text{if } V_{ho}(v_t) = \emptyset \\ e^{-\alpha\ell}, & \text{otherwise} \end{cases} \tag{3}$$

$$R_{jump}(v_i) = \begin{cases} \left\{ r \mid r \in T, V_{Skip}^r(v_i) \neq \emptyset \right\}, & \text{if nonempty} \\ \left\{ r \mid r \in S, V_{Skip}^r(v_i) \neq \emptyset \right\}, & \text{otherwise} \end{cases} \tag{4}$$

Algorithm 1 shows the details of the random walk with the above-mentioned strategies to sample from the heterogeneous graph [18]. For each node $v \in V$, we initialize a random walk starting from v_i until the maximum length L_{max} is reached.

4.2 Graph-Node Embedding Learning

Based on the number of generated random walk nodes, we use the SkipGram model [16] to train graph node embeddings. The SkipGram model can maximize the co-occurrence probability between the current node and its context nodes. The probability of co-occurrence in the SkipGram model is defined by Eq. (5).

$$P\left((v_i, v_j) \in T_{walk}\right) = \delta\left(\vec{v}_i \cdot \vec{v}_j\right) \tag{5}$$

Let $\delta(\bullet)$ be a sigmoid function, $\vec{v}_i$ and $\vec{v}_j$ be the vectors of v_i and v_j, we can define the negative sampling probability by Eq. (6), where V_F represents the negative sampling node randomly selected. In summary, the model needs to maximize the loss function, as shown in Eq. (7).

$$P((v_i, v_F) \notin T_{walk}) = 1 - P((v_i, v_F) \in T_{walk}) \tag{6}$$

$$\log P\left((v_i, v_j) \in T_{walk}\right) + \beta \cdot E_{v_F}\left(P\left(v_i, v_j\right) \notin T_{walk}\right) \tag{7}$$

Then, we use the generated graph-node vector to concatenate the text vector that is pre-trained with *gensim* (https://radimrehurek.com/gensim/). With this mechanism, we can integrate the graph feature of a node with its text feature. Finally, we take the concatenated vector as input and measure the graph-node similarity based on the cosine similarity, as shown in Eq. (8).

$$sim(v_i, v_j) = \frac{\vec{v}_i \cdot \vec{v}_j}{\|\vec{v}_i\| \|\vec{v}_j\|} \tag{8}$$

4.3 Textual Relevance

In addition to the graph-node similarity between nodes in the graph, we also consider the textual relevance that can reflect the similarity of the documents between companies crawled from web encyclopedia.

In Doc2vec [17], each sentence is represented by a unique vector which is expressed as a column of matrix D. Each word is also represented by a unique vector which is represented by a certain column of the matrix W. The word vector corresponding to the input word and the sentence vector corresponding to the sentence are used as the input of the input layer. The vector of the sentence and the word vector of the sample are added to form a new vector X. Then, we can use this vector X to predict the predicted words in the window. Doc2vec differs from Word2vec in that Doc2vec introduces a new sentence vector to the input layer. Paragraph vectors can also be regarded as another kind of word vectors. As a result, we use the concatenation of the vector representing three contextual words to predict the fourth word. The paragraph vector representing the paragraph topic is used to avoid missing information in the current context.

Based on the above representation of documents, we further employ the cosine similarity to compute the textual relevance between two companies. Basically, for a given company, the companies with high similarity scores will be regarded as the main competitors of the given company. In our implementation, we return the top-k companies as potential competitors for a given company.

5 Performance Evaluation

We conduct experiments on data sets crawled from Wikipedia and DBPedia to evaluate the performance of the proposed graph-embedding-based competitor mining algorithm. All the experiments are performed on a computer with an Intel Core i7-7700 3.6 GHz CPU, 16 GB DRAM, and a 512 GB SSD. All algorithms are implemented by Python 3.5.

5.1 Settings

Data Set. The data set was crawled from Wikipedia and DBPedia. We especially focus on the competitive relationships in Wikipedia and DBPedia. As a result, we prepare a data set of companies that contain 2616 companies from different industries. As shown in Table 1, the heterogeneous graph constructed from the data set has 12692 entity nodes, and 32898 edges, in which 16403 are homogenous edges, and the rest are heterogeneous edges.

Table 1. Characteristics of the graph constructed from the data set

Dataset	Company Graphs
$|V|$	12692
$|E|$	32898
$|E_{ho}|$	16403
$|E_{he}|$	16495

Among the extracted entities from Wikipedia and DBPedia, we choose 50 companies that are listed in Fortune 500 to measure the effectiveness of the graph embedding based competitor mining algorithm. One problem is the lack of annotated ground truth. The best way is to conduct a substantial survey over major companies but this is time-consuming. Therefore, we take the evaluation method used in the literature [19], which uses the recommendation results of Google search when we input a specific company name as the search criteria.

Compared Methods. In our experiments, we mainly compare our graph embedding approach with four previous algorithms, as described below.

(1) *Pathsim* [18]. This method computes node similarity in a heterogeneous graph by meta-paths. We use two kinds of meta-paths to extract competitive relationships from the data set, which are "C(company) - I(industry)-C(company)" and "C(Company) - P(Product) - PT(Product Type) - P(Product) - C(Company)". If there is a meta-path connected to a given company in the graph which is within the above two kinds of meta-paths, we output the company entities on the meta-path as the competitors of the company.

(2) *Doc2vec* [17]. This method is a text mining algorithm, which learns the text vector of a company based on the text corpus crawled from Wikipedia. We use *gensim* to train document vectors and set the dimension number to 200 by default.

(3) *Metapath2vec* [12]. This is also a heterogeneous graph embedding method which adopts meta-path-based random walks to construct the heterogeneous neighborhood of a node. Then, it uses a skip-gram model to perform node embeddings. We use this model to train node embeddings and set the dimension number to 200.

(4) *Hin2vec* [20]. This method applies the neural network model to carry out multiple prediction training tasks for a target set of relationships. It finally learns the latent vectors of nodes and meta-paths in a heterogeneous information network.

Parameters Setting. The settings of the parameters in the graph embedding are as follows: $r = 30$, $L_{max} = 30$, $k = 10$, $and\ d = 200$.

5.2 Results

Figures 2, 3, and 4 show the comparison of precision, recall, and F1-measure, respectively, for the five methods. The X-axis represents the top-n competitors returned by each method. As companies mainly focus on their major competitors in the market, we measure the performance of each method for extracting top-5, top-10, and top-15 competitors.

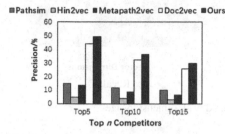

Fig. 2. Precision comparison of the five methods

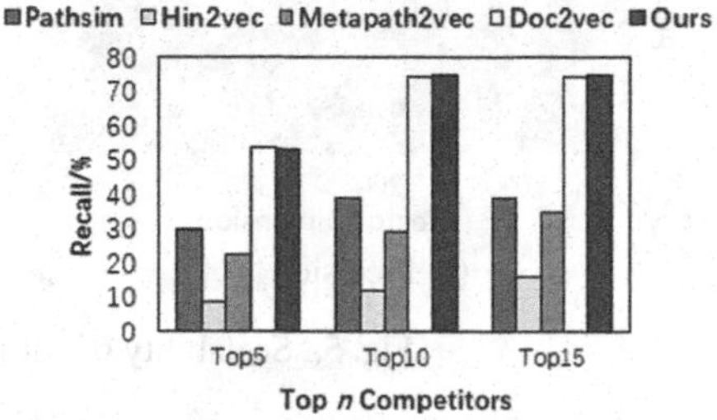

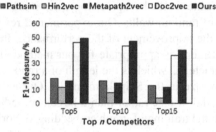

Fig. 3. Recall comparison of the five methods

Fig. 4. F1-measure comparison of the five methods

We can see that all figures report a very similar trend. Pathsim, Metapath2vec, and Hin2vec all get the worst performance in all experiments, indicating that it is not a good choice to only consider the graph structure in competitor mining. In addition, the text-only method Doc2vec performs better than Pathsim, Metapath2vec, and Hin2vec, meaning that textual information is more important for competitor mining from web encyclopedia. This is because the documents in Wikipedia and DBpedia are semi-structured and with good quality. Our method that combines the graph embedding and textual relevance outperforms all the other four methods. This is mainly due to the combination of graph embedding and textual relevance. Thus, we can conclude that it is better to consider both graph structural information and textual relevance for mining competitors from web encyclopedia.

Next, we vary the dimension size to measure the scalability of our method concerning the dimension size. The results in terms of precision and recall are shown in Fig. 5. We can see that the precision of our method is relatively stable when the dimension size varies from 100 to 400. Therefore, our method can be extended to larger documents that involve more dimensions. Note that the documents in web encyclopedia like Wikipedia and DBPedia are generally smaller than free web pages, e.g., news web pages. Thus, we believe that our method can also be applied to free web pages in case we devise an effective approach to remove the incredible information from free web pages.

Finally, we measure the effect of parameters on the performance of our method. The results are shown in Fig. 6, where four parameters including α, k, L, and w are measured.

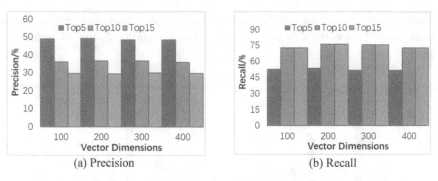

(a) Precision   (b) Recall

Fig. 5. Scalability of our method w.r.t. dimension size

The parameter α defines the probability of node traversing during the random walk. We can see that our method achieves the best precision when it is set to 0.4. The parameter k represents the number of random walks. Our experimental result shows that higher k will not always result in the improvement of the performance. Instead, a middle value, which is 30 in our experiment, is appropriate for our method. This is also applicable to the setting of the parameter L, which is the length of a random walk. The parameter w represents the window size in the SkipGram model. Our experiment shows that it is better to use a small window size rather than a large one, indicating only the words nearby are meaningful in the training of graph embedding vectors.

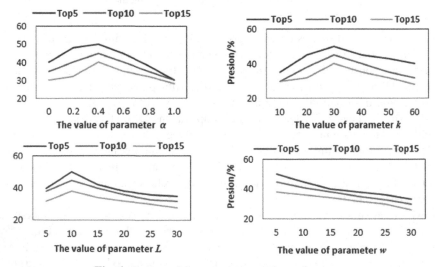

Fig. 6. Impact of the parameters on the performance

6 Conclusions

In this paper, we study the problem of mining competitive relationships by learning heterogeneous graphs constructed from web encyclopedia including Wikipedia and DBPedia. We develop a general framework for mining competitors from web encyclopedia. Particularly, we use the similarity among graph nodes to measure competitive relationships and propose to extract competitors based on graph-node similarity and textual relevance. We evaluate the proposed method on data sets crawled from Wikipedia and DBpedia, and compare our algorithm with four existing algorithms. The experimental results suggest the effectiveness of our proposal.

Some future works are worth further investigating. First, companies may has different competitive relationships [21], which need to be further distinguished. Second, we will study the credibility evaluation [22] of Wikipedia information to enhance the validity of the detected competitive relationships.

Acknowledgement. This study is supported by the National Key Research and Development Program of China (2018YFB0704404) and the National Science Foundation of China (61672479). Peiquan Jin is the corresponding author.

References

1. Bao, S., Li, R., Yu, Y., Cao, Y.: Competitor Mining with the Web. IEEE Trans. Knowl. Data Eng. **20**(10), 1297–1310 (2008)
2. Bondarenko, A., et al.: Comparative web search questions. WSDM, 52–60 (2020)
3. Zhao, J., Jin, P.: Conceptual modeling for competitive intelligence hiding in the internet. J. Softw. **5**(4), 378–386 (2010)
4. Zhao, J., Jin, P.: Towards the extraction of intelligence about competitor from the web. In: Lytras, M.D., et al. (eds.) Visioning and Engineering the Knowledge Society. A Web Science Perspective. Lecture Notes in Computer Science, vol. 5736, pp. 118–127. Springer, Berlin, Heidelberg (2009). https://doi.org/10.1007/978-3-642-04754-1_13
5. Chen, X., Wu, Y.: Web mining from competitors' websites. In: KDD, pp. 550–555 (2005)
6. Li, S., Lin, C., Song, Y., Li, Z.: Comparable Entity mining from comparative questions. In: ACL, pp. 650–658 (2010)
7. Ruan, T., Xue, L., Wang, H., Pan, J.: Bootstrapping yahoo! finance by wikipedia for competitor mining. In: Qi, G., Kozaki, K., Pan, J., Yu, S. (eds.) Semantic Technology. Lecture Notes in Computer Science, vol. 9544, pp. 108–126. Springer, Cham (2015). https://doi.org/10.1007/978-3-319-31676-5_8
8. Lange, D., Böhm, C., Naumann, F.: Extracting Structured Information from Wikipedia Articles to Populate Infoboxes. CIKM, pp. 1661–1664 (2010)
9. Haidar-Ahmad, L., Zouaq, A., Gagnon, M.: Automatic extraction of axioms from wikipedia using SPARQL. In: Sack, H., et al. (eds.) The Semantic Web. ESWC 2016. Lecture Notes in Computer Science, vol. 9989, pp. 60–64. Springer, Cham (2016). https://doi.org/10.1007/978-3-319-47602-5_13
10. Perozzi, B., Al-Rfou, R., Skiena, S.: DeepWalk: online learning of social representations. KDD, 701–710 (2014)
11. Grover, A., Leskovec, J.: node2vec: scalable feature learning for networks. KDD, 855–864 (2016)

12. Dong, Y., Chawla, N., Swami, A.: metapath2vec: scalable representation learning for heterogeneous networks. KDD, 135–144 (2017)
13. Haghighat, M., Li, J.: Toward fast regex pattern matching using simple patterns. In: ICPADS, pp. 662–670 (2018)
14. Hill, B., Shaw, A.: Consider the redirect: a missing dimension of wikipedia research. OpenSym **28**(1–28), 4 (2014)
15. Tamir, R.: A random walk through human associations. In: ICDM, pp. 442–449 (2005)
16. Pickhardt, R., et al.: A generalized language model as the combination of skipped n-grams and modified Kneser Ney smoothing. In: ACL, pp. 1145–1154 (2014)
17. Le, Q., Mikolov, T.: Distributed representations of sentences and documents. In: ICML, pp. 1188–1196 (2014)
18. Sun, Y., Han, J., Yan, X., Yu, P., Wu, T.: PathSim: meta path-based top-K similarity search in heterogeneous information networks. PVLDB **4**(11), 992–1003 (2011)
19. Ni, C., Liu, K., Torzec, N.: Layered graph embedding for entity recommendation using wikipedia in the yahoo! knowledge graph. In: WWW, pp. 811–818 (2020)
20. Fu, T., Lee, W., Lei, Z.: HIN2Vec: explore meta-paths in heterogeneous information networks for representation learning. In: CIKM, pp. 1786–1806 (2017)
21. Zhao, J., Jin, P., Liu, Y.: Business relations in the web: semantics and a case study. J. Softw. **5**(8), 826–833 (2010)
22. Zhao, J., Jin, P.: Extraction and credibility evaluation of web-based competitive intelligence. J. Softw. **6**(8), 1513–1520 (2011)

Graph Neural Network

Fine-Grained Semantics-Aware Heterogeneous Graph Neural Networks

Yubin Wang[1,2], Zhenyu Zhang[1,2], Tingwen Liu[1,2(✉)], Hongbo Xu[1], Jingjing Wang[1], and Li Guo[1,2]

[1] Institute of Information Engineering,
Chinese Academy of Sciences, Beijing, China
{wangyubin,zhangzhenyu1996,liutingwen,hbxu,wangjingjing,guoli}@iie.ac.cn
[2] School of Cyber Security,
University of Chinese Academy of Sciences, Beijing, China

Abstract. Designing a graph neural network for heterogeneous graph which contains different types of nodes and links have attracted increasing attention in recent years. Most existing methods leverage meta-paths to capture the rich semantics in heterogeneous graph. However, in some applications, meta-path fails to capture more subtle semantic differences among different pairs of nodes connected by the same meta-path. In this paper, we propose Fine-grained Semantics-aware Graph Neural Networks (FS-GNN) to learn the node representations by preserving both meta-path level and fine-grained semantics in heterogeneous graph. Specifically, we first use multi-layer graph convolutional networks to capture meta-path level semantics via convolution on edge type-specific weighted adjacent matrices. Then we use the learned meta-path level semantics-aware node representations as guidance to capture the fine-grained semantics via the coarse-to-fine grained attention mechanism. Experimental results semi-supervised node classification show that FS-GNN achieves state-of-the-art performance.

Keywords: Graph neural network · Heterogeneous graph · Fine-grained semantics · Meta-path

1 Introduction

Graph neural networks (GNNs), which can learn from graph-structured data, have been successfully applied in various tasks, such as node classification [6,8], graph classification [10,26], link prediction [14] and recommendation [16]. Most of the existing GNNs perform on homogeneous graphs, where all objects and relations are of the same type. However, real-world data tends to be presented as a heterogeneous graph that contains multiple types of objects and relations.

A heterogeneous graph combines different aspects of information. Figure 1 illustrates a toy example of heterogeneous graph, including three types of objects (author, paper, and conference) and six types of relations (cite/cited,

© Springer Nature Switzerland AG 2020
Z. Huang et al. (Eds.): WISE 2020, LNCS 12342, pp. 71–82, 2020.
https://doi.org/10.1007/978-3-030-62005-9_6

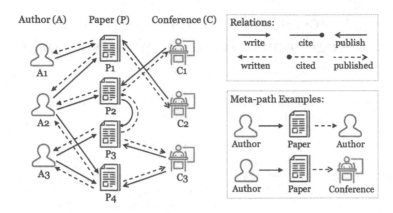

Fig. 1. A toy example of heterogeneous graph.

write/written, and publish/published). Due to the heterogeneity of nodes and edges, a heterogeneous graph contains more comprehensive information and rich semantics. Meta-path [20], a composite relation connecting two nodes, is a widely used structure to capture the semantics. For example, the "Author $\xrightarrow{write}$ Paper $\xrightarrow{written}$ Author" path means authors collaborating on the same papers, while "Author $\xrightarrow{write}$ Paper $\xrightarrow{published}$ Conference" path means authors publishing papers on conferences. Due to the complexity of heterogeneous graph, traditional graph neural networks cannot be directly applied to heterogeneous graph.

Designing a graph neural network for heterogeneous graph have attracted increasing attention in recent years. [24] proposed HAN which transforms a heterogeneous graph into homogeneous graphs by predefined meta-paths and generates nodes representations by fusing the representations learned on each constructed homogeneous graph. This approach requires hand-crafted meta-paths for each problem, but it is hard to exhaustively enumerate and select valuable meta-paths manually. To address this issue, [15] proposed GTN to learn to transform a heterogeneous graph into useful meta-path graphs for each task without any predefined meta-paths and generate node representations via convolution on the learned meta-path graphs.

The key idea of these methods is to identify useful meta-paths to capture the rich semantics in heterogeneous graph. However, in some applications, meta-path fails to capture more subtle semantics. It is because that the information passing by the meta-path, such as features of heterogeneous nodes, is lost in the process of generating meta-path based neighbors. For example, the "Author $\xrightarrow{write}$ Paper $\xrightarrow{written}$ Author" path describes the collaboration relation among authors. However, it cannot depict the fact that *Philip S. Yu* and *Jiawei Han* have many collaborations in data mining field but they seldom collaborate in information retrieval field. This motivates us to design a novel graph neural network to capture the fine-grained semantics under the same meta-path.

In this paper, we propose Fine-grained Semantics-aware Graph Neural Networks (FS-GNN) to learn the node representations preserving both meta-path-level and fine-grained semantics in heterogeneous graph without any predefined meta-paths. Specifically, given the node features as input, we first use the type-specific transformation matrix to project different types of node features into the same space. Then we use a pre-trained meta-path level semantics-aware network to learn the node representations for preserving meta-path level semantics. Afterward, we use the learned node representations to guide the fine-grained semantics-aware network to capture more subtle semantics between pairs of nodes connected by the same relation via the coarse-to-fine grained attention mechanism. It enforces the model to learn the optimal combination of the features of heterogeneous neighbors and lead to better node representations.

We conduct extensive experiments on node classification to evaluate our proposed model. And experimental results on several datasets show that FS-GNN achieves state-of-the-art performance. The source code of this paper can be obtained from https://github.com/jadbin/FS-GNN.

2 Related Work

We briefly mention related work in the field of heterogeneous graph mining, graph embedding, and graph neural networks.

Heterogeneous Graph Mining. In recent years, heterogeneous graph analysis attracts much attention because of rich structural and semantic information in this kind of network [18]. As a general information modeling method, the heterogeneous graph has been widely applied to many data mining tasks such as link prediction [28], classification [1] and recommendation [17]. Meta-path [20] can effectively capture subtle semantics among objects, and many works have exploited meta-path based mining tasks. For example, [27] proposed a meta-path based deep convolutional classification model for collective classification in heterogeneous graph.

Graph Embedding. Graph embedding is proposed to embed graph-structured data into a low dimensional space while preserving the graph structure and property so that the learned embeddings can be applied to the downstream tasks. For example, inspired by word2vec [12,13] proposed DeepWalk which uses SkipGram to learn node embeddings from node sequences generated via random walks over the graph. To address the heterogeneity of graph, [3] introduced meta-path guided random walks and proposed metapath2vec for representation learning in heterogeneous graph. HIN2Vec [4] learns representations of nodes and meta-paths simultaneously via multiple prediction training tasks. Besides these random the random walk based methods, many other approaches have been proposed, such as the deep neural network based methods [22], the matrix factorization based methods [23], etc. Furthermore, attributed graph embedding models [9,11,25] have been proposed to leverages both graph structure and node attributes for learning node embeddings.

Graph Neural Networks. Graph neural networks (GNNs) extend the deep neural network to deal with graph-structured data. Recently, many works generalizing convolutional operation on the graph, and they are often categorized as spectral methods and spatial methods. Spectral methods define graph convolution based on the spectral graph theory. [2] developed convolution operation based graph Fourier transformation. [8] then simplified the previous method by using a linear filter to operate one-hop neighboring nodes. GTN [15] generates node representations via convolution on the learned meta-path graphs. Spatial methods which define convolution directly on the graph, operating on groups of spatially close neighbors. For instance, GraphSAGE [6] performs various aggregators such as mean-pooling over a fixed-size neighborhood of each node. GAT [21] adopts attention mechanisms to learn the relative weights between two connected nodes. HAN [24] leverages node-level attention and semantic-level attention to model the importance of nodes.

Unlike the existing GNNs designed for heterogeneous network which learn the node representations preserving meta-path level semantics, we introduce coarse-to-fine grained attention to capture more subtle semantics difference among the different pairs of nodes connected by the same meta-path. And experiments show that this can further improve the effectiveness of the learned node representations.

3   Preliminaries

A heterogeneous graph, denoted as $\mathcal{G} = (\mathcal{V}, \mathcal{E})$, consists of a node set $\mathcal{V}$ and a link set $\mathcal{E}$. A heterogeneous graph is also associated with a node type mapping function $\phi : \mathcal{V} \to \mathcal{A}$ and a link type mapping function $\psi : \mathcal{E} \to \mathcal{R}$. $\mathcal{A}$ and $\mathcal{R}$ denote the sets of predefined node types and link types. Figure 1 shows a toy example of heterogeneous graph. It consists of three types of objects (author, paper, and conference) and six types of relations (cite/cited, write/written, and publish/published).

In heterogeneous graph, two nodes can be connected via different semantic paths, which are called meta-paths. A meta-path is defined as a path in the form of $A_1 \xrightarrow{R_1} A_2 \xrightarrow{R_2} \cdots \xrightarrow{R_l} A_{l+1}$, which describes a composite relation $R = R_1 \circ R_2 \circ \cdots \circ R_l$ between objects A_1 and A_{l+1}, where $\circ$ denotes the composition operator on relations. Different meta-paths always reveal different semantics. For example, as shown in Fig. 1, the "Author $\xrightarrow{write}$ Paper $\xrightarrow{written}$ Author" path means authors collaborating on the same papers, while "Author $\xrightarrow{write}$ Paper $\xrightarrow{published}$ Conference" path means authors publishing papers on conferences.

Existing graph neural networks designed for heterogeneous graph [15,24] leverage meta-paths to capture the rich semantics. However, due to the information passing by the meta-path, such as features of heterogeneous nodes, is lost in the process of aggregating features of meta-path based neighbors, these models fail to capture more subtle semantic difference among different pairs of

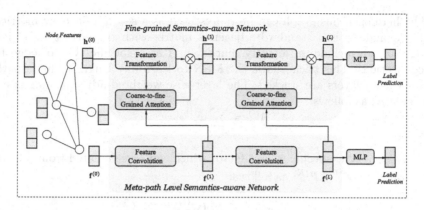

Fig. 2. The overall architecture of FS-GNN.

nodes connected by the same meta-path. In this paper, we propose a graph neural network, namely FS-GNN, to exploit fine-grained semantics under the same meta-path in heterogeneous graph.

4 Proposed Model

In this section, we will give more details of FS-GNN. The overall structure of FS-GNN is shown in Fig. 2. FS-GNN consists of a Meta-path Level Semantics-aware Network (MSN) and a Fine-grained Semantics-aware Network (FSN). We first use MSN to capture meta-path level semantics without any predefined meta-paths. We then use the node representations learned by the MSN as guidance for FSN to capture fine-grained semantics via the coarse-to-fine grained attention. Finally, we present the objectives for model training.

4.1 Meta-path Level Semantics-Aware Network

The goal of MSN is to learn the node representations preserving meta-path level semantics without any predefined meta-paths. Following [15], we view the meta-path as a combination of a list of edges with specific types and assign a type-specific weight to each edge to evaluate the importance of the meta-path.

Specifically, we use a multi-layer graph convolutional network [8] to aggregate the features of neighboring nodes on the weighted adjacent matrices. The layer-wise propagation rule of feature convolution can be defined as follows:

$$\mathbf{f}^{(l+1)} = \sigma \left(\widetilde{D}^{(l)^{-1}} \widetilde{A}^{(l)} \mathbf{f}^{(l)} W_f^{(l)} \right) \tag{1}$$

Here, $\widetilde{A}^{(l)}$ is the layer-wise weighted adjacency matrix as explained in the following paragraph. $\widetilde{D}^{(l)^{-1}}$ is the degree matrix of $\widetilde{A}^{(l)}$, and $\widetilde{D}_{ii}^{(l)} = \sum_j \widetilde{A}_{ij}^{(l)}$. $\sigma(\cdot)$ denotes an activation function such as $\text{ReLU}(\cdot) = \max(0, \cdot)$. $\mathbf{f}^{(l)} \in \mathbb{R}^{|\mathcal{V}| \times d_f^{(l)}}$ denotes the hidden representations of node i in the lth layer.

The heterogeneous graph can be represented as a set of adjacency matrices $\{A_r\}_{r\in\mathcal{R}}$, and A_r is a standard adjacency matrix with the edge type r. Following [15], we include an identity matrix in A as a specific type of edge for self-connections. This trick allows the MSN to learn any length of meta-paths up to l when l layers are stacked. The layer-wise weighted adjacent matrix can be calculated as follows:

$$\widetilde{A}^{(l)} = \sum_{r\in\mathcal{R}} \omega_r^{(l)} A_r \tag{2}$$

where $\omega_r^{(l)}$ is a type-specific weight factor which can be calculated from a learnable parameter $\mathbf{w}_r^{(l)} \in \mathbb{R}^{|\mathcal{R}|}$ as follows:

$$\omega_r^{(l)} = \frac{\exp(\mathbf{w}_r^{(l)})}{\sum_{k\in\mathcal{R}} \exp(\mathbf{w}_k^{(l)})} \tag{3}$$

Due to the heterogeneity of nodes in heterogeneous graph, different types of nodes have different feature spaces. Therefore, initially, we apply a type-specific transformation $M_{\phi(i)} \in \mathbb{R}^{d_f^{(0)} \times d_{in}^{\phi(i)}}$ to project the features of different types of nodes into the same feature space:

$$\mathbf{f}_i^{(0)} = M_{\phi(i)}\mathbf{x}_i \tag{4}$$

where $\mathbf{x}$ can be either attribute features of nodes or one-hot vector for the nodes without attributes.

The node representations $\mathbf{f}^{(l+1)}$ are obtained by aggregating information from the features of their neighborhoods $\mathbf{f}^{(l)}$. After going through L layers of feature convolution, the output representation of node i learned by MSN can be represented as $\mathbf{f}_i = \mathbf{f}_i^{(L)}$. We use the output $\mathbf{f}$ as meta-path level semantics-aware node representations to guide FSN to capture the subtle semantics difference under the same meta-path.

To stabilize the learning process of MSN, we have found extending our method to employ multi-head mechanism to be beneficial, similarly to [21]. Specifically, with K heads executing the procedure of Eq. 1, we concatenate the low-dimensional vectors of different heads and output the representation of each node in the layer $l + 1$ as follows:

$$\mathbf{f}_i^{(l+1)} = \mathop{\Big\|}_{k=1}^{K} \mathbf{f}_i^{(l+1,k)} \tag{5}$$

where $\|$ is the concatenation operator and M denotes the number of heads.

4.2   Fine-Grained Semantics-Aware Network

The goal of FSN is to exploit the fine-grained semantics in heterogeneous graph to learn better node representations under the guidance of meta-path level semantics-ware node representations.

FSN consists of several stacked layers. In each layer l, we would like the output $\mathbf{h}_i^{(l+1)}$ to be composed of K independent components, i.e., $\mathbf{h}_i^{(l+1)} = [\mathbf{c}_i^{(l+1,1)}, \mathbf{c}_i^{(l+1,2)}, \cdots, \mathbf{c}_i^{(l+1,K)}]$ where $\mathbf{c}_k \in \mathbb{R}^{\frac{d_h^{(l+1)}}{K}}$ $(1 \leq k \leq K)$. Each component $\mathbf{c}_k$ is for describing the aspect of node i with fine-grained semantics. The key challenge is to identify which neighbors actually affect the kth aspect of node i. To this end, we propose the coarse-to-fine grained attention mechanism which is presented as follows.

In each layer, similar to MSN, we first apply a type-specific linear transformation $W_c^{(l,k)} \in \mathbb{R}^{\frac{d_h^{(l+1)}}{K} \times \frac{d_h^{(l)}}{K}}$ to each node i:

$$\widehat{\mathbf{c}}_i^{(l,k)} = W_c^{(l,k)} \mathbf{c}_i^{(l,k)} \tag{6}$$

where $\mathbf{h}_i^{(l)} \in \mathbb{R}^{d_h^{(l)}}$ denotes the hidden representations of node i in the lth layer.

We use the meta-path level semantics-aware node representations learned from MSN to guide FSN to capture the fine-grained semantics via the coarse-to-fine grained attention. The attention scores between connected nodes can be calculated as follows:

$$a_{ij}^{(l,k)} = \mathbf{v}^{(l)\top} \tanh(W_1^{(l)} \mathbf{f}_i^{(l+1)} + W_2^{(l)} \mathbf{f}_j^{(l+1)} + W_3^{(l)} \widehat{\mathbf{c}}_i^{(l,k)} + W_4^{(l)} \widehat{\mathbf{c}}_j^{(l,k)}) \tag{7}$$

Here $\mathbf{v}^{(l)} \in \mathbb{R}^{d_h^{(l)}}$ is a learnable attention vector. $W_1^{(l)} \in \mathbb{R}^{d_h^{(l+1)} \times d_f^{(l+1)}}$, $W_2^{(l)} \in \mathbb{R}^{d_h^{(l+1)} \times d_f^{(l+1)}}$, $W_3^{(l)} \in \mathbb{R}^{d_h^{(l+1)} \times \frac{d_h^{(l+1)}}{K}}$ and $W_4^{(l)} \in \mathbb{R}^{d_h^{(l+1)} \times \frac{d_h^{(l+1)}}{K}}$ are trainable transformation matrices. $\cdot^\top$ represents transposition. Then we can obtain the attention weights by normalizing the attention scores with the softmax function:

$$\alpha_{ij}^{(l)} = \frac{\exp(\omega_{\psi_{i,j}}^{(l)} a_{ij})}{\sum_{k \in \mathcal{N}_i} \exp(\omega_{\psi_{i,k}}^{(l)} a_{ik})} \tag{8}$$

where $\mathcal{N}_i$ is the neighborhood of node i in the graph. $\omega_S^{(l)}$ is a edge type-specific weight factor which can be calculated from a learnable parameter $\mathbf{w}_S^{(l)} \in \mathbb{R}^{|\mathcal{R}|}$:

$$\omega_r^{(l)} = \frac{\exp(\mathbf{w}_r^{(l)})}{\sum_{k \in \mathcal{R}} \exp(\mathbf{w}_k^{(l)})} \tag{9}$$

The representation of node i in aspect k can be aggregated by the projected features with the corresponding coefficients as follows:

$$\mathbf{z}_i^{(l+1,k)} = \sigma \left(\sum_{j \in \mathcal{N}_i} \alpha_{ij} \widehat{\mathbf{c}}_j^{(l)} \right) \tag{10}$$

Finally, we concatenate the low-dimensional vectors of different aspects and output the representation of each node in the layer $l + 1$ as follows:

$$\mathbf{h}_i^{(l+1)} = \mathop{\Big\|}_{k=1}^{K} \mathbf{z}_i^{(l+1,k)} \tag{11}$$

Similar to MSN, initially, we apply a type-specific transformation $M_{\phi(i)} \in \mathbb{R}^{d_h^{(0)} \times d_{in}^{\phi(i)}}$ to project the features of different types of nodes into the same feature space as follows:

$$\mathbf{h}_i^{(0)} = M_{\phi(i)} \mathbf{x}_i \tag{12}$$

The node representations $\mathbf{h}^{(l+1)}$ are obtained by aggregating information from the features of their neighborhoods $\mathbf{h}^{(l)}$. After going through L attention layers, the output representation of node i learned by FSN can be represented as $\mathbf{h}_i = \mathbf{h}_i^{(L)}$. And $\mathbf{h}_i$ can be transferred to downstream tasks such as node classification.

4.3   Model Training

We train MSN and FSN separately and here we give the objective functions. To optimize the representations toward semi-supervised node classification task, we integrate the node representations learned by the MSN or FSN into a classifier implemented with a one-layer MLP with softmax function to predict the labels of nodes as follows:

$$\widehat{\mathbf{y}}_i = \mathrm{softmax}(\mathrm{MLP}_\theta(\mathbf{o}_i)) \tag{13}$$

where θ is the trainable parameters of the classifier, $\mathbf{o}_i$ is output representation of node i from MSN or FSN, i.e., $\mathbf{o}_i = \mathbf{f}_i$ or $\mathbf{o}_i = \mathbf{h}_i$. Then we can minimize the cross-entropy loss over all labeled nodes between the ground-truth and the prediction:

$$\mathcal{L} = -\frac{1}{|\mathcal{V}_L|} \sum_{i \in \mathcal{V}_L} \sum_{j=1}^{C} \mathbf{y}_{ij} \cdot \log \widehat{\mathbf{y}}_{ij}^T \tag{14}$$

where $\mathcal{V}_L$ is the set of labeled nodes and C denotes the number of classes.

5   Experiments

This section first introduces datasets and experimental settings, and then presents performance comparison results with baselines in order to validate the effectiveness of FS-GNN.

5.1   Datasets and Baselines

We follow existing studies [15,24] and use three heterogeneous graph benchmark datasets for evaluation, including two citation network datasets ACM, DBLP and one movie dataset IMDB. The statistics of datasets are summarized in Table 1.

ACM contains three types of nodes (paper (P), author (A), subject (S)) and four types of edges (PA, AP, PS, SP). The papers are labeled according to the conference they published and divided into three classes (*Database, Wireless Communication, Data Mining*). Paper features correspond to elements of a bag-of-words represented of keywords.

Table 1. The statistics of datasets.

Dataset	ACM	DBLP	IMDB
# Nodes	8,994	18,405	12,772
# Edges	25,922	67,946	37,288
# Edge Type	4	4	4
# Features	1,902	334	1,256
# Classes	3	4	3
# Training	600	800	300
# Validation	300	400	300
# Test	2,125	2,857	2,339

DBLP contains three types of nodes (paper (P), author (A), conference (C)) and four types of edges (PA, AP, PC, CP). The authors are labeled research area according to the conferences they submitted, and they are divided into four areas (*Database, Data Mining, Machine Learning, Information Retrieval*). Author features are the elements of a bag-of-words represented of keywords.

IMDB contains three types of nodes (movie (M), actor (A), director (D)) and four types of edges (MA, AM, MD, DM). The movies are divided into three classes (*Action, Comedy, Drama*) according to their genre. Movie features correspond to elements of a bag-of-words represented of plots.

To evaluate the performance of our proposed FS-GNN, we compare against several state-of-the-art baselines as specified in [15], including the graph embedding methods and graph neural network based methods.

DeepWalk [13] A graph embedding method that learns node embeddings from the node sequences generated via random walks.

metapath2vec [3] A heterogeneous graph embedding method which performs meta-path based random walk and utilizes skip-gram with negative sampling technique to generate node embeddings.

GCN [8] A graph convolutional network which utilizes a localized first-order approximation of the spectral graph convolution.

GAT [21] A graph neural network which uses attention mechanism to model the differences between the node and its one-hop neighbors.

HAN [24] A graph neural network that exploits manually selected meta-paths and leverages node-level attention and semantic-level attention to model the importance of nodes and meta-paths respectively.

GTN [15] A graph neural network transforms a heterogeneous graph into multiple new graphs defined by meta-paths and generates node representations via convolution on the learned meta-path graphs.

5.2   Experimental Setup

In our experiments, we conduct the semi-supervised node classification task to evaluate the performance of our proposed model. The partition of datasets is the same as the previous studies [15,24].

We train a two-layer FS-GNN for IMDB, and a three-layer FS-GNN for ACM and DBLP. We initialized parameters $\mathbf{w}^{(l)}$ in each layer of MSN and FSN with a constant value. We randomly initialize other parameters following [5]. We adopt the Adam optimizer [7] for parameter optimization with weight decay as 0.0005. We set the learning rate as 0.005 for DBLP and 0.001 for ACM and IMDB. We apply dropout [19] with $p = 0.5$ to both layers' inputs, as well as to the normalized attention coefficients. For a fair comparison, we set the output dimension to 64 following [15,24]. We set the number of heads M in MSN as 8, and this means the output dimension of each head is 8. The number of different aspects of each node in FSN is set as 8.

5.3   Node Classification Results

The results of the semi-supervised node classification task are summarized in Table 2, where the best results are highlighted in bold. Following [15], we use the Macro-F1 metric for quantitative evaluation. We present the mean F1 score over 10 runs of our method and reuse the results already reported in [15] for baselines. FS-GNN-M denotes that we only use the output of MSN to predict the labels of nodes.

We can observe that: (1) Our FS-GNN outperforms all the baselines and achieve state-of-the-art results. The performance gain is from two folds. First, the guidance from MSN enforces the model to capture the meta-path level semantics. Second, the model successfully learns to capture fine-grained semantics under the same meta-path via the coarse-to-fine grained attention. (2) FS-GNN-M consistently outperforms GCN and GAT which are designed for homogeneous graph.

Table 2. Results of node classification in terms of F1 score (%). Bold marks highest number among all models. * marks statistically significant improvements over GTN with $p < 0.01$ under a student t-test.

Method	ACM	DBLP	IMDB
DeepWalk [13]	67.42	63.18	32.08
metapath2vec [3]	87.61	85.53	35.21
GCN [8]	91.60	87.30	56.89
GAT [21]	92.33	93.71	58.14
HAN [24]	90.96	92.83	56.77
GTN [15]	92.68	94.18	60.92
FS-GNN-M (ours)	92.46	93.81	60.54
FS-GNN (ours)	**93.57***	**94.72***	**63.20***

This proves that capturing rich semantics in heterogeneous graph leads to effective node representations to achieve better performance. (3) FS-GNN-M achieves competitive results compared with GTN. It demonstrates that the multi-layer convolution on type-specific weighted adjacent matrices is capable of learning the importance of meta-path.

6   Conclusion

In this paper, we propose FS-GNN to learn the node representations of a heterogeneous network without any predefined meta-paths. It is able to learn meta-path level semantics-aware node representations via multi-layer convolution on type-specific weighted adjacent matrices. And it learns fine-grained semantics-aware node representations under the guidance of the meta-path level semantics-aware node representations via the coarse-to-fine grained attention mechanism. Experimental results demonstrate that our FS-GNN achieves state-of-the-art performance on several semi-supervised node classification benchmarks. The future directions include studying the efficacy of coarse-to-fine grained attention layers combined with other GNNs such as GTN [15]. Also, we will try to extend FS-GNN and apply it to other tasks such as graph classification [10,26].

Acknowledgment. This work is supported by the National Key Research and Development Program of China (grant No. 2016YFB0801003) and the Strategic Priority Research Program of Chinese Academy of Sciences (grant No. XDC02040400).

References

1. Bangcharoensap, P., Murata, T., Kobayashi, H., Shimizu, N.: Transductive classification on heterogeneous information networks with edge betweenness-based normalization. In: WSDM, pp. 437–446 (2016)
2. Bruna, J., Zaremba, W., Szlam, A., LeCun, Y.: Spectral networks and locally connected networks on graphs. In: ICLR (2013)
3. Dong, Y., Chawla, N.V., Swami, A.: metapath2vec: scalable representation learning for heterogeneous networks. In: KDD, pp. 135–144 (2017)
4. Fu, T.Y., Lee, W.C., Lei, Z.: HIN2Vec: explore meta-paths in heterogeneous information networks for representation learning. In: CIKM, pp. 1797–1806 (2017)
5. Glorot, X., Bengio, Y.: Understanding the difficulty of training deep feed for leveraging graph wavelet transform to address the short-comings of previous spectral graphrd neural networks. In: AISTATS, pp. 249–256 (2010)
6. Hamilton, W., Ying, Z., Leskovec, J.: Inductive representation learning on large graphs. In: NIPS, pp. 1024–1034 (2017)
7. Kingma, D.P., Ba, J.: Adam: a method for stochastic optimization. arXiv preprint arXiv:1412.6980 (2014)
8. Kipf, T.N., Welling, M.: Semi-supervised classification with graph convolutional networks. In: ICLR (2017)
9. Li, J., Dani, H., Hu, X., Tang, J., Chang, Y., Liu, H.: Attributed network embedding for learning in a dynamic environment. In: CIKM, pp. 387–396 (2017)

10. Li, R., Wang, S., Zhu, F., Huang, J.: Adaptive graph convolutional neural networks. In: AAAI (2018)
11. Liao, L., He, X., Zhang, H., Chua, T.S.: Attributed social network embedding. TKDE **30**(12), 2257–2270 (2018)
12. Mikolov, T., Sutskever, I., Chen, K., Corrado, G.S., Dean, J.: Distributed representations of words and phrases and their compositionality. In: NIPS, pp. 3111–3119 (2013)
13. Perozzi, B., Al-Rfou, R., Skiena, S.: DeepWalk: online learning of social representations. In: KDD, pp. 701–710 (2014)
14. Schütt, K., Kindermans, P.J., Felix, H.E.S., Chmiela, S., Tkatchenko, A., Müller, K.R.: SchNet: a continuous-filter convolutional neural network for modeling quantum interactions. In: NIPS, pp. 991–1001 (2017)
15. Seongjun, Y., Jeong, M., Kim, R., Kang, J., Kim, H.: Graph transformer networks. In: NIPS, pp. 11960–11970 (2019)
16. Shi, C., et al.: Deep collaborative filtering with multi-aspect information in heterogeneous networks. TKDE (2019)
17. Shi, C., Hu, B., Zhao, W.X., Philip, S.Y.: Heterogeneous information network embedding for recommendation. TKDE **31**(2), 357–370 (2018)
18. Shi, C., Li, Y., Zhang, J., Sun, Y., Philip, S.Y.: A survey of heterogeneous information network analysis. TKDE **29**(1), 17–37 (2016)
19. Srivastava, N., Hinton, G., Krizhevsky, A., Sutskever, I., Salakhutdinov, R.: Dropout: a simple way to prevent neural networks from overfitting. JMLR **15**(1), 1929–1958 (2014)
20. Sun, Y., Han, J., Yan, X., Yu, P.S., Wu, T.: PathSim: meta path-based top-k similarity search in heterogeneous information networks. VLDB **4**(11), 992–1003 (2011)
21. Veličković P., Cucurull, G., Casanova, A., Romero, A., Liò, P., Bengio, Y.: Graph attention networks. In: ICLR (2018)
22. Wang, D., Cui, P., Zhu, W.: Structural deep network embedding. In: KDD, pp. 1225–1234 (2016)
23. Wang, X., Cui, P., Wang, J., Pei, J., Zhu, W., Yang, S.: Community preserving network embedding. In: AAAI (2017)
24. Wang, X., et al.: Heterogeneous graph attention network. In: WWW, pp. 2022–2032 (2019)
25. Zhang, C., Swami, A., Chawla, N.V.: SHNE: representation learning for semantic-associated heterogeneous networks. In: WSDM, pp. 690–698 (2019)
26. Zhang, M., Cui, Z., Neumann, M., Chen, Y.: An end-to-end deep learning architecture for graph classification. In: AAAI (2018)
27. Zhang, Y., Xiong, Y., Kong, X., Li, S., Mi, J., Zhu, Y.: Deep collective classification in heterogeneous information networks. In: WWW, pp. 399–408 (2018)
28. Zhang, Y., Tang, J., Yang, Z., Pei, J., Yu, P.S.: COSNET: connecting heterogeneous social networks with local and global consistency. In: KDD, pp. 1485–1494 (2015)

DynGCN: A Dynamic Graph Convolutional Network Based on Spatial-Temporal Modeling

Jing Li[1], Yu Liu[1], and Lei Zou[1,2(✉)]

[1] Peking University, Beijing, China
zoulei@pku.edu.cn
[2] National Engineering Laboratory for Big Data Analysis Technology
and Application (PKU), Beijing, China

Abstract. Representation learning on graphs has recently attracted a lot of interest with graph convolutional networks (GCN) achieving state-of-the-art performance in many graph mining tasks. However, most of existing methods mainly focus on static graphs while ignoring the fact that real-world graphs may be dynamic in nature. Although a few recent studies have gone a step further to incorporate sequence modeling (e.g., RNN) with the GCN framework, they fail to capture the dynamism of graph structural (i.e., spatial) information over time. In this paper, we propose a **Dynamic Graph Convolutional Network** (**DynGCN**) that performs spatial and temporal convolutions in an interleaving manner along with a model adapting mechanism that updates model parameters to adapt to new graph snapshots. The model is able to extract both structural dynamism and temporal dynamism on dynamic graphs. We conduct extensive experiments on several real-world datasets for link prediction and edge classification tasks. Results show that DynGCN outperforms state-of-the-art methods.

Keywords: Dynamic graphs · GCN · Representation learning

1 Introduction

Owing to the success of convolutional neural networks (CNN) in fields such as computer vision, natural language processing and speech recognition, researchers have showed a lot of interest in the topic of graph neural networks (GNN). However, compared with the Euclidean data structure of images, natural languages and speech signals, graphs are non-Euclidean data which makes it unsuitable for us to apply convolutions designed in CNN to graphs directly.

Therefore, graph convolutional networks (GCN) are specially designed to extract topological structures on graphs. They extend the concept of *convolution* to graph domain by designing operations that aggregate neighborhood information. Since topological structures and local information are extracted by

© Springer Nature Switzerland AG 2020
Z. Huang et al. (Eds.): WISE 2020, LNCS 12342, pp. 83–95, 2020.
https://doi.org/10.1007/978-3-030-62005-9_7

the graph convolution operation, GCN is a powerful architecture in graph representation learning. Some existing studies [7,9] have proved that GCN achieves state-of-the-art performance in a lot of graph mining tasks such as node classification, edge classification, link prediction, clustering, etc.

It is worthwhile to further investigate that most of existing graph convolutional networks are designed for static graphs. However, real-world graphs are dynamic in nature with insertion/deletion of nodes and edges or changing of properties all the time. For example, users of a financial network are constantly trading with each other, users of a social network may develop friendship over time and authors of a citation network are publishing new papers all the time, thus resulting in the dynamism of these networks. Under these dynamic scenarios, it requires dynamic models that aim at capturing not only structural information but also historical information on dynamic graphs.

A few recent studies have gone a step further by combining GCN with recurrent neural networks (RNN) since GCN is designed for structural information extraction and RNN is designed for sequence modeling which makes them a natural match in dynamic graph representation learning. Among them, some models use GCN to aggregate neighborhood information and then feed the output embeddings into RNN architectures to extract sequence information [11,17]. While EvolveGCN [15] combines them in a different manner. It utilize RNN to evolve GCN parameters which results in a dynamic evolving GCN model for every snapshot along time axis. However, these existing works either model graph dynamism only on node embeddings that lacks adaptive learning or on model parameters that lacks structural dynamism extraction.

In this paper, we focus on both spatial and temporal dynamism on dynamic graphs and propose a **Dyn**amic **G**raph **C**onvolutional **N**etwork (**DynGCN**) that performs spatial-temporal convolutions in an interleaving manner with a model adapting mechanism. We conduct several experiments for DynGCN model on link prediction and edge classification tasks. DynGCN outperforms dynamic graph convolution baselines.

Our main contributions can be summarized as follows:

- We propose DynGCN, a self-adapting, spatial-temporal graph convolutional network that learns rich latent graph representations on dynamic graphs.
- We consider the problem of dynamic graph representation learning as a combination of spatial dynamism and temporal dynamism where either spatial or temporal dynamism is often ignored by existing methods.
- We give a comprehensive study for representation learning under extreme class imbalance situations. Our model shows superiority in identifying minority class with preserving total accuracy which indicates the necessity of spatial-temporal modeling.
- We conduct several experiments based on real-world scenarios. DynGCN outperforms state-of-the-art baselines thus demonstrates the efficacy of DynGCN in learning rich and high-level representations.

The remainder of the paper is organized as follows: we present related works for representation learning in Sect. 2 and then give a detailed description of our proposed DynGCN model in Sect. 3. In order to prove the efficacy of DynGCN, we show the experimental results and give a detailed analysis in Sect. 4. Last but not least, we conclude our work in Sect. 5.

2   Related Work

2.1   Static Graph Representation Learning

Representation learning aims to learn node embeddings into low dimensional vector space. A traditional way on static graphs is to perform Singular Vector Decomposition (SVD) on the similarity matrix computed from the adjacency matrix of the input graph [3,14]. Despite their straightforward intuition, matrix factorization-based algorithms require more scalability. Inspired by Word2Vec [12] that maps words into low dimensional space, researchers developed some random walk-based approaches (e.g., DeepWalk [16] and Node2Vec [6]) that perform random walks [13] over the input graph and then feed these paths (regraded as *sentences*) into a SkipGram model [12] to learn node embeddings. What's more, with the success of deep learning, GNN has become an efficient tool in static graph representation learning [4,5,7,9,18]. One typical idea among these GNNs lies in graph convolution operation which adopts the concept of neighborhood aggregation to extract structural information. Yet, all methods mentioned above focus on static graphs and lack the ability to capture dynamism in some applications where graphs may change dynamically.

2.2   Dynamic Graph Representation Learning

It is a straightforward idea to extend static methods to dynamic ones with extra updating mechanism or sequence modeling architectures. Since GNN (e.g. GCN) is designed for structural extraction and RNN (e.g. LSTM or GRU) is designed for sequence modeling, it makes them a natural match to form a dynamic graph neural network.

Seo et al. proposed two GCRN [17] architectures that naturally use GCN to learn node embeddings and then feed them into LSTM to learn sequence dynamism along time axis. The only difference between these two architectures is that the second model uses a modified LSTM that replaces the Euclidean 2D convolution operator by graph convolution operator. Similarly, Manessia et al. proposed WD-GCN/CD-GCN [11] to learn dynamic graph embeddings by combining a modified LSTM and an extension of graph convolution operations to learn long short-term dependencies together with graph structures. WD-GCN takes as input a graph sequence while CD-GCN takes as input the corresponding ordered sequence of vertex features.

Furthermore, GCN-GAN [10] models the temporal link prediction tasks in weighted dynamic networks by combining GCN and RNN with generative adversarial networks (GAN). They use GCN to capture topological characteristics of each snapshot and employ LSTM to characterize the evolving features with a

GAN followed to generate the next snapshot thus to learn a more effective representation. STAR [20] adds a dual attention mechanism on top of the GCN and RNN architectures. On the other side, Aldo Pareja et al. models dynamism is a different way, they proposed EvolveGCN [15] that employs RNN architecture (e.g. GRU or LSTM) to evolve the GCN weight parameters along time dimension to adapt models to graph dynamism.

Although these methods combine GCN with RNN from different aspects, they model dynamism either on output node embeddings that lacks adaptive learning or on the weight parameters that lacks the extraction of graph strctural dynamism.

3   Method

3.1   Problem Definition

A *graph* is denoted as $G = (V, E)$ where V is the set of vertices and E is the set of edges. The adjacency matrix of graph G is denoted as $A \in R^{N \times N}$, where $A_{i,j} = 1$ if $v_i, v_j \in V$ and $(v_i, v_j) \in E$ otherwise $A_{i,j} = 0$. Given a graph $G = (V, E)$, *graph representation learning* is a mapping function $f : V \to R^d$, where $d \ll N$. The mapping function f embeds every node into low-dimensional space while preserving original information at the same time, namely $y_v = f(v), \forall v \in V$. The more similar two nodes are in the graph G, the closer their representation vectors y_u and y_v are in the embedding space.

Dynamic graphs may be divided into two types according to the active time type for an edge, either time instances or time intervals, which results in continuous-time or discrete-time dynamic graphs [8,19]. In our settings, we formulate dynamic graphs under discrete time and define *dynamic graph* and *dynamic graph representation learning* as follows:

Definition 1 (Dynamic Graph). *A dynamic graph is denoted as a set of graphs, i.e.* $\mathcal{G} = \{G_1, G_2, \ldots, G_T\}$, *where each* $G_t = (V_t, E_t)$ *is a snapshot at time step* $t, t \in \{1, 2, \ldots, T\}$. *For graph* $G_t, t \in \{1, 2, \ldots, T\}$, *the adjacency matrix* $A_t \in R^{N \times N}$ *is defined as follows:* $(A_t)_{i,j} = 1$ *if* $(v_t)_i, (v_t)_j \in V_t$ *and* $((v_t)_i, (v_t)_j) \in E_t$, *otherwise,* $(A_t)_{i,j} = 0$.

Definition 2 (Dynamic Graph representation learning). *Dynamic graph representation learning is to learn a set of mappings* $\mathcal{F} = \{f_1, f_2, \ldots, f_T\}$ *where each mapping function* $f_t, \forall t \in \{1, 2, \ldots, T\}$ *maps nodes of a dynamic graph* G_t *at time t into low-dimensional space while preserving origin information at the same time. The More similar two nodes are in the origin dynamic graph, the closer their representation vectors are in the embedding space.*

3.2   Architecture Overview

As shown in Fig. 1, DynGCN performs spatial and temporal convolutions in an interleaving manner. The first layer of the model is a spatial convolution layer where the GCN model is updated by a GRU cell to self-adapt to graph dynamism. The spatial convolution layer aggregates neighborhood structural information with dynamism awareness. The second layer is a temporal convolution layer that aggregates information from current and historical time steps with dilated convolutions. After that, every node is represented by his current neighborhood information along with his historical neighborhood information. Then we add another spatial convolution layer on the top to aggregate spatial temporal information from neighborhoods.

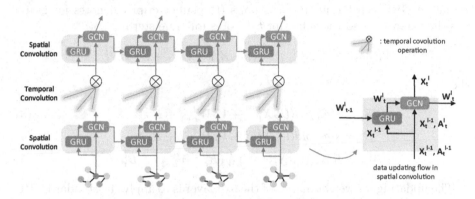

Fig. 1. An overview of DynGCN architecture

3.3 Spatial Convolution Layer

GCN has showed its superiority in learning graph topological structures, we utilize GCN unit to learn the structural information of every snapshot in dynamic graphs. Formally, given a graph $G_t = (V_t, E_t)$ at time step t, the adjacency matrix is denoted by $A_t \in R^{N \times N}$. At the l-th layer of GCN, it takes output node embedding vectors X_t^{l-1} of the $(l-1)$-th layer and the adjacency matrix A_t as input, and output the updated node embedding vectors X_t^l. We can write the operation of a GCN layer as follows:

$$X_t^l = \mathcal{F}(X_t^{l-1}, A_t, W_t^l) = \sigma(\hat{D}_t^{-1/2} \hat{A}_t \hat{D}_t^{-1/2} X_t^{l-1} W_t^l) \tag{1}$$

where $\hat{A}_t = A_t + I$ and I is an identity matrix. $\hat{D}_t = diag(\sum_{j=1}^N \hat{A}_{t(ij)})$ and $\hat{D}_t^{-1/2} \hat{A}_t \hat{D}_t^{-1/2}$ is a normalization of A_t served as a approximated graph convolution filter. W_t^l is the weight matrix of the l-th layer at time t and function σ is an activation function (e.g. ReLU). The initial input X_t^0 for the first layer is the node features matrix $Z_t \in R^{N \times K}$ where each row represents a K dimensional feature vector for each of N nodes. After L layers of graph convolution operations, the output matrix contains aggregated neighborhood information for every node in every single graph.

In consideration of graph dynamism, the spatial convolution layer further extends static GCN architecture with self-adapting mechanisms which is first introduced in EvolveGCN. The spatial convolution layer utilizes a recurrent network to update weight matrix of GCN unit. Although GRU and LSTM both work in updating GCN parameters, we choose GRU for our implementation since the input of GRU unit consists of hidden state and current embedding while LSTM only takes hidden state into consideration. By updating model parameters, GCN architecture is self-adapted to every time step.

For weight matrix W_t^l of l-th layer at time t, we obtain it by W_{t-1}^l and X_t^{l-1} through a GRU cell formulated as follows (the superscript l denotes for graph convolution layer, and the subscript t denotes for time step):

$$W_t^l = \mathcal{G}(X_t^{l-1}, W_{t-1}^l) = (1 - Z_t^l) \circ W_{t-1}^l + Z_t^l \circ \widetilde{W}_t^l \tag{2}$$

where:

$$Z_t^l = sigmoid(U_Z^l X_t^{l-1} + V_Z^l W_{t-1}^l + B_Z^l) \tag{3}$$

$$R_t^l = sigmoid(U_R^l X_t^{l-1} + V_R^l W_{t-1}^l + B_R^l) \tag{4}$$

$$\widetilde{W}_t^l = tanh(U_W^l X_t^{l-1} + V_W^l (R_t^l \circ W_{t-1}^l) + B_W^l) \tag{5}$$

The updating for weight matrix of the l-th layer is to apply the standard GRU operation on each column of the involved matrices independently since standard GRU maps vectors to vectors but we have matrices here. We treat W_{t-1}^l as the hidden state of GRU. Embedding vector X_t^{l-1} is chosen as the input of GRU at every time step to represent current information. $Z_t^l, R_t^l, \widetilde{W}_t^l$ are the update gate output, the reset gate output, and the pre-output, respectively. To deal with the inequality between the columns of weight matrix W_{t-1}^l and embedding matrix X_t^{l-1}, a summarization [2] on X_t^{l-1} is further added to the evolving graph convolution layer to transfer X_t^{l-1} to have the same columns as W_{t-1}^l .

GCN unit aggregates embedding vectors from $(l - 1)$-th layer to l-th layer and GRU unit updates weight matrix from time step $t - 1$ to t. The detailed data updating flow in spatial convolution layer is illustrated in Fig. 1. Formally, the spatial convolution architecture can be written as follows:

$$X_t^l = \mathcal{F}(X_t^{l-1}, A_t, \mathcal{G}(X_t^{l-1}, W_{t-1}^l)) \tag{6}$$

Where functions $\mathcal{F}$ and $\mathcal{G}$ are graph convolution operation and weight evolving operation respectively as declared above.

3.4   Temporal Convolution Layer

It is a key issue to capture temporal information along time dimension in dynamic graph embedding problems. A lot of existing models employ RNN architectures for sequence modeling. However, RNN-based methods are memory consuming and time consuming because of the complex gate mechanisms.

Besides, standard RNN suffers from gradient disappear and can only obtain short-term memory.

Although variant architectures like LSTM and GRU fix these problems to some extent, they are still at a disadvantage compared to CNN-based architectures (specially, TCN [1]) which require less memory and training time along with a flexible receptive field size. What's more, RNN-based methods model temporal dynamism by gate mechanism where historical information is considered only in hidden state of every time step. While in CNN-based methods, information of every historical time steps are aggregated by convolution operations which guarantees rich information and at the same time, unifies the idea of spatial and temporal convolutions.

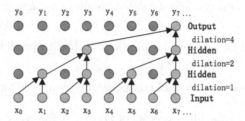

Fig. 2. An example of dilated convolution in temporal convolution layer with $dilation = 1, 2, 4$ and $filter\_size = 2$. This guarantees that the aggregated information covers all historical time steps.

We employ TCN architecture to capture historical information in the proposed DynGCN model. The temporal convolution layer uses a 1D fully convolutional unit and a causal convolutional network unit. The 1D fully-convolutional unit guarantees that the output has the same length as the input and the causal convolutional unit guarantees that an output at time step t_k is convoluted with input of time steps $t \leq t_k$, which means aggregating information from current and historical time steps. Further, the causal convolution is equipped with dilation to enable a larger and more flexible receptive field. An example of dilated convolution is shown in Fig. 2. Formally, given $X^l \in R^{T \times M}$, a length-T sequence of node embedding vectors in the l-th layer with M channels, and a filter $f : \{0, 1, \ldots k - 1\}$, the temporal convolution operation $\mathcal{H}$ on element x of X^l is formalized as follow:

$$\mathcal{H}(x) = (X^l *_d f)(x) = \sum_{i=0}^{k-1} f(i) \cdot X^l_{x-d \cdot i} \tag{7}$$

d is the dilation factor, k is the filter size and $x - d \cdot i$ represents the direction of the past. Benefit from this, we can obtain a larger receptive field by either increasing filter size k or increasing the dilation factor d. What's more, convolution operations provide parallelism in processing a sequence thus is efficient.

Node embedding vectors of the l-th layer are aggregations of neighborhood information. By performing temporal convolution operations on node embedding vectors, we can then aggregate historical information along time axis. With another evolving graph convolution layer at the top, we get representations with current and historical information for both a node and its neighborhood. Therefore, performing spatial and temporal convolutions in an interleaving manner ensures both topological and historical information to be included in high-level node embedding vectors learned from our DynGCN architecture.

3.5   Tasks and Model Training

In order to show the capacity in representation learning of DynGCN, we further train the model based on two specific downstream tasks: *link prediction* and *edge classification.*

Link Prediction. The task of link prediction aims to predict the existence of an edge (u, v) at a future time step t.

Edge Classification. The task of edge classification is to classify the label of an edge (u, v) at time step t.

Given the output embedding vectors of DynGCN for every node, we form the prediction of an edge label or edge existence based on concatenating embedding vectors of two nodes of this edge. Assuming that the output embedding vector of node u at time t is denoted by X_t^u, and P is the parameter matrix, the prediction can be written as $y_t^{uv} = softmax(P[X_t^u; X_t^v])$. The loss function is defined as $L = -\sum_{t=1}^{T} \sum_{(u,v)} \alpha_{uv} \sum_{i=1}^{N} (z_t^{uv})_i log(y_t^{uv})_i$, where z_t^{uv} is the ground-truth one-hot label vector and the nonuniform weight α_{uv} serves as hyperparameter to balance class distribution since datasets we use in our experiments suffer from serious class imbalance problems.

4   Experiments

4.1   Datasets

We verify the effectiveness of the DynGCN model on three data sets. All statistics about these data sets are summarized in Tabel 1.

Table 1. Dataset statistics

Datasets	Nodes	Edges	Positive Edges	Train/Test/Val (Edge classification)	Train/Test/Val (Link prediction)
OTC	5881	35592	89%	96/21/21	96/14/28
Alpha	3783	24186	93%	96/22/22	96/13/28
AS	6474	13895	—	—	70/10/20

- **OTC.**[1] This network of bitcoin users rate each other in a scale from -10 (total distrust) to $+10$ (trust) and every rating comes with a time stamp. We generate a sequence of 138 time steps and split it into training, testing and validation sets as shown in Table 1. Specially, this data set suffers from class imbalance probelms since there are 89% of positive edges.
- **Alpha.**[2] Alpha is similar to OTC but extracted from a different bitcoin platform. We extract a sequence of 140 time steps. Alpha has a higher ratio of 93% positive ratings than OTC.
- **Autonomous Systems (AS).**[3] The graph of routers comprising the Internet are organized into sub-graphs called autonomous systems and each exchanges traffic flows with some neighbors. AS dataset is a who-talks-to-whom network constructed from the BGP (Border Gateway Protocol) logs. The dataset spans an interval of 785 days we form a sequence of 100 time steps.

4.2 Baselines

We compare DynGCN with both static and dynamic baselines where dynamism is considered in different manners. By conducting these comprehensive comparisons, we verify that our architecture design makes sense and has the ability to model rich information.

- **GCN** [9]: This is a classical method for static graph representation learning that learns node embeddings by aggregating neighborhood information. For dynamic graph, we use the same GCN model for all snapshots.
- **GCN-GRU**: This baseline is a combination of graph convolution and sequence modeling. The output node embeddings of GCN unit are feed into a GRU architecture to model dynamism on node representations. This method comes from the Method 1 proposed by Seo et al. [17] and we further adjust their GNN from ChebNet to GCN and their RNN from LSTM to GRU for a better comparison.
- **EvolveGCN** [15]: Different from the above mentioned GCN-GRU method, EvolveGCN uses RNN to model dynamism on GCN parameters. The authors proposed two types of EvolveGCN model, namely EvolveGCN-H and EvolveGCN-O. The -H version is implemented by using GRU architectures to evolve GCN parameters, and the $-$O version is implemented by using LSTM architectures to evolve GCN parameters.

4.3 Experimental Results

Link Prediction. The evaluation metric for link prediction is Mean Average Precision (MAP). MAP averages the predictions for all nodes and higher MAP

---

[1] http://snap.standford.edu/data/soc-sign-bitcoin-otc.html.

[2] http://snap.standford.edu/data/soc-sign-bitcoin-alpha.html.

[3] http://snap.stanford.edu/data/as-733.html.

shows that the model can predict well for more nodes. Figure 3 shows the performance comparisons of all methods in link prediction tasks. Although all methods perform badly because of the data distribution and class imbalance, DynGCN still outperforms all other baselines in all three datasets.

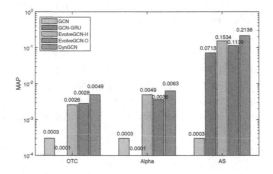

Fig. 3. MAP results for link prediction (the y-axis is in log scale).

Edge Classification. Table 2 shows the performance comparisons between DynGCN and all baselines on edge classification task. The accuracy and weighted F1 score are for all classes and F1 score and the corresponding precision and recall are for the negative edges. Since both datasets suffer from serious class imbalance problems, classification tasks on them are of great challenges. Our proposed DynGCN model achieves the best performance by a significant margin.

Table 2. Experimental results for edge classification tasks

Datasets	Methods	Accuracy	Weighted F1	F1 (Precision/Recall)
OTC	GCN	0.5535	0.6155	0.3186 (0.2050/0.7148)
	GCN-GRU	0.5981	0.6551	0.3386 (0.2256/0.6788)
	EvolveGCN-H	0.6839	0.7118	0.2171 (0.1754/0.2848)
	EvolveGCN-O	0.6873	0.7168	0.2446 (0.1962/0.3246)
	DynGCN	**0.7949**	**0.8019**	**0.3583** (0.3309/0.3905)
Alpha	GCN	0.6495	0.6814	0.0992 (0.0802/0.1303)
	GCN-GRU	0.6340	0.6865	0.3433 (0.2353/0.6347)
	EvolveGCN-H	0.7569	0.7668	0.3160 (0.2738/0.3735)
	EvolveGCN-O	0.6773	0.7126	0.2971 (0.2288/0.4234)
	DynGCN	**0.8011**	**0.8027**	**0.3519** (0.3549/0.3489)

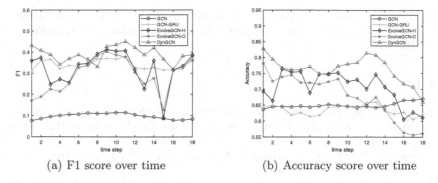

(a) F1 score over time   (b) Accuracy score over time

Fig. 4. Performance over time for dataset Alpha on edge classification.

Advantage of Dynamic Modeling. We further plot the accuracy score and F1 score for the minority class over time for Alpha on edge classification. Although the same results are also observed on OTC, we omit figures due to space limitation. As shown in Fig. 4 (a), there is an obvious gap between GCN method and all other methods while DynGCN still outperforms all baselines along time steps. Besides, the accuracy score for GCN method is also relatively small as shown in Fig. 4 (b). Intuitively, GCN is designed for static graphs and no dynamism is considered in GCN model. The superior performance of other methods against GCN indicates the advantage of dynamic modeling.

Advantage of Spatial-Temporal Modeling. By further analyzing Fig. 4, we can observe that DynGCN outperforms all baselines in F1 score of minority class and accuracy score of all classes over time. Especially, at time step 15, all baselines show extremely low performance, but DynGCN performs relatively stable and still keeps an absolute advantage in F1 score. This is benefit from the special architecture of DynGCN designed to capture both spatial and temporal information from current and the past.

5   Conclusion

In this paper, we proposed DynGCN for representation learning on dynamic graphs. By conducting spatial and temporal convolutions in an interleaving manner with a model adapting architecture, we obtain rich information aggregated along both time dimension and graph dimension. The extended experiments show that DynGCN outperforms all baselines. The study of dynamic graph representation learning opens up a lot of future works. For example, the model scalability remains more improvements and we plan to extend this work to other graph mining tasks like node classification, clustering. Besides, it is also an interesting research topic on modeling continuous time for dynamic graphs.

Acknowledgements. This work was supported by NSFC under grant 61932001, 61961130390. This work was also supported by Beijing Academy of Artificial Intelligence (BAAI).

References

1. Bai, S., Kolter, J.Z., Koltun, V.: An empirical evaluation of generic convolutional and recurrent networks for sequence modeling. CoRR abs/1803.01271 (2018). http://arxiv.org/abs/1803.01271
2. Cangea, C., Veličković, P., Jovanović, N., Kipf, T., Liò, P.: Towards sparse hierarchical graph classifiers (2018)
3. Cao, S., Lu, W., Xu, Q.: Grarep: Learning graph representations with global structural information. In: Proceedings of the 24th ACM International on Conference on Information and Knowledge Management, pp. 891–900. CIKM 2015, ACM, New York, USA (2015). https://doi.org/10.1145/2806416.2806512
4. Chen, J., Ma, T., Xiao, C.: Fastgcn: Fast learning with graph convolutional networks via importance sampling (2018)
5. Defferrard, M., Bresson, X., Vandergheynst, P.: Convolutional neural networks on graphs with fast localized spectral filtering. In: Lee, D.D., Sugiyama, M., Luxburg, U.V., Guyon, I., Garnett, R. (eds.) Advances in Neural Information Processing Systems, vol. 29, pp. 38vol. 44–3852. Curran Associates, Inc. (2016). http://papers.nips.cc/paper/6081-convolutional-neural-networks-on-graphs-with-fast-localized-spectral-filtering.pdf
6. Grover, A., Leskovec, J.: Node2vec: scalable feature learning for networks. In: Proceedings of the 22Nd ACM SIGKDD International Conference on Knowledge Discovery and Data Mining, pp. 855–864. KDD 2016, ACM, New York, USA (2016). https://doi.org/10.1145/2939672.2939754
7. Hamilton, W., Ying, Z., Leskovec, J.: Inductive representation learning on large graphs. In: Guyon, I., et al. (eds.) Advances in Neural Information Processing Systems, vol. 30, pp. 1024–1034. Curran Associates, Inc. (2017). http://papers.nips.cc/paper/6703-inductive-representation-learning-on-large-graphs.pdf
8. Kazemi, S.M., Goel, R., Jain, K., Kobyzev, I., Sethi, A., Forsyth, P., Poupart, P.: Relational representation learning for dynamic (knowledge) graphs: A survey. CoRR abs/1905.11485 (2019). http://arxiv.org/abs/1905.11485
9. Kipf, T.N., Welling, M.: Semi-supervised classification with graph convolutional networks. CoRR abs/1609.02907 (2016). http://arxiv.org/abs/1609.02907
10. Lei, K., Qin, M., Bai, B., Zhang, G., Yang, M.: GCN-GAN: a non-linear temporal link prediction model for weighted dynamic networks. CoRR abs/1901.09165 (2019). http://arxiv.org/abs/1901.09165
11. Manessi, F., Rozza, A., Manzo, M.: Dynamic graph convolutional networks. Pattern Recogn. **97**, 107000 (2020). https://doi.org/10.1016/j.patcog.2019.107000, http://www.sciencedirect.com/science/article/pii/S0031320319303036
12. Mikolov, T., Chen, K., Corrado, G., Dean, J.: Efficient estimation of word representations in vector space (2013)
13. Noh, J.D., Rieger, H.: Random walks on complex networks. Phys. Rev. Lett. **92**(11) 3 (2004). https://doi.org/10.1103/physrevlett.92.118701
14. Ou, M., Cui, P., Pei, J., Zhang, Z., Zhu, W.: Asymmetric transitivity preserving graph embedding. In: Proceedings of the 22nd ACM SIGKDD International Conference on Knowledge Discovery and Data Mining, pp. 1105–1114. KDD 22016, ACM, New York, USA (2016). https://doi.org/10.1145/2939672.2939751

15. Pareja, A., et al.: Evolvegcn: evolving graph convolutional networks for dynamic graphs. CoRR abs/1902.10191 (2019). http://arxiv.org/abs/1902.10191
16. Perozzi, B., Al-Rfou, R., Skiena, S.: Deepwalk: Online learning of social representations. In: Proceedings of the 20th ACM SIGKDD International Conference on Knowledge Discovery and Data Mining, pp. 701–710. KDD 22014, ACM, New York, USA (2014). https://doi.org/10.1145/2623330.2623732
17. Seo, Y., Defferrard, M., Vandergheynst, P., Bresson, X.: Structured sequence modeling with graph convolutional recurrent networks (2016)
18. Veličković, P., Cucurull, G., Casanova, A., Romero, A., Liò, P., Bengio, Y.: Graph attention networks (2017)
19. Wang, Y., Yuan, Y., Ma, Y., Wang, G.: Time-dependent graphs: definitions, applications, and algorithms. Data Sci. Eng. 4(4), 352–366 (2019). https://doi.org/10.1007/s41019-019-00105-0
20. Xu, D., Cheng, W., Luo, D., Liu, X., Zhang, X.: Spatio-temporal attentive RNN for node classification in temporal attributed graphs. In: Proceedings of the 28th International Joint Conference on Artificial Intelligence, pp. 3947–3953. IJCAI 2019, AAAI Press (2019)

NeuLP: An End-to-End Deep-Learning Model for Link Prediction

Zhiqiang Zhong[1], Yang Zhang[2], and Jun Pang[1,3(✉)]

[1] Faculty of Science, Technology and Medicine, University of Luxembourg,
Esch-sur-Alzette, Luxembourg
jun.pang@uni.lu
[2] CISPA Helmholtz Center for Information Security, Saarland Informatics Campus,
Saarbrücken, Germany
[3] Interdisciplinary Centre for Security, Reliability and Trust,
University of Luxembourg, Esch-sur-Alzette, Luxembourg

Abstract. Graph neural networks (GNNs) have been proved useful for link prediction in online social networks. However, existing GNNs can only adopt shallow architectures as too many layers will lead to over-smoothing and vanishing gradient during training. It causes nodes to have indistinguishable embeddings if locally they have similar structural positions, and this further leads to inaccurate link prediction. In this paper, we propose a unified end-to-end deep learning model, namely Neural Link Prediction (NeuLP), which can integrate the linearity and non-linearity user interactions to overcome the limitation of GNNs. The experimental evaluation demonstrates our model's significant improvement over several baseline models. Moreover, NeuLP achieves a reliable link prediction given two users' different types of attributes and it can be applied to other pairwise tasks. We further perform in-depth analyses on the relation between prediction performance and users' geodesic distance and show that NeuLP still can make accurate link prediction while two users are far apart in the networks.

1 Introduction

Link prediction is one of the key problems in online social networks (OSNs) mining, which aims to estimate the likelihood of the existence of an edge (relationship) between two nodes (users) [14]. It has been widely applied for friendship recommendation in OSNs, which is important to increase user engagement and leads to more satisfactory user experience. Besides OSNs, link prediction has drawn the attention from many different application domains and researches in various fields, such as in the study of protein-protein interaction networks [10], and identifying hidden or missing criminals in terrorist networks [3].

The existing methods for link prediction in OSNs can be classified into three categories, ranging from the early methods focusing on hand-crafted features (e.g., see methods based on user profiles [2] and graph structure [14,15]); shallow network embedding-based methods, such as LINE [22] and node2vec [4]; to

© Springer Nature Switzerland AG 2020
Z. Huang et al. (Eds.): WISE 2020, LNCS 12342, pp. 96–108, 2020.
https://doi.org/10.1007/978-3-030-62005-9_8

recently emerged graph neural network (GNN)-based methods [5,9,23]. Among these, GNNs is currently the most popular paradigm, largely owing to their efficiency learning capability. In contrast, hand-crafted features methods are limited in extracting latent information in the OSNs, and shallow network embedding-based methods cannot incorporate user attributes.

GNN-based methods follow a common scheme that explores user information from either social graph or user attributes as node embeddings and then calculates user similarity for making link prediction. However, we argue that it is insufficient since the classic GNNs cannot capture the global information for a given user pair in an OSN. Current GNNs generally follow message-passing mechanism and assume that with the use of enough GNN layers (times of interactive aggregation), the model can capture long-range dependencies. However, recent research shows that GNNs are not robust to multiple layers since they would become over-smoothing and lead to vanishing gradient during training [11,13]. Therefore, existing GNNs have shallow architecture, 2–4 layers. It causes nodes to have indistinguishable embedding vectors if locally they are at similar positions of an OSN. To fill this gap, You et al. propose a novel GNN model (P-GNNs) which could capture the position of a given node with respect to all other nodes in the graph [24]. However, the introduced anchor-set sampling brings with high computational costs and performs less well on attributed networks.

In this paper, we propose a unified end-to-end deep learning model, namely Neural Link Prediction (NEULP), which could efficiently integrate the linearity and non-linearity user interactions to overcome the above limitations, and NEULP is flexible to handle different types of user attributes for link prediction. We firstly use two shallow GNNs to transform user information (their social relations and attributes) to embeddings which interactively aggregate user features from neighbours. We then add a fusion layer to merge embeddings of two users into one vector, which performs element-wise product on two user embeddings to capture the linearity interactions between two users. However, a simple element-wise product does not account for the non-linearity interactions between users, so we further apply multilayer perceptron (MLP) on the vector to map it to low dimensions. The MLP can explore the non-linearity interactions between two users, where each layer performs a non-linear transformation with an activation function. This architecture is inspired by the recent progress of recommendation systems [6] which is useful for mining online user behaviours. In the end, we utilise a linear transformation layer to summarise the extracted features of a pair of users as the possibility of existing a link between them.

We conduct extensive experiments on five benchmark datasets and two Instagram datasets in various pairwise prediction tasks to evaluate the performance of NEULP with respect to several state-of-the-art baselines, including (1) link prediction with only network structures and (2) link prediction on attributed social network datasets with one/two type(s) of user attributes. The experimental results demonstrate our model's significant improvement over the baselines, with up to 5.8% improvement in terms of the AUC score, and its capability in leveraging user attributes and interactions when compared with GNN-based baselines. Moreover, NEULP achieves a strong link prediction when given users' two different types of attributes. We further perform in-depth analyses on the relation between prediction performance and the graph geodesic of two users in OSNs and this demonstrates that our model successfully used user interactions to improve GNN-based baselines.

2   Related Work

We briefly review the state-of-the-art on link prediction in OSNs, including methods based on hand-crafted features, shallow network embedding and GNNs [19]. The hand-crafted features-based methods extract the feature hidden inside the user attributed and edges' structures. A heuristic score is used to measure the similarity and connectivity between two users [2,14,15]. These methods are pellucid and efficient, but cannot explore the latent information in OSNs.

Shallow network embedding-based methods learn a link's formation mechanism from the network other than assuming a particular mechanism (e.g., common neighbours), and they can be summarised into two groups, i.e., DeepWalk-based methods and matrix factorisation-based methods. DeepWalk [20] pioneers network embedding methods by considering the node paths traversed by random walks over the network as sentences and leveraging Skip-gram model [18] for learning node representations. On the other hand, some works adopt the idea of matrix factorisation for network embedding [16,21]. However, these methods cannot deal with user attributes naturally and cannot optimise parameters for a specific task.

In the past few years, many studies have adopted GNNs to aggregate node information in arbitrary graph-structured data for link prediction. Most of the existing GNN models (e.g., see [5,9,23]) rely on a series of graph message-passing architectures that perform the convolution in the graph domain by aggregating node feature messages from its neighbours in the graph and stacked multiple GNN layers can capture the long-range node dependencies. However, existing GNNs can only adopt the shallow architecture, with 2−4 layers, which is not enough to get global network structure information. You et al. [24] recently proposed a global position-aware information capture mechanism to overcome this limitation. However, the introduced anchor-set sampling incurs high computational costs and makes the approach perform less well on attributed datasets.

3   Problem Definition

A social network is denoted as $G = (U, E)$, where $U = \{u_1, \ldots, u_n\}$ represents the set of users and $E \subseteq U \times U$ is the set of links between users. Besides, we can use the adjacency matrix $A = (a_{ij}) \in \mathbb{R}^{n \times n}$ to present the social graph, where we have $a_{ij} = 1$ if $(u_i, u_j) \in E$, otherwise $a_{ij} = 0$. We formally define the social network with user-shared information (i.e., attributes) as $G = (U, E, H)$, where $H \in \mathbb{R}^{n \times k}$, k is the dimension of user attributes. Each node $u_i \in U$ is associated with some types of features vectors and a feature type mapping

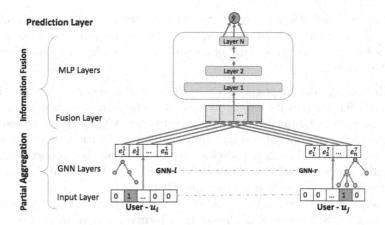

Fig. 1. NEuLP's model architecture. The two input user feature vectors (left: u_i, right: u_j) are firstly transformed with two GNNs and further fused with multiple propagation layers, and the output is the predicted possibility of existing friendship between u_i and u_j.

function. $H = \{x_i \mid u_i \in U\}$ is the set of node features for all nodes, where x_i is the associated node feature of the node u_i.

Given a social network $G = (U, E)$ or an attributed social network $G = (U, E, H)$, the goal of the link prediction problem is to learn a scoring function $s\colon U \times U \longrightarrow \mathbb{R}$ for predicting a new link between an unlabelled pair of users in a set $E_p := (U \times U) \setminus E$.

4   Proposed Approach

4.1   Model Overview

Our proposed model NEuLP has three main components as shown in Fig. 1: partial aggregation, information fusion and model prediction. In the partial aggregation part, we use the input layer to get the encoded user attributes and feed them into GNN layers. Then, the GNN could iteratively aggregate information from a user's local neighbourhoods in the social network and update the user presentation (embedding). After obtaining user embeddings through partial aggregation simultaneously, we add a fusion layer to merge embeddings of two users into one vector and further apply MLP on the vector to explore the non-linearity interactions between these two users. At last, we adopt a prediction layer to linearly summarise the extracted feature of two users as the possibility of existing an edge between them.

4.2   Partial Aggregation

In this subsection, we describe the details of adopting GNNs to iteratively aggregate the partial information from a user's local network neighbourhood and update its embedding.

Input Layer. The bottom input layer consists of two feature vectors h_i and h_j, where $h_i, h_j \in H$, that describe user u_i and user u_j, respectively. They can be customised to support a wide range of user attributes, such as location check-ins and posted hashtags.

GNN Encoding. We adopt the GNN to iteratively aggregate neighbour information for each user in its sub-network and update their embeddings. More specifically, we use Graph Convolutional Networks (GCN) [9] as an example GNN, due to its widespread use and remarkable contributions. GCN operates directly on a graph and induces node feature vectors from the properties of their neighbourhoods. The model can be built by stacking multiple convolutional layers to involve information from farther neighbours.

Formally, given an attributed social network $G = (U, E, H)$, and its adjacent matrix A, a GCN layer is a nonlinear transformation that maps from H to $H^{(1)} \in \mathbb{R}^{n \times k'}$:

$$H^{(1)} = \phi(\hat{A} H W^{(0)} + b^{(0)}) \tag{1}$$

where $W^{(0)} \in \mathbb{R}^{k \times k'}$, $b^0 \in \mathbb{R}^{k'}$ are model parameters. ϕ is a non-linear activation function. k' is the output feature dimension and $\hat{A}$ is the normalisd symmetric adjacency matrix which is defined as:

$$\hat{A} = \tilde{D}^{-\frac{1}{2}} \tilde{A} \tilde{D}^{-\frac{1}{2}} \tag{2}$$

Here, $\tilde{A} = A + I_N$ and $\tilde{D}$ is the degree matrix. I_N is an identity matrix with dimension of n and $\tilde{D}_{ij} = \sum_j \tilde{A}_{ij}$.

With multiple convolutional layers, the GCN model is then described as:

$$H^{(\ell)} = \phi(\tilde{A} H^{(\ell-1)} W^{(\ell-1)} + b^{\ell-1}) \tag{3}$$

where ℓ denotes the layer number, and $H^{(\ell)}$ is the node embedding matrix passing by layers. The intuition is that at each layer node aggregates information from their local neighbours. Therefore, we get two embedding matrices $H_1^{(\ell)}$ and $H_2^{(\ell)}$ from the left and the right GCNs, respectively.

4.3   Information Fusion

We continue to describe the information fusion layer and how to explore nonlinearity user interactions with the MLP layers.

Fusion Layer. In this layer, we merge embeddings of two users ($h_i^{(\ell)} \in H_1^{(\ell)}, h_j^{(\ell)} \in H_2^{(\ell)}$) into one vector. In detail, inspired by the Neural Matrix Factorisation model, NeuMF [6], we define the fusion function as:

$$z_0 = \phi(W^T h_i \odot h_j) \tag{4}$$

where $\odot$, ϕ and W denote the element-wise product of vectors, activation function and edge weights of the output layer, respectively. The benefit of adopting the fusion layer instead of concatenating two embeddings together is that we can extract the linearity pair-wise interaction between two users. This compensates the insufficiency of GCNs to acquire global user interactions.

MLP Layers. Since the fusion layer analogues the matrix factorisation model to capture the linear interaction between two users. However, simply capturing linearity user interaction is not enough. To address this issue, we propose to add hidden layers on the fused vector, using a standard MLP to extract the non-linear interactions for pairs of users. In this sense, we can endow the interactions between u_i and u_j, rather than the way of fusion layer that uses only a fixed element-wise product on them. More precisely, the MLP layers in our NEULP model is defined as:

$$z_1 = \phi_1(W_1^T z_0 + c_1)$$
$$z_2 = \phi_2(W_2^T z_1 + c_2)$$
$$\dots$$
$$z_m = \phi_m(W_m^T z_{m-1} + c_m)$$

(5)

where W_m, c_m, and ϕ_m denote the weight matrix, bias vector, and activation function for the m-th layer's perceptron, respectively.

4.4   Model Prediction

After the partial aggregation and information fusion layers, we obtain a representation vector $r_{ij} = z_m$ for each pair of users (u_i and u_j), which presents the latent features between them. As such, in the last prediction layer, we summarise the latent features as a predicted score which represents the possibility of existing a link between these two users:

$$\hat{y}_{ij} = \sigma(W_f^T \phi_{m+1}(z_m))$$

(6)

where $\sigma(x) = \frac{1}{1+e^{-x}}$ to scale the output into $(0, 1)$. With this, we successfully sum up the comprehensive information between u_i and u_j into a similarity score $\hat{y}_{ij}$ for link prediction.

4.5   Optimisation

To learn model parameters, we optimise the Binary Cross Entropy Loss (BCELoss) which has been intensively used in link prediction models. We adopt the mini-batch Adam [8] to optimise the prediction model and update the model parameters. In particular, for a batch of randomly sampled user tuple (u_i, u_j), we compute the possibility $\hat{y}_{ij}$ of existing a link between them after partial aggregation and information fusion, and then update model parameters by using the gradient of the loss function.

5   Experiments

In this section, we conducted intensive experiments in order to answer the following research questions:

- RQ1: How effective is NEULP when compared with the state-of-the-art baselines focusing on network structures?
- RQ2: How does NEULP perform when compared with the state-of-the-art baselines for attributed social network datasets with only one type of user attributes?
- RQ3: Can we utilise NEULP for link prediction with two different types of user attributes? How effective will it be?

Table 1. Statistics summary of the seven datasets.

Statistics	USAir	Yeast	Cora	Email	Citeseer	Instageam-1	Instagram-2
# of Nodes	332	2,375	2,708	799	3,312	6,815	12,944
# of Edges	2,126	11,693	5,429	10,182	4,660	36,232	61,963
Node features	No	No	Yes	No	Yes	Yes	Yes
Node label	No	No	Yes	Yes	Yes	No	No

5.1   Dataset Description

We evaluate our model on five widely used datasets for graph-related machine learning tasks: USAir, Yeast, Cora, Email and Citeseer. They are publicly accessible on the websites, together with their dataset descriptions.[1] We further use an Instagram dataset, which we collected from Instagram relying on its public API.[2] Our data collection follows a similar strategy as the one proposed by Zhang et al. [26]. Concretely, we sample users from New York by their geotagged posts. Then, for each user, we collected all her/his posted hashtags, and further performed following preprocessing to filter out the users matching any of the following criteria: (i) users whose number of followers are above the 90th percentile (celebrities) or below the 10th percentile (bots); (ii) users without location check-ins or posted hashtags. Then we generate two datasets Instagram-1 and Instagram-2 according to different filter conditions:

- Instagram-1: users with no less than 100 location check-ins and with no less than 10 friends.
- Instagram-2: users with no less than 100 location check-ins and with no less than 5 friends.

---

[1] https://linqs.soe.ucsc.edu/data, and http://snap.stanford.edu/data/index.html.
[2] The dataset was collected in 01/2016 when Instagram's API was publicly available.

Table 2. Link prediction performance comparison with baseline methods on network datasets (AUC). OOM: out of memory.

	USAir	Yeast	Cora	Instagram-1	Instagram-2
MF	0.831 ± 0.010	0.903 ± 0.003	0.816 ± 0.005	0.740 ± 0.013	0.714 ± 0.008
NeuMF	0.801 ± 0.015	0.907 ± 0.005	0.686 ± 0.009	0.744 ± 0.001	0.727 ± 0.001
node2vec	0.805 ± 0.002	0.905 ± 0.008	0.770 ± 0.005	0.785 ± 0.004	0.738 ± 0.002
GCN	0.903 ± 0.006	0.938 ± 0.003	0.819 ± 0.006	0.804 ± 0.004	0.773 ± 0.003
GraphSAGE	0.897 ± 0.007	0.933 ± 0.004	0.838 ± 0.008	0.802 ± 0.005	0.770 ± 0.002
GAT	0.902 ± 0.006	0.935 ± 0.004	0.839 ± 0.008	0.796 ± 0.005	0.767 ± 0.002
P-GNNs	0.911 ± 0.018	0.940 ± 0.006	0.852 ± 0.008	0.734 ± 0.007	OOM
HGANE	0.901 ± 0.004	0.932 ± 0.006	0.845 ± 0.010	0.797 ± 0.007	0.771 ± 0.008
NeuLP	$\mathbf{0.952} \pm 0.009$	$\mathbf{0.965} \pm 0.003$	$\mathbf{0.894} \pm 0.006$	$\mathbf{0.813} \pm 0.001$	$\mathbf{0.790} \pm 0.003$

The reason for generating these two Instagram datasets is that we need datasets with different types of user attributes to discuss the possibility of applying NeuLP to attributed social networks with different types of user attributes (RQ3). The statistics of the seven datasets are given in Table 1.

For the experiments, we use two sets of 10% existing links and an equal number of nonexistent links as test and validation sets. We use the left 80% existing links and an equal number of nonexistent links as the training sets.

5.2   Experimental Settings

Evaluation Metrics. Like existing link prediction studies [4], we adopt the most frequently-used metrics "Area under the receiver operating characteristic" (AUC) to measure the performance.

Baselines. To demonstrate the effectiveness of NeuLP, we compare its performance with several baseline methods: MF [17], NeuMF [6], node2vec [4] and several state-of-the-art GNN variants, including GCN [9], GraphSAGE [5], GAT [23], and P-GNNs [24]. Besides, we also adopt HGANE [7], which is a collective network embedding framework with a hierarchical graph attention mechanism.

In order to study the performances of NeuLP on the attributed datasets (link prediction using one same type or two different types of user attributes), we additionally compare our model with walk2friend [1] and tag2friend [25].

Implementation Details. For the node2vec related approaches, i.e., node2vec, walk2friend, tag2friend, we use the default settings as in [4]. All the neural network-related baselines follow their original paper's code at their GitHub pages if available; otherwise, we implement it by ourself following their original paper, e.g., HGANE. For NeuLP, we use a simple GNN model, GCN, for partial aggregation and the layer numbers of GCN and MLP are 2 and 4, respectively.[3]

---

[3] Code and datasets are available at https://github.com/zhiqiangzhongddu/NeuLP.

Table 3. Link prediction performance comparison between NEuLP and baseline methods on attributed social network datasets with one user attribute type (AUC). OOM: out of memory.

Attributes	Methods	Datasets	
		Instagram-1	Instagram-2
Location check-in	walk2friend	0.831 ± 0.007	0.820 ± 0.003
	GCN	0.808 ± 0.005	0.781 ± 0.003
	GraphSAGE	0.801 ± 0.006	0.775 ± 0.003
	GAT	0.784 ± 0.006	0.758 ± 0.008
	P-GNNs	0.686 ± 0.012	OOM
	HGANE	0.830 ± 0.007	0.799 ± 0.002
	NEuLP	0.832 ± 0.002	0.828 ± 0.001
	NEuLP*	$\mathbf{0.885 \pm 0.001}$	$\mathbf{0.880 \pm 0.001}$
Hashtag	tag2friend	0.724 ± 0.013	0.696 ± 0.011
	GCN	0.812 ± 0.004	0.781 ± 0.003
	GraphSAGE	0.806 ± 0.004	0.775 ± 0.003
	GAT	0.774 ± 0.012	0.763 ± 0.009
	P-GNNs	0.717 ± 0.010	OOM
	HGANE	0.819 ± 0.005	0.781 ± 0.004
	NEuLP	0.836 ± 0.002	0.826 ± 0.002
	NEuLP*	$\mathbf{0.880 \pm 0.002}$	$\mathbf{0.875 \pm 0.001}$

To make a fair comparison, all methods are set to have a similar number of parameters and trained for the same number of epochs.

5.3   Performance Comparison (RQ1)

We first compare the link prediction performance of approaches on the network datasets, which do not include any user-shared information. Here, we use the adjacent matrix A as user features set. We compare the performance of NEuLP with the baselines, except for walk2friend and tag2friend, since the latter two require user-shared information.

From the results shown in Table 2, we first observe that our model consistently outperforms all the baselines with remarkable improvements. It indicates the effectiveness of NEuLP for link prediction with only network structures. In particular, there is 4.9% AUC improvement on the Cora dataset. NEuLP gets 30.3% AUC improvement over NeuMF (i.e., on the Cora dataset), this indicates the observed network plays an important role in link prediction, as NeuMF does not include the network topology. NEuLP also outperforms P-GNNs on the Instagram-1 dataset (with 13.5% AUC improvement), it indicates the advantage of our model on real OSN datasets to capture their global information.

5.4   Attributed OSNs (RQ2 & RQ3)

We continue to discuss the performance of various methods on attributed social networks. More specifically we present the performance of NEuLP with one type of attributes (RQ2) and discuss the possibility of applying NEuLP on attributed social networks with different types of attributes (RQ3). We use walk2friend & tag2friend as the baselines for OSN datasets with location check-ins and hashtags, respectively, and several variant GNNs as baselines for both datasets.

Link Prediction with One Attribute Type. Here we use the encoded user-attributes matrix as user features set (i.e., one-hot encoding), where 1 means the user visited a location (or published a hashtag), and 0 otherwise. NEuLP* means we adopt the embeddings generated by random walk-based methods, e.g., walk2friend, tag2friend, as input user features.

From Table 3, we can see that NEuLP outperforms walk2friend, tag2friend and other GNN models with up to 5.8% AUC improvement. Moreover, NEuLP* gets improvements over NEuLP on all three datasets with up to 6.4% AUC improvement, it means the information reflected by location check-ins or hashtag is a good supplement to the social network structure and our model can well capture such information.

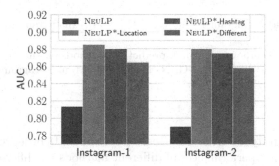

Fig. 2. Link prediction performance comparison of NEuLP on the Instagram datasets. NEuLP: without user attributes. NEuLP*-feature: adopting random-walk related methods generated embeddings as user features.

Link Prediction with Two Different Attribute Types. To answer the research question RQ3, we design experiments on attributed social networks with two different types of user attributes, which means performing link prediction given two user's different types of information, e.g., one user shares location check-ins, and the other user shares hashtags (see NEuLP*-Different in Fig. 2). Following the previous settings, we randomly sample two groups of users where the first group users share their location check-ins and the second group sharing hashtags. These two groups account for half of all users. We organise two graphs where users with location check-ins are connected through locations and users with hashtags are connected through hashtags. Then we apply walk2friend and tag2fiend to these two graphs to generate user embeddings (model settings are as same as default settings). We use these two generated user embeddings matrix as two user features sets for left and right input, respectively.

From the experimental results are shown in Fig. 2, we can see that NEULP on attributed social networks with different types of user attributes gets significantly better performances compared with the results without using user attributes. NEULP*-Different gets similar prediction performance on attributed social networks with the ones using one user attribute type, i.e., NEULP*-Location and NEULP*-Hashtag. It demonstrates the flexibility of our model on making link prediction, even when utilising different types of information that users share in OSNs.

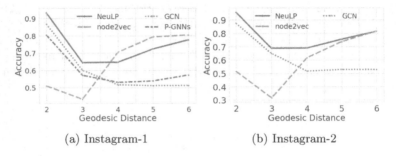

(a) Instagram-1   (b) Instagram-2

Fig. 3. Prediction accuracy of four link prediction methods on the Instagram datasets with different geodesic distances. P-GNNs leads to OOM on Instagram-2 dataset; thus not showing in figure (b).

Performance Analyses with Geodesic Distances. In order to further investigate the prediction performances of different methods for different users pairs, we perform in-depth analyses on the relation between prediction performance and the graph geodesic of two users in OSNs. The results are drawn in Fig. 3, which has two sub-figures to plot the prediction accuracy of the methods (node2vec, GCN, P-GNNs and NEULP) with different user-pair geodesic distances (i.e., the number of edges in a shortest path connecting two nodes in the network) for the two Instagram datasets. We see that the accuracy of GCN and P-GNNs significantly decreases when increasing the distance. This means GNN models cannot capture latent information for user pairs of large distances. In contrast, node2vec behaves in an opposite way: it achieves higher accuracy when increasing the distance and it does not perform well when the distance is small. The reason might be that (long) random walks can capture long-distance information. However, NEULP gets the most balanced performances, this demonstrates that it successfully uses both linearity and non-linearity user interactions to improve the performance of GNN-based models for link prediction.

6   Conclusion and Future Work

In this paper, we have presented NEULP as an end-to-end deep learning model for link prediction in OSNs. NEULP incorporates linearity and non-linearity user interactions to overcome the limitations of current GNN-based link prediction methods. Compared with the state-of-the-art methods, NEULP can not only achieve significant performance improvements but also have the flexibility of utilising different types of user attributes for link prediction. In future, we plan to extend NEULP for link prediction in temporal (or time-varying) networks [12].

Acknowledgements. This work is partially supported by the Luxembourg National Research Fund through grant PRIDE15/10621687/SPsquared.

References

1. Backes, M., Humbert, M., Pang, J., Zhang, Y.: walk2friends: inferring social links from mobility profiles. In: CCS, pp. 1943–1957. ACM (2017)
2. Bhattacharyya, P., Garg, A., Wu, S.F.: Analysis of user keyword similarity in online social networks. Social Network Anal. Min. **1**, 143–158 (2011)
3. Chen, A., Gao, S., Karampelas, P., Alhajj, R., Rokne, J.: Finding hidden links in terrorist networks by checking indirect links of different sub-networks. In: Wiil, U.K. (ed.) Counterterrorism and Open Source Intelligence. LNSN, pp. 143–158. Springer, Vienna (2011). https://doi.org/10.1007/978-3-7091-0388-3_8
4. Grover, A., Leskovec, J.: node2vec: scalable feature learning for networks. In: KDD, pp. 855–864. ACM (2016)
5. Hamilton, W.L., Ying, Z., Leskovec, J.: Inductive representation learning on large graphs. In: NIPS, pp. 1025–1035. NIPS (2017)
6. He, X., Liao, L., Zhang, H., Nie, L., Hu, X., Chua, T.: Neural collaborative filtering, In: WWW. pp. 173–182. ACM (2017)
7. Jiao, Y., Xiong, Y., Zhang, J., Zhu, Y.: Collective link prediction oriented network embedding with hierarchical graph attention. In: CIKM, pp. 419–428. ACM (2019)
8. Kingma, D.P., Ba, J.: Adam: a method for stochastic optimization. In: ICLR (2015)
9. Kipf, T.N., Welling, M.: Semi-supervised classification with graph convolutional networks. In: ICLR (2017)
10. Lei, C., Ruan, J.: A novel link prediction algorithm for reconstructing protein-protein interaction networks by topological similarity. Bioinformatics **29**(3), 355–364 (2012)
11. Li, G., Müller, M., Qian, G., Delgadillo, I.C., Abualshour, A., Thabet, A.K., Ghanem, B.: DeepGCNs: Making GCNs Go as Deep as CNNs. CoRR (2019). abs/1910.06849
12. Li, J., Cheng, K., Wu, L., Liu, H.: Streaming link prediction on dynamic attributed networks. In: WSDM, pp. 369–377. ACM (2018)
13. Li, Q., Han, Z., Wu, X.: Deeper insights into graph convolutional networks for semi-supervised learning. In: AAAI, pp. 3538–3545. AAAI (2018)
14. Liben-Nowell, D., Kleinberg, J.: The link-prediction problem for social networks. J. Am. Soc. Inf. Sci. Technol. **58**(7), 1019–1031 (2007)
15. Lichtenwalter, R.N., Lussier, J.T., Chawla, N.V.: New perspectives and methods in link prediction. In: SIGKDD, pp. 243–252. ACM (2010)

16. Liu, X., Murata, T., Kim, K., Kotarasu, C., Zhuang, C.: A general view for network embedding as matrix factorization. In: WSDM, pp. 375–383. ACM (2019)
17. Menon, A.K., Elkan, C.: Link prediction via matrix factorization. In: Gunopulos, D., Hofmann, T., Malerba, D., Vazirgiannis, M. (eds.) ECML PKDD 2011. LNCS (LNAI), vol. 6912, pp. 437–452. Springer, Heidelberg (2011). https://doi.org/10.1007/978-3-642-23783-6_28
18. Mikolov, T., Chen, K., Corrado, G., Dean, J.: Efficient estimation of word representations in vector space. In: ICLR (2013)
19. Mutlu, E.C., Oghaz, T.A.: Review on Graph Feature Learning and Feature Extraction Techniques for Link Prediction. CoRR (2019). abs/1901.03425
20. Perozzi, B., Al-Rfou, R., Skiena, S.: DeepWalk: online learning of social representations. In: KDD, pp. 701–710. ACM (2014)
21. Qiu, J., Dong, Y., Ma, H., Li, J., Wang, C., Wang, K., Tang, J.: NetSMF: large-scale network embedding as sparse matrix factorization. In: WWW, pp. 1509–1520. ACM (2019)
22. Tang, J., Qu, M., Wang, M., Zhang, M., Yan, J., Mei, Q.: LINE: large-scale information network embedding. In: WWW. pp. 1067–1077. ACM (2015)
23. Velickovic, P., Cucurull, G., Casanova, A., Romero, A., Lio, P., Bengio, Y.: Graph attention networks. In: ICLR (2018)
24. You, J., Ying, R., Leskovec, J.: Position-aware graph neural networks. In: ICML. JMLR (2019)
25. Zhang, Y.: Language in our time: an empirical analysis of hashtags. In: WWW, pp. 2378–2389. ACM (2019)
26. Zhang, Y., Humbert, M., Rahman, T., Li, C.T., Pang, J., Backes, M.: Tagvisor: a privacy advisor for sharing hashtags. In: WWW, pp. 287–296. ACM (2018)

PLSGAN: A Power-Law-modified Sequential Generative Adversarial Network for Graph Generation

Qijie Bai[1], Yanting Yin[1], Yining Lian[1], Haiwei Zhang[1,2(✉)],
and Xiaojie Yuan[1,2]

[1] College of Computer Science, Nankai University, Tianjin, China
{baiqijie,lianyining}@dbis.nankai.edu.cn, yanting.yin@mail.nankai.edu.cn,
{zhhaiwei,yuanxj}@nankai.edu.cn
[2] College of Cyber Science, Nankai Univeristy, Tianjin, China

Abstract. Characterizing and generating graphs is essential for modeling Internet and social network, constructing knowledge graph and discovering new chemical compound molecule. However, the non-unique and high-dimensional nature of graphs, as well as complex community structures and global node/edge-dependent relationships, prevent graphs from being generated directly from observed graphs. As a well-known deep learning framework, generative adversarial networks (GANs) provide a feasible way, and have been applied to graph generation. In this paper, we propose PLSGAN, a Power-Law-modified Sequential Generative Adversarial Network to address aforementioned challenges of graph generation. First, PLSGAN coverts graph topology to node-edge sequences by sampling based on biased random walk. Fake sequences are produced by trained generator and assembled to get target graph. Second, power law distribution of node degree is taken into consideration to modify the learning procedure of GANs. Last, PLSGAN is evaluated on various datasets. Experimental results show that PLSGAN can generate graphs with topological features of observed graphs, exhibit strong generalization properties and outperform state-of-the-art methods.

Keywords: Graph generation · Generative adversarial networks · Power law distribution

1 Introduction

Graph generation has an important significance in many domains, including Internet and social network modeling, new chemical compound molecule discovery, and knowledge graph construction. For chemical compound discovery, generated graphs can be seen as new molecules, where nodes are corresponding to atoms and edges are corresponding to chemical bonds [4]. For financial fraud detection, generated synthetic financial networks can be used in empirical studies under the premise of protecting private information [8]. For graph-related

© Springer Nature Switzerland AG 2020
Z. Huang et al. (Eds.): WISE 2020, LNCS 12342, pp. 109–121, 2020.
https://doi.org/10.1007/978-3-030-62005-9_9

academic research, generative methods can be used to synthesize and augment experimental datasets. Diverse methods have been proposed in the history of graph generation. Many methods for graph generation have been proposed based on graph theories [23] and statistical theorems [5], which provide the basis for evaluating the characteristics of graphs.

Traditional studies mainly focus on Internet structure. Based on Autonomous System connectivity in the Internet, Inet [13] has been presented as a topology generator and further been improved to Inet-3.0 [24]. Both of Inet and Inet3.0 believe node degree distribution of Internet topology obeys some certain mathematical laws. So they perform well in degree distribution but for other features, evaluating results are unsatisfying. The key challenge of graph generation is how to directly produce graphs with similar features to the observed graphs. Mathematical graph theories and statistical theorems can only measure artificially designated but not general properties, so it is difficult to generate graphs similar to observed ones in all aspects by traditional methods.

To date, with the development and widely usage of deep learning, new approaches have appeared and provided a possible way to graph generation. Generative methods on deep learning for graphs can be categorized into four group: recurrent graph neural networks, graph autoencoders, graph generative adversarial networks and hybrid generative models. Recurrent graph neural networks decompose the topology of graphs into sequences, and transform graph generation into sequence generation [16,25]. Graph autoencoders embed graphs into a low-dimensional space, and then reconstruct them as generated graphs [11,14,17,22]. Graph generative adversarial networks (Graph GANs) consist of a generator used to produce fake graphs and a discriminator used to distinguish fake graphs from the real data [4,7]. Some methods put recurrent graph neural networks or other graph neural networks into GANs and modify the generator so that it produces fake samples instead of the whole graph [3,26]. All these methods can directly produce graphs from observed ones but still have limitations. For example, [16,22] are only applicable to the small graphs. [3,25] focus too much on the generalization ability of all graph features, resulting in no outstanding performance on some certain characteristics like traditional methods.

In this paper, we propose PLSGAN, a power-law-modified sequential generative adversarial network, for graph generation. We balance our model between the generalization ability of all graph features and the specific characteristic node degree distribution. Like other recurrent graph neural networks, we regard the topological characteristics of graphs as the concentrated embodiment of the sequence characteristics of nodes and edges. We leverage biased random walk to sample sequences from observed graphs. Like other graph generative adversarial networks, the generator is designed to produce fake samples and the discriminator is designed to distinguish them from real samples. Note that there are a number of real-life networks, such as Internet and social networks, whose node degrees conform to the power law distribution. Thus, we use the power law regressed on the observed graphs to modify the method.

To prove the effectiveness of our method, we compare PLSGAN to other state-of-the-art methods with diverse evaluations. Experimental results show that our method is not inferior to other methods in the generalization ability of graph features and is superior to the counterparts in most characteristics.

The main contributions are summarized as follows:

- We introduce PLSGAN which generates new graphs directly from observed ones. PLSGAN captures characteristics of a graph by learning samples from the original graph via biased random walk and assembles the target graph.
- We have a special optimization for node degree distribution. We regress power law parameters on the observed graph and modify the node degree distribution of generated graph based on these parameters.
- We design extensive experiments to compare PLSGAN with state-of-the-art models on diverse evaluations. Experimental results prove the effectiveness of our method.

The rest of this paper is organized as follows. In Sect. 2, we briefly review related studies, including traditional methods and deep learning methods. Section 3 presents the proposed PLSGAN in details. Experiments are described in Sect. 4. In Sect. 5, we conclude this paper.

2 Related Work

This work aims at topological graph generation, which has been proposed to simulate real-life networks. In this section, we summarize the existing studies based on two typical kinds of solutions.

2.1 Traditional Graph Generation

Traditional graph generation methods refer to using mathematical graph theories and statistical theorems to capture specific topological features of the observed graph, and then add nodes and edges to the generated newly graph according to these features. Faloutsos et al. [6] point out that node degree of the Internet topology obeys the power law distribution. Jin et al. [13] apply this rule to the method Inet for graph generation. Inet generates a new network topology by the power law function counted from the degree distribution of the observed network. Subsequently, Winick et al. [24] find that the degree distribution of the Internet topology does not strictly obey the power law, and propose the more consistent complementary cumulative distribution function (CCDF) instead. Based on CCDF, they take connectivity into account and improve Inet to Inet-3.0. Albert et al. [1] present a random graph model based on statistical theorem. All these rules and methods are built on real Internet topology, but later researches and experiments [26] show that they are also applicable to other graph structure data such as social networks and citation relationships.

2.2 GNN-based Graph Generation

With the rapid development of computer technology in recent years, neural network methods, which extremely rely on computation, have been developed by leaps and bounds. Franco et al. [20] first proposed general neural network method for different types of graphs, such as directed, undirected, cyclic and acyclic ones. Then researchers begin to leverage GNN to graph generation issues and its generalization of features has attracted the attention of many researchers. To learn more complex topological features of graphs, one idea is to convert graph structure to sequences by proposing nodes and edges. You et al. [25] present GraphRNN, which models graph generation process in two levels, graph-level and edge-level. In graph level, recurrent neural network adds one node to an existed node sequence each time step. In edge level, recurrent neural network produces one edge between the new node and previous nodes. Simonovsky et al. [22] use graph autoencoder(GraphVAE) to capture the topological features of the observed graph and reconstruct the generated graph. Ma et al. [17] add the regularization to GraphVAE and propose RGVAE model.

Since Goodfellow et al. [9] propose generative adversarial networks, the architecture has been widely used in data generative models. Bojchevski et al. [3]'s NetGAN combines LSTM [12] with GANs to capture the topological features of the observed graph by sequently sampling. It uses Wasserstein GANs [2] to train the model and get fake samples. Assemble the adjacency matrix with these samples and achieve the final generated graph. De Cao and Kipf [4] present a molecular generative adversarial network which combines the R-GCNs [21], GANs, and reinforcement learning to achieve small molecular graphs. For labeled graph generation, Fan et al. [7] propose LGGAN that feeds the labels of graphs as extra inputs for generator according to conditional GAN [18] and as outputs of discriminator according to auxiliary conditional GAN [19]. Misc-GAN [26] is a multi-scale graph generative model and divides the observed graph into different levels of granularity. Each level is generated by GANs independently and then all parts coalesce into the generated graph.

All these existing methods cannot address the generalization ability of all graph features and specific characteristics at the same time. So in the rest of this paper, we introduce PLSGAN, which balances both of them.

3 Methodology

In this section, we introduce PLSGAN in details. The pipeline of our method can be described as follows. There are three key points in our method: sampling sequences from observed graphs, capturing topological characteristics and modifying node degree distribution. The architecture is shown in Fig. 1.

3.1 Sampling Sequences from Observed Graphs

In our method, sampling is the first step and determines whether topological characteristics of observed graphs can be detected by subsequent steps. Here we adopt biased second-order random walk [10] as sampling strategy to our method.

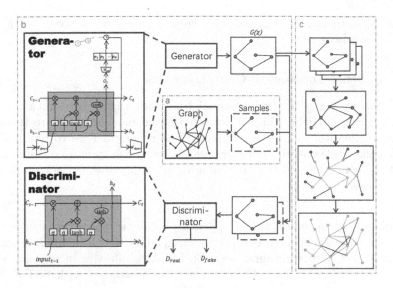

Fig. 1. Architecture of PLSGAN proposed in this paper. Three key points of the model corresponding to part a (sampling sequences from observed graphs), part b (capturing topological characteristics) and part c (modifying node degree distribution).

To capture characteristics better, the probability that a random walk jumps from one node to another is biased based on the history of the random walk. We use E to denote the edge set of observed graph and c_i to denote the i-th node in one random walk. Then we have:

$$P\left(c_i = x | c_{i-1} = v\right) = \begin{cases} \frac{\alpha_{vx}}{\sum_{(v,v') \in E} \alpha_{vv'}} & if \ (v, x) \in E \\ 0 & otherwise \end{cases} \quad (1)$$

where α_{vx} denotes unnormalized transition probability from node v to x. We calculate the value of α_{vx} as follows. Assuming that a random walk has just jumped from node t to v, and currently it is staying at node v. Let x_i denote the adjacent nodes of v. Then

$$\alpha_{pq}(v, x_i) = \begin{cases} \frac{1}{p} & if \ d_{tx_i} = 0 \\ 1 & if \ d_{tx_i} = 1 \\ \frac{1}{q} & if \ d_{tx_i} = 2 \end{cases} \quad (2)$$

where d_{tx_i} denotes the length of the shortest path between node t and x_i in the range of $\{0, 1, 2\}$. The variable $\alpha_{pq}(v, x)$ in Eq. 2 corresponds to α_{vx} in Eq. 1. The parameters p and q are used to control the exploration of random walks. If we set p to a large value, the probability of jumping back to node t in next hop will be low and this is benefit for further exploration. Meanwhile, the parameter q controls whether the next node is nearer to or further from last hop node t. A higher value for q makes random walks jump away from source node c_0 rapidly while a lower value leaves random walk around c_0. In this way, we can capture both local and global topological features of observed graph from sampling sequences.

3.2 Capturing Topological Characteristics

In this part, we use generative adversarial network architecture to capture the topological characteristics. Like other GAN-based models, there are two components: a generator and a discriminator. The generator tries to produce fake random walks while the discriminator tries to distinguish them from real samples. Both of the generator and the discriminator are trained using backpropagation in this process.

Generator. The generator is an implicit probabilistic model that can produce fake random walk sequences like $(v_1, v_2, ..., v_n)$(The length of random walk is n). As a typical recurrent neural network, LSTM is effective in sequential processing, so we adopt it here. The generator is constructed with LSTM f_θ, in which θ denotes the parameters. At each time step t, f_θ updates its cell states C_t and hidden states h_t. The hidden states h_t correspond to the probabilistic distribution of next node. Let $\sigma(\cdot)$ denote softmax function. Sampling from the distribution $v_t \sim Cat(\sigma(h_t))$ determines the next node for current fake random walk. At time step $t + 1$, next node v_{t+1} and cell states C_t will be passed to LSTM cells recurrently. For the first node, a latent code z and initial function $g_{\theta'}$ replace C_t and h_t.

$$\begin{cases} z \sim \mathcal{N}(0, I_d) \\ C_0 = g_{\theta'}(z) \\ (h_t, C_t) = f_\theta(m_{t-1}, 0) \\ v_t \sim Cat(\sigma(h_t)) \end{cases} \quad (t = 1, 2, ..., n) \quad (3)$$

Discriminator. In our method, the discriminator is only used to guide the training of generator. A simple model structure can work well enough and preserve a low level of extra computing. With this in consideration, we use original LSTM as the discriminator. For a real or fake random walk, the discriminator takes a one-hot vector of one node as input at each time step and outputs the probability that this random walk is real after processing the whole sequence.

Details for GAN. In each iteration of the generator, a discrete probabilistic distribution h_t of length N(number of nodes) is produced. It takes a large amount of computation to calculate on such high dimensional data. To optimize this model, the hidden states of LSTM cells are described as vectors with far lower dimension than N, and when needed to output, they are up-projected to $\mathbb{R}^N$ using W_{up} matrix. Naturally, sampled nodes v_t need to be down-projected from the format of one-hot vector to low dimensional vector using corresponding W_{down} matrix to be the input of each iteration.

Early stopping is another important issue for model training. We expect to generate graphs with similar features to the observed graphs rather than copies of the observed ones. So we set the max value of edge overlap(EO), which means the ratio of edges appeared in both generated fake random walks and the observed graphs. When the value of EO reaches the max, training will stop.

To prevent mode collapse in training, we use Wasserstein GAN [2]. After training, the generator achieves the ability of producing almost "real" random walks. Till current step, we have captured the topological characteristics indeed and stored them in these random walks.

3.3   Modifying Node Degree Distribution

The outputs of above GAN are a series of random walks. In order to modify node degree distribution, the topological graph structure need to be constructed first. A score matrix $S \in \mathbb{R}^{N \times N}$ counts the frequencies of edges appeared in generated random walks. Set $s_{ij} = s_{ji} = max\{s_{ij}, s_{ji}\}$ so that S is symmetrized. To make the constructed graph reflect the general characteristics of generated random walks better, we sampled the edges as follows.

$$p_{ij} = \frac{s_{ij}}{\sum_{u,v} s_{uv}} \qquad (4)$$

where p_{ij} denotes the probability that the edge (i, j) is sampled. Keep sampling until the amount of sampled edges reaches a certain percentage of the edge number of observed graph.

Power-law-based generative models have been proposed in [13] and [24] to produce new Internet topology. As is shown in Fig. 2, modifying mainly depends on adding nodes and edges to current generated graph and this will increase the degree of existing nodes. To minimize this effect, we modify in ascending order of node degrees. Regress the parameter γ and scaling factor k on the observed graph in the following equation:

$$P(d) = k \bullet d^{\gamma} \qquad (5)$$

Using regressed parameters and the amount of nodes, the specific number of nodes with any certain degree can be calculated. Then add nodes that have not yet appeared in generated graph and connect them to existing connected subgraph. After all due nodes have appeared in generated graph, we make a statistics to achieve current lacking degrees for all nodes. Connect pairs of nodes, both of which don't have enough neighbors, until the node degree distribution of generated graph has conformed to the power law.

When adding edges to generated graph, an important question is which pairs of nodes we should choose to connect. Assuming that 50 nodes of generated graph lacking degrees of 80 in total, then how should we choose to add 40 edges among these nodes? In PLSGAN, we consider the connecting probability between nodes of different degrees. Calculate on the observed graph as follows:

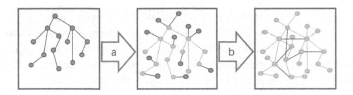

Fig. 2. Partial process of modifying node degree distribution. (a) Add nodes that have not appeared in generated graph. (b) Add edges to fit power-law node degree distribution.

$$w_i^j = MAX \left(1, \sqrt{\left(log\frac{d_i}{d_j}\right)^2 + \left(log\frac{f(d_i)}{f(d_j)}\right)^2}\right) \bullet d_j \tag{6}$$

where d_i denotes the degree of node i, and $f(d_i)$ denotes the number of nodes with degree of d_i. The value of w_i^j means the weight of d_j with respect of d_i. If we set node i as one end of an edge, then the probability p_i^j of the other end being node j is defined as:

$$p_i^j = \frac{w_i^j}{\sum_{k \in G} w_i^k} \tag{7}$$

where G refers to the node set of generated graph. In this way, node pairs are chosen and connected in order to make node degree distribution of the generated graph conform to power law.

4   Experiment

In this section, we design the experiments to evaluate PLSGAN. We compare PLSGAN with state-of-the-art baselines by calculating various graph characteristics provided by graph theories and statistical theorems. Experimental results demonstrate the generative ability for graph of our method.

4.1   Datasets

Internet topology is the main experimental data for early graph generation research. Consider Autonomous Systems (ASs) as nodes and the communications among them as edges, then a network can be constructed. The addition and deletion of nodes and edges in this network happen over time. The dataset AS733 [15] consists of snapshots of ASs sampled once a day for 733 times and we select 4 of them for our experiments. The statistical information is shown in Table 1.

Table 1. The statistics of experimental datasets.

Graph ID	Node number	Edge number
1	3015	5156
2	3109	5341
3	3213	5624
4	3311	5864

4.2 Baselines

- **Inet-3.0** [24] is a traditional method, which is based on the rule that it is the CCDF rather than node degree distribution which obeys the power law. So it regresses the power law of CCDF from observed graph and creates a new graph from nothing.
- **NetGAN** [3] is a deep learning model for graph generation. The core idea is that topological features can be captured via random walks. It relies on generative adversarial networks and recurrent neural networks to produce fake random walks which would be used to assemble the new graph.

4.3 Measures

To evaluate PLSGAN and above baselines, we measure the following characteristics on generated graphs of these methods.

- **Complementary cumulative distribution**. CCDF is defined as the probability of randomly selecting a node with a degree greater than k. CCDF for real-life graphs is often more in line with power law.
- **Clustering coefficient**. Clustering coefficient is used to measure the connections among each node's adjacent nodes.
- **Coordination coefficient**. Coordination coefficient characterizes the interconnections between nodes of different degrees. It can be used to measure the tendency of larger-degree nodes in graph G to connect with other larger-degree nodes.
- **Richclub coefficient**. Richclub coefficient measures the kernel and peripheral layered structure of the graph. The kernel is composed of large-degree nodes in the graph, and nodes in the kernel have a high-density connection with each other.
- **Average length of shortest paths**. The property describes the average length of the shortest paths for all node pairs in the graph.

4.4 Evaluation

Based on the above experimental settings, the results are shown in Table 2, Fig. 3 and Fig. 4.

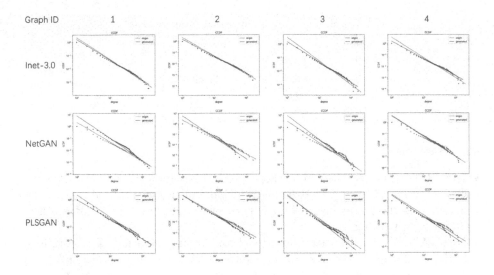

Fig. 3. CCDF scatter and regressed power law plots on the datasets for PLSGAN and the baselines. The red parts of plots are of observed graphs and the blue parts are of generated graphs. (Color figure online)

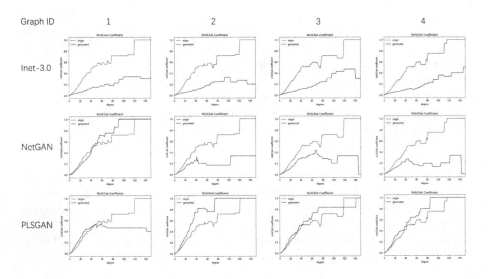

Fig. 4. Richclub coefficient plots on the datasets for PLSGAN and the baselines. The red parts of plots are of observed graphs and the blue parts are of generated graphs. (Color figure online)

Table 2. Numerical metrics of evaluation results

Graph ID	Method	Node number	Edge number	Avg. Clustering	Coordination	Avg. shortest path Len
1	Observed	3015	5156	0.1815	−0.2289	3.7615
	Inet-3.0	3015	4824	0.0176	**−0.2197**	4.1209
	NetGAN	2108	6051	**0.1695**	−0.1737	3.3410
	PLSGAN	2693	4629	0.1412	−0.2528	**3.9949**
2	Observed	3109	5341	0.1804	−0.2246	3.8047
	Inet-3.0	3109	4997	0.0193	−0.2173	4.1382
	NetGAN	1740	3522	0.0631	−0.1628	4.0126
	PLSGAN	2765	4784	**0.1248**	**−0.2204**	**3.9078**
3	Observed	3213	5624	0.1755	−0.2194	3.7701
	Inet-3.0	3213	5272	0.0213	**−0.2127**	4.0915
	NetGAN	2245	5827	0.0948	−0.1671	3.6072
	PLSGAN	2870	5047	**0.1453**	−0.2363	**3.6937**
4	Observed	3311	5864	0.1817	−0.2153	3.7639
	Inet-3.0	3311	5555	0.0287	−0.2078	4.0330
	NetGAN	2375	5239	0.0864	−0.1710	**3.7440**
	PLSGAN	2954	5261	**0.0940**	−0.2201	4.1470

Table 2 mainly summarizes the numerical metrics, including basic statistical properties, the average of clustering coefficient, coordination coefficient and the average length of shortest paths, evaluated on the datasets for PLSGAN and the baselines. It is clear that our method PLSGAN outperforms these baselines for most datasets on most metrics, especially the average of clustering coefficient, which reveals the distribution rules and interrelationships of node clusters in the topological graph. Figure 3 and Fig. 4 show two functional metrics, complementary cumulative distribution function and richclub coefficient. For CCDF, our method doesn't show too much superiority but meanwhile it is not inferior to the baselines. However, for the metric Richclub coefficient, the polylines of the generated graphs by PLSGAN fit the ones by the observed graph better. Above experimental results clearly show the superiority of PLSGAN.

5   Conclusion

In this paper, we propose PLSGAN, which captures the topological characteristics from node sequences sampled by biased random walks, and trains a LSTM-based GANs to generate fake node sequences that characterize the topology. After assembling, a power-law-based modifying process is performed. Experimental results on real-life datasets show that our method performs better than state-of-the-art models for the issue of graph generation.

Acknowledgements. This work is supported by NSFC (NO. U1936206, U1836109) and "the Fundamental Research Funds for the Central Universities", Nankai University (No. 63201209).

References

1. Albert, R., Barabási, A.L.: Statistical mechanics of complex networks. Rev. Modern Phys. **74**(1), 47 (2002)
2. Arjovsky, M., Chintala, S., Bottou, L.: Wasserstein GAN. arXiv preprint arXiv:1701.07875 (2017)
3. Bojchevski, A., Shchur, O., Zügner, D., Günnemann, S.: Netgan: generating graphs via random walks. arXiv preprint arXiv:1803.00816 (2018)
4. De Cao, N., Kipf, T.: Molgan: an implicit generative model for small molecular graphs. arXiv preprint arXiv:1805.11973 (2018)
5. Dixon, W.J., Massey, Jr., F.J.: Introduction to statistical analysis (1951)
6. Faloutsos, M., Faloutsos, P., Faloutsos, C.: On power-law relationships of the internet topology. In: ACM SIGCOMM Computer Communication Review, vol. 29, pp. 251–262. ACM (1999)
7. Fan, S., Huang, B.: Deep generative models for generating labeled graphs (2019)
8. Fich, E.M., Shivdasani, A.: Financial fraud, director reputation, and shareholder wealth. J. Financ. Econ. **86**(2), 306–336 (2007)
9. Goodfellow, I., et al.: Generative adversarial nets. In: Advances in neural Information Processing Systems, pp. 2672–2680 (2014)
10. Grover, A., Leskovec, J.: node2vec: scalable feature learning for networks. In: Proceedings of the 22nd ACM SIGKDD International Conference on Knowledge Discovery and Data Mining, pp. 855–864. ACM (2016)
11. Grover, A., Zweig, A., Ermon, S.: Graphite: iterative generative modeling of graphs. arXiv preprint arXiv:1803.10459 (2018)
12. Hochreiter, S., Schmidhuber, J.: Long short-term memory. Neural Comput. **9**(8), 1735–1780 (1997)
13. Jin, C., Chen, Q., Jamin, S.: Inet: internet topology generator (2000)
14. Kipf, T.N., Welling, M.: Variational graph auto-encoders. arXiv preprint arXiv:1611.07308 (2016)
15. Leskovec, J., Kleinberg, J., Faloutsos, C.: Graphs over time: densification laws, shrinking diameters and possible explanations. In: Proceedings of the Eleventh ACM SIGKDD International Conference on Knowledge Discovery in Data Mining, pp. 177–187. ACM (2005)
16. Li, Y., Vinyals, O., Dyer, C., Pascanu, R., Battaglia, P.: Learning deep generative models of graphs. arXiv preprint arXiv:1803.03324 (2018)
17. Ma, T., Chen, J., Xiao, C.: Constrained generation of semantically valid graphs via regularizing variational autoencoders. In: Advances in Neural Information Processing Systems, pp. 7113–7124 (2018)
18. Mirza, M., Osindero, S.: Conditional generative adversarial nets. arXiv preprint arXiv:1411.1784 (2014)
19. Odena, A., Olah, C., Shlens, J.: Conditional image synthesis with auxiliary classifier GANs. In: Proceedings of the 34th International Conference on Machine Learning, vol. 70. pp. 2642–2651. JMLR. org (2017)

20. Scarselli, F., Gori, M., Tsoi, A.C., Hagenbuchner, M., Monfardini, G.: The graph neural network model. IEEE Trans. Neural Networks **20**(1), 61–80 (2008)
21. Schlichtkrull, M., Kipf, T.N., Bloem, P., van den Berg, R., Titov, I., Welling, M.: Modeling relational data with graph convolutional networks. In: Gangemi, A., et al. (eds.) ESWC 2018. LNCS, vol. 10843, pp. 593–607. Springer, Cham (2018). https://doi.org/10.1007/978-3-319-93417-4_38
22. Simonovsky, M., Komodakis, N.: GraphVAE: towards generation of small graphs using variational autoencoders. In: Kůrková, V., Manolopoulos, Y., Hammer, B., Iliadis, L., Maglogiannis, I. (eds.) ICANN 2018. LNCS, vol. 11139, pp. 412–422. Springer, Cham (2018). https://doi.org/10.1007/978-3-030-01418-6_41
23. West, D.B., et al.: Introduction to Graph Theory, vol. 2. Prentice Hall, Upper Saddle River (1996)
24. Winick, J., Jamin, S.: Inet-3.0: Internet topology generator. Technical report, Technical Report CSE-TR-456-02, University of Michigan (2002)
25. You, J., Ying, R., Ren, X., Hamilton, W.L., Leskovec, J.: GraphRNN: a deep generative model for graphs. arXiv preprint arXiv:1802.08773 (2018)
26. Zhou, D., Zheng, L., Xu, J., He, J.: Misc-GAN: a multi-scale generative model for graphs. Front. Big Data **2**, 3 (2019). https://doi.org/10.3389/fdata

Social Network

User Profile Linkage Across Multiple Social Platforms

Manman Wang, Wei Chen, Jiajie Xu, Pengpeng Zhao, and Lei Zhao[✉]

Institute of Artificial Intelligence, School of Compute Science and Technology,
Soochow University, Suzhou, China
mmwang9@stu.suda.edu.cn, {robertchen,xujj,ppzhao,zhaol}@suda.edu.cn

Abstract. Linking user profiles belonging to the same people across multiple social networks underlines a wide range of applications, such as cross-platform prediction, cross-platform recommendation, and advertisement. Most of existing approaches focus on pairwise user profile linkage between two platforms, which can not effectively piece up information from three or more social platforms. Different from the previous work, we investigate the user profile linkage across multiple social platforms by proposing an effective and efficient model called MCULK. The model contains two key components: 1) Generating a similarity graph based on user profile matching candidates. To speed up the generation, we employ the locality sensitive hashing (LSH) to block user profiles and only measure the similarity for the ones within the same bucket. 2) Linking user profiles based on similarity graph. Extensive experiments are conducted on two real-world datasets, and the results demonstrate the superiority of our proposed model MCULK compared with the state-of-art methods.

Keywords: User profile linkage · Social network · Clustering

1 Introduction

Nowadays, social networks (SNs) have greatly enriched people's daily life. It's common that a person participates in multiple social networks (e.g., Twitter[1] and Instagram[2]) to meet different needs due to limited services provided by a single SN. According to a report from GlobalWebIndex, the average number of social media accounts of each internet user is more than 3 in 2019[3]. Since each SN only reflects some aspects of a user's interests, we can have a better and deeper understanding of a user's behaviors after linking all profiles belonging to him/her. User profile linkage (UPL) is of critical importance for many applications, such as cross-platform recommendation [1], knowledge graph refinement [13], and fake user information verification [7].

---

[1] https://twitter.com/.
[2] https://www.instagram.com/.
[3] https://www.globalwebindex.com/reports/social.

Z. Huang et al. (Eds.): WISE 2020, LNCS 12342, pp. 125–140, 2020.
https://doi.org/10.1007/978-3-030-62005-9_10

Considering the great practical significance of UPL, many researchers have paid much attention to this topic [7,11,14,19]. Despite of the great contributions made by them, most of their proposed approaches focus on linking user profiles across two platforms, which have so far fallen short of successfully addressing following problem.

In real applications, given m different social platforms, some users may have profiles on each of them while others only have profiles on partial platforms, we call this phenomenon *"platform uncertainty"*. For instance, given four platforms F(Foursquare), W(WhatsApp), I(Instagram), and T(Twitter) in Fig. 1, Jim has a profile on each platform, Tom only has profiles (u_1 and u_2) on F and W, and Lee has profiles on T and I. The *"uncertainty"* is a common phenomenon and will bring following two challenges.

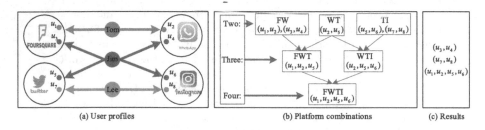

| (a) User profiles | (b) Platform combinations | (c) Results |

Fig. 1. An example of platform uncertainty

- **High Computation Cost**. Given m different platforms, the first step is to train models for calculating user profile similarity between any two platforms, if traditional methods [7,17,20] are directly applied to this problem. Assuming the number of user profiles on each platform is n, the upper bound of similarity measurement is $\frac{m \cdot (m-1)}{2} n^2$. Next, the results of the first step are integrated to find actual linked user profiles, and upper bound of platform combinations to be considered is $\sum_{i=3}^{m} C_m^i$. Obviously, such approach will lead to huge time cost if given a large m or n.
- **Incorrect Linkage Chain**. During the integration of the results of pairwise linkage, there may exist "incorrect linkage chain". In Fig. 1, assuming that pairwise sets $\{(u_1, u_2), (u_2, u_5), (u_3, u_4), (u_5, u_6), (u_7, u_8)\}$ are obtained after computing similarity for 24 ($\frac{4 \cdot (4-1)}{2} \cdot 2^2$) pairwise user profiles. Note that, the assumption is reasonable since we cannot achieve user profile linkage with 100% precision and recall in most of real cases. Next, cross-platform combinations are integrated and final results $\{(u_3, u_4), (u_7, u_8), (u_1, u_2, u_5, u_6)\}$ are returned. All chains like $(u_2, u_5) \rightarrow \{(u_1, u_2, u_5), (u_2, u_5, u_6)\} \rightarrow (u_1, u_2, u_5, u_6)$ are called *"incorrect linkage chain"* in this work.

To address above challenges, an effective and efficient model MCULK is proposed, which includes two phases: (1) similarity graph generation; (2) clustering on the similarity graph. In the first phase, a blocking technique is first used to reduce the number of user profile pairs to be considered, with the goal of achieving high efficiency. Next, features are extracted from user profile pairs. Finally, a pre-trained classifier Logistic Regression is applied to predict whether two profiles belonging to the same user, and the output of this step is stored as a similarity graph. Note that, each vertice of the graph represents a user profile, and the edge between two vertices denotes the probability that they belong to the same user. In the second phase, clustering is performed on the similarity graph and the disjoint user profile clusters are returned. By this way, the *"incorrect linkage chain"* problem can be addressed, since a user profile only belongs to one cluster.

To sum up, we make following main contributions in this study.

- We propose a novel model MCULK to link user profiles across multiple platforms by considering effectiveness and efficiency simultaneously.
- To reduce the high computation cost of UPL and avoid "incorrect linkage chain", we construct a similarity graph and perform clustering on the graph.
- We conduct extensive experiments on two real-world datasets, and the results demonstrate that our proposed model can achieve better performance than the state-of-art methods.

The remainder of this paper is organized as follows. Section 2 summarizes related work and Sect. 3 formulates the UPL problem. The details of the proposed model are introduced in Sect. 4. The time complexity is discussed in Sect 5, and the experimental evaluation is reported in Sect. 6. The paper is concluded in Sect. 7.

2   Related Work

The goal of UPL is to piece up information across multiple social platforms by employing the similarities between profiles. A lot of efforts have been devoted to address UPL problem. Zafarani et al. [19] analyzed user behaviors from human limitation, exogenous factors to endogenous factors when selecting usernames across multiple SNs, and exploited such redundancies to match user profiles. To improve performance, Li et al. [7] used username and name to link user profiles and employed Gale-Shapley algorithm to address matching confusion like one-to-many or many-to-many relationships between user profiles. However, people may select completely different username and name in different SNs, so it's difficult to have a high UPL precision only employing username or name. Zhang et al. [20] extracted seven features from user profile to improve linkage precision, e.g., username, language, description, url, popularity, location, avatar. Mu et al. [11] proposed ULINK and projected all user profiles from different social platforms into the latent user space by minimizing the distance between match pairs and maximizing the distance between unmatched pairs in the latent user space. Raad et al. [14] proposed a matching framework to address UPL problem, which considered all the profile's attributes and allowed users to specify the weight of the attribute.

As UPL is a subproblem of user identity linkage (UIL), UIL approaches can also address UPL problem and can improve accuracy since the more information we have, the further we distinguish the user profile from others. Goga et al.[3] extracted three features from user generated content (UGC), i.e., location, timestamp, user's writing style, and used these features to match user accounts across SNs, which found that location is the most powerful feature. Li et al. [8] was similar to [3], but which focused on spatial, temporal and text information extracted from UGC and applied three-level classifier to match user accounts. Zhou et al. [22] focused on friendship structure in SNs and proposed the Friend Relationship-Based User Identification algorithm, which calculated a match degree of all candidate matched pairs and only considered top rank pairs as identical users. To further improve identification performance, researchers attempted hybrid approaches to address UIL problem. Kong et al. [4] took user's social into account to improve UIL performance, with features extracted from user's social, spatial, temporal and text information and one-one constraint, they could effectively match user accounts. Gao et al. [2] proposed CNL, which incorporates heterogeneous user attributes and social features and learned model parameters by EM algorithm. Liu et al. [10] proposed a multi-objective semi-supervised framework by jointly modeling heterogeneous behaviors (e.g., profile features and content features) and structure consistency. Xie et al. [18] combined user profile information and network structure, and employed a network embedding approach to link user identities. But in fact few people would like to public their friend network structure in SNs due to privacy protection.

Compared to these works, our approach mainly focuses on three or more social platforms user profile linkage, and aims to return a set of user profile clusters where user profiles within a cluster match with each other and different clusters belong to different person.

3   Problem Formulation

In this section, we present several definitions used throughout the paper and formulate the problem.

Definition 1. *User Profile Cluster.* *A user profile cluster $C = \{u_1, u_2, \cdots, u_k\}$ is a set of user profiles that are assumed to belong to the same user. If each user profile of the cluster is from a different social platform, the cluster is called "source-consistent" [12], otherwise "source-inconsistent". A "complete cluster" [15] is "source-consistent" and contains profiles from all social platforms.*

Definition 2. *Similarity Graph.* *Given an undirected weighted similarity graph $G = (V, E)$, where V is set of vertices with each vertice representing a user profile, E is the set of edges, and the weight of each edge represents the probability of two profiles belonging to the same user.*

Definition 3. *Weak Edge* [16]. *Given an edge (u_s, u_t), assume u_s and u_t are from platforms A and B respectively. u_s have several edges E_{st} to user profiles of B, u_t have several edges E_{ts} to user profiles of A, (u_s, u_t) is defined e*

Problem Formulation. Given m social platforms, let $U = \{u_1, u_2, ..., u_n\}$ denote the set of all user profiles. We aim to divide U into mutually exclusive clusters $\{C_1, C_2, ..., C_M\}$ with the goal of putting all profiles belonging to the same user into a cluster C_i, i.e., maximizing the following objective function:

$$\max \sum_{j=1}^{M} \sum_{i=1, e_i \in C_j}^{N} w(e_i)$$

where e_i is an edge in cluster C_j, $w(e_i)$ denotes the weight of e_i, N represents the number of edges in C_j, and the number of clusters M is determined by Algorithm 2, which is introduced in Subsect. 4.2.

Here we assume that there is no user profile duplication in all platforms, i.e. there does not exist any two user profiles inside one platform which belong to one real-world person [21].

4 Approaches

Observed from Fig. 2, our proposed model MCULK consists of two main phases: (1) generating a similarity graph based on user profile similarity; (2) performing clustering on the similarity graph to divide the user profile set U into mutually exclusive clusters $\{C_1, C_2, ..., C_M\}$.

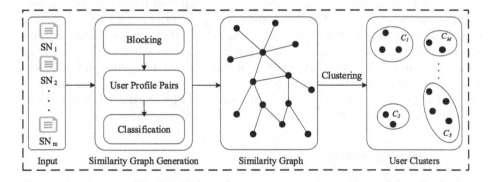

Fig. 2. Overview of the proposed model

Algorithm 1: Similarity Graph Generation

Input: a set of user profiles U, a pre-trained logistic regression classifier $\mathcal{L}_c$, a jacard
similarity threshold S_{jac};
Output: an undirected similarity graph G
1 $V \leftarrow \emptyset, E \leftarrow \emptyset, G \leftarrow (V, E)$;
2 $B \leftarrow$ MinHashLSH(U, S_{jac}); // B is a set of buckets;
3 $P \leftarrow \emptyset$;
4 **for** *each bucket* $b \in B$ **do**
5 $\quad\lfloor$ get all user profile pairs from b, and add these pairs into P;

6 $\{(u_i, u_j, w(u_i, u_j))\} \leftarrow \mathcal{L}_c(P)$; //Compute probability $w(u_i, u_j)$;
7 add the edge (u_i, u_j) with weight $w(u_i, u_j)$ into graph G;
8 **return** G

4.1   Similarity Graph Generation

The steps involved in the similarity graph generation are described in
Algorithm 1. The upper bound of user profile pairs to be considered across m
social platforms is $\frac{m \cdot (m-1)}{2} \cdot n^2$. Obviously, the time cost will be unacceptable if
we compute the similarity for all pairs. To address the problem, MinHashLSH
[5] is employed to locate similar user profiles and a jaccard similarity threshold
S_{jac} is used to optimize MinHashLSH by minimizing the false positive and false
negative. And the information that is used to block user profiles is the q-gram of
username and name because the previous studies [9, 19] have shown that a person
prefers to select similar usernames or names across different social platforms.

Algorithm 1 starts with initializing an empty graph, followed by indexing
all user profiles through MinHashLSH in a set B(line 2), which consists of a
set of buckets containing many user profiles. In line 5, user profile pairs are
generated from these buckets where each user profile of a pair is from different
platforms, i.e. there are no direct links between user profiles from the same
platform. Next, all possible matching pairs are detected from P, where a pre-
trained logistic regression classifier is applied to judge whether a pair is matching
or not. During the process, the logistic regression label 1 means matched and
label 0 means unmatched, and $w(u_i, u_j)$ denotes the probability of u_i and u_j to
be a true match. The algorithm returns all possible matched pairs and use them
to generate an undirected graph (line 6–7), where u_i and u_j are two adjacent
vertices in the graph and $w(u_i, u_j)$ is their weight. Finally, the similarity graph
G is returned, which will be further processed in the next phase.

4.2   Clustering on Similarity Graph

A clustering method is applied to find all user profiles belonging to the same
person, which is based on the ideas of link strength proposed by Saeedi et al.
[16]. Generally, there are two main errors for any clustering method. First, a single cluster in fact may contain more than one user profiles that needed to be put
as separate clusters (splitting cluster). Second, multiple clusters that actually
belong to same user should be merged to create a single bigger cluster (merging clusters). To address above problems, splitting and merging are taken into

account during clustering. Algorithm 2 outlines the clustering approach, which consists of three main steps: (1) partitioning the similarity graph to generate initial clusters; (2) splitting clusters with low intra-cluster similarity; (3) merging high similar clusters. Figure 3 illustrates the algorithm with an example, and user profiles are from four platforms A, B, C and D and belong to same person with same index, e.g. a_0, b_0, c_0.

Initial Clustering. This step aims to create source-consistent clusters CS_{temp} on similarity graph. All weak edges (e.g. $(b_0, c_0), (c_1, d_2)$) are first removed from similarity graph, then the C_{init} is obtained by generating connected components on similarity graph, and these components may be source-inconsistent because user profiles from same platform may be indirectly linked. In Fig. 3, the largest component is source-inconsistent as the edges $(a_0, b_0), (a_1, b_0)$ cause source-inconsistent for platform A. To determine which edge of a component should be removed, the element $Assoc(u)$, which initially only contains u (line 8), is used to record all associated profiles of u. Then all edges of a component are sorted in descending order by edge weights and each edge (u_s, u_t) will be checked to see whether it causes source-inconsistent. If source-inconsistent, the edge will be removed. Otherwise, the related sets will be updated with the union of $Assoc(u_s)$ and $Assoc(u_t)$. In our example, assume the edge (a_0, b_0) is first checked and vertices a_0 and b_0 are updated by setting $Assoc(a_0) = Assoc(b_0) = \{a_0, b_0\}$. Then we check the edge (a_0, c_0) and set $Assoc(a_0) = Assoc(b_0) = Assoc(c_0) = \{a_0, b_0, c_0\}$. When processing (b_0, a_1), which causes inconsistent, and the edge (b_0, a_1) will be removed. After eliminating all source-inconsistent edges of a component, CS_{temp} is updated by performing connected components again. Finally, four clusters are obtained in Fig. 3.

Split Clusters. This step aims to separate low similarity user profiles from a cluster to ensure high intra-cluster similarity. Before determining a possible split of each cluster from CS_{temp}, vertices' similarities are needed to compute. Here we define vertice v_i similarity in a cluster C as follows:

$$sim(v_i) = \frac{\sum_{v_j \in adj(v_i)} w(v_i, v_j)}{|adj(v_i)|}$$

where $adj(v_i)$ is a set of vertices that have edges to v_i in cluster C, and $w(v_i, v_j)$ is the weight between v_i and v_j.

After computing vertices' similarities of a cluster, vertices are sorted in ascending order and the vertice v_i is separated from the current cluster to form a new cluster, if $sim(v_i)$ is the lowest and less than the given splitting threshold t_s. Then this splitting process is repeated until the similarities of all vertices in current cluster are larger than the threshold t_s. In our example, the vertice c_3 is separated from the cluster (a_1, b_1, c_3). After processing all initial clusters, we will check whether the resulted clusters should be added to the CS or merged to create a bigger cluster. For each resulted cluster, if the cluster is a complete cluster, it will be directly added to CS to reduce the computation in the merge step. Otherwise, a representative will be created for the cluster by combining the

---

Algorithm 2: Clustering on Similarity Graph

Input: similarity graph $G = (V, E)$, the pre-trained classifier $\mathcal{L}_c$, the blocking threshold K,
splitting threshold t_s, merging threshold t_m

Output: user profile cluster sets CS

1 $CS \leftarrow \emptyset, CS_{temp} \leftarrow \emptyset$;
2 // initial clustering;
3 $E' \leftarrow$ removeWeakEdges(E);
4 $C_{init} \leftarrow$ getConnectedComponents(V, E');
5 **for** $C_i(V_i, E_i) \in C_{init}$ **do**
6   **if** $sourceConsistent(C_i)$ **then** $CS_{temp} \leftarrow CS_{temp} \cup C_i$;
7   **else**
8        $V_i \leftarrow$ initAssocUser(V_i);
9        $E_{sorted} \leftarrow$ sortEdges(E_i);
10   **for** $(u_s, u_t) \in E_{sorted}$ **do**
11   **if** $sourceConsistent(Assoc(u_s), Assoc(u_t))$ **then**
12             $V_i \leftarrow$ updateSets$(Assoc(u_s), Assoc(u_t))$;
13   **else** $E_i \leftarrow$ removeEdge(u_s, u_t) ;
14       $CS_{temp} \leftarrow CS_{temp} \cup$ getConnectedComponents(C_i);

15 // splitting ;
16 $C_{split} \leftarrow \emptyset, C_{repre} \leftarrow \emptyset$;
17 **for** $C_i \in CS_{temp}$ **do** $C_{split} \leftarrow C_{split} \cup$ splitCluster(C_i, t_s) ;
18 **for** $C_i \in C_{split}$ **do**
19   **if** $isComplete(C_i)$ **then** $CS \leftarrow CS \cup C_i$;
20   **else** $C_{repre} \leftarrow C_{repre} \cup$ createRepresentative(C_i) ;

21 // merging ;
22 $CM \leftarrow$ computeClusterSim$(C_{repre}, \mathcal{L}_c, t_m, K)$;
23 **while** $CM \neq \emptyset$ **do**
24    $(r_1, r_2) \leftarrow$ getBestMatch(CM);
25    $cm \leftarrow$ merge(r_1, r_2);
26   remove r_1, r_2 from the index table, C_{repre} and CM;
27   **if** $isComplete(cm)$ **then** $CS \leftarrow CS \cup cm$;
28   **else**
29       $C_{repre} \leftarrow C_{repre} \cup$ createRepresentive(cm);
30   index cm;
31       $CM \leftarrow$ adaptMapping$(CM, C_{repre}, cm, \mathcal{L}_c, t_m, K)$;

32 $CS \leftarrow CS \cup$ getCluster$(C_{repre}, 2)$;
33 **return** CS

---

properties from user profiles of the cluster to simplify the similarity computation between clusters. For each property with multiple values of the representative, a preferred value is selected by pre-determined rules, e.g. longest string for string values. And user profiles and their platforms are also tracked in each representative, which will be helpful in the merge step to avoid source-inconsistent between clusters.

Merge Clusters. This step aims to detect and merge high similar clusters that actually belong to same user. The process of generating similar clusters is computationally expensive because the number of cluster pairs increases quadratically. To speed up the process, a blocking technique is applied to prune unnecessary pairwise comparisons. Here we directly use inverted index where the keys are the union of q-gram of the username and name from all cluster representatives and consider top K candidate clusters. To be more clearly, for each cluster in C_{repre}, the top K clusters are returned as candidates by the number of common tokens in descending order. And platforms tracked in the representatives are also

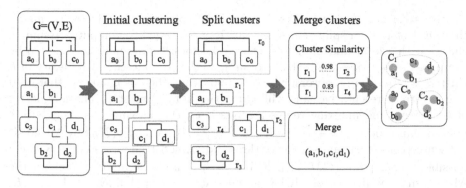

Fig. 3. An example of clustering

used to further reduce the number of cluster pairs that need to be compared by source-consistent constraint. In Fig. 3, there are two merge candidates after similarity computing. The resulted cluster pairs are recorded in CM(line 22) and the similarity of each pair in CM is larger than the threshold t_m. While CM is not empty (line 23–31), a highest similar cluster pair (r_1, r_2) is first obtained, and then the pair will be merged to create a bigger cluster. Next, we need to update CM by removing pairs related to r_1 or r_2 in the CM and remove r_1, r_2 from cluster representatives C_{repre} and index tables. Lastly, the bigger cluster will be added into CS if it is complete. AS shown in Fig. 3, (a_1, b_1, c_1, d_1) is obtained. Otherwise, we create a representative for this bigger cluster cm, find clusters that are similar to cm, and update CM with these new cluster pairs. After processing CM, CS is updated with cluster representatives C_{repre} by a constraint that the size of a cluster should be more than one. In line 33, the final cluster sets CS are returned. In our example, three user clusters (a_0, b_0, c_0), (a_1, b_1, c_1, d_1) and (b_2, d_2) are finally obtained.

5 Time Complexity Analysis

In this section, we analyze the time complexity of Algorithm 2, which contains three main steps: initial clustering, splitting and merging.

There are m social platforms participating in the linkage process. Assume the graph G resulted in Algorithm 1 with V nodes and E edges. In the first step, the type of each edge is first determined, which has a complexity of $O(E)$. Then all weak edges are removed from G in time $O(E - E')$ where $E - E'$ is the number of weak edges. And the connected components are generated by BFS, which has a complexity of $O(V + E')$. The step of removing all inconsistent edges has a complexity of $O(3V + E'(1 + m + logE') - E'')$ in the worst case that all nodes are within a single component where line 6 is $O(V)$, line 8 is $O(V)$, line 9 is $O(E'logE')$, line 10–13 is $O(mE')$ for updating at most m nodes each time and line 14 is $O(V + E' - E'')$, E'' is the number of inconsistent edges. Assume the first step produce $h(h \in [\frac{V}{m}, V])$ clusters. In the splitting step, the size of each cluster is smaller than $m + 1$, so the times of split for each cluster is smaller than m. Before each split, nodes' similarities are needed to compute. Assume a cluster with size m and need to split $m - 1$ times, the number of nodes to be computed is $m + m - 1 + ... + 2 = O(m^2)$, so the computation complexity of h clusters is $O(hm^3)$ in the worst case. After splitting, there is n clusters that need to generate representatives and will participate in merging step. The complexity of generating and indexing representatives is $O(n)$. Indexing n clusters with b buckets can produce $O(\frac{n^2}{b})$ cluster pairs by MCUL while $O(Kn)$ by MCULK. By source-consistent and similarity threshold t_m constraints, only a very small possible merge pairs $(CM$ with $|CM| \ll \frac{n^2}{b})$ are generated. Therefore, the Algorithm 2 complexity of MCUL is quadratic in n and linear in V and E, and for MCULK is linear in V, E and n with acceptable performance loss.

By constructing the similarity graph in Algorithm 1 and conducting clustering in Algorithm 2, we reduce the complexity of user profile linkage across multiple social platforms, and avoid "*incorrect linkage chain*" introduced in introduction. The high performance of our proposed model is demonstrated in Sect. 6.

6   Experiment

6.1   Experiment Settings

Two real-world datasets are used to conduct experiments, the features of datasets and functions used to compute user profile similarity are presented in Table 1.

DS1. This dataset[4] is collected by [17] and contains real-world user profiles from three different social platforms(Google+, Twitter, Instagram). We remove some user profiles from DS1 where username and name are both null and chose a 40–60 split in the DS1 for training(14886 user profiles) and testing (21840 user profiles). The test sets contain 21840 user profiles that needed to be linked, 1541 users have profiles on only one of three platforms, 4385 users have profiles on two of three platforms, 3843 users have profiles on three platforms, i.e., $1541 + (4385 \times 2) + (3843 \times 3) = 21840$.

DS2. This dataset[5] is provided by [15], where each user profile contains following attributes: name, surname, postcode, and suburb. In our experiments, we select 102044 user profiles across four different platforms from the dataset to generate DS2 and chose a 20–80 split in the DS2 for training (19606) and testing (82438). The test sets contain 82438 user profiles that needed to be linked, 5877 users have profiles on only one of four platform, 5062 users have profiles on two of four platforms, 7771 users have profiles on three of four platforms, and 10781 users have profiles on four platforms, i.e., $5877 + (5062 \times 2) + (7771 \times 3) + (10781 \times 4) = 82438$. Note that, such selection is significant for evaluating the better performance of our proposed model compared with existing studies, where the original datasets have different number of user profiles.

---

[4] http://vishalshar.github.io/data/.
[5] https://dbs.uni-leipzig.de.

Table 1. The features of datasets and similarity measurement functions

Dataset	Feature	Similarity measurement function
DS1	f1: username f2: name f3: bio f4: username+name	jaro-winkler(f1, f2, f4), edit-distance(f1, f2) longest common substring/subsequence(f1, f2) tf-idf cosine(f3), bigrams jaccard(f4) average of best match [6] (f4) jensen-shannon(f4)
DS2	f1: name+surname f2: postcode f3: suburb	jaro-winkler(f1, f2, f3), edit-distance(f1) longest common substring/subsequence(f1) bigrams jaccard(f1), jensen-shannon(f1) average of best match(f1)

Compared Algorithms. To evaluate the performance of MCULK, we compare the results of MCULK with that of following algorithms.

OPL [20]: A supervised method which employs a probabilistic classifier that integrates prior knowledge into Naive Bayes to improve UPL performance and uses token-based canopies technique to reduce unnecessary pairwise comparisons, and tokens above frequency threshold θ are discarded.

CNL [2]: An unsupervised method which links user profiles from different social platforms by constructing a generative probabilistic model via exponential family and employing EM algorithm to infer parameters of the generative model.

LINKSOCIAL [17]: A supervised method based on profile similarity, which uses clustering to reduce computation cost. This is also a pairwise linkage method.

UISN-UD [7]: The model employs a two-stage supervised learning classifier to address UPL problem and applys the one-to-one constaint to optimize the pairwise matching results by Gale-Shapley algorithm.

MCUL: This is a variant of MCULK where all possible merging cluster pairs will be considered from each bucket.

MCULK: Top-K candidate clusters are taken into account to reduce the time cost in the merge step.

Evaluation Metrics. The recall, precision and F1 are employed as the metrics to evaluate our model performance. Recall is the fraction of the number of correct user profile clusters that have been found among the number of all real user profile clusters, and precision is the fraction of the number of correct user profile clusters that have been found among the number of result clusters.

$$Recall = \frac{|CorrectClusters|}{|AllClusters|} \quad Precision = \frac{|CorrectClusters|}{|ResultClusters|} \quad F1 = \frac{2 \cdot Recall \cdot Precision}{Recall + Precision}$$

6.2 Performance Evaluation

This section illustrates the results of UPL across multiple social platforms on real-world data sets. Figure 4 shows the recall, precision, and F1 of all algorithms

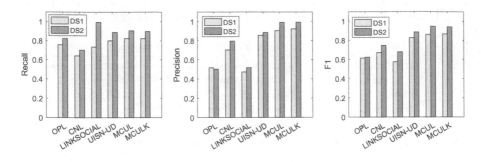

Fig. 4. Performances of all approaches about complete clusters on different datasets

on DS1 and DS2 about complete clusters, i.e. the size of resulted clusters is 3 on DS1 and 4 on DS2. Since OPL, CNL, LINKSOCIAL and UISN-UD are pairwise linkage methods, they cannot effectively address the *"uncertainty"* problem introduced in introduction, where some users only have profiles on partial platforms. After integrating the results of two pairwise linkage for three platforms in DS1 (i.e. $A \rightarrow B, B \rightarrow C$) and three pairwise linkage for four platforms in DS2 (i.e. $A \rightarrow B, B \rightarrow C, C \rightarrow D$) [11], the results are obtained in Fig. 4. Observed from which, CNL has lower recall compared to other methods and the reason is that CNL is an unsupervised method and the similarity vectors do not fit exponential family distributions well. As expected, UISN-UD has better performance than OPL and LINKSOCIAL in DS1 and DS2 since UISN-UD employs the Gale-Shapley algoritm to eliminate the one-to-many or many-to-many relationships existed in the pairwise matching results. And we observe that OPL has better performance than LINKSOCIAL in DS1 since OPL employs prior knowledge about the rarity of tokens and embeds it directly within its probabilistic model, and users usually select fairly unique usernames that may incorporate rarer tokens. And we also find that LINKSOCIAL has high recall on DS2 as it directly return the target profile with maximum score for a source profile between two platforms without further judgments, and the maximum score may be very small (e.g. 0.1), this is the reason for its low precision. Overall, the model MCUL and MCULK both outperform other methods, and the reason is that our model can eliminate the *"incorrect linkage chain"* caused by the integration of pairwise linkage results. Furthermore, the performance of MCULK ($K = 1$) is close to MCUL, which means considering the number of common tokens between user profile clusters to select candidates is feasible.

Table 2 presents the results of MCUL and MCULK while taking all clusters into account (i.e., complete clusters and incomplete clusters). The high precision, recall, and F1 demonstrate that our model can tackle the *"uncertainty"* problem.

Efficiency. Table 3 presents the average running time of different methods. Obviously, UISN-UD is the most time consuming method, as it considers all pairs from different platforms while other methods employ different approaches to reduce computation cost. Without surprise, MCULK has the least time cost.

Table 2. Performances of MCUL, MCULK on different datasets

Dataset	|cluster|	Platforms	Method	Recall	Precision	F1
DS1	8228	3	MCUL	0.786	0.8568	0.8199
			MCULK	0.7904	0.869	0.8278
DS2	23614	4	MCUL	0.9114	0.9216	0.9165
			MCULK	0.905	0.9169	0.9109

Table 3. Average running time(s) of different methods

Dataset	OPL	CNL	LINKSOCIAL	UISN-UD	MCUL	MCULK
DS1	2.634	0.246	5.967	69.039	0.277	**0.149**
DS2	4.312	0.118	0.939	49.929	0.136	**0.092**

This is because, it employs MinHashLSH to reduce the number of user profile pairs to be considered during similarity graph generation, and just takes top-K candidates into account while merging clusters.

Performance w.r.t. Varied t_s, t_m. Splitting and merging clusters are necessary due to the inevitable errors caused by clustering algorithms. Figure 5 shows the performance of MCULK while varying thresholds t_s, t_m. The increase of them both lead to the increase of precision, as clusters with high intra-cluster similarity are more likely to be correct. To balance the recall and precision, we set $t_s = 0.97, t_m = 0.96$ in DS1 and $t_s = 0.98, t_m = 0.98$ in DS2.

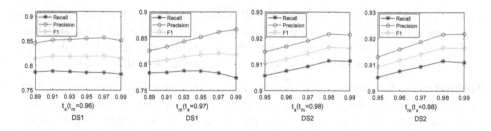

Fig. 5. Performance of MCULK in different datasets w.r.t. varied t_s, t_m

Performance w.r.t. Varied K. Since MCUL has high computation cost in merging step due to the upper bound of cluster pairs is $O(\frac{n^2}{b})$, top-K candidate clusters for each cluster in C_{repre} are considered to reduce the time cost to $O(Kn)$ with MCULK. As shown in Fig. 6, the performance of MCULK has little change with K increasing and is close to MCUL, which means MCULK is effective.

Scalability. To investigate the scalability of MCULK, we expand DS2 to five sources and change the number of platforms that participate in linkage process.

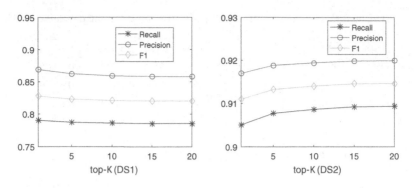

Fig. 6. Performance of MCULK w.r.t. varied K

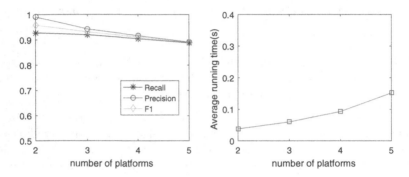

Fig. 7. Scalability of MCULK

Generally, the increase of platforms will lead to decrease of performance. In Fig. 7, we obverse that the precision is slowly reducing. But MCULK still has high recall, precision and F1 while taking linear time consuming with the increase of platforms, these results demonstrate the high scalability of MCULK.

Robustness. To study the robustness of MCULK, we adjust the proportion of users that have profiles only in one platform. Obviously, the proportion of such users can affect the linkage performance. In general, the larger proportion of such users will lead to the smaller precision, recall, and F1. As shown in Table 4, we observe that the performance of MCULK is promised even the proportion is 50%, which means MCULK has high robustness. Note that, 1:1953 denotes the number of users that only have profiles on one platform and 2:4092 represents the number of users that have profiles on two platforms.

Table 4. Robustness of MCULK

Dataset	All users	Proportion	Recall	Precision	F1
DS1	9769	20%(**1:1953**, 2:4092, 3:3724)	0.7886	0.8623	0.8238
		30%(**1:2930**, 2:3603, 3:3236)	0.7833	0.8422	0.8117
		40%(**1:3907**, 2:3114, 3:2748)	0.7753	0.8247	0.7992
		50%(**1:4884**, 2:2625, 3:2260)	0.776	0.801	0.7883
DS2	29491	20%(**1:5877**, 2:5062, 3:7771, 4:10781)	0.905	0.9169	0.9109
		30%(**1:8847**, 2:4178, 3:6887, 4:9579)	0.9028	0.9107	0.9067
		40%(**1:11796**, 2:3294, 3:6003, 4:8398)	0.8991	0.9002	0.8996
		50%(**1:14745**, 2:2410, 3:5119, 4:7217)	0.8954	0.8865	0.8909

7   Conclusion and Future Work

In this paper, we study the problem of user profile linkage across multiple social platforms and propose a model to address this problem, which contains two main components: blocking and clustering. Unlike the existing studies, we return user profile clusters and the size of these clusters can be from 2 to the number of platforms. The extensive experiments on real-world datasets have demonstrated the effectiveness, efficiency, scalability, and robustness of the proposed model.

In the future work, we can further improve the performance of our method by involving more user information, such as user generated content and user friendship network structure. And we can also explore the scalability of the proposed model on larger datasets by performing it in a parallel way.

Acknowledgments. This work was supported by the Major Program of Natural Science Foundation of Jiangsu Higher Education Institutions of China under Grant No. 19KJA610002 and 19KJB520050, and the National Natural Science Foundation of China under Grant No. 61902270, a project funded by the Priority Academic Program Development of Jiangsu Higher Education Institutions.

References

1. Cao, D., et al.: Cross-platform app recommendation by jointly modeling ratings and texts. ACM Trans. Inf. Syst. **35**(4), 37:1–37:27 (2017)
2. Gao, M., Lim, E., Lo, D., Zhu, F., Prasetyo, P.K., Zhou, A.: CNL: collective network linkage across heterogeneous social platforms. In: ICDM, pp. 757–762. IEEE Computer Society (2015)
3. Goga, O., Lei, H., Parthasarathi, S.H.K., Friedland, G., Sommer, R., Teixeira, R.: Exploiting innocuous activity for correlating users across sites. In: WWW, pp. 447–458. ACM (2013)
4. Kong, X., Zhang, J., Yu, P.S.: Inferring anchor links across multiple heterogeneous social networks. In: CIKM, pp. 179–188. ACM (2013)
5. Leskovec, J., Rajaraman, A., Ullman, J.D.: Mining of Massive Datasets, 2nd edn. Cambridge University Press, Cambridge (2014)

6. Li, Y., Peng, Y., Ji, W., Zhang, Z., Xu, Q.: User identification based on display names across online social networks. IEEE Access **5**, 17342–17353 (2017)
7. Li, Y., Peng, Y., Zhang, Z., Yin, H., Xu, Q.: Matching user accounts across social networks based on username and display name. World Wide Web **22**(3), 1075–1097 (2018). https://doi.org/10.1007/s11280-018-0571-4
8. Li, Y., Zhang, Z., Peng, Y., Yin, H., Xu, Q.: Matching user accounts based on user generated content across social networks. Future Gener. Comput. Syst. **83**, 104–115 (2018)
9. Liu, J., Zhang, F., Song, X., Song, Y., Lin, C., Hon, H.: What's in a name?: an unsupervised approach to link users across communities. In: WSDM, pp. 495–504. ACM (2013)
10. Liu, S., Wang, S., Zhu, F., Zhang, J., Krishnan, R.: HYDRA: large-scale social identity linkage via heterogeneous behavior modeling. In: SIGMOD, pp. 51–62. ACM (2014)
11. Mu, X., Zhu, F., Lim, E., Xiao, J., Wang, J., Zhou, Z.: User identity linkage by latent user space modelling. In: KDD. pp. 1775–1784. ACM (2016)
12. Nentwig, M., Rahm, E.: Incremental clustering on linked data. In: ICDM, pp. 531–538. IEEE (2018)
13. Paulheim, H.: Knowledge graph refinement: a survey of approaches and evaluation methods. Semant. Web **8**(3), 489–508 (2017)
14. Raad, E., Chbeir, R., Dipanda, A.: User profile matching in social networks. In: NBiS, pp. 297–304. IEEE Computer Society (2010)
15. Saeedi, A., Nentwig, M., Peukert, E., Rahm, E.: Scalable matching and clustering of entities with FAMER. CSIMQ **16**, 61–83 (2018)
16. Saeedi, A., Peukert, E., Rahm, E.: Using link features for entity clustering in knowledge graphs. In: Gangemi, A., Navigli, R., Vidal, M.-E., Hitzler, P., Troncy, R., Hollink, L., Tordai, A., Alam, M. (eds.) ESWC 2018. LNCS, vol. 10843, pp. 576–592. Springer, Cham (2018). https://doi.org/10.1007/978-3-319-93417-4_37
17. Sharma, V., Dyreson, C.E.: LINKSOCIAL: linking user profiles across multiple social media platforms. In: ICBK, pp. 260–267. IEEE Computer Society (2018)
18. Xie, W., Mu, X., Lee, R.K., Zhu, F., Lim, E.: Unsupervised user identity linkage via factoid embedding. In: ICDM, pp. 1338–1343. IEEE Computer Society (2018)
19. Zafarani, R., Liu, H.: Connecting users across social media sites: a behavioral-modeling approach. In: KDD, pp. 41–49. ACM (2013)
20. Zhang, H., Kan, M.-Y., Liu, Y., Ma, S.: Online social network profile linkage. In: Jaafar, A., et al. (eds.) AIRS 2014. LNCS, vol. 8870, pp. 197–208. Springer, Cham (2014). https://doi.org/10.1007/978-3-319-12844-3_17
21. Zhong, Z., Cao, Y., Guo, M., Nie, Z.: Colink: an unsupervised framework for user identity linkage. In: AAAI, pp. 5714–5721 (2018)
22. Zhou, X., Liang, X., Zhang, H., Ma, Y.: Cross-platform identification of anonymous identical users in multiple social media networks. IEEE Trans. Knowl. Data Eng. **28**(2), 411–424 (2016)

Budgeted Influence Maximization
with Tags in Social Networks

Suman Banerjee[1]([⊠]), Bithika Pal[2], and Mamata Jenamani[2]

[1] Indian Institute of Technology Gandhinagar, Gandhinagar, India
suman.b@iitgn.ac.in
[2] Indian Institute of Technology Kharagpur, Kharagpur, India
bithikapal@iitkgp.ac.in, mj@iem.iitkgp.ac.in

Abstract. Given a social network, where each user is associated with
a selection cost, the problem of *Budgeted Influence Maximization* (*BIM
Problem*) asks to choose a subset of them (known as seed users) within
the allocated budget whose initial activation leads to the maximum number of influenced nodes. In reality, the influence probability between two
users depends upon the context (i.e., tags). However, existing studies
on this problem do not consider the tag specific influence probability. To
address this issue, in this paper we introduce the TAG-BASED BUDGETED
INFLUENCE MAXIMIZATION Problem (*TBIM Problem*), where along with
the other inputs, a tag set (each of them is also associated with a selection cost) is given, each edge of the network has the tag specific influence probability, and here the goal is to select influential users as well as
influential tags within the allocated budget to maximize the influence.
Considering the fact that different tag has different popularity across
the communities of the same network, we propose three methodologies
that work based on *effective marginal influence gain computation*. The
proposed methodologies have been analyzed for their time and space
requirements. We evaluate the methodologies with three datasets, and
observe, that these can select seed nodes and influential tags, which leads
to more number of influenced nodes compared to the baseline methods.

Keywords: Social network · BIM problem · Seed nodes · Tags

1 Introduction

A social network is an interconnected structure among a group of agents. One
of the important phenomenon of social networks is the diffusion of information
[5]. Based on the diffusion process, a well studied problem is the *Social Influence
Maximization* (*SIM Problem*), which has an immediate application in the context
of *viral marketing*. The goal here is to get wider publicity for a product by initially
distributing a limited number of free samples to highly influential users. For a

The work of the first author is supported by the Institute Post Doctoral Fellowship
Grant of IIT Gandhinagar (MIS/IITGN/PD-SCH/201415/006).

Z. Huang et al. (Eds.): WISE 2020, LNCS 12342, pp. 141–152, 2020.
https://doi.org/10.1007/978-3-030-62005-9_11

given social network and a positive integer k, the SIM Problem asks to select k users for initial activation to maximize the influence in the network. Due to potential application of this problem in viral marketing [4], different solution methodologies have been developed. Look into [2] for recent survey.

Recently, a variant of this problem has been introduced by Nguyen and Zheng [7], where the users of the network are associated with a selection cost and the seed set selection is to be done within an allocated budget. There are a few approaches to solve the problem [1,7]. In all these studies, it is implicitly assumed that irrespective of the context, influence the probability between two users will be the same, i.e., there is a single influence probability associated with every edge. However, in reality, the scenario is different. It is natural that a sportsman can influence his friend in any sports related news with more probability compared to political news. This means the influence probability between any two users are context specific, and hence, in Twitter a follower will re-tweet if the tweet contains some specific hash tags. To address this issue, we introduce the *Tag-based Budgeted Influence Maximization (TBIM) Problem*, which considers the tag specific influence probability assigned with every edge.

Ke et al. [6] studied the problem of finding k seed nodes and r influential tags in a social network. However, their study has two drawbacks. First, in reality most of the social networks are formed by rational human beings. Hence, once a node is selected as seed then incentivization is required (e.g., free or discounted sample of the item to be advertised). Also, the cost of displaying a viral marketing message in any media platforms such as *Vonag*[1] is associated with a cost. As the message is constituted by the tags, hence it is important to consider the individual cost for tags. Their study does not consider these issues. Secondly, in their study, they have done the tag selection process at the network level. However, in reality, popular tags may vary from one community to another in the same network. Figure 1 shows community wise distribution of Top 5 tags for the Last.fm dataset. From the figure, it is observed that Tag No. 16 has the highest popularity in Community 3. However, its popularity is very less in Community 4. This signifies that tag selection in network level may not be always helpful to spread influence in each community of the network. To mitigate this issues, we propose three solution methodologies for the TBIM Problem, where the tag selection is done community wise. To the best of our knowledge, this is the second study in this direction. The main contributions of this paper are as follows:

- Considering the tag specific influence probability, in this paper we introduce the Tag-based Budgeted Influence Maximization Problem (TBIM Problem).
- Two iterative methods have been proposed with their detailed analysis.
- To increase the scalability, an efficient pruning technique has been developed.
- The proposed methodologies have been implemented with three publicly available datasets and a set of experiments have been performed to show their effectiveness.

---

[1] https://www.vonage.com.

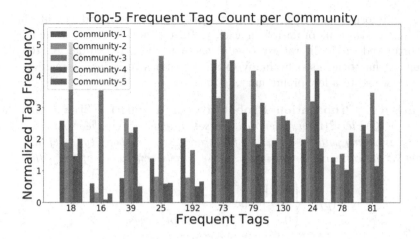

Fig. 1. Community specific distribution top 5 tags in 'Last.fm' dataset

Rest of the paper is arranged as follows: Sect. 2 contains some background material and defines the TBIM Problem formally. The proposed solution methodologies for this problem have been described in Sect. 3. In Sect. 4, we report the experimental results of the proposed methodologies.

2   Background and Problem Definition

The social network is represented as a directed and edge weighted graph $\mathcal{G}(\mathcal{V}, \mathcal{E}, \mathcal{P})$, where the vertex set, $\mathcal{V}(\mathcal{G}) = \{u_1, u_2, \ldots, u_n\}$ is the set of n users, the edge set $\mathcal{E}(\mathcal{G}) = \{e_1, e_2, \ldots, e_m\}$ is the set of m social ties among the users. Along with $\mathcal{G}$, we are also given with a tag set $T = \{t_1, t_2, \ldots, t_a\}$ relevant to the users of the network. $\mathcal{P}$ is the edge weight function that assigns each edge to its tag specific influence probability, i.e., $\mathcal{P} : \mathcal{E}(\mathcal{G}) \longrightarrow (0, 1]^{|T|}$. This means that each edge of the network is associated with a influence probability vector, whose each dimension is for a particular tag. For all $(u_i u_j) \in \mathcal{E}(\mathcal{G})$, we denote its corresponding influence probability vector as $\mathcal{P}_{u_i \to u_j}$. Also, for a particular tag $t \in T$ and an edge $(u_i u_j) \in \mathcal{E}(\mathcal{G})$, we denote the influence probability of the edge $(u_i u_j)$ for the tag t as $\mathcal{P}^t_{u_i \to u_j}$. Now, a subset of the available tags $T' \subseteq T$ which are relevant to the campaign may be used. It is important how to compute the effective probability for each edge and this depends upon how the selected tags are aggregated. In this study, we perform the *independent tag aggregation* shown in Eq. 1.

$$\mathcal{P}^{T'}_{u_i \to u_j} = 1 - \prod_{t \in T'} (1 - \mathcal{P}^t_{u_i \to u_j}) \tag{1}$$

Diffusion in Social Networks. To conduct a campaign using a social network, a subset of the users need to be selected initially as seed nodes (denoted by $\mathcal{S}$). The users in the set $\mathcal{S}$ are informed initially, and the others are ignorant about the information. These seed users start the diffusion, and the information is diffused by the rule of an information diffusion model. There are many such rules proposed in the literature. One of them is the MIA Model [8]. Recently, this model has been used by many existing studies in influence maximization [9]. Now, we state a few preliminary definitions.

Definition 1 (Propagation Probability of a Path). *Given two vertices* $u_i, u_j \in V(\mathcal{G})$, *let* $\mathbb{P}(u_i, u_j)$ *denotes the set of paths from the vertex* u_i *to* u_j. *For any arbitrary path* $p \in \mathbb{P}(u_i, u_j)$ *the propagation probability is defined as the product of the influence probabilities of the constituent edges of the path.*

$$\mathcal{P}(p) = \begin{cases} \prod_{(u_i u_j) \in \mathcal{E}(p)} \mathcal{P}_{u_i \to u_j}^{T'} & \text{if } \mathbb{P}(u_i, u_j) \neq \phi \\ 0 & \text{otherwise} \end{cases} \tag{2}$$

Here, $\mathcal{E}(p)$ *denotes the edges that constitute the path* p.

Definition 2 (Maximum Probabilistic Path). *Given two vertices* $u_i, u_j \in V(\mathcal{G})$, *the maximum probabilistic path is the path with the maximum propagation probability and denoted as* $p_{(u_i u_j)}^{max}$. *Hence,*

$$p_{(u_i u_j)}^{max} = \underset{p \in \mathbb{P}(u_i, u_j)}{argmax} \ \mathcal{P}(p) \tag{3}$$

Definition 3 (Maximum Influence in Arborescence). *For a given threshold* θ, *the maximum influence in-arborescence of a node* v *is defined as*

$$MIIA(v, \theta) = \bigcup_{u \in V(\mathcal{G}), \mathcal{P}(p_{(uv)}^{max}) \geq \theta} p_{(uv)}^{max} \tag{4}$$

Given a seed set $\mathcal{S}$ and a node $v \notin \mathcal{S}$, in MIA Model, the influence from $\mathcal{S}$ to v is approximated by the rule that for any $u \in \mathcal{S}$ can influence v through the paths in $p_{(uv)}^{max}$. The influence probability of a node $u \in MIIA(v, \theta)$ is denoted as $ap(u, \mathcal{S}, MIIA(v, \theta))$, which is the probability that the node u will be influenced by the nodes in $\mathcal{S}$ and influence is propagated through the paths in $MIIA(v, \theta)$. This can be computed by the Algorithm 2 of [8]. Hence, the influence spread obtained by the seed set $\mathcal{S}$ is given by the Eq. 5.

$$\sigma(\mathcal{S}) = \sum_{v \in V(\mathcal{G})} ap(v, \mathcal{S}, MIIA(v, \theta)) \tag{5}$$

Problem Definition. To study the BIM Problem along with the input social network, we are given with the selection costs of the users which is characterized by the cost function $\mathcal{C} : V(\mathcal{G}) \to \mathbb{R}^+$, and a fixed budget $\mathcal{B}$. For any user $u \in V(\mathcal{G})$, its selection cost is denoted as $\mathcal{C}(u)$. The BIM Problem asks to choose a subset of users $\mathcal{S}$ such that $\sigma(\mathcal{S})$ is maximized and $\sum_{u \in \mathcal{S}} \mathcal{C}(u) \leq \mathcal{B}$. Let, $\mathcal{K} = \{\mathcal{K}_1, \mathcal{K}_2, \ldots, \mathcal{K}_\ell\}$ denotes the set of communities of the network. Naturally, all the tags that are considered in a specific context (i.e., T) may not be relevant to each of the communities. We denote the relevant tags of the community $\mathcal{K}_i$ as $T_{\mathcal{K}_i}$. It is important to observe that displaying a tag in any on-line platform may associate some cost, which can be characterized by the tag cost function $\mathcal{C}^T : T \to \mathbb{R}^+$. Now, for a set of given tags and seed nodes what will be the number of influenced nodes in the network? This can be defined as the tag-based influence function. For a given seed set $\mathcal{S}$ and tag set T', the tag-based influence function $\sigma(\mathcal{S}, T')$ returns the number of influenced nodes, which is defined next.

Definition 4 (Tag-Based Influence Function). *Given a social network* $\mathcal{G}(\mathcal{V}, \mathcal{E}, \mathcal{P})$, *a seed set* $\mathcal{S} \subseteq \mathcal{V}(\mathcal{G})$, *tag set* $T' \subseteq T$, *the tag-based influence function* σ^T *that maps each combination of subset of the nodes and tags to the number of influenced nodes, i.e.,* $\sigma^T : 2^{\mathcal{V}(\mathcal{G})} \times 2^T \longrightarrow \mathbb{R}_0$.

Finally, we define the Tag-based Budgeted Influence Maximization Problem.

Definition 5 (TBIM Problem). *Given a social network* $\mathcal{G}(\mathcal{V}, \mathcal{E}, \mathcal{P})$, *Tag set* T, *seed cost function* $\mathcal{C}^S : \mathcal{V}(\mathcal{G}) \to \mathbb{R}^+$, *tag cost function* $\mathcal{C}^T : T \to \mathbb{R}^+$ *and the budget* $\mathcal{B}$, *the TBIM Problem asks to select a subset of the tags from the communities, i.e.,* $T'_{\mathcal{K}_i} \subseteq T_{\mathcal{K}_i}$, $\forall i \in [\ell]$ *(here,* $T'_{\mathcal{K}_i} \cap T'_{\mathcal{K}_j} = \emptyset$, $\forall i \neq j$ *and* $T' = \bigcup_{i \in [\ell]} T_{\mathcal{K}_i}$*), and nodes* $\mathcal{S} \subseteq \mathcal{V}(\mathcal{G})$ *to maximize* $\sigma^T(\mathcal{S}, T')$ *such that* $\sum_{u \in \mathcal{S}} \mathcal{C}^S(u) + \sum_{i \in [\ell]} \sum_{t \in T_{\mathcal{K}_i}} \mathcal{C}^T(t) \leq \mathcal{B}$.

3   Proposed Methodologies

Here, we describe two different approaches and one subsequent improvement to select tags and seed users for initiating the diffusion process. Before stating the proposed approaches, we first define the *Effective Marginal Influence Gain*.

Definition 6 (Effective Marginal Influence Gain). *Given a seed set* $\mathcal{S} \subset \mathcal{V}(\mathcal{G})$, *tag set* $T' \subset T$, *the effective marginal influence gain (EMIG) of the node* $v \in \mathcal{V}(\mathcal{G}) \backslash \mathcal{S}$ *(denoted as δ_v) with respect to the seed set $\mathcal{S}$ and tag set T' is defined as the ratio between the marginal influence gain to its selection cost, i.e.,*

$$\delta_v = \frac{\sigma^T(\mathcal{S} \cup \{v\}, T') - \sigma^T(\mathcal{S}, T')}{\mathcal{C}^S(v)}. \tag{6}$$

In the similar way, for any tag $t \in T \backslash T'$, *its EMIG is defied as*

$$\delta_t = \frac{\sigma^T(\mathcal{S}, T' \cup \{t\}) - \sigma^T(\mathcal{S}, T')}{\mathcal{C}^T(t)}. \tag{7}$$

For the user-tag pair (v, t), $v \in \mathcal{V}(\mathcal{G}) \backslash \mathcal{S}$, *and* $t \in T \backslash T'$, *its EMIG is defined as*

$$\delta_{(v,t)} = \frac{\sigma^T(\mathcal{S} \cup \{v\}, T' \cup \{t\}) - \sigma^T(\mathcal{S}, T')}{\mathcal{C}^S(v) + \mathcal{C}^T(t)}. \tag{8}$$

3.1 Methodologies Based on Effective Marginal Influence Gain Computation of User-Tag Pairs (EMIG-UT)

In this method, first the community structure of the network is detected, and the total budget is divided among the communities based on its size. In each community, the shared budget is divided into two halves to be utilized to select tags and seed nodes, respectively. Next, we sort the communities based on its size in ascending order. Now, we take the smallest community first and select the most frequent tag which is less than or equal to the budget. Next, each community from smallest to the largest is processed for tag and seed node selection in the following way. Until the budget for both tag and seed node selection is exhausted, in each iteration the user-tag pair that causes maximum EMIG value is chosen and kept into the seed set and tag set, respectively. The extra budget is transferred to the largest community. Algorithm 1 describe the procedure.

Algorithm 1: Effective Marginal Influence Gain Computation of User-Tag Pairs (EMIG-UT)

Data: Social Network $\mathcal{G}$, Tag Set T, Node Cost Function $\mathcal{C}^S$, User-Tag Count Matrix $\mathcal{M}$, Tag Cost Function $\mathcal{C}^T$, Budget $\mathcal{B}$

Result: Seed set $\mathcal{S} \subseteq \mathcal{V}(\mathcal{G})$, and Tag set $T' \subseteq T$

1   $\mathcal{S} = \emptyset$; $T' = \emptyset$; $Community = Community\_Detection(\mathcal{G})$;

2   $\mathcal{K} \longleftarrow \{\mathcal{K}_1, \mathcal{K}_2, \ldots, \mathcal{K}_l\}$; $\mathcal{K}_{max} = Largest\_Community(\mathcal{K})$;

3   $T_{\mathcal{K}} = Crete\_Matrix(|\mathcal{K}|, |T|, 0)$;

4   $T_{\mathcal{K}} = $ Count the tag frequencies in each communities;

5   Sort row of $T_{\mathcal{K}}$ corresponding to the smallest community ;

6   $Create\_Vector(\mathcal{B}_S^k, \ell, 0)$; $Create\_Vector(\mathcal{B}_T^k, \ell, 0)$;

7   **for** $i = 1$ *to* $|\mathcal{K}|$ **do**

8    $\left| \quad \mathcal{B}^k = \frac{|\mathcal{V}_{\mathcal{K}_i}|}{n}.\mathcal{B}$; $\mathcal{B}_S^k[i] = \frac{\mathcal{B}^k}{2}$; $\mathcal{B}_T^k[i] = \frac{\mathcal{B}^k}{2}$; \right.$

9   **end**

10   $\mathcal{K} = Sort(\mathcal{K})$; $T' = T' \cup \{t' : t'$ is most frequent in $\mathcal{K}_1$ and $\mathcal{C}(t) \leq \mathcal{B}_T^k[i]\}$;

11   $\mathcal{B}_T^k[1] = \mathcal{B}_T^k[1] - \mathcal{C}^T(t')$;

12   **for** $i = 1$ *to* $|\mathcal{K}|$ **do**

13   **while** $\mathcal{B}_S^k[i] > 0$ *and* $\mathcal{B}_T^k[i] > 0$ **do**

14     $(u,t) = \underset{\substack{v \in \mathcal{V}_{\mathcal{K}_i}(\mathcal{G}) \setminus \mathcal{S}, \mathcal{C}(v) \leq \mathcal{B}_S^k[i]; \\ t'' \in T \setminus T', \mathcal{C}(t'') \leq \mathcal{B}_T^k[i]}}{argmax} \delta_{(v,t'')}$; $\mathcal{S} = \mathcal{S} \cup \{u\}$; $T' = T' \cup \{t\}$;

15     $\mathcal{B}_S^k[i] = \mathcal{B}_S^k[i] - \mathcal{C}^S(u)$; $\mathcal{B}_T^k[i] = \mathcal{B}_T^k[i] - \mathcal{C}^T(t)$;

16   **end**

17    $\mathcal{B}_S^k[max] = \mathcal{B}_S^k[max] + \mathcal{B}_S^k[i]$; $\mathcal{B}_T^k[max] = \mathcal{B}_T^k[max] + \mathcal{B}_S^k[i]$;

18   **end**

Now, we analyze Algorithm 1. For detecting communities using Louvian Method requires $\mathcal{O}(n \log n)$ time. Computing the tag count in each communities requires $\mathcal{O}(n|T|)$ time. Community is the array that contains the community number of the user in which they belong to, i.e., `Community[i]=x` means the user u_i belongs to Community $\mathcal{K}_x$. From this array, computing the size of each communities and finding out the maximum one requires $\mathcal{O}(n)$ time. Dividing the budget among the communities for seed node and tag selection and Sorting the communities require $\mathcal{O}(\ell)$ and $\mathcal{O}(\ell \log \ell)$ time, respectively. From the smallest community, choosing the highest frequency tag requires $\mathcal{O}(|T| \log |T|)$ time. Time requirement for selecting tags and seed nodes in different communities will be different. For any arbitrary community $\mathcal{K}_x$, let, $\mathcal{C}(\mathcal{S}_{\mathcal{K}_x}^{min})$ and $\mathcal{C}(T_{\mathcal{K}_x}^{min})$ denote the minimum seed and tag selection cost of this community, respectively. Hence, $\mathcal{C}(\mathcal{S}_{\mathcal{K}_x}^{min}) = \min\limits_{u \in V_{\mathcal{K}_x}} \mathcal{C}^S(u)$ and $\mathcal{C}(T_{\mathcal{K}_x}^{min}) = \min\limits_{t \in T_{\mathcal{K}_x}} \mathcal{C}^T(t)$. Here, $V_{\mathcal{K}_x}$ and $T_{\mathcal{K}_x}$ denote the nodes and relevant tags in community $\mathcal{K}_x$, respectively. Also, $\mathcal{B}_{\mathcal{K}_x}^S$ and $\mathcal{B}_{\mathcal{K}_x}^T$ denotes the budget for selecting seed nodes and tags for the community $\mathcal{K}_x$, respectively. Now, it can be observed that, the number of times while loop (Line number 13 to 16) runs for the community $\mathcal{K}_x$ is $\min\{\frac{\mathcal{B}_{\mathcal{K}_x}^S}{\mathcal{C}(\mathcal{S}_{\mathcal{K}_x}^{min})}, \frac{\mathcal{B}_{\mathcal{K}_x}^T}{\mathcal{C}(T_{\mathcal{K}_x}^{min})}\}$ and it is denoted as $r_{\mathcal{K}_x}$. Let, $r_{max} = \max\limits_{\mathcal{K}_x \in \{\mathcal{K}_1, \mathcal{K}_2, \dots, \mathcal{K}_\ell\}} r_{\mathcal{K}_x}$. The number of times the marginal influence gain needs to be computed is of $\mathcal{O}(\ell r_{max} n |T|)$. Assuming the time requirement for computing the MIIA for a single node with threshold θ is of $\mathcal{O}(t_\theta)$ [8]. Hence, computation of $\sigma(\mathcal{S})$ requires $\mathcal{O}(n t_\theta)$ time. Also, after updating the tag set in each iteration updating the aggregated influence probability requires $\mathcal{O}(m)$ time. Hence, execution from Line 17 to 24 of Algorithm 1 requires $\mathcal{O}(\ell r_{max} n |T|(n t_\theta + m))$. Hence, the total time requirement for Algorithm 1 of $\mathcal{O}(n \log n + n|T| + n + \ell + \ell \log \ell + \ell r_{max} n |T|(n t_\theta + m)) = \mathcal{O}(n \log n + \ell \log \ell + \ell r_{max} n |T|(n t_\theta + m))$. Additional space requirement for Algorithm 1 is to store the `Community` array which requires $\mathcal{O}(n)$ space, for $\mathcal{B}_S^k$ and $\mathcal{B}_T^k$ require $\mathcal{O}(\ell)$, for $T_{\mathcal{K}}$ requires $\mathcal{O}(\ell|T|)$, for storing MIIA path $\mathcal{O}(n(n_{i\theta} + n_{o\theta}))$ [8], for aggregated influence probability $\mathcal{O}(m)$, for $\mathcal{S}$ and T' require $\mathcal{O}(n)$ and $\mathcal{O}(|T'|)$, respectively. Formal statement is presented in Theorem 1.

Theorem 1. *Time and space requirement of Algorithm 1 is of $\mathcal{O}(n \log n + |T| \log |T| + \ell \log \ell + \ell r_{max} n |T|(n t_\theta + m))$ and $\mathcal{O}(n(n_{i\theta} + n_{o\theta}) + \ell|T| + m)$, respectively.*

3.2 Methodology Based on Effective Marginal Influence Gain Computation of Users (EMIG-U)

As observed in the experiments, computational time requirement of Algorithm 1 is very high. To resolve this problem, Algorithm 2 describes the *Effective Marginal Influence Gain Computation of Users (EMIG-U)* approach, where after community detection and budget distribution, high frequency tags from the communities are chosen (Line 3 to 11), and effective influence probability for each of the edges are computed. Next, from each of the communities until their respective budget is exhausted, in each iteration the node that causes maximum EMIG value are chosen as seed nodes. As described previously, time requirement

---

Algorithm 2: Effective Marginal Influence Gain Computation of User (EMIG-U)

Data: Social Network $\mathcal{G}$, Tag Set T, Node Cost Function $\mathcal{C}^S$, User-Tag Count Matrix $\mathcal{M}$, Tag Cost Function $\mathcal{C}^T$, Budget $\mathcal{B}$

Result: Seed set $\mathcal{S} \subseteq \mathcal{V}(\mathcal{G})$, and Tag set $T' \subseteq T$

1 Execute Line Number 1 to 14 of Algorithm 1;
2 Sort each row of the $T_{\mathcal{K}}$ matrix;
3 **for** $i = 1$ *to* $|\mathcal{K}|$ **do**
4 | **for** $j = 1$ *to* $|T|$ **do**
5 | | **if** $\mathcal{C}(t_j) \leq \mathcal{B}_T^k[i]$ *and* $t_j \notin T'$ **then**
6 | | | $T' = T' \cup \{t_j\}$; $\mathcal{B}_T^k[i] = \mathcal{B}_T^k[i] - \mathcal{C}^T(t_j)$;
7 | | **end**
8 | **end**
9 | $\mathcal{B}_T^k[max] = \mathcal{B}_T^k[max] + \mathcal{B}_T^k[i]$;
10 **end**
11 **for** *All* $(u_i u_j) \in \mathcal{E}(\mathcal{G})$ **do**
12 | Compute aggregated influence using Equation 1;
13 **end**
14 **for** $i = 1$ *to* $|\mathcal{K}|$ **do**
15 | **while** $\mathcal{B}_S^k[i] > 0$ **do**
16 | | $u = \underset{v \in \mathcal{V}_{\mathcal{K}_i}(\mathcal{G}) \backslash \mathcal{S}, \mathcal{C}(v) \leq \mathcal{B}_S^k[i]}{argmax} \delta_v$; $\mathcal{S} = \mathcal{S} \cup \{u\}$;
17 | | $\mathcal{B}_S^k[i] = \mathcal{B}_S^k[i] - \mathcal{C}^S(u)$;
18 | **end**
19 | $\mathcal{B}_S^k[max] = \mathcal{B}_S^k[max] + \mathcal{B}_S^k[i]$;
20 **end**

---

for executing Line 1 to 14 is $\mathcal{O}(n \log n + |T| \log |T| + n|T| + n + \ell + \ell \log \ell) = \mathcal{O}(n \log n + |T| \log |T| + n|T| + \ell \log \ell)$. Sorting each row of the matrix $T_{\mathcal{K}}$ requires $\mathcal{O}(\ell |T| \log |T|)$ time. For any arbitrary community $\mathcal{K}_x$, the number of times the `for` loop will run in the worst case is of $\mathcal{O}(\frac{\mathcal{B}_{\mathcal{K}_x}^T}{\mathcal{C}(T_{\mathcal{K}_x}^{min})})$. Let, $t_{max} = \underset{\mathcal{K}_x \in \{\mathcal{K}_1, \mathcal{K}_2, ..., \mathcal{K}_\ell\}}{max} \frac{\mathcal{B}_{\mathcal{K}_x}^T}{\mathcal{C}(T_{\mathcal{K}_x}^{min})}$. Also, in every iteration, it is to be checked whether the selected tag is already in T' or not. Hence, the worst case running time from Line 3 to 11 will be of $\mathcal{O}(\ell t_{max} |T'|)$. Computing aggregated influence probabilities for all the edges (Line 12 to 14) requires $\mathcal{O}(m |T'|)$ time. For the community $\mathcal{K}_x$, the number of times the `while` loop in Line 16 will run in worst case is of $\mathcal{O}(\frac{\mathcal{B}_{\mathcal{K}_x}^S}{\mathcal{C}(\mathcal{S}_{\mathcal{K}_x}^{min})})$.

Let, $s_{max} = \underset{\mathcal{K}_x \in \{\mathcal{K}_1, \mathcal{K}_2, ..., \mathcal{K}_\ell\}}{max} \frac{\mathcal{B}_{\mathcal{K}_x}^S}{\mathcal{C}(\mathcal{S}_{\mathcal{K}_x}^{min})}$. Hence, the number of times the marginal influence gain will be computed is of $\mathcal{O}(\ell s_{max} n)$. Contrary to Algorithm 1, in this case MIIA path needs to be computed only once after the tag probability aggregation is done. Hence, worst case running time from Line 15 to 22 is of $\mathcal{O}(\ell s_{max} n + n t_\theta)$. The worst case running time of Algorithm 2 is of $\mathcal{O}(n \log n +$

$|T|\log|T| + n|T| + \ell\log\ell + \ell|T|\log|T| + \ell t_{max}|T'| + m|T'| + \ell s_{max}n + nt_\theta) =$
$\mathcal{O}(n\log n + n|T| + \ell\log\ell + \ell|T|\log|T| + \ell t_{max}|T'| + m|T'| + \ell s_{max}n + nt_\theta).$
It is easy to verify that the space requirement of Algorithm 2 will be same as
Algorithm 1. Hence, Theorem 2 holds.

Theorem 2. *Running time and space requirement of Algorithm 2 is of*
$\mathcal{O}(n\log n + n|T| + \ell\log\ell + \ell|T|\log|T| + \ell t_{max}|T'| + m|T'| + \ell s_{max}n + nt_\theta)$ *and*
$\mathcal{O}(n(n_{i\theta} + n_{o\theta}) + \ell|T| + m)$, *respectively.*

3.3   Efficient Pruning Technique (EMIG-U-Pru)

Though, Algorithm 2 has better scalability compared to Algorithm 1, still it is
quite huge. The main performance bottleneck of Algorithm 2 is the excessive
number of EMIG computations. Hence, it will be beneficial, if we can prune off
some of the nodes, in such a way that even if we don't perform this computation
for these nodes, still it does not affect much on the influence spread. We propose
the following pruning strategy. Let, $\mathcal{S}^i$ denotes the seed set after the ith iteration.
$\forall u \in V(\mathcal{G})\backslash\mathcal{S}^i$, if the outdegree of u, i.e. $outdeg(u)$ will be decremented by
$|\mathcal{N}^{in}(u) \cap \mathcal{S}^i|$, where $\mathcal{N}^{in}$ denotes the set of incoming neighbors of u. All the
nodes in $V(\mathcal{G})\backslash\mathcal{S}^i$ are sorted based on the computed outdegree to cost ratio
and top-k of them are chosen for the EMIG computation. We have stated for
the ith iteration. However, the same is performed in every iteration. Due to the
space limitation, we are unable to present the entire algorithm and its analysis.
However, we state the final result in Theorem 3.

Theorem 3. *Running time and space requirement of the proposed pruning*
strategy is of $\mathcal{O}(n\log n + \ell|T|\log|T| + n|T| + \ell\log\ell + \ell t_{max}|T'| + m|T'| +$
$\ell s_{max}(n_{max}|\mathcal{S}| + n_{max}\log n_{max} + k) + nt_\theta)$ *and* $\mathcal{O}(n(n_{i\theta} + n_{o\theta}) + \ell|T| + m)$,
respectively.

4   Experimental Evalution

We use three datasets, namely, **Last.fm** [3], **Delicious** [3], and **LibraryThing**
[10]. No. of nodes, edges and tags in these datasets are 1288, 11678, and 11250;
1839, 25324, and 9749; and 15557, 108987, and 17228, respectively. For all the
three datasets it have been observed that the frequency of the tags decreases
exponentially. Hence, instead of dealing with all the tags, we have selected 1000
tags in each datasets using most frequent tags per community. Now, we describe
the experimental setup. Initially, we start with the influence probability setting.

- **Trivalency Setting**: In this case, for each edge and for all the tags influence
  probabilities are randomly assigned from the set $\{0.1, 0.01, 0.001\}$.
- **Count Probability Setting**: By this rule, for each edge (u_iu_j) its influence
  probability vector is computed as follows. First, element wise subtraction
  from $\mathcal{M}^{u_i}$ to $\mathcal{M}^{u_j}$ is performed. If there are some negative entries, they
  are changed to 0. We call the obtained vector as $\mathcal{M}^{u_i-u_j}$. Next, 1 is added

with each entries of the vector $\mathcal{M}^{u_i}$. We call this vector as $\mathcal{M}^{u_i+1}$. Now, the element wise division of $\mathcal{M}^{u_i-u_j}$ is performed by $\mathcal{M}^{u_i+1}$. The resultant vector is basically the influence probability vector for the edge $(u_j u_i)$. Here 1 is added with each of the entries of $\mathcal{M}^{u_i}$ before the division just to avoid infinite values in the influence probability vector.

- **Weighted Cascade Setting**: Let, $\mathcal{N}^{in}(u_i)$ denotes the set of incoming neighbors for the node u_i. In standard weighted cascade setting, $\forall u_j \in \mathcal{N}^{in}(u_i)$, the influence probability for the edges $(u_j u_i)$ is equal to $\frac{1}{deg^{in}(u_i)}$. Here, we have adopted this setting in a little different way. Let, $\mathcal{M}^{u_i}$ denotes the tag count vector of the user u_i (i^{th} row of the matrix $\mathcal{M}$). Now, $\forall u_j \in \mathcal{N}^{in}(u_i)$, we select the corresponding rows from $\mathcal{M}$, apply column-wise sum on the tag-frequency entries and perform the element wise division of the vector $\mathcal{M}^{u_i}$ by the summed up vector. The resultant vector is assigned as the influence probability for all the edges from $\forall u_j \in \mathcal{N}^{in}(u_i)$ to u_i.

Cost and Budget. We have adopted the *random setting* for assigning selection cost to each user and tag as mentioned in [7]. Selection cost for each user and tag are selected from the intervals $[50, 100]$ and $[25, 50]$, respectively uniformly at random. We have experimented with fixed budget values starting with 1000, continued until 8000, incremented each time by 1000, i.e., $\mathcal{B} = \{1000, 2000, \ldots, 8000\}$.

The following baseline methods have been used for comparison.

- **Random Nodes and Random Tags (RN+RT):** According to this method, the allocated budget is divided into two equal halves. One half will be spent for selecting seed nodes and the other one for selecting tags. Now seed nodes and tags are chosen randomly until their respective budgets are exhausted.
- **High Degree Nodes and High Frequency Tags (HN+HT):** In this method, after dividing the budget into two equal halves, high degree nodes and high frequency tags are chosen until their respective budget is exhausted.
- **High Degree Nodes and High Frequency Tags with Communities (HN+HT+COM):** In this method, after dividing the budget into two equal halves, first the community structure of the network is detected. Both of these divided budgets are further divided among the communities based on the community size. Then apply **HN+HT** for each community.

The implementations have been done with *Python 3.5 + NetworkX 2.1* on a HPC Cluster with 5 nodes each of them having 64 cores and 160 GB of memory and available at https://github.com/BITHIKA1992/BIM_with_Tag.

Figure 2 shows the Budget vs. Influence plot for all the datasets. From the figures, it has been observed that the seed set selected by proposed methodologies leads to more influence compared to the baseline methods. For the 'Delicious' dataset, for $\mathcal{B} = 8000$, under weighted cascade setting, among the baseline methods, **HN+HT+COMM** leads to the expected influence of 744, whereas the same for **EMIG-UT**, **EMIG-U**, **EMIG-U-Prunn** methods are 804, 805, and 805, respectively, which is approximately 8% more compared to

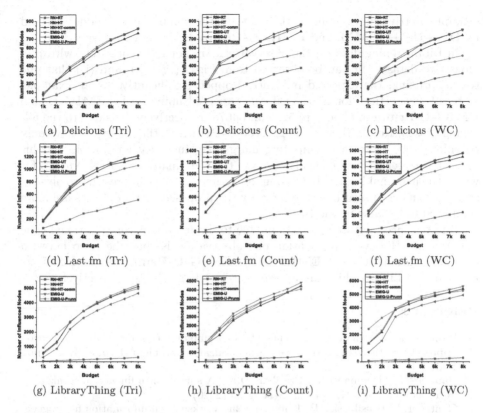

(a) Delicious (Tri)   (b) Delicious (Count)   (c) Delicious (WC)

(d) Last.fm (Tri)   (e) Last.fm (Count)   (f) Last.fm (WC)

(g) LibraryThing (Tri)   (h) LibraryThing (Count)   (i) LibraryThing (WC)

Fig. 2. Budget Vs. influence plot for **Delicious**, **Last.fm**, and **LibraryThing** datasets under the trivalency, weighted cascade and count probability settings.

HN+HT+COMM. The influence due to the seed set selected by **EMIG-U-Prunn** under Weighted Cascade, trivalency, and, Count setting are 805, 816, and 861 which are 62.5%, 63.2% and 66.85% of the number of nodes of the network, respectively. Also, we observe that for a given budget, the number of seed nodes selected by the proposed methodologies are always more compared to baseline methods. For example, under the trivalency setting with $\mathcal{B} = 8000$, the number of seed nodes selected by **RN+RT**, **HN+HT**, and **HN+HT+COMM** methods are 54. The same for **EMIG-UT**, **EMIG-U**, and **EMIG-U-Prunn** are 59, 62, and 62, respectively.

In case of 'Last.fm' dataset also, similar observations are made. For example, under trivalency setting with $\mathcal{B} = 8000$, the expected influence by the proposed methodologies **EMIG-UT**, **EMIG-U**, and **EMIG-U-Prunn** are 1230, 1226, and 1219, respectively. The same by **RN+RT**, **HN+HT**, and **HN+HT+COMM** methods are 515, 1067, and 1184, respectively. In this dataset also, it has been observed that the number of seed nodes selected by the proposed methodologies are more compared to the baseline methods. For example, in trivalency setting, with $\mathcal{B} = 8000$, the number of seed nodes selected by **HN+HT+COMM**, and **EMIG-U-Prunn** are 51 and 59, respectively.

In LibraryThing dataset, the observations are not fully consistent with previous two datasets. It can be observed from Fig. 2 ((g), (h), and (i)) that due to the pruning, the expected influence dropped significantly. As an example, in trivalency setting, for $\mathcal{B} = 8000$, the expected influence by **EMIG-U** and **EMIG-U-Prunn** methods are 5265 and 4673, respectively. It is due to the following reason. Recall, that in the **EMIG-U-Prunn** methodology, we have only considered 200 nodes for computing marginal gain in each iteration. As this dataset is larger than previous two, hence their are many prospective nodes for which the marginal has not been computed. However, it is interesting to observe still the number of seed nodes selected by the proposed methodologies are more compared to baseline methods.

Due to space limitation, we are unable to discuss about computational time requirement. However, we mention one observation is that the ratio between the computational time of **EMIG-U** and **EMIG-U-Prunn** for the Delicious, Last.fm, and LibraryThing are approximately 1.1, 2, and 10, respectively.

References

1. Banerjee, S., Jenamani, M., Pratihar, D.K.: Combim: a community-based solution approach for the budgeted influence maximization problem. Expert Syst. Appl. **125**, 1–13 (2019)
2. Banerjee, S., Jenamani, M., Pratihar, D.K.: A survey on influence maximization in a social network. Knowl. Inf. Syst. 1–39 (2020)
3. Cantador, I., Brusilovsky, P., Kuflik, T.: 2nd workshop on information heterogeneity and fusion in recommender systems (HetRec 2011). In: Proceedings of the 5th ACM Conference on Recommender Systems, RecSys 2011. ACM, New York (2011)
4. Chen, W., Wang, C., Wang, Y.: Scalable influence maximization for prevalent viral marketing in large-scale social networks. In: Proceedings of the 16th ACM SIGKDD International Conference on Knowledge Discovery and Data Mining, pp. 1029–1038. ACM (2010)
5. Guille, A., Hacid, H., Favre, C., Zighed, D.A.: Information diffusion in online social networks: a survey. ACM Sigmod Rec. **42**(2), 17–28 (2013)
6. Ke, X., Khan, A., Cong, G.: Finding seeds and relevant tags jointly: for targeted influence maximization in social networks. In: Proceedings of the 2018 International Conference on Management of Data, pp. 1097–1111. ACM (2018)
7. Nguyen, H., Zheng, R.: On budgeted influence maximization in social networks. IEEE J. Sel. Areas Commun. **31**(6), 1084–1094 (2013)
8. Wang, C., Chen, W., Wang, Y.: Scalable influence maximization for independent cascade model in large-scale social networks. Data Min. Knowl. Discov. **25**(3), 545–576 (2012)
9. Yalavarthi, V.K., Khan, A.: Steering top-k influencers in dynamic graphs via local updates. In: IEEE International Conference on Big Data, Big Data 2018, Seattle, WA, USA, December 10–13, 2018. pp. 576–583 (2018)
10. Zhao, T., McAuley, J., King, I.: Improving latent factor models via personalized feature projection for one class recommendation. In: Proceedings of the 24th ACM International on Conference on Information and Knowledge Management, pp. 821–830. ACM (2015)

Modeling Implicit Communities from Geo-Tagged Event Traces Using Spatio-Temporal Point Processes

Ankita Likhyani[1(✉)], Vinayak Gupta[2], P. K. Srijith[3], Deepak P.[4], and Srikanta Bedathur[2]

[1] Indraprastha Institute of Information Technology Delhi, Delhi, India
`ankital@iiitd.ac.in`
[2] Indian Institute of Technology Delhi, Delhi, India
`{vinayak.gupta,srikanta}@cse.iitd.ac.in`
[3] Indian Institute of Technology Hyderabad, Hyderabad, India
`srijith@iith.ac.in`
[4] Queen's University Belfast, Belfast, UK
`deepaksp@acm.org`

Abstract. The location check-ins of users through various location-based services such as Foursquare, Twitter and Facebook Places, generate large traces of geo-tagged events. These event-traces often manifest in hidden (possibly overlapping) communities of users with similar interests. Inferring these implicit communities is crucial for forming user profiles for improvements in recommendation and prediction tasks. Given only time-stamped geo-tagged traces of users, can we find out these implicit communities, and characteristics of the underlying influence network? Can we use this network to improve the next location prediction task? In this paper, we focus on the problem of community detection as well as capturing the underlying diffusion process. We propose CoLAB, based on spatio-temporal point processes for information diffusion in continuous time but discrete space of locations. It simultaneously models the implicit communities of users based on their check-in activities, without making use of their social network connections. CoLAB captures the semantic features of the location, user-to-user influence along with spatial and temporal preferences of users. The latent community of users and model parameters are learnt through stochastic variational inference. To the best of our knowledge, this is the first attempt at jointly modeling the diffusion process with activity-driven implicit communities. We demonstrate CoLAB achieves upto 27% improvements in location prediction task over recent deep point-process based methods on geo-tagged event traces collected from Foursquare check-ins.

A. Likhyani and V. Gupta are contributed equally.
The original version of this chapter was revised: the author's name was corrected to Deepak P. The correction to this chapter is available at
https://doi.org/10.1007/978-3-030-62005-9_41

Keywords: Location based social networks · Community detection · Information diffusion · Spatio-temporal point process

1   Introduction

Proliferation of smartphone usage and pervasive data connectivity have made it possible to collect enormous amounts of mobility information of users with relative ease. Foursquare announced in early 2018 that it collects more than 3 billion events every month from its 25 million users[1]. These events generate a location information diffusion process through an underlying –possibly hidden– network of users that determines an *location adoption* behavior among users. Location adoption primarily depends upon user's spatial, temporal and categorical preferences. For instance, one user's check-in at a newly opened jazz club could inspire another user to visit the same club or a similar club in her vicinity depending upon the distance from the club and time of the day/week. These users may not have a social connection, but share an implicit connection because of similar activity choices. This often leads to the formation of –possibly overlapping– communities of users with similar behavior. Detecting a community of such like-minded people from large geo-tagged events can benefit applications in various settings such as targeted advertisements and friend recommendation. Prior work [41] also suggests that social connections alone are not as effective for prediction tasks.

Example 1. We illustrate the spatio-temporal activity-based network and communities that CoLAB derives using the US dataset we collected, shown in Fig. 1. We observe that explicit social connections, shown using black edges, are very sparse and are unable to capture the implicit influence network. On the other hand, the latent network derived by CoLAB, shown using gray edges, can identify significantly higher number of relations in the latent influence network. It can be observed that it does not only captures the clusters but also identify

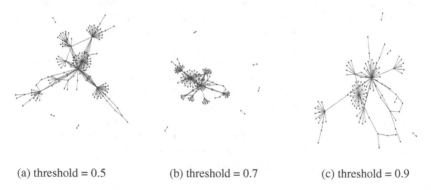

(a) threshold = 0.5   (b) threshold = 0.7   (c) threshold = 0.9

Fig. 1. CoLAB extracts underlying diffusion network over US data using geotagged checkin traces. This Maximum Weighted Spanning Forest (MWSF) is constructed by varying the threshold for edge weights (i.e. influence score computed using CoLAB). The inferred influence network depicts a tree-like structure of influence.

---

[1] https://bit.ly/2BdhnnP (accessed in February 2019).

potential influencers (and their influence networks) which can critically help in prediction tasks.

In this paper we determine that the location adoption process and *community formation* among users can be explained by the same latent factors underlying the observed behavior of users without considering their social network information. We propose **COLAB** (Communities of Location Adoption Behaviour) that focuses on jointly inferring location specific influence scores between users and the communities they belong to, based solely on their activity traces. **COLAB** completely disregards the social network information, which makes it suitable for scenarios where only activity traces are available. Note that if social network is available, it can be used as a prior or a regularization over the influence matrix we derive. Unlike the current best-performing models for location prediction task (e.g., [38]), CoLAB avoids the community formation from being biased only by the availability of social connections. Thus we generate better communities even for users with few or no social connections as shown in Fig. 1.

Further, we generate *overlapping communities* of users that take into account their spatio-temporal patterns, and the shared special interests over location categories. Although a user may be part of multiple communities with different special interests, we enforce each event or check-in to be associated with only one of those communities. Prior works have focused on modeling temporal + textual together [41] or temporal + spatial features together [37]. None of the techniques, to the best of our knowledge, have modeled the entire combination of temporal, spatial and location semantics in a spatio-temporal point process to infer the underlying influence network.

Contributions. Our main contributions are as follows:

1. We propose a novel model called CoLAB to model activity patterns over geo-tagged event traces. It leverages *spatio-temporal Hawkes process* [5,28] to not only construct an information diffusion-based latent network but also recover overlapping community structures within it.
2. We develop a novel *stochastic variational inference technique* to infer the latent communities and model parameters.
3. As our target is to identify communities that comprise users who share interests as opposed to just being socially connected, there is unfortunately no gold-standard community information to evaluate our results. Therefore, we first empirically evaluate our method over synthetic data; further, results show our inference algorithm can accurately recover the model parameters i.e. the matrix that contains community affinity score for each user and the influence matrix between users. For community evaluation on real data, we make use of a joint loss function that evaluates on the basis intra-community properties defined in terms of users' affinity to venue category and the spatial distribution of their checkins.
4. We evaluate on two real-world geo-tagged event traces collected from two countries – viz., SA (Saudi Arabia) and US. The experimental results demonstrates that we achieve upto 27% improvement over neural network based models.

2   Related Work

In this section, we briefly review existing literature on community detection and characterization of location adoption.

2.1   Community Detection

Within general social networks, there has been much work on detecting communities as overlapping or non-overlapping clusters of users such that there is a high degree of connectedness between them. Techniques have largely considered social-network connectedness as the main driver in forming communities. Community detection techniques could adopt a top-down approach starting from the entire graph and form communities by separating them into more coherent subsets that would eventually become communities. Methods from this school include those that use graph-based methods [27] and filtering out edges based on local informations such as number of mutual friends [42]. Analogously, community discovery could proceed bottom-up by aggregating proximal nodes to form communities. Techniques within this category have explored methods such as merging proximal cliques [16] or by grouping nodes based on affinities to 'leader' nodes [15]. Point processes [6], which are popular models for sequential and temporal data, have been recently explored for community detection in general social networks [29]. NetCodec, the method proposed therein, targets to simultaneously detect community structure and network infectivity among individuals from a trace of their activities. The key idea is to leverage multi-dimensional Hawkes process in order to model the relationship between network infectivity vis-a-vis community memberships and user popularity, in order to address the community discovery and network infectivity estimation tasks simultaneously. Our task of community detection within LBSNs is understandably more complex due to the primacy of spatial information in determining LBSN community structures. Accordingly, we leverage a spatio-temporal variant of multi-dimensional Hawkes process, in devising our solution.

LBSN Community Detection: There has been existing work [19,23,25,32] on community detection in LBSNs that leverage features such as venue categories, temporal features, geo-span, social-status, structure based and location based features in determining community structure within clustering formulations. They have used standard ML based techniques such as Spectral clustering [25], M^2 [31,32]: a k-means based clustering ([31] uses M^2 clustering with meta information of user), Entropy-based [23], LDA (Latent Dirichlet Allocation) on Foursquare tips data [1] and Sequence-Mining [19] based techniques. Our method, as illustrated therein, models a wider variety of features, incorporating both spatio-temporal check-in information as well as venue category information, in inferring communities and user-user influence.

2.2   Location Adoption Characterization

With each LBSN check-in being associated with a location, the location information is central to LBSNs. There has been much research into modelling user-location correlations in various forms, which may be referred to as *location adoption* in general.

[14] is the closest work to our work, it is a Hawkes Process based model. Hawkes Process [12] It is a point process with self triggering property, has been explored to model temporal occurrences of events in social media [34]. In [14] authors determine the patterns from geo located posts from twitter. Our model deals with discrete locations unlike [14] which simply considers sampling of spatial locations from a continuous distribution which can not be used for modeling check-in locations. The intensity function in CoLAB seamlessly integrates temporal and spatial components of the model and jointly learns the parameters whereas in [14] location and time are modeled separately. Moreover, we use multi-dimensional Hawkes Process as intensities differs across users. The parameter estimation of multi-dimensional Hawkes Process is much harder and Monte Carlo Simulations based methods used in [14] are very slow. We introduce the use of stochastic variational inference (SVI) to overcome this. [11] and [10] exploit the role of social correlations and temporal cyclical behaviors in improving upon the task of predicting the next location that a user would check-in. Location categories (e.g., restaurant, pub etc.) has also been seen to be useful for next location prediction [22]. [36] exploit geographical influence across locations in recommending POIs (points of interest) to LBSN users.

Another task that has attracted significant attention within LBSNs is that of quantifying user influence in LBSNs. [39] consider using geo-social correlation between users to estimate mutual influence. [33] leverage the observation that mutual influence is central to successful recommendation to identify a set of influential seed nodes using information cascade models. [3] adopt influence propagation models to address a slight variant, that of identifying region-specific influential users. [18] and [30] propose other kinds of models for the influence maximization task within LBSNs. Location Promotion [21,44] and Trip Purpose Detection [13] form other les addressed tasks within the context of LBSNs. Mobility models that mine spatial patterns based on generative models [11], Gaussian distributions [4] and kernel density based estimations [20] have been particularly popular in modeling location adoption behavior within LBSNs.

3   Problem Statement

Consider a geo-tagged event trace S over a set of L locations $\mathbb{L} = \{\ell_k\}_{k=1}^{L} = \{(x_k, y_k)\}_{k=1}^{L}$, a set of I users $\mathbb{U} = \{i_k\}_{k=1}^{I}$ and V categories (restaurants, entertainments etc.). Let us consider there are N events, with the nth check-in denoted as $E_n = (t_n, \ell_n, c_n, i_n, g_n)$ and $\mathbb{E} = \{E_n\}_{n=1}^{N}$. The notation denotes that E_n is the check-in event involving the user i_n checks-in to location ℓ_n at time t_n, with the category associated with the location being c_n and the *latent community* of the user associated with the check-in is g_n. The task is to learn the latent community associated with the users and effectively model the diffusion of information among users. Towards this, we aim to learn a matrix ϕ of size $|U|$ x $|M|$ where i^{th} row represents community participation for the ith user (assuming M communities). In addition, to model the diffusion process, we estimate matrix A_{ij}, where an element a_{ij} represents the influence of ith user on j^{th} user.

4   CoLAB Model

In this section we describe our model to infer the ϕ matrix i.e. the communities vector inferred from check-in activities and information diffusion over the users in the network.

4.1   Spatio-Temporal Data Modeling

Given the check-in events E as defined earlier, for modeling time and spatial components w.r.t. communities, we define the Hawkes process based model as follows:

Intensity Function. We model the user's community-specific intensity using the multi-dimensional spatio-temporal Hawkes process. Multi-dimensional because influence from other users also contribute in the intensity of a user [43]. Consider the task of estimating the *community-specific* intensity of a user i_n towards generating a check-in $E_n = (t_n, \ell_n, c_n, i_n, g_n)$; the multi-dimensional spatio-temporal Hawkes process formulation yields the following:

$$\lambda_{i_n, g_n}(t_n, \ell_n) = \mu_{i_n}\eta_{g_n} + \sum_{t_k < t_n} A_{i_k i_n} \kappa(t_n - t_k, \ell_n - \ell_k)\mathbb{I}(g_k = g_n) \qquad (1)$$

where μ_{i_n} is the base intensity of user i_n and η_{g_n} is weight associated to g_n towards a community g_n, and $\boldsymbol{\eta} = \{\eta_g | g = 1, \ldots, M\}$ with $\eta_g \geq 0$. $\lambda_{i_n, g_n}(t_n, \ell_n)$ is community specific intensity of user i at the nth instance. We allow historical check-ins to contribute to the intensity - proportionate to their temporal and spatial proximity to t_n and ℓ_n respectively, and weighted using the influence between user i_n and i_k (i.e., $A_{i_k i_n}$) - as long as they belong to the same community, enforced by the indicator function $\mathbb{I}(g_k = g_n)$. Here, $\kappa(t_n - t_k, \ell_n - \ell_k)$ is the triggering exponential kernel which factorises over time and location.

$$\kappa(t_n - t_k, \ell_n - \ell_k) = \kappa(t_n - t_k) * \kappa(\ell_n - \ell_k), \qquad (2)$$

where, $\kappa(t_n - t_k) = \exp(-\nu(t_n - t_k))$ is the time specific triggering kernel with ν decay and $\kappa(\ell_n - \ell_k) = \frac{1}{2\pi h}exp\left(-\frac{\|\ell_n - \ell_k\|}{2h}\right)$ is the location specific triggering kernel with h bandwidth. When the decay parameter is low, the influence of the previous events is high and similarly when the bandwidth parameter is high the influence of previous locations is high.

In general, the intensity of a particular user i at some time t and location ℓ, is given as the sum of intensities that are estimated at the level of each community i.e. total intensity $\lambda_i(t, \ell) = \sum_g \lambda_{i,g}(t, \ell)$.

4.2   Category Distribution

The category c associated with a check-in is represented as a $|V|$-length vector and it represents one of V possible categories[2] such as restaurant, entertainment etc. associated with the check-in. Also, we assume that the category depends on the underlying latent community associated with this check-in. For example, some community may be more inclined towards restaurants while another community is oriented towards sports. The category is modelled as a sample from a Multinomial (categorical) distribution,

$$c \sim Multinomial(\boldsymbol{\theta_g}) \tag{3}$$

where $\boldsymbol{\theta_g}$ is a $|V|$-length vector whose elements encode probability of each category and which depends on the community g that the check-in belongs to. We assume a prior over $\boldsymbol{\theta_g}$ as a sample from Dirichlet distribution with parameters $\boldsymbol{\theta_0}$. We write the conditional distribution $p(c, \boldsymbol{\theta_g}|\boldsymbol{\theta_0})$ as:

$$p(c, \boldsymbol{\theta_g}|\boldsymbol{\theta_0}) = p(c|\boldsymbol{\theta_g})p(\boldsymbol{\theta_g}|\boldsymbol{\theta_0}) = \theta_{g,c} \frac{\Gamma(\sum_j \theta_{0,j})}{\prod_j \Gamma(\theta_{0,j})} \prod_j \theta_{g,j}^{\theta_{0,j}-1} \tag{4}$$

j runs over the V categories and $p(c|\theta_g)$ is given as $\theta_{g,c}$.

4.3   Distribution over Communities

We assume the latent variable g (the communities) associated with the user for some check-in, is distributed as multinomial distribution parameterized by π_i for a user i. π_{ig} represents the probability user i belongs to community g.

$$g \sim Multinomial(\boldsymbol{\pi_i}) \tag{5}$$

4.4   Generative Process

Note that, for sampling (t, x, y), the thinning algorithm proposed in [17] is modified in order to sample location coordinates from discrete "venue" set rather the continuous space. First we consider a discrete set of locations L for the user based on her region. We sample (x', y') at n^{th} iteration, from a Gaussian distribution centered at the previous coordinates in the $(n-1)^{th}$ iteration: (x_{n-1}, y_{n-1}). Once (x', y') is sampled, the nearest coordinate in the L is determined and returned as (x_n, y_n).

---

[2] We overload the notation c to also represent a scalar categorical value in the set $\{1, \ldots, V\}$.

Algorithm 1: Generative Process

---

1 Initialize the number of communities M, and number of checkins N_i for each user;
2 Set μ_i proptional to N_i;
3 Initialize A_{ij} as column normalized matrix;
4 Initialize π, η and θ as Dirichlet-Multinomial distribution ;
5 Initialize $\lambda_i(t_0, x_0, y_0) = \mu_i \quad \forall i = 1, \ldots, U$;
6 **for** $n = 1$ to N **do**
7   Sample (t_n, ℓ_n) from $\sum_{i=1}^{U} \lambda_i(t, \ell)$;
8   Sample i_n from Multinomial $(\lambda_1(t_n, \ell_n), \lambda_2(t_n, \ell_n) , \ldots , \lambda_U(t_n, \ell_n))$;
9   Sample g_n from a Multinomial (π_{i_n});
10   Sample c_n from $Multinomial(\theta_{g_n})$ (θ_g is defined in section 4.2);
11 **end**

---

5 Estimation and Inference

Given the multi-dimensional Hawkes process model defined above, the joint probability density function over the check-in events $\mathbb{E}$ is given as:

$$\prod_{n=1}^{N} p(t_n, l_n, c_n, g_n | i_n) = \prod_{n=1}^{N} \Big((p(t_n, l_n | i_n, g_n) \times p(g_n | i_n)) \times p(c_n | g_n, \theta) \Big) \quad (6)$$

Here, $p(c_n | g_n, \theta) = \theta_{g_n, c_n}$, where c_n represents the category associated with the nth check-in, and $p(g_n | i_n) = \pi_{i_n, g_n}$ the probability that user i_n belong to the community g_n.

$$\prod_{n=1}^{N} p(t_n, \ell_n | i_n, g_n) = \prod_{n=1}^{N} \lambda_{i_n, g_n}(t_n, \ell_n) \exp\Big(-\sum_{i=1}^{U} \int_0^T \int_{\ell_{min}}^{\ell_{max}} \lambda_i(t, \ell) dt d\ell\Big) \quad (7)$$

is the likelihood (event density) of generating the observations given the community and users in the interval $[(0, \ell_{min}), (T, \ell_{max})]$. The first term in (7) provides the instantaneous probability of occurrence of the observed events and the second term provides the probability that no event happens outside these observations (survival probability) [6]. Thus, the complete joint log likelihood is:

$$\mathcal{LL} = \sum_{n=1}^{N} \Big(\log \lambda_{i_n, g_n}(t_n, \ell_n) + \log \pi_{i_n, g_n} + \log \theta_{g_n, c_n} \Big)$$

$$-\sum_{i=1}^{U} \int_0^T \int_{\ell_{min}}^{\ell_{max}} \lambda_i(t, \ell) dt d\ell \quad (8)$$

Assuming communities are known, we can estimate the model parameters μ, η, A, θ_g's and π's by maximum likelihood estimation. We treat the kernel parameters, and the Dirichlet parameters as the hyper-parameters which we initialize to some fixed values. However, the communities are latent and the maximum likelihood estimation cannot be applied directly. This calls for the expectation maximization algorithm, where the parameters are estimated after integrating out the latent variables from the joint likelihood using the posterior distribution over the latent variables. In our case, the posterior distribution over latent communities is given as

$$
p(g_1, \ldots, g_n | \{t_n, \ell_n, c_n, i_n\}_{n=1}^{N})
$$
$$
= \frac{\prod_{n=1}^{N} p(t_n, \ell_n | i_n, g_n) \times p(c_n | g_n, \theta) \times p(g_n | i_n)}{\sum_{g_1, \ldots, g_n} \prod_{n=1}^{N} p(t_n, \ell_n | i_n, g_n) \times p(c_n | g_n, \theta) \times p(g_n | i_n)} \quad (9)
$$

The posterior distribution over the latent communities cannot be obtained in closed form due to the intractable normalization constant (denominator term) which involves an exponential number of summation terms. Markov chain Monte Carlo methods [2] can be used to obtain samples from the posterior. However, these approaches are not scalable to large datasets and becomes computationally expensive for use in LBSNs. To overcome this, we use a variational expectation maximization algorithm where we approximate the posterior over communities using a variational distribution and estimate the model parameters and variational parameters by maximizing a variational lower bound [40].

5.1   Variational Expectation Maximization

The latent variable g_n is dependent on different types of feature set i.e. space, time through $p(t_n, \ell_n | i_n, g_n)$ and semantics through $p(c_n | g_n, \theta)$. Though the prior over g_n is conjugate to $p(c_n | g_n, \theta)$, it is not with respect to $p(t_n, \ell_n | i_n, g_n)$ and hence the posterior over g_n cannot be computed in closed form. Moreover, g_n is inter-dependent i.e. at current step g_n is dependent on history from g_1 to g_{n-1} as well as the future ones i.e. g_{n+1}. Thus marginalizing out over such interconnected latent variables to compute the normalization constant for the posterior is intractable. To this end we assume a variational distribution over g_n's conditioned on the user i_n. The conditional variational distribution over g_n is considered to be a multinomial distribution with parameters ϕ_{i_n}. The variational parameter ϕ_i for a user i represents the posterior probability distributions over the communities for the user as observed from the data.

$$
q(g_n | i_n) = Multinomial(g_n | \phi_{i_n}) \quad (10)
$$

The variational parameters can be learnt by minimizing the KL divergence between the variational posterior (10) and the exact posterior (9). However, a direct minimization of KL divergence is not possible due to the intractable posterior. Following variational inference approach, the variational parameters are learnt by maximizing a variational lower bound, Evidence Lower Bound (ELBO), which indirectly minimizes the KL divergence. ELBO is obtained by considering an expected value of the complete joint log likelihood w.r.t. the variational distribution and acts as a lower bound to the marginal likelihood or evidence (normalization constant of the posterior). Hence, ELBO is useful to learn the model parameters also in addition to the variational parameters. Using the variational distribution defined in (10) and the complete joint log likelihood (8), we obtain the ELBO as:

$$\mathcal{L} = \sum_{n=1}^{N} \left(\mathbb{E}_q[\log \lambda_{i_n,g_n}(t_n,\ell_n)] + \sum_{m=1}^{M} \phi_{i_n,m} \log \pi_{i_n,m} + \sum_{m=1}^{M} \phi_{i_n,m} \log \theta_{m,c_n} \right)$$

$$- \sum_{i=1}^{U} \int_0^T \int_{\ell_{min}}^{\ell_{max}} \mathbb{E}_q[\lambda_i(t,\ell)] d\ell dt - \mathbb{E}_q[\log q] \quad (11)$$

Here, $\mathbb{E}_q$ represents the expectation with respect to the variational distribution q defined in (10). We learn the variational parameters and the model parameters by maximizing the ELBO. Table 1 lists the model parameters and variational parameters to be learnt using ELBO. All the terms in the ELBO except the first term can be computed in closed form.

Table 1. Parameters to be estimated and whether a Hyperparameter

Par	Description	H
μ	Base intensity	
η	Weight associated towards community	
A_{ij}	Influence matrix	
h	Bandwidth (KDE)	✓
ν	Temporal decay parameter	✓
θ_0	Dirichlet prior: category	✓
θ_g	Multinomial prior: categories/community	
π	Multinomial prior: communities (All users)	
ϕ	Variational parameters: communities (All users)	

Since the first term in (11) cannot be computed in closed form, we approximate it using the samples from the variational posterior (10) (Monte-Carlo approximation).

$$\mathbb{E}_q[\log \lambda_{i_n,g_n}(t_n,\ell_n)] \approx \frac{1}{S} \sum_{s=1}^{S} \log \lambda_{i_n,g^{(s)}}(t_n,\ell_n) \quad (12)$$

where $g^{(s)}$ represent the vector of N samples sampled from the joint variational distribution over all the g_n's, i.e. $q(g^{(s)}) = \prod_{i=1}^{N} q(g_n^{(s)}|\phi_{i_n})$. This results in a stochastic variational lower bound where the stochasticity arises due to the approximation of expectation using Monte Carlo sampling [40]. We learn the model parameters μ, η, A_{ij}, θ by maximizing the stochastic variational lower bound [26] using gradient based methods. However learning the variational parameters is problematic as the variational parameters does not appear explicitly in the stochastic term but only through the samples. For determining gradient w.r.t. ϕ we apply the Reinforcement trick to the stochastic term and compute the gradient as follows [9]:

$$\nabla_\phi \mathbb{E}_q[\log \lambda_{i_n,g_n}(t_n,\ell_n)] \approx \frac{1}{S}\sum_{s=1}^{S} \log \lambda_{i_n,g^{(s)}}(t_n,\ell_n) \nabla_\phi \log q(g^{(s)}) \tag{13}$$

6 Experiments

Compared Baselines. We empirically evaluate the performance of CoLAB over synthetic and real datasets with the following baselines:

- **STHP:** Spatio-Temporal Hawkes Process models the diffusion process across spatial and temporal dimensions but ignores the *category* associated location feature. This is the baseline derived from CoLAB ignoring the categories.
- **Sequence Mining** [19]: First extracts frequent occuring venue category sequences and assigns communities based on clusters with similar patterns.
- **DH** [8]: Dirichlet-Hawkes, clusters continuous time event streams using a modified Hawkes model with preferential cluster assignment through Dirichlet Process.
- **RMTPP** [7]: Recurrent Marked Temporal Point Process model the time and the marker information by learning a general representation of the non-linear dependency over the history based on recurrent neural networks. In this model, event history is embedded into a compact vector representation which is then used for predicting the next event time and marker type.

For an even comparison, we feed the check-in events to all the baselines and evaluate the community quality formed by DH and Sequence Mining along with location prediction for STHP and RMTPP.

6.1 Synthetic Data

We generate synthetic data using Algorithm 1 (statistics in Table 2). The set of locations, #users to categories i.e. (U:V) and to the number of checkins i.e. (U:N) are kept similar to the real data collected from Brazil with ν (temporal decay parameter) is set to 0.01 [35] and h is picked up from the bandwidth values learned from users' checkins. During inference, we use *true values* for all the parameters, except for the influence matrix A_{ij} and the user-community posterior ϕ which are estimated. We use $RelErr(A_{ij}, \hat{A}_{ij}) = \frac{1}{I^2}\sum_{i,j=1}^{I} \frac{|a_{ij}-\hat{a}_{ij}|}{|a_{ij}|}$ (and similarly for ϕ), as the metric to evaluate the ability to recover the true values. Table 3 indicates that the stochastic variational inference technique offers considerably better reconstruction of the parameters, recording significant reductions in the error.

Table 2. Dataset properties

Property	Synthetic	SA (Real)	US (Real)
#Users(U)	100	95	133
#Communities(M)	10	–	–
#Categories(V)	200	314	524
#Check-ins(N)	9777	15110	22059

Table 3. *RelErr* on A and ϕ and true positives for location prediction results at Top-K on synthetic data

Technique		STHP	CoLAB
RelErr	A	0.99174	**0.04813**
	ϕ	0.13807	**0.07216**
Top K	5	681	**972**
	10	1206	**1677**

Table 4. Comparison of CoLAB with other baselines and CoLAB without A_{ij} and μ

Dataset	K	STHP	RMTPP	CoLAB w/o Aij	CoLAB w/o μ	CoLAB
SA (#testcases = 2805)	5	287	250	279	331	**339**
	10	455	499	478	589	**593**
	20	950	951	911	1038	**1043**
	50	1539	1539	1520	1664	**1666**
US (#testcases = 4395)	5	153	172	172	231	**237**
	10	456	467	445	507	**512**
	20	870	888	919	920	**927**
	50	1700	1691	**1759**	1668	1673

6.2   Real Data

For real data, we use our crawls over Foursquare conducted between January-2015 and March-2016 and construct two collections consisting of check-ins from Saudi Arabia (SA) and United States (US), with details given in Table 2. We allocated first 80% (as per check-in timestamp) of each dataset to *training* and remaining for *testing*. Here, we also use temporal decay parameter as 0.01 [35] and h is learnt for each user based on Silverman's rule of thumb in kernel density estimation [2] and is fixed during the joint estimation.

Location Prediction with COLAB. For location prediction task, we predict the next location from the previously seen locations in the training set at $M(\#communities) = 10$ and at various top-K ranked (Eq. 8) cutoffs. Since, DH is for clustering event streams, therefore we make use of STHP and RMTPP as baselines for the Location Prediction task. Table 4 shows that CoLAB is able to offer significant improvements (18–37% in the top-5). As we increase K, these gains diminish because the number of candidate locations saturates. We also study the effect of A_{ij} (influence matrix) and μ (base intensity) over CoLAB's performance. It can be observed that without A_{ij}, the CoLAB's performance degrades signifying CoLAB's ability to capture the underlying diffusion process well.

Impact of #Communities. In Fig. 2 we study the impact of M, over US data we observe that with increasing M the prediction accuracy improves and then diminishes, signifying optimal value of M for better predictions. CoLAB performs significantly better than STHP, primarily due to the better estimation of A_{ij} (influence matrix), because of the presence of category information in the CoLAB model. For brevity SA results are omitted, but similar patterns were observed.

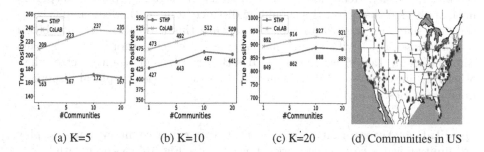

(a) K=5   (b) K=10   (c) K=20   (d) Communities in US

Fig. 2. Effect of Varying M over US data: (a)–(c) and the plot of communities obtained by CoLAB (Color figure online)

6.3   Community Assessment

We plot communities over US data in Fig. 2d, where colored dots represents a check-in. In Fig. 2d, it can be observed that; (i) *Overlap* between communities due to data concentration in cities. (ii) CoLAB is able to capture communities across cities. As a consequence of this behavior, clustering quality metrics such as Silhouette score[3] that penalize the overlapping clusters are not meaningful indicators of the resulting communities. For brevity SA results are omitted, but similar patterns were observed.

Unfortunately, we lack the community ground truth for users, making communities assessment a non-trivial task. Thus, we use a metric a joint loss function for the intra-community properties through a mixture of (i) *category loss* ($\mathcal{L}_{cat}$) and (ii) *location loss* ($\mathcal{L}_{loc}$). **Category Loss:** We consider all categories associated to locations as independent *marks* of a point process. Hence to estimate the category affinity in a community, we consider similarity among the check-in categories using pre-trained word embeddings [24] and devise a *loss* function.

$$\mathcal{L}_{cat} = \frac{1}{|\mathbf{T}|} \sum_{E_n \in \mathbf{T}} \sum_{g \in \mathcal{M}} \left\{ 1 - \frac{\mathbf{v}_{E_n} \cdot \mu_g}{||\mathbf{v}_{E_n}||_2 ||\mu_g||_2} \right\} \cdot \Phi(E_n, g) \tag{14}$$

where $\mathbf{T}$ represents test data, μ_g is *category mean* for a community g using K_{cat} frequent categories, $\mathbf{v}_{E_n}$ is the category vector for event E_n; $\Phi(E_n, g)$ indicates whether E_n is assigned to community g, with $\mathcal{M}$ as all communities. Table 5 demonstrates CoLAB's ability to capture category dynamics across communities. Although, it can be observed that at $K_{cat} = 10$ Dirichlet Hawkes performs

---

[3] https://en.wikipedia.org/wiki/Silhouette_(clustering).

better because with most frequent categories like restaurant, coffee shop etc. DH assigns it most of the communities. Note that, DH is unable to capture communities with varied categories as seen at $K_{cat} = 50$ and 100, that CoLAB is able to capture.

Location Loss: [4] show that users tend to visit nearby locations given we ignore the bias of loyalty. Hence ideally, a community of users not spatially dispersed in their checkin characteristics should be distinct from another community with checkins spanning large distances. We capture this through a distance based *k-means* loss ($\mathcal{L}_{loc}$) with cluster means (μ_l) for checkin coordinates for each community. In Table 6 we can see CoLAB performs significantly better than other baselines because CoLAB can better capture the geographical dispersion in communities.

Qualitative Assessment. We claim that a user in a community will display an affinity towards certain categories. For US data, Fig. 3 shows even with highly overlapping venue categories, our model finds the intricate differences between a community (a) with affinity to music and a community (b) with affinity towards food/bar joints. The word clouds for SA dataset shows similar properties and have been avoided for brevity.

Table 5. Results for category loss

Dataset	K_{cat}	Sequence mining		DH	STHP	CoLAB
		Daily	Weekly			
SA	10	250.24	236.19	**103.47**	125.17	119.41
	50	1118.23	1089.27	862.05	842.63	**826.72**
	100	2007.93	2120.76	1983.61	1784.04	**1749.37**
US	10	248.45	217.56	**98.37**	118.32	113.86
	50	956.87	990.45	781.83	793.07	**771.15**
	100	1907.84	2020.49	1901.37	1605.64	**1583.02**

Table 6. Results for location loss

Datasets	Sequence mining		DH	STHP	CoLAB
	Daily	Weekly			
SA	600.67	547.56	413.16	306.73	**298.09**
US	1127.34	1067.50	1039.08	849.92	**834.64**

(a) Community A   (b) Community B

Fig. 3. Word cloud of categories in two communities from US

7   Conclusion

In this paper, we presented CoLAB that uses spatio-temporal Hawkes process to infer the implicit communities using a novel stochastic variational inference technique. Empirical evaluations over synthetic as well as real-world datasets highlight its prowess with significant improvements in location and community detection tasks. This illustrates the effectiveness of the modelling used in CoLAB in generating user communities even in the absence of social connectedness information. In future work, we would like to explore scalability of CoLAB through sample-based inference techniques.

References

1. Alrumayyan, N., Bawazeer, S., AlJurayyad, R., Al-Razgan, M.: Analyzing user behaviors: a study of tips in foursquare. In: Alenezi, M., Qureshi, B. (eds.) 5th International Symposium on Data Mining Applications. AISC, vol. 753, pp. 153–168. Springer, Cham (2018). https://doi.org/10.1007/978-3-319-78753-4_12
2. Bishop, C.M.: Pattern Recognition and Machine Learning. Springer, New York (2006)
3. Bouros, P., Sacharidis, D., Bikakis, N.: Regionally influential users in location-aware social networks. In: SIGSPATIAL (2014)
4. Cho, E., Myers, S.A., Leskovec, J.: Friendship and mobility: user movement in location-based social networks. In: SIGKDD (2011)
5. Cho, Y.S., Galstyan, A., Brantingham, P.J., Tita, G.: Latent self-exciting point process model for spatial-temporal networks, vol. 19 (2014)
6. Daley, D.J., Vere-Jones, D.: An Introduction to the Theory of Point Processes. Volume I: Elementary Theory and Methods. Probability and Its Applications, 2nd edn. Springer, New York (2003). https://doi.org/10.1007/b97277
7. Du, N., Dai, H., Trivedi, R., Upadhyay, U., Gomez-Rodriguez, M., Song, L.: Recurrent marked temporal point processes: embedding event history to vector. In: KDD (2016)
8. Du, N., Farajtabar, M., Ahmed, A., Smola, A.J., Song, L.: Dirichlet-Hawkes processes with applications to clustering continuous-time document streams. In: SIGKDD (2015)
9. Gal, Y.: Uncertainty in deep learning. Ph.D. thesis. University of Cambridge (2016)
10. Gao, H., Tang, J., Hu, X., Liu, H.: Modeling temporal effects of human mobile behavior on location-based social networks. In: CIKM (2013)

11. Gao, H., Tang, J., Liu, H.: Exploring social-historical ties on location-based social networks. In: ICWSM (2012)
12. Hawkes, A.G., Oakes, D.: A cluster process representation of a self-exciting process. J. Appl. Probab. **11**(3), 493–503 (1974)
13. Hu, W., Jin, P.J.: An adaptive Hawkes process formulation for estimating time-of-day zonal trip arrivals with location-based social networking check-in data. Transp. Res. Part C: Emerg. Technol. **79**, 136–155 (2017)
14. Jankowiak, M., Gomez-Rodriguez, M.: Uncovering the spatiotemporal patterns of collective social activity. In: SDM (2017)
15. Khorasgani, R.R., Chen, J., Zaïane, O.R.: Top leaders community detection approach in information networks. In: 4th SNA-KDD Workshop on Social Network Mining and Analysis. Citeseer (2010)
16. Kumpula, J.M., Kivelä, M., Kaski, K., Saramäki, J.: Sequential algorithm for fast clique percolation. Phys. Rev. E **78**(2), 026109 (2008)
17. Lewis, P.A.W., Shedler, G.S.: Simulation of nonhomogeneous poisson processes by thinning. Nav. Res. Logist. Q. **26**(3), 403–413 (1979)
18. Li, G., Chen, S., Feng, J., Tan, K.L., Li, W.: Efficient location-aware influence maximization. In: SIGMOD (2014)
19. Li, H., Deng, K., Cui, J., Dong, Z., Ma, J., Huang, J.: Hidden community identification in location-based social network via probabilistic venue sequences. Inf. Sci. **422**, 188–203 (2018)
20. Lichman, M., Smyth, P.: Modeling human location data with mixtures of kernel densities. In: SIGKDD (2014)
21. Likhyani, A., Bedathur, S., Deepak, P.: LoCaTe: influence quantification for location promotion in location-based social networks. In: IJCAI (2017)
22. Likhyani, A., Padmanabhan, D., Bedathur, S., Mehta, S.: Inferring and exploiting categories for next location prediction. In: WWW (2015)
23. Liu, J., Li, Y., Ling, G., Li, R., Zheng, Z.: Community detection in location-based social networks: an entropy-based approach. In: IEEE CIT (2016)
24. Mikolov, T., Grave, E., Bojanowski, P., Puhrsch, C., Joulin, A.: Advances in pre-training distributed word representations. In: LREC (2018)
25. Noulas, A., Scellato, S., Mascolo, C., Pontil, M.: Exploiting semantic annotations for clustering geographic areas and users in location-based social networks. In: The Social Mobile Web, ICWSM Workshop (2011)
26. Paisley, J.W., Blei, D.M., Jordan, M.I.: Variational Bayesian inference with stochastic search. In: ICML (2012)
27. Prat-Pérez, A., Dominguez-Sal, D., Larriba-Pey, J.L.: High quality, scalable and parallel community detection for large real graphs. In: Proceedings of the 23rd International Conference on World Wide Web, pp. 225–236. ACM (2014)
28. Reinhart, A.: A review of self-exciting spatio-temporal point processes and their applications. Stat. Sci. **33**, 299–318 (2018)
29. Tran, L.Q., Farajtabar, M., Song, L., Zha, H.: NetCodec: community detection from individual activities. In: SDM (2015)
30. Wang, X., Zhang, Y., Zhang, W., Lin, X.: Distance-aware influence maximization in geo-social network. In: ICDE (2016)
31. Wang, Z., Zhang, D., Zhou, X., Yang, D., Yu, Z., Yu, Z.: Discovering and profiling overlapping communities in location-based social networks. IEEE Trans. Syst. Man Cybern.: Syst. **44**(4), 499–509 (2014)

32. Wang, Z., Zhang, D., Yang, D., Yu, Z., Zhou, X.: Detecting overlapping communities in location-based social networks. In: Aberer, K., Flache, A., Jager, W., Liu, L., Tang, J., Guéret, C. (eds.) SocInfo 2012. LNCS, vol. 7710, pp. 110–123. Springer, Heidelberg (2012). https://doi.org/10.1007/978-3-642-35386-4_9

33. Wu, H.-H., Yeh, M.-Y.: Influential nodes in a one-wave diffusion model for location-based social networks. In: Pei, J., Tseng, V.S., Cao, L., Motoda, H., Xu, G. (eds.) PAKDD 2013. LNCS (LNAI), vol. 7819, pp. 61–72. Springer, Heidelberg (2013). https://doi.org/10.1007/978-3-642-37456-2_6

34. Yang, S.H., Zha, H.: Mixture of mutually exciting processes for viral diffusion. In: Proceedings of the 30th International Conference on International Conference on Machine Learning, ICML 2013, vol. 28, pp. II-1–II-9. JMLR.org (2013)

35. Yang, S.H., Zha, H.: Mixture of mutually exciting processes for viral diffusion. In: ICML, ICML 2013, pp. II-1–II-9. JMLR.org (2013). http://dl.acm.org/citation.cfm?id=3042817.3042894

36. Ye, M., Yin, P., Lee, W.C., Lee, D.L.: Exploiting geographical influence for collaborative point-of-interest recommendation. In: SIGIR (2011)

37. Yuan, B., Li, H., Bertozzi, A.L., Brantingham, P.J., Porter, M.A.: Multivariate spatiotemporal Hawkes processes and network reconstruction. SIAM J. Math. Data Sci. 1, 356–382 (2019)

38. Zarezade, A., Jafarzadeh, S., Rabiee, H.R.: Recurrent spatio-temporal modeling of check-ins in location-based social networks. PLoS ONE 13(5), 1–20 (2018)

39. Zhang, C., Shou, L., Chen, K., Chen, G., Bei, Y.: Evaluating geo-social influence in location-based social networks. In: CIKM (2012)

40. Zhang, C., Bütepage, J., Kjellström, H., Mandt, S.: Advances in variational inference. CoRR abs/1711.05597 (2017)

41. Zhang, P., Wang, X., Li, B.: On predicting twitter trend: factors and models. In: Proceedings of the 2013 IEEE/ACM International Conference on Advances in Social Networks Analysis and Mining, ASONAM 2013, pp. 1427–1429. Association for Computing Machinery, New York (2013). https://doi.org/10.1145/2492517.2492576

42. Zhao, F., Tung, A.K.: Large scale cohesive subgraphs discovery for social network visual analysis. Proc. VLDB Endow. 6(2), 85–96 (2012)

43. Zhou, K., Zha, H., Song, L.: Learning triggering kernels for multi-dimensional Hawkes processes. In: ICML, vol. 28 (2013)

44. Zhu, W.Y., Peng, W.C., Chen, L.J., Zheng, K., Zhou, X.: Modeling user mobility for location promotion in location-based social networks. In: SIGKDD (2015)

Detecting Social Spammers in Sina Weibo Using Extreme Deep Factorization Machine

Yuhao Wu[1], Yuzhou Fang[1], Shuaikang Shang[2], Lai Wei[1], Jing Jin[2], and Haizhou Wang[1(✉)]

[1] College of Cybersecurity, Sichuan University,
Chengdu 610065, People's Republic of China
{wuyuhao,fangyuzhou,weilai@scu.edu.cn}@stu.scu.edu.cn,
whzh.nc@scu.edu.cn
[2] College of Computer Science, Sichuan University,
Chengdu 610065, People's Republic of China
{shangshuaikang,jinjing.0306}@stu.scu.edu.cn

Abstract. Online social networks (OSNs) are a form of social media that allow users to obtain news and information as well as connect with others to share content. However, the emergence of social spammers disrupts the normal order of OSNs significantly. As one of the most popular Chinese OSNs in the world, Sina Weibo is seriously affected by social spammers. With the continuous evolution of social spammers in Sina Weibo, they are gradually indistinguishable from benign users. In this paper, we propose a novel approach for social spammer detection in Sina Weibo using extreme deep factorization machine (xDeepFM). Specifically, we extract thirty features from four categories, namely profile-based, interaction-based, content-based, and temporal-based features to distinguish between social spammers and benign users. Furthermore, we build a detection model based on xDeepFM to implement the effective detection of social spammers. The proposed approach is empirically validated on the real-world data collected from Sina Weibo. The experimental results show that this approach can detect social spammers in Sina Weibo more effectively than most of the existing approaches.

Keywords: Online social networks · Social spammers · Sina Weibo · Extreme deep factorization machine

1 Introduction

With the rapid development of information technology, online social networks (OSNs) have had a significant impact on public life, which enable users to conduct massive-scale and real-time communication [2]. However, OSNs have gradually emerged social spammers, namely users with malicious purposes in OSNs. As social spammers continue to evolve, their behaviors including guidance of

© Springer Nature Switzerland AG 2020
Z. Huang et al. (Eds.): WISE 2020, LNCS 12342, pp. 170–182, 2020.
https://doi.org/10.1007/978-3-030-62005-9_13

online public opinion, malicious commentary, defamation, and ideology infiltration have posed huge damage to normal social order and even national stability [7]. For instance, during the presidential election in the U.S. in 2016, social spammers spread a large number of fake tweets in Twitter, and the decisions of many voters were affected by such tweets [19]. Accordingly, the detection of social spammers in OSNs is of great significance.

Currently, most of the existing research of social spammer detection is carried out on Twitter [1,11,17] and Facebook [4,6,18], and there are relatively few studies based on the Chinese OSNs, such as Sina Weibo [8,13]. Sina Weibo is one of the largest Chinese microblogging services in the world, which has a significant influence on the Chinese social communication. Meanwhile, Sina Weibo has also become one of the most active OSNs of social spammers for its popularity. Spammers widely conduct social spamming, which threaten the quality of Sina Weibo's social network [8]. The research of social spammer detection in Sina Weibo needs to be undertaken more comprehensive and in-depth.

This paper proposes a novel approach for detecting social spammers in Sina Weibo using extreme deep factorization machine (xDeepFM) [12], which consists of three components: data collection component, feature extraction component, and detection component. To begin with, the data collection component is responsible for collecting user data from Sina Weibo. Next, the feature extraction component is used to extract features of social spammers and benign users. Eventually, in the detection component, a xDeepFM model is employed for detection.

Contributions. The main contributions of the paper are summarized as follows:

- A novel deep learning-based approach for detecting social spammers in Sina Weibo is proposed, which mainly comprises three combined components, namely, a data collection component, a feature extraction component, and a detection component.
- A total of thirty features are extracted to identify social spammers in Sina Weibo accurately. These features can be divided into four categories: profile-based features, interaction-based features, content-based features as well as temporal-based features.
- A deep learning model, xDeepFM, is employed for detection. The evaluation results show that it significantly outperforms the widely used models.

Paper Organization. The rest of this paper is organized as follows. In Sect. 2, related work in the field of social spammer detection is introduced. The proposed social spammer detection approach is elaborated in Sect. 3. Furthermore, Sect. 4 describes the experimental setup and evaluation results. Finally, Sect. 5 concludes the research and plans for future work.

2   Related Work

In this section, we summarize some important studies on social spammer detection using machine learning approaches in recent years. Machine learning approaches are effective in detecting social spammers. Generally, the machine learning approaches can be categorized into classical machine learning approaches and deep learning approaches, which are introduced separately below.

The classical machine learning approach performs social spammer detection in OSNs by training a classical machine learning classification model. Chu et al. [9] measured and characterized the behaviors of humans, spammers, and cyborgs on Twitter. Also, an automated classification system was designed for detecting social spammers. After that, Yang et al. [21] made an analysis of the evasion tactics utilized by spammers and further designed several new features to detect more spammers. In their work, random forest (RF), decision tree (DT), and some other classical machine learning models were applied. In [1], an integrated social media content analysis platform was proposed, which leverages three aspects of features including user-generated content, social graph connections, and user profile activities to identify social spammers. Meanwhile, supervised machine learning models such as support vector machine (SVM), and RF etc. were used for detection, and RF model got the highest accuracy of 96.07%. Alghamdi et al. [2] utilized focused interest patterns of users and combined unsupervised and supervised machine learning to detect social spammers. In [3], supervised machine learning algorithms including RF, SVM, and logistic regression (LR) were employed to detect organized behaviors. Meanwhile, user-based features, temporal-based features, and features of collective behavior were used to distinguish users. Fazil et al. [10] defined six new features and redefined two features in their work. They focused on four categories features, namely metadata-based, content-based, interaction-based, and community-based features. These features were fed to RF, DT, and Bayesian network (BN) machine learning classifiers to detect social spammers.

In recent studies of social spammer detection, deep learning approaches begin to be more and more wildly used. Comparing with classical machine learning approaches, deep learning approaches have better generalization performance, especially when dealing with big data. Cai et al. [5] proposed an extreme learning machine (ELM)-based approach for social spammer detection in Sina Weibo. In their work, features were extracted from message content and behavior, and ELM is used to identify social spammers. In [11], a deep neural network based on contextual long short-term memory (LSTM) architecture that exploits both content and metadata to detect social spammers was proposed. The results showed that the approach had a high area under curve value. Moreover, a deep learning-based social spammer detection scheme was proposed in [17], i.e. DeBD, which utilizes tweet joint features, tweet metadata temporal features, and feature fusing.

3 The Proposed Social Spammer Detection Approach

In this section, we describe the proposed approach for detecting social spammers in Sina Weibo, the architecture of which is shown in Fig. 1. It is composed of data collection component, feature extraction component, and detection component. The details of each component of the proposed approach are described below.

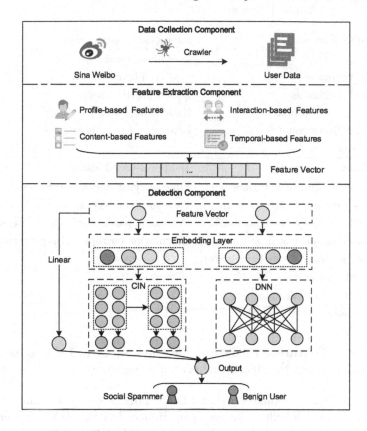

Fig. 1. The architecture of the proposed approach.

3.1 Data Collection

Sina Weibo provides developer APIs for data accessing, and these APIs are great ways for researchers and developers to collect user data from Sina Weibo at no cost. However, these APIs have some strict restrictions on data collection. To meet the needs of research, a high-performance and multi-threaded web crawler is developed. The crawler can create multi-tasks with multiple proxy IP to cycle through them and build a series of API requests to download raw HTML data from the web. Then, the valid data such as user profiles and posts of users is extracted, which is stored in a database next.

3.2 Feature Extraction

In this component, a total of thirty features of social spammers and benign users are extracted and divided into profile-based, interaction-based, content-based, and temporal-based features. Table 1 summarizes all the considered features.

Table 1. Summary of the extracted features.

Profile-based features	Interaction-based features
Default avatar	# of likes (mean)
Default nickname	# of comments (mean)
Length of nickname	# of reposting (mean)
Completeness of profile	Ratio of reposted posts
Ratio of follower count	Diversity of posting sources
Active level	
Content-based features	Temporal-based features
# of mentions (mean & variance)	Time interval (mean & variance)
# of hashtags (mean & variance)	Time interval (maximum & minimum)
# of URLs (mean & variance)	Burstiness parameter of time interval
# of punctuations (mean & variance)	Information entropy of time interval
# of interjections (mean & variance)	
# of words (variance)	
# of pictures (variance)	
Post sentiment score (mean)	

Profile-Based Features. Profile-based features are extracted from profiles of users, which can reveal the differences between benign users and social spammers.

Default avatar and nickname (DA and DN): Lots of social spammers use the default avatars and default nicknames [20]. Hence, DA and DN are considered in our work. The value of $DA(u)$ is 1 if user u uses the default avatar, otherwise it is 0, and the calculation of $DN(u)$ is similar to $DA(u)$.

Length of nickname (LN): The LN feature was used and achieved good results in [20]. Also, we adopt LN in our work. Since there are strict restrictions on LN in Sina Weibo, the value range of $LN(u)$ of user u is $\{LN(u)|2 \leq LN(u) \leq 30\}$.

Completeness of profile (CP): Users of Sina Weibo can fill in or change their profiles. Benign users have real friend-making demands, so they tend to fill in their profiles carefully. However, the profiles of social spammers are usually incomplete. Thus, CP is used in our work, like the way of using this feature in [21].

Ratio of follower count (RF): The RF feature is widely used [8–10]. To compute this feature, we follow the method used in these works.

Active level (AL): The AL feature is to measure how active a user is. In our work, we define AL of user u as

$$AL(u) = \sum_{i=1}^{M} \beta_i \times \varphi_i, 0 < \beta_i < 1, \tag{1}$$

where β_i is the value of the ith level, φ_i is the weight of the ith level, and M is the number of used levels. In our work, whether verified (verification is 1, otherwise 0) and the normalized user level are weighted to calculate a user's AL feature. The value range of $AL(u)$ is $\{AL(u) \mid 0 < AL(u) \le 1\}$.

Interaction-Based Features. The posts of users can be commented, reposted, and liked by others. These interactions often reflect the difference between benign users and social spammers. Therefore, interaction-based features are extracted.

Mean of the number of likes, comments, and reposting (ML, MC, and MR): The number of likes, comments, and reposting on a user's posts can quantify the popularity of the user, and most of posts of social spammers are illogical, so they have few likes, comments, or reposting. Hence, ML, MC, and MR are employed, which can be computed using the method in [15].

Diversity of posting sources (DS): Generally, posts come with posting sources, such as computer, mobile, etc. Benign users' posts tend to have different posting sources, while social spammers' posts usually have few posting sources. Therefore, we consider the DS feature and use the Margalef diversity index to calculate it, which is given by

$$DS(u) = \frac{\gamma - 1}{\ln K}, \tag{2}$$

where γ denotes the number of types of posting sources of the user, K represents the number of posts of the user u.

Ratio of reposted posts (RR): A number of posts of social spammers are reposted from other users, or generated using probabilistic methods. We thereby use RR feature to distinguish users, which is defined as the ratio of the number of reposted posts to the total number of posts [10,13].

Content-Based Features. Generally, the content of different posts of social spammers is similar and the writing habits of social spammers are illogical. Hence, content-based features are employed to identify social spammers.

Mean and variance of the number of mentions (MM and VM): In Sina Weibo, users use "@" to mention other users when posting. The number of mentions in posts can significantly distinguish users [10]. Hence, MM and VM of user u are defined as

$$MM(u) = \frac{1}{K} \sum_{i=1}^{K} \eta_i, \text{ and } VM(u) = \frac{1}{K} \sum_{i=1}^{K} (\eta_i - MM(u))^2, \tag{3}$$

where η_i is the number of mentions in the ith post.

Mean and variance of the number of hashtags (MH and VH): In Sina Weibo, "#" is used by users to participate in the discussion of topics while posting posts [9,10]. Accordingly, MH and VH are taken as two features, which can be computed in the same way as MM and VM.

Mean and variance of the number of URLs (MU and VU): URLs are usually used by social spammers for advertising, monetization, etc. [9]. The number of URLs can judge the quality of users' posts [10]. Thus, MU and VU are employed, which can be computed like MM and VM.

Mean and variance of the number of punctuations (MP and VP): The use of punctuations in posts can reflects a user's writing habits. In the posts of social spammers, the use of punctuations is often unreasonable. For this reason, MP and VP are used, which can be calculated like MM and VM.

Mean and variance of the number of interjections (MI and VI): An interjection is a word or expression that occurs as an utterance on its own and expresses a spontaneous feeling or reaction, such as "oh", "ah", "o", "ha", etc. These words can reflect a user's writing style. Thus, MI and VI are employed. To compute them, we follow the method used in MM and VM.

Variance of the number of words (VW): The word counts of different posts of a social spammer are usually similar [20]. As such, VW is given to identify social spammers in our work.

Variance of the number of pictures (VNP): When posting, users can add pictures to their posts. Generally, the number of pictures on each post of a social spammer is similar. Thus, VNP is used.

Mean of the post sentiment score (MS): Features of sentiment are the features extracted through sentiment analysis of posts [14,20]. In our work, sentiment scores of posts are calculated, and the MSS feature is employed. The value range of MS is $\{MS(u) \mid 0 \leq MS(u) \leq 1\}$.

Temporal-Based Features. Temporal-based features are extracted from the time of posts. In our work, the series of time intervals between each post of a user is defined as $\theta = [\chi_1, \chi_2, \ldots, \chi_{K-1}]$, where K is the number of posts of the user. The following temporal-based features are used to distinguish users.

Mean and variance of the time interval (MT and VT): In [8], the regularity of the time of user's posts is considered, and the variance of the post time was employed. In our work, MT and VT of user u are defined by

$$MT(u) = \frac{1}{K-1} \sum_{i=1}^{K-1} \chi_i, \text{ and } VT(u) = \frac{1}{K-1} \sum_{i=1}^{K-1} (\chi_i - MT(u))^2, \quad (4)$$

where χ_i is the time interval of two consecutive posts.

Maximum and minimum of the time interval (MAT and MIT): A number of social spammers do not post for a long time after posting a large number of posts in a short time. Thus, the features MAT and MIT are used. We sort the series of time interval to get a new series: $\theta' = [\chi'_1, \chi'_2, \ldots\ldots, \chi'_{K-1}]$ ($\chi'_i \leq \chi'_{i+1}, 1 \leq i \leq K-1$). Then, MAT and MIT of the user u are computed by

$$MAT(u) = \frac{1}{\mu}\sum_{i=1}^{\mu} \chi'_i, \text{ and } MIT(u) = \frac{1}{\mu}\sum_{i=K-\mu}^{K-1} \chi'_i. \tag{5}$$

After analysis, when $\mu = 5$, this pair of features can distinguish users well.

Burstiness parameters of time interval (BT): BT can distinguish users as well. There are three special values for BT: $\varepsilon - 1$, ε, $\varepsilon + 1$, which can be understood as a completely regular behavior, a completely Poisson behavior, and the most bursty behavior, respectively [16]. Generally, the values of BT of social spammers are close to $\varepsilon - 1$ and $\varepsilon + 1$.

Information entropy of time interval (IT): The Shannon entropy can be applied to quantify the regularity of the posting time interval series of users [16]. Thus, IT is employed in our work, and the smaller the IT, the greater the probability that the user is a social spammer.

3.3   xDeepFM for Detection

The detection component is based on the xDeepFM model [12], which combines a compressed interaction network (CIN) and a deep neural network (DNN) into a unified model. xDeepFM can learn certain bounded-degree feature interactions explicitly, and learn arbitrary low-order and high-order feature interactions implicitly as well. The characteristics of xDeepFM, being able to learn patterns of combinatorial features automatically and generalize to unseen features, have a great effect on the detection of social spammers.

After completing extracting features of users and forming feature vectors, each feature vector is divided into two types of features: continues and categorical features. Continuous features are those that have numerical values, while categorical features are features that can take on one of a limited number of possible values. This component transforms categorical features and continuous features by one-hot encoding and normalization, respectively. A user's feature vector can be represented as x_i (i denotes the ith user among all users), the output of xDeepFM is the probability that the user is a social spammer, which is given by

$$\hat{y}_i = \sigma(w_1^T x_i + w_2^T d_i + w_3^T c_i + b), \tag{6}$$

where σ is the sigmoid function, d_i and c_i are the outputs of the DNN and CIN, respectively. w_1, w_2, w_3, and b denote learnable parameters. Moreover, the cross-entropy cost function is used for overcoming the learning slowdown problem of quadratic cost function, which is defined by

$$\mathcal{L} = -\frac{1}{n}\sum_{i=1}^{n} y_i log \hat{y}_i + (1 - y_i)log(1 - \hat{y}_i), \tag{7}$$

where n is the total number of training feature vectors of users, and the value range of i is $\{i|1 \leq i \leq n\}$. In addition, y_i is the true category of the ith user (social spammer or benign user). Furthermore, the optimization process is to minimize a objective function, it is defined as

$$\mathcal{J} = \mathcal{L} + \lambda\|\varTheta\|, \tag{8}$$

where λ is the regularization term, $\varTheta$ represents the parameter set of the model.

4 Experiments and Evaluation

In this section, we evaluate the performance of the proposed approach for detecting social spammers. Our experiments are conducted on a Ubuntu 18.04.3 LTS platform with an Intel Xeon E5-2618L v3 CPU and a NVIDIA GeForce RTX 2080TI GPU (64 GB RAM). In the numerical results analysis, four metrics are taken into consideration to evaluate the performance of detection models, namely accuracy, precision, recall, and F-score. Each experiment is repeated ten times independently, and the average results are shown.

4.1 Dataset

In order to train our detection model, we collect a mass of user data. Then, five researchers in related fields are invited to manually annotate the collected user data. To be specific, the five researchers first independently annotate user data, and then mutually verify the results of annotation. Eventually, a dataset with 10,000 social spammers and 10,000 benign users is constructed after completing annotation, data balance processing, and feature extraction. Further, the dataset is divided into 60% for training, 20% for validation, and 20% for testing. The brief description of the dataset is shown in Table 2.

Table 2. Overview of the dataset.

Category	Training	Validation	Testing	Posts
Benign user	6,000	2,000	2,000	118,199
Social spammer	6,000	2,000	2,000	96,307
Total	12,000	4,000	4,000	214,506

4.2   Performance Evaluation

To evaluate the effectiveness of the proposed approach, first, we complete model training on the constructed dataset, and the training process of our detection model is shown in Fig. 2. Notably, the xDeepFM model that we used is stable and convergent. Moreover, overfitting is clearly well suppressed by using dropout technology. This yields better performance for training and testing. Further, we compare the xDeepFM model with baseline models that are widely used in this field including SVM [1,3], LR [3], DT [10,21], and RF [1,10]. Table 3 and Fig. 3 show the experimental results. It can be seen that the classical machine learning models of LR, SVM, and DT have almost no advantages over the ensemble machine learning models of RF in this case. Among all detection models, our detection model xDeepFM has the highest accuracy, precision, recall, and F-score. Such results indicate that the characteristics of xDeepFM, the ability to learn and generalize patterns of combinatorial features, have obvious advantages in detecting social spammers.

Table 3. Performance of different detection models.

Model	Accuracy	Precision	Recall	F-score
SVM [1,3]	0.9544	0.9365	0.9752	0.9554
LR [3]	0.9673	0.9564	0.9791	0.9676
DT [10,21]	0.9724	0.9693	0.9757	0.9725
RF [1,10]	0.9794	0.9790	0.9801	0.9795
xDeepFM	**0.9816**	**0.9829**	**0.9802**	**0.9815**

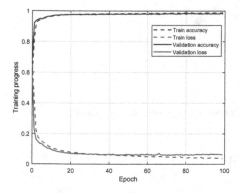

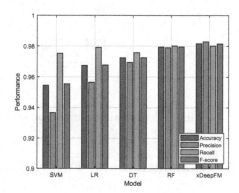

Fig. 2. Training progress of our detection model xDeepFM.

Fig. 3. Performance comparison between xDeepFM and other detection models.

4.3   Feature Ablation Study

In order to evaluate the validity of each category of feature, we conduct feature ablation test on the full feature set and four subsets of the full feature set using xDeepFM model. The subsets of the feature set can be represented by the set-difference function as

$$F \backslash F' = \{x | x \in F \wedge x \notin F'\}, \tag{9}$$

where F is the set with all features, F' is the subset of F with a particular category of features, and x is all user data of a feature. Therefore, $F \backslash Profile$, $F \backslash Interaction$, $F \backslash Content$, and $F \backslash Temporal$ represent the feature set with profile-based, interaction-based, content-based, and temporal-based features removed from F, respectively.

The results of the feature ablation test are shown in Fig. 4. Firstly, the detection model xDeepFM performs much better on the feature set F that contains all the features than on other feature sets, which proves that each category of extracted features has the distinguishability between social spammers and benign users. Moreover, xDeepFM performs worst using feature set of $F \backslash Content$, which indicates that the distinguishability of content-based features is the greatest. Whereas the performance of xDeepFM using feature set of $F \backslash Temporal$ is second only to that of F, which means that the distinguishability of temporal-based features is the worst.

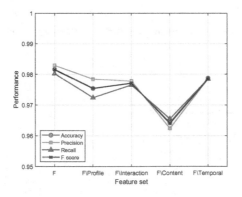

Fig. 4. Performance comparison on different feature sets using xDeepFM.

5   Conclusion and Future Work

In this paper, we have proposed a novel approach for detecting social spammers in Sina Weibo. The proposed approach mainly includes three components: data collection component, feature extraction component, and detection component. Specifically, we extracted thirty features to distinguish between social spammers and benign users, which can be divided into profile-based, interaction-based, content-based, and temporal-based features. Furthermore, this paper built a xDeepFM model to detect social spammers. Extensive experiments on the real-world data collected from Sina Weibo showed that, compared with the widely used detection models, the proposed xDeepFM-based detection model is superior in terms of the accuracy, precision, recall, and F-score.

Future work will focus on more online social networks to further verify and improve the proposed social spammer detection approach.

Acknowledgements. This work is supported by the National Natural Science Foundation of China (NSFC) under grant nos. 61802270, 61802271, 81602935, and 81773548. Haizhou Wang is the corresponding author. The authors thank anonymous reviewers for their helpful comments to improve the paper.

References

1. Al-Qurishi, M., Hossain, M.S., Alrubaian, M., Rahman, S.M.M., Alamri, A.: Leveraging analysis of user behavior to identify malicious activities in large-scale social networks. IEEE Trans. Ind. Inform. **14**(2), 799–813 (2017)
2. Alghamdi, B., Xu, Y., Watson, J.: A hybrid approach for detecting spammers in online social networks. In: Hacid, H., Cellary, W., Wang, H., Paik, H.-Y., Zhou, R. (eds.) WISE 2018. LNCS, vol. 11233, pp. 189–198. Springer, Cham (2018). https://doi.org/10.1007/978-3-030-02922-7_13
3. Beğenilmiş, E., Uskudarli, S.: Organized behavior classification of tweet sets using supervised learning methods. In: Proceedings of the 8th International Conference on Web Intelligence, Mining and Semantics, pp. 1–9. ACM (2018)
4. Boshmaf, Y., et al.: Íntegro: leveraging victim prediction for robust fake account detection in large scale OSNs. Comput. Secur. **61**, 142–168 (2016)
5. Cai, C., Li, L., Zeng, D.: Detecting social bots by jointly modeling deep behavior and content information. In: Proceedings of the 26th ACM Conference on Information and Knowledge Management, pp. 1995–1998. ACM (2017)
6. Cao, Q., Yang, X., Yu, J., Palow, C.: Uncovering large groups of active malicious accounts in online social networks. In: Proceedings of the 21st ACM Conference on Computer and Communications Security, pp. 477–488. ACM (2014)
7. Chakraborty, M., Pal, S., Pramanik, R., Ravindranath Chowdary, C.: Recent developments in social spam detection and combating techniques: a survey. Inf. Process. Manag. **52**(6), 1053–1073 (2016)
8. Chen, H., Liu, J., Lv, Y., Li, M.H., Liu, M., Zheng, Q.: Semi-supervised clue fusion for spammer detection in Sina Weibo. Inf. Fusion **44**, 22–32 (2018)
9. Chu, Z., Gianvecchio, S., Wang, H., Jajodia, S.: Detecting automation of Twitter accounts: are you a human, bot, or cyborg? IEEE Trans. Dependable Secur. Comput. **9**(6), 811–824 (2012)
10. Fazil, M., Abulaish, M.: A hybrid approach for detecting automated spammers in Twitter. IEEE Trans. Inf. Forensics Secur. **13**(11), 2707–2719 (2018)
11. Kudugunta, S., Ferrara, E.: Deep neural networks for bot detection. Inf. Sci. **467**, 312–322 (2018)

12. Lian, J., Zhou, X., Zhang, F., Chen, Z., Xie, X., Sun, G.: XDeepFM: combining explicit and implicit feature interactions for recommender systems. In: Proceedings of the 24th ACM Conference on Knowledge Discovery and Data Mining, pp. 1754–1763. ACM (2018)
13. Lian, Y., Dong, X., Chi, Y., Tang, X., Liu, Y.: An internet water army detection supernetwork model. IEEE Access **7**, 55108–55120 (2019)
14. Loyola-González, O., López-Cuevas, A., Medina-Pérez, M.A., Camiña, B., Ramírez-Márquez, J.E., Monroy, R.: Fusing pattern discovery and visual analytics approaches in tweet propagation. Inf. Fusion **46**, 91–101 (2019)
15. Mohammad, S., Khan, M.U., Ali, M., Liu, L., Shardlow, M., Nawaz, R.: Bot detection using a single post on social media. In: Proceedings of the 3rd World Conference on Smart Trends in Systems, Security and Sustainability, pp. 215–220. IEEE (2019)
16. Pan, J., Liu, Y., Liu, X., Hu, H.: Discriminating bot accounts based solely on temporal features of microblog behavior. Phys. A **450**, 193–204 (2016)
17. Ping, H., Qin, S.: A social bots detection model based on deep learning algorithm. In: Proceedings of the 18th IEEE International Conference on Communication Technology, pp. 1435–1439. IEEE (2018)
18. Santia, G.C., Mujib, M.I., Williams, J.R.: Detecting social bots on Facebook in an information veracity context. In: Proceedings of the 13th International AAAI Conference on Web and Social Media, pp. 463–472. AAAI (2019)
19. Shao, C., Ciampaglia, G.L., Varol, O., Yang, K.C., Flammini, A., Menczer, F.: The spread of low-credibility content by social bots. Nat. Commun. **9**(1), 1–9 (2018)
20. Varol, O., Ferrara, E., Davis, C.A., Menczer, F., Flammini, A.: Online human-bot interactions: detection, estimation, and characterization. In: Proceedings of the 11th International AAAI Conference on Web and Social Media, pp. 280–289. AAAI (2017)
21. Yang, C., Harkreader, R., Gu, G.: Empirical evaluation and new design for fighting evolving Twitter spammers. IEEE Trans. Inf. Forensics Secur. **8**(8), 1280–1293 (2013)

Clustering Hashtags Using Temporal Patterns

Borui Cai[1], Guangyan Huang[1(✉)], Shuiqiao Yang[2], Yong Xiang[1],
and Chi-Hung Chi[3]

[1] School of Information Technology, Deakin University, Melbourne, Australia
{bcai,guangyan.huang,yong.xiang}@deakin.edu.au
[2] School of Computer Science, University of Technology Sydney, Ultimo, Australia
shuiqiao.yang@uts.edu.au
[3] Data61, CSIRO, Sydney, Australia
chihung.chi@csiro.au

Abstract. Twitter hashtags provide a high-level summary of tweets, while cluster hashtags have many applications. Existing text-based methods (relying on explicit words in tweets) are greatly affected by the sparsity of the short tweet texts and the low co-occurrence rates of hashtags in tweets. Meanwhile, semantically related hashtags but using different text-expressions may show similar temporal patterns (i.e., the frequencies of hashtag usages changing with the time), which can help capture events, opinions and synonyms. In this paper, we propose a novel clustering hashtags by their temporal patterns (CHTP) method as a complement to text-based methods. In CHTP, hashtags are represented as hashtag time series that show their temporal patterns, so, hashtag clusters can be discovered by clustering hashtag time series. Density-based clustering algorithms are suitable to discover naturally shaped hashtag clusters but they are not fine enough (use one distance threshold to define density) to differentiate clusters of various density levels. Therefore, we develop a new parameter-free Density-Sensitive Clustering (DSC) algorithm to discover clusters of different density levels and use it in CHTP to group hashtags by temporal patterns. DSC recursively partitions the dataset from coarse-grained to fine-grained (using adaptive distance thresholds) to discover hashtag clusters of different density levels. Experiments conducted on Twitter datasets show that the DSC algorithm finds hashtag clusters of different densities more effectively than counterpart methods, and CHTP (using DSC) can discover meaningful hashtag clusters, 36% of which cannot be found by the text-based approaches.

Keywords: Hashtag · Time series clustering · Cluster density

1 Introduction

Twitter is a popular web microblogging and social networking service, on which users interact and share information with short messages (tweets). A twitter

© Springer Nature Switzerland AG 2020
Z. Huang et al. (Eds.): WISE 2020, LNCS 12342, pp. 183–195, 2020.
https://doi.org/10.1007/978-3-030-62005-9_14

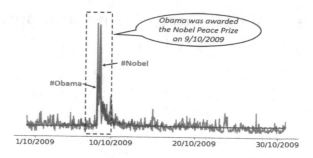

Fig. 1. The similar time series of #Obama and #Nobel can help detect the event that Obama was awarded the Nobel Peace Prize on 9/10/2009.

hashtag is a meta-tag that the user creates for classifying tweets based on their meanings and subjects. For example, tweets tagged with the hashtag #Jobs are related to job opportunities. This high-level summary/label information explicitly provided by users makes hashtag promising in finding dynamic relationships of tweets [2]. Hashtag clustering that finds semantically related hashtags can be used for event discovery [13,16], opinion extraction and synonym detection.

Most existing methods discover hashtag clusters using the tweet texts [13,14]. These text-based methods represent hashtags using explicit words in tweet texts. For example, with the bag-of-words (BoW) representation [13], one hashtag is represented by the multiplicity of the co-occurring words in the tweets. However, the short tweet texts may not reveal the whole hashtag relationships. First, many tweets may report the same event but use totally different words. Second, hashtag co-occurrence is not always reliable since many tweets have no more than two hashtags. Affected by these problems, text-based methods only can extract part of the information revealed by hashtags. The recent method [12] uses the temporal pattern of hashtags related to known events, i.e., time series that record the frequencies of hashtag usages changing with the time, to discover other related hashtags. Inspired by this, we use hashtag time series (which show temporal pattern) to cluster hashtags, because the frequencies of certain hashtag usages are highly correlated with the popularity of the corresponding events/topics. For example, in Fig. 1, the hashtag time series of #Obama and #Nobel can help discover the event that Obama was awarded the Nobel Peace Prize.

In this paper, we propose a novel clustering hashtags by their temporal patterns (CHTP) method. In CHTP, hashtags are represented as hashtag time series, so, clusters can be discovered by clustering hashtag time series. However, clustering hashtag time series faces two challenges. First, it is impractical to manually determine the number of hashtag clusters, because a Twitter dataset includes world-wide events/topics. Second, these hashtag clusters are naturally shaped and follow no specific distributions. Therefore, density-based time series clustering is preferred in CHTP since it can find arbitrarily shaped clusters. Many density-based clustering methods require single distance thresholds to define density of data objects. Unfortunately, specifying a suitable distance threshold is difficult because there are clusters of different density levels in the complex Twitter datasets. Thus, we propose a novel parameter-free Density-Sensitive Clustering (DSC) algorithm for CHTP to find hashtag clusters of different density levels. DSC is a density-based recursive partition process, and it discovers a hashtag time series group as a hashtag cluster if it cannot be split into multiple subgroups in lower recursions. Comprehensive experiments have been conducted on Twitter datasets to demonstrate the effectiveness of the proposed CHTP method, which uses DSC for hashtag clustering. Therefore, the contributions of this paper are listed as follows:

1) We provide a novel CHTP method for hashtag clustering, which groups hashtags using the common temporal patterns, rather than the tweet texts.
2) We develop a new parameter-free Density-Sensitive Clustering (DSC) algorithm for CHTP to discover hashtag clusters of different densities.
3) Comprehensive experiments conducted with four Twitter datasets show that averagely 36% of meaningful hashtag clusters discovered by CHTP (using DSC) cannot be found by text-based approaches; and the proposed DSC algorithm is more effective than the counterpart algorithms to find hashtag clusters.

The rest of this paper is organized as follows. In Sect. 2, we review the related work. We detail the CHTP method in Sect. 3, and then develop the DSC algorithm in Sect. 4. The proposed method is evaluated in Sect. 5. The paper is concluded in Sect. 6.

2   Related Work

Hashtag is widely used in tweets and can represent summary information of tweets to extract useful information. Most hashtag clustering methods discover hashtag clusters by the tweet texts tagged with the hashtags (text-based methods). SMSC [13] uses the bag-of-words (BoW) representation of co-occurred tweet texts with Kmeans to find hashtag clusters. Meanwhile, topic model is used in HGTM [14] to cluster hashtags by analyzing a hashtag graph built with the co-occurrence information of hashtags. The method in [7] further integrates lexical and contextual text information to improve clustering performance. In addition to tweet texts, external knowledge, i.e. the semantics of hashtags obtained from WordNet [8] or Wikipedia [6], is utilized to improve the accuracy of hashtag clustering.

Other than the text-based methods, SAX* [12] uses temporal similarity as the semantic relatedness for hashtags, and it detects hashtag clusters with string patterns deduced from external sources (e.g. Wikipedia) [9]. We extend this idea and discover hashtag clusters from their temporal patterns by clustering hashtag time series. We briefly review the recent time series clustering methods, upon which the proposed method is built developed. Partition-based time series clustering, such as kshape [10] and KSC [15], find cluster representatives and minimize the distances of time series to nearest representatives. However, they adopt

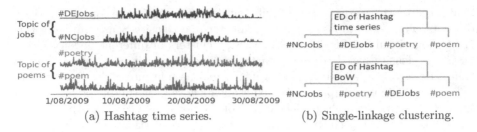

(a) Hashtag time series.   (b) Single-linkage clustering.

Fig. 2. Four hashtag time series (as shown in (a)) obtained from a Twitter corpus [15]. (b) shows single-linkage clustering with Euclidean distance on hashtag time series (top) and hashtag BoW representation (bottom), respectively.

spherical-shape clusters that are sensitive to outliers and noise in the complex Twitter datasets. Model-based methods. e.g. Gaussian Mixture Model [4] and Gaussian Inverse Covariance [5], cluster time series by the optimizations of specific models, but a model that well-explain the large Twitter datasets may have impractical complexity. Density-based clustering methods are preferred since they can find natural-shaped clusters. For example, YADING [3] hierarchically adopts DBSCAN to find time series clusters of different densities; however, the inflection points (on the distance-to-nearest-neighbours curve) YADING used to determine densities levels are not significant in the sparse Twitter datasets. TADPole [1] groups time series into clusters using the density and the distance by DPC [11]; but that makes TADPole hard to find clusters of different density levels since a global distance threshold is applied in the entire dataset.

3   Problem Definition

In this section, we present the CHTP method, and it comprises two steps:

1) Represent hashtags as hashtag time series. (see Sect. 3.1)
2) Discover hashtag clusters by clustering hashtag time series. (see Sect. 3.2)

3.1   Hashtag Time Series

CHTP discovers hashtag clusters by hashtag temporal patterns, and each hashtag is represented as a time series denoted as $Z = \{z_1, z_2, ..., z_m\}$, where z_j is the frequency of a hashtag at time bin j (for example, one hour [15]). A hashtag time series dataset DS is a collection of hashtag time series denoted as $DS = \{Z_1, Z_2, ...Z_n\}$. Each hashtag time series is preprocessed to be scale-invariant by applying z-normalization as follows:

$$x_j = \frac{z_j - \mu}{\delta} \ (1 \leq j \leq m),\tag{1}$$

where $\mu = \frac{1}{m}\sum z_j$ and $\delta = \sqrt{\frac{\sum(z_j - \mu)^2}{m-1}}$. The distance of two normalized hashtag time series, X and Y, is measured by Euclidean distance ($ED(X,Y)$).

We show four hashtag time series as an example in Fig. 2 (a). By applying single-linkage clustering with Euclidean distance, two clusters are correctly discovered (the top in Fig. 2 (b)). However, with text-based hashtag BoW representation, unrelated hashtags are grouped together (the bottom in Fig. 2 (b)).

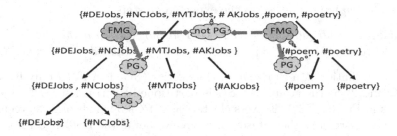

Fig. 3. The depth-first recursive partition of example hashtag time series.

3.2   Clustering by Hashtag Time Series

Specifying the number or the shape of clusters in a Twitter dataset is difficult since tweets posted world-widely cover enormous events/topics, and thus density-based clustering that discovers natural-shaped clusters is favored. Existing density-based time series clustering algorithms, such as TADPole [1], face a major challenge to cluster hashtag time series, i.e. the global distance threshold they used to define density cannot differentiate density levels. That may undermine the clustering accuracy, because a large distance threshold may group irrelevant hashtags into clusters; while a small threshold cannot discover sparse clusters with small densities. Therefore, CHTP demands a density-based clustering algorithm that is adaptable to different densities of hashtag clusters.

4   The DSC Algorithm

In this section, we present the DSC algorithm used in CHTP to cluster hashtag time series. DSC discovers clusters of different density levels by adaptive distance thresholds. In general, DSC partitions the dataset recursively from the coarse-grained to the fine-grained (in a depth-first manner), with adaptive distance thresholds, and clusters of different density levels are discovered as *full pure group*s. A *full pure group* contains highly correlated time series that are less relevant with time series in other *full pure group*s. Before explaining *full pure group*, we first define *pure group*.

Definition 1. *Pure group (PG): a group of time series that cannot be split into multiple subgroups by density, i.e., each partition produces at most one subgroup (contains more than one time series) during the recursive partitioning.*

Examples of *PG* are shown in Fig. 3, which records the recursive partition process of six hashtags. Apparently, groups with 2 or 3 time series are *PGs*. Based on the definition, the partitioned subgroups of a *PG* are also *PGs* but locates at different hierarchies on the partition tree. *PGs* on the top hierarchies (as *full pure group*) are regarded as the expected clusters.

Definition 2. *Full pure group (FPG): full pure group is the pure group of the top hierarchy, i.e., full pure group must be partitioned from a non-PG.*

Based on Definition 2, a subgroup of time series (SG) is a FPG only if 1) SG is a PG and 2) S (the group that SG is partitioned from) is not a PG (see the examples in Fig. 3). DSC discovers clusters as $FPGs$ by analyzing the recursive process that partitions the dataset, and now we detail the DSC algorithm.

4.1   Data Partition

In DSC, a recursive *DataPartition* algorithm is used to partition the dataset (DS) by density, and the input dataset for each partition process in the recursion is denoted as S. *DataPartition* has two components, i.e. the group forming function ($FormGroup$) that partitions S by forming subgroups and the cluster detector that indicates whether a subgroup is a FPG. $FormGroup$ forms subgroups by density, and one time series, X, finds a set of neighbours (N_X) as follows:

$$N_X = \{Y : ED(X,Y) < d_c, Y \in S\}, \tag{2}$$

where d_c is the adaptive distance threshold (will be discussed later). The density of X, ρ_X, is calculated as follows:

$$\rho_X = \sum \left(1 - \frac{ED(X,Y)^2}{d_c^2}\right), Y \in N_X. \tag{3}$$

We borrow the idea of DPC [11] to group time series by density, that is, X is density connected to a specific neighbour (n_X) as follows:

$$n_X = \underset{Y:\rho_Y > \rho_X, Y \in N_X}{\arg\min} ED(X,Y), \tag{4}$$

X is the centre (C) of a group if $n_X = null$, which means X has the local maximum density among its neighbours. Groups are developed by two steps: 1) each C is assigned a unique group label; 2) starting from centers, the group labels are spread from n_X to X in the decreasing order of density. After all time series acquire group labels from relative centers, S is partitioned into subgroups (SG) for further partitioning.

4.2 Adaptive Distance Threshold

We especially expect the adaptive d_c to satisfy the following two requirements:

1) *FormGroup* can always partition S into subgroups or individuals, which ensures the termination of the recursive partition.
2) Partitions are conducted from the coarse-grained to the fine-grained. This ensures sparse clusters are not omitted.

We use Minimum Spanning Tree (MST) built with S to find the adaptive d_c. Nodes of MST are the time series in S and the edges are the time series distances. Then, d_c of *FormGroup* is assigned as the largest edge on the MST. We show that d_c satisfies the first requirement by proving that at least two time series are split after partitioning. Assume $E_{ij} = ED(X_i, X_j)$ is the longest edge of MST built with S, and we assign $d_c = E_{ij}$ for *FormGroup*.

Algorithm 1. DataPartition.

Input: Hashtag time series dataset/subset S, *Clusters*
1: $\text{MST} = BuildMST(S)$
2: d_c = largest edge of MST (based on Lemma 1 and Lemma 2)
3: $Groups = FormGroup(S, d_c)$, $Subgroups = \{\}$, PG_S = False
4: **for** each $SG \in Groups$ **do**
5:     **if** $|SG| > 1$ **then** $Subgroups = Subgroups \cup \{SG\}$
6: **if** $|Subgroups| = 0$ or $|Subgroups| = 1$ **then** PG_S = True
7: **for** each $SG \in Subgroups$ **do**
8:     PG_{SG} = DataPartition(SG, *Clusters*)
9:     **if** PG_{SG} = False **then** PG_S = False
10:    **if** PG_{SG} = True and PG_S = False **then** $Clusters = Clusters \cup SG$
Output: PG_S

Lemma 1. X_i and X_j do not stay in the same subgroup after S is partitioned by FormGroup, with $d_c = E_{ij}$.

Proof. MST is split into $LMST = \{X_{l_1}, ..., X_{l_n}\}$ and $RMST = \{X_{r_1}, ..., X_{r_n}\}$ after removing E_{ij}, and $X_i \in LMST$ and $X_j \in RMST$. Now assume X_i and X_j still stay in the same subgroup. Since $d_c = E_{ij}$, $X_j \notin N_{X_i}$, therefore, X_i and X_j must be connected through $\{X_{\eta_0} = X_i, X_{\eta_1}, ..., X_{\eta_{k-1}}, X_{\eta_k} = X_j\}$, in which $E_{\eta_{t-1}, \eta_t} < d_c, \forall t \in \{1, ..., k\}$. Since $X_l \in LMST$ and $X_r \in RMST$, $\exists t \in \{1, ..., k\}$ s.t. $X_{\eta_{t-1}} \in LMST$ and $X_{\eta_t} \in RMST$, and thus the spanning tree by connecting $LMST$ with $RMST$ with $E_{\eta_{t-1}\eta_t}$ ($< E_{ij}$) has a smaller weights than MST, which is impossible and the assumption is wrong. It is proved.

Now we show that the second requirement is also satisfied due to Lemma 2.

Lemma 2. *The largest edge on MST of SG (one subgroup of S grouped by FormGroup) is always smaller than the largest edge on MST of S.*

Proof. MST_{SG} (the MST of SG) comprises several subtrees, i.e., $\mathrm{MST}_{SG} = \{ST_1, ..., ST_k\}$ ($ST_i \in \mathrm{MST}_{SG}$ and $ST_i \cap ST_j = \emptyset, \forall i, j \in [1, k]$). In MST_{SG}, $\{ST_1, ..., ST_k\}$ are connected with the $k - 1$ edges among them. Assume the largest edge on MST_{SG} (E_{ab}, which is also d_c^{SG}) connects ST_a and ST_b, thus $E_{ab} = \min_{X_l \in ST_a, X_r \in ST_b} ED(X_l, X_r)$. Since SG is grouped by *FormGroup* on S with d_c^S (the largest edge on MST_S), $\exists ED(X_l, X_r) < d_c^S$ s.t. $X_l \in ST_a, X_r \in ST_b$. Therefore, $d_c^{SG} = E_{ab} \leq ED(X_l, X_r) < d_c^S$. It is proved.

The pseudo-code of *DataPartition* is shown in Algorithm 1. d_c is assigned as the longest edge on MST at lines 1–2. S is partitioned at line 3, and the obtained *subgroup*s are further partitioned at line 8. The clusters are discovered as *FPG*s is shown at lines 8–10.

4.3   Complexity Analysis

The complexity of *DataPartition* for the initial dataset (DS) is $O(n \log n + 2n^2)$, including building MST from DS for $O(n^2)$ and adopting *FormGroup* for $O(n \log n + n^2)$. Meanwhile, the recursion depth (γ) is usually much smaller than n, because *FormGroup* can partition the dataset efficiently with a well-designed d_c. Therefore, the complexity of DSC is $O(\gamma n \log n + \gamma n^2)$.

5   Evaluation

In this section, we evaluate the performance of the proposed CHTP method (hashtag time series clustering by DSC) and the DSC clustering algorithm by answering the following two questions:

Table 1. Statistics of the 4 Twitter datasets.

Dataset	Hashtags	Tweets (million)	Date of Tweets
Aug/2009	1875	10.6	1/8/2009–31/8/2009
Sep/2009	1462	7.0	1/9/2009–30/9/2009
Oct/2009	1181	5.4	1/10/2009–31/10/2009
Nov/2009	791	3.4	1/11/2009–30/11/2009

1) Q1: Can DSC find hashtag clusters more effectively than the counterpart TADPole and YADING? (Sect. 5.1)
2) Q2: Is it necessary to use CHTP to cluster hashtag when the text-based approaches exist? (Sect. 5.2)

All algorithms are implemented with python 2.7, and the experiments are run on a Windows 10 platform with 3.4 GHz CPU and 16 GB RAM.

Datasets: We use a Twitter corpus [15] that comprises tweets posted from August to November 2009, for our experiments. These tweets are split into 4 subsets by the months they were posted. In each subset, we select important hashtags (with frequency larger than 1000) along with the tweets they appeared to generate the corresponding dataset. The time bin is specified as one hour following [15]. The statistics of the datasets are shown in Table 1.

Counterparts for DSC: TADPole [1] and YADING [3], two density-based time series algorithms as discussed in Sect. 2, are compared with DSC in experiment 1 (see Sect. 5.1) to answer Q1. To be fair, distance measurements used are unified as Euclidean distance in TADPole, YADING and DSC. **Counterparts for CHTP**: We explicitly choose SMSC [13], which adopts the hashtag BoW (text based) and Kmeans, as the counterpart of CHTP in experiment 2 to answer Q2. We also use the BoW representation with DSC, i.e., TextDSC, to directly compare with CHTP. Besides DSC, TextDSC and YADING, which are parameter-free, the results of TADPole and SMSC are obtained under the optimal parameters that maximize Silhouette Coefficient. Clusters that contain multiple hashtags are used for comparison.

Table 2. Clustering accuracy.

Dataset	TADPole	YADING	DSC
Aug/2009	0.53	0.52	**0.79**
Sep/2009	0.62	0.51	**0.78**
Oct/2009	0.67	0.54	**0.76**
Nov/2009	0.71	0.54	**0.74**
Average	0.63	0.53	**0.77**

5.1 Accuracy and Effectiveness of DSC on Hashtag Time Series

In this experiment, we compare DSC with TADPole and YADING for clustering hashtag time series. We evaluate the clustering accuracy by validating each hashtag cluster by its contained hashtags since manually labeling the hashtags is impractical. A cluster is *valid* only if the contained hashtags are synonyms/abbreviations or represent an event/topic (see Fig. 1) in the search results from Google. The accuracy of hashtag clustering is measured by F_1 score.

The clustering accuracy results are shown in Table 2. DSC, TADPole and YADING achieve fair clustering results in the datasets and the average accuracy are 0.77, 0.63 and 0.53, respectively. Specifically, DSC out-performs TADPole and YADING in all the 4 datasets, and the average improvements of accuracy to TADPole and YADING are around 22% and 45%, respectively.

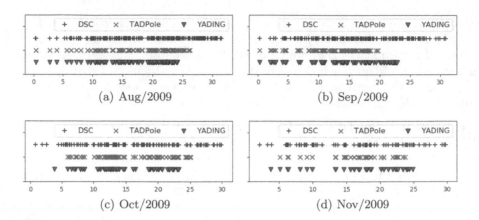

Fig. 4. Density distribution of hashtag clusters discovered by DSC, TADPole and YADING, respectively.

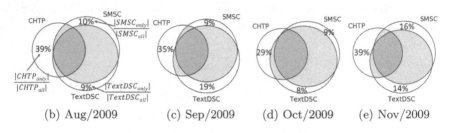

Fig. 5. Relationships of hashtag clusters discovered by CHTP, SMSC and TextDSC, and shaded area indicates similar clusters shared by multiple methods.

We compare the densities of *valid* hashtag clusters discovered by DSC, TADPole and YADING, with the density of a cluster being the average nearest-neighbour-distance of the contained hashtags. The larger the average nearest-neighbour-distance, the smaller the density, and the results are shown in Fig. 4. Generally, DSC finds more *valid* hashtag clusters than TADPole and YADING in all the 4 Twitter datasets, and YADING discovers the least *valid* hashtag clusters, which results in its low *recall*.

5.2 CHTP vs. Text-Based Approaches

In this experiment, we analyze the significance of hashtag clustering by temporal pattern through showing whether the text-based methods (TextDSC and SMSC) can discover similar *valid* hashtag clusters as CHTP. We regard two *valid* hashtag clusters (discovered by different methods) are similar if they shares more than three fourth hashtags. Then we summarize the relationship of *valid* hashtag clusters discovered by CHTP, TextDSC and SMSC as shown in Fig. 5.

The results show that, in all the 4 Twitter datasets, around 36% of the *valid* hashtag clusters discovered by CHTP cannot be discovered by TextDSC/SMSC,

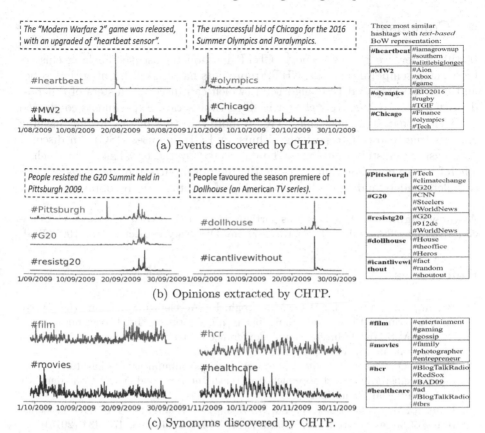

(a) Events discovered by CHTP.

(b) Opinions extracted by CHTP.

(c) Synonyms discovered by CHTP.

Fig. 6. Examples of events (a), opinions (b) and synonyms (c) discovered by CHTP, but SMSC and TextDSC fail to find. The right column shows the nearest hashtags with hashtag BoW representation.

even when using the same clustering algorithm (DSC). This result suggests that the hashtag temporal pattern partially represent distinctive correlations of hashtags compared with the tweet texts, and CHTP can supplement hashtag clustering that only uses tweet texts. In contrast, TextDSC and SMSC share many hashtag clusters since they use the same clue, i.e., the tweet texts.

We show some examples of hashtag clusters only discovered by CHTP to further understand the hashtag clusters discovered by temporal patterns in Fig. 6. That includes two events discovered only by CHTP (Fig. 6 (a)), two examples of extracted opinion (Fig. 6 (a)) and two examples of discovered synonyms (Fig. 6 (c)). To compare with the text-based BoW representation, we show the three most similar hashtags (with BoW) of the hashtags in the right-side column of Fig. 6. The result shows that hashtags having similar temporal patterns are not quite similar in their BoW representations, and that is the reason that TextDSC/SMSC fails to discover the hashtag clusters.

6   Conclusion

In this paper, we propose a novel CHTP method to discover hashtag clusters by hashtag temporal patterns. CHTP represents hashtags as hashtag time series and uses the proposed DSC algorithm (which can discover clusters of different density levels) to effectively cluster hashtag time series. Experiments conducted on Twitter datasets show that DSC is more effective in discovering hashtag clusters than two counterpart algorithms, and CHTP (uses DSC) can discover 36% hashtag clusters that cannot be discovered by the text-based approaches. Therefore, we conclude that by using the temporal pattern of hashtags, a more complete understanding of the relationship of hashtags can be obtained.

Acknowledgments. This work was partially supported by Australia Research Council (ARC) DECRA Project (DE140100387) and Discovery Project (DP190100587).

References

1. Begum, N., Ulanova, L., Wang, J., Keogh, E.: Accelerating dynamic time warping clustering with a novel admissible pruning strategy. In: Proceedings of the 21th ACM SIGKDD International Conference on Knowledge Discovery and Data Mining, pp. 49–58. ACM (2015)
2. DeMasi, O., Mason, D., Ma, J.: Understanding communities via hashtag engagement: a clustering based approach. In: Tenth International AAAI Conference on Web and Social Media (2016)
3. Ding, R., Wang, Q., Dang, Y., Fu, Q., Zhang, H., Zhang, D.: YADING: fast clustering of large-scale time series data. Proc. VLDB Endow. **8**, 473–484 (2015)
4. Hallac, D., Nystrup, P., Boyd, S.: Greedy Gaussian segmentation of multivariate time series. Adv. Data Anal. Classif. **13**, 727–751 (2019)
5. Hallac, D., Vare, S., Boyd, S., Leskovec, J.: Toeplitz inverse covariance-based clustering of multivariate time series data. In: Proceedings of the 23rd ACM SIGKDD International Conference on Knowledge Discovery and Data Mining, pp. 215–223. ACM (2017)
6. Javed, A., Lee, B.S.: Sense-level semantic clustering of hashtags. In: Lossio-Ventura, J.A., Alatrista-Salas, H. (eds.) SIMBig 2015–2016. CCIS, vol. 656, pp. 1–16. Springer, Cham (2017). https://doi.org/10.1007/978-3-319-55209-5_1
7. Javed, A., Lee, B.S.: Hybrid semantic clustering of hashtags. Online Soc. Netw. Media **5**, 23–36 (2018)
8. Li, T., Wu, Y., Zhang, Y.: Twitter hash tag prediction algorithm. In: Proceedings on the International Conference on Internet Computing (ICOMP), pp. 1–5 (2011)
9. Lin, J., Keogh, E., Wei, L., Lonardi, S.: Experiencing sax: a novel symbolic representation of time series. Data Min. Knowl. Discov. **15**, 107–144 (2007)
10. Paparrizos, J., Gravano, L.: Fast and accurate time-series clustering. ACM Trans. Database Syst. (TODS) **42**, 1–49 (2017)
11. Rodriguez, A., Laio, A.: Clustering by fast search and find of density peaks. Science **344**, 1492–1496 (2014)
12. Stilo, G., Velardi, P.: Hashtag sense clustering based on temporal similarity. Comput. Linguist. **43**, 181–200 (2017)

13. Tsur, O., Littman, A., Rappoport, A.: Efficient clustering of short messages into general domains. In: Proceedings of the 7th International Conference on Weblogs and Social Media, ICWSM 2013, pp. 621–630 (2013)
14. Wang, Y., Liu, J., Qu, J., Huang, Y., Chen, J., Feng, X.: Hashtag graph based topic model for tweet mining. In: 2014 IEEE International Conference on Data Mining (ICDM), pp. 1025–1030. IEEE (2014)
15. Yang, J., Leskovec, J.: Patterns of temporal variation in online media. In: Proceedings of the Fourth ACM International Conference on Web Search and Data Mining, pp. 177–186. ACM (2011)
16. Zhong, Z., Zhang, Y., Pang, J.: A graph-based approach to explore relationship between hashtags and images. In: Cheng, R., Mamoulis, N., Sun, Y., Huang, X. (eds.) WISE 2020. LNCS, vol. 11881, pp. 473–488. Springer, Cham (2019). https://doi.org/10.1007/978-3-030-34223-4_30

Nonnegative Residual Matrix Factorization for Community Detection

Yulong Pei[1], Cong Liu[2(✉)], Chuanyang Zheng[3], and Long Cheng[4]

[1] Eindhoven University of Technology, Eindhoven, Netherlands
y.pei.1@tue.nl
[2] Shandong University of Technology, Zibo, China
liucongchina@163.com
[3] Hong Kong Baptist University, Kowloon Tong, Hong Kong
17251311@life.hkbu.edu.hk
[4] Dublin City University, Dublin, Ireland
long.cheng@dcu.ie

Abstract. Community detection is one of the most important and challenging problems in graph mining and social network analysis. Nonnegative Matrix Factorization (NMF) based methods have been proved to be effective in the task of community detection. However, real-world networks could be noisy and existing NMF based community detection methods are sensitive to the outliers and noise due to the utilization of the squared loss function to measure the quality of graph regularization and network reconstruction. In this paper, we propose a framework based on the nonnegative residual matrix factorization (NRMF) to overcome this limitation. In this method, a residual matrix, represented by the matrix reconstruction error, is explicitly introduced to capture the impact of outliers and noise. The residual matrix should be sparse intuitively so some sparse regularization can be used to model the sparsity. Specifically, three different types of sparse regularization, i.e., L_0, L_1 and $L_{2,1}$, have been studied. Multiplicative update rules and different thresholding operators are used to learn these lower-rank matrices. Extensive experiments on benchmark networks with and without known communities demonstrate that our framework is more robust so that it outperforms state-of-the-art NMF based approaches in community detection task.

Keywords: Nonnegative residual matrix factorization · Community detection · L_p regularization

1 Introduction

Network data is ubiquitous in our daily life, for example, social networks, road networks, and Internet networks. Analyzing these networks is of both theoretical and practical values. Recent years have witnessed numerous network analysis tasks. Among these tasks community detection is one of the most important and challenging problems. A community can be defined as a group of users that

© Springer Nature Switzerland AG 2020
Z. Huang et al. (Eds.): WISE 2020, LNCS 12342, pp. 196–209, 2020.
https://doi.org/10.1007/978-3-030-62005-9_15

(1) interact with each other more frequently than with those outside the group and (2) are more similar to each other than to those outside the group [15]. The research on community detection is beneficial for a variety of real-world applications such as online marketing and recommendation systems.

A variety of approaches have been proposed to solve the problem of community detection in different types of networks, e.g., homogeneous networks, attributed networks, and heterogeneous networks. More details about community detection can be found in the survey paper [6]. Among these methods, Nonnegative Matrix Factorization (NMF) based methods have attracted increasing attention since it has been proved to be effective in detecting communities and has powerful interpretability in clustering. NMF based community detection approaches identify the hidden communities by decomposing the adjacency matrix into lower-rank matrices [22,24]. From the perspective of statistics, NMF can be viewed as probabilistic [10] and Bayesian models [20] by factorizing the input data into distributions instead of real-value matrices. Nonnegative matrix tri-factorization (NMTF), which extends the standard NMF, factorizes the input matrix into three low-rank matrices so that it can provide more information about the interaction between lower-rank representations. Therefore, it has been utilized to community detection task to identify communities and learn community interaction simultaneously [15,28]. However, real-world networks could be noisy. Existing NMF based community detection methods are sensitive to the outliers and noise due to the utilization of the squared loss function to measure the quality of graph regularization and network reconstruction [7]. The community detection performance may be degraded by the inevitable noise.

To deal with this issue, we propose a framework based on the nonnegative residual matrix factorization (NRMF) in this study. A residual matrix, represented by the matrix reconstruction error, is explicitly introduced to capture the impact of outliers and noise. The residual matrix should be sparse intuitively so some sparse regularization can be used to model the sparsity. Three different types of norms, i.e., L_0, L_1 and $L_{2,1}$, have been studied for the sparsity modeling. To optimize the *NRMF* framework with different types of regularization, multiplicative update rules [11] are used to learn the lower-rank matrices and different thresholding operators are used to learn the residual matrix corresponding to different regularization types. To evaluate the performance of the proposed *NRMF*, we conduct community detection experiments on two types of real-world networks from different domains: networks with and without known communities.

Our contributions can be summarized as follows:

- We propose a nonnegative residual matrix factorization based framework *NRMF* to detect communities. A residual matrix, which is represented by the matrix reconstruction error, is explicitly introduced to model the impact of outliers and noise. The residual matrix is regularized to capture the sparsity.

– We explore three different regularization types to model the sparsity, i.e., L_0, L_1 and $L_{2,1}$ norms, on the residual matrix. We also propose the multiplicative update rules and different thresholding operators to optimize the objectives.
– We evaluate the effectiveness of *NRMF* using several real-world networks. The experimental results demonstrate that our method outperforms state-of-the-art NMF based methods in community detection.

The rest of the paper is organized as follows. Section 2 provides an overview of the related work. Notations and problem formulation are given in Sect. 3. Section 4 explains the proposed *NRMF* framework. In Sect. 5 we then discuss our experimental study. Finally, in Sect. 6 we draw conclusions and outline directions for future work.

2   Related Work

Communities are groups of vertices which probably share common properties and/or play similar roles within the graph [6]. Traditional community detection methods aim to partition nodes into different groups such that the number of edges between groups is minimal. For example, cut-based graph partition [9], k-core decomposition [5]. There are some methods aiming at explore the graph structures. Modularity maximization aims to find communities which can optimize a predefined measures [13]. Spectral graph clustering makes use of the eigenvectors of Laplacian matrices to group nodes [6].

Recently, NMF-based clustering approaches have also been applied in community detection. NMF as well as NMTF techniques have been used for community detection in networks in [22] where it aims to factorize the adjacency matrix of the given network. Bayesian version of NMF for community detection has been proposed in [19] to identify overlapping communities. Since NMTF explicitly models data interactions through an extra latent factor, it provides better interpretability and then has been employed in community detection. NMTF with specific bounds has been proposed in [28] to detect overlapping communities. NMTF with graph regularization has been used in [15] to discovery communities in attributed social networks. REACT [17] employs two NMTF components for community detection and role discovery and a regularization component for modeling the relations between communities and roles. Deep Autoencoder-like NMF [26] consists of encoder and decoder, where the encoder component attempts to transform the original network into the community membership space and the decoder seeks to reconstruct the original network from the community membership space with the aid of the hierarchical mappings learnt in the encoder component. Local community detection problem has been studied in [8]. Community detection in multi-layer networks using NMF has been studied in [12]. To autonomously determine the number for communities, an adaptive NMF model has been proposed in [25].

With the rapid development of deep learning techniques, community preserving network embedding approaches which are based on deep learning have attracted enormous attention from machine learning and graph mining communities recently. Existing network embedding methods have reported promising results in community detection. For example, M-NMF [23] exploits the consensus relationship between the representations of nodes and community structure, and then jointly optimize NMF based representation learning model and modularity based community detection model. Community Embedding (ComE) [3] is an embedding based method for joint node embedding and community detection. DNGE [16] uses Gaussian embedding to detect communities in dynamic networks.

Due to the space limitation, we refer the reader to [6] for more details in community detection research.

3 Preliminary

We first summarize the notations used in this study in Table 1 and then introduce the backgrounds of the techniques we will use in this paper.

Table 1. Summary of the notations.

Notation	Description
n	Number of nodes.
e	Number of edges.
c	Number of communities.
$A_{n \times n}$	Adjacency matrix of the given network.
$C_{n \times c}$	Community membership matrix.
$S_{n \times n}$	Residual matrix.
$M_{c \times c}$	Community interaction matrix.
$L_{n \times n}$	Graph Laplacian matrix.
α	Trade-off parameter for the residual matrix.
β	Trade-off parameter for graph regularization.

Data could be noisy and there may exist some entries of the data corrupted arbitrarily. To capture such noisy information in the framework of NMF, in [7,18] a sparse error matrix S has been introduced to capture the sparse corruption. Thus, a robust factorization to approximate the input data matrix A can be defined as:

$$A \approx UV^T + S, \tag{1}$$

where S is supposed to be sparse. To guarantee the sparsity, [18] proposed to use L_0 norm to regularize S and [7] utilized L_1 norm as the sparseness constraint. Therefore, the objective function can be defined as

$$\min_{U,V,S} \|A - UV^T - S\|_F^2. \tag{2}$$

Then, we will introduce the formal definitions of different norms:

- L_0 **Norm**: L_0 norm is not really a norm and it is defined as the number of nonzero elements in the given matrix.
- L_1 **Norm**: L_1 norm is the sum of the absolute values of the columns. Given a matrix $A_{n \times n}$, formally

$$L_1(A) = \max_{1 \le j \le n} \sum_{i=1}^{n} |A_{ij}|. \tag{3}$$

L_1 Norm is robust to noise and outliers because it considers the sparsity in rows.
- $L_{2,1}$ **Norm**: $L_{2,1}$ Norm combines L_1 and L_2 norms. Given a matrix $A_{n \times n}$, $L_{2,1}(A)$ is defined as

$$L_{2,1}(A) = \sum_{i=1}^{N} \left(\sum_{j=1}^{N} |A_{ij}|^2 \right)^{1/2}. \tag{4}$$

$L_{2,1}$ norm controls the capacity of A and also ensures A to be sparse in rows so it is robust to noise and outliers.

4   The Proposed Method

In this paper, motivated by the robust NMF [7,18], we propose a general framework, which is less sensitive to noise and outliers, to detect communities. The framework can be formulated as:

$$\min_{C,M,S} \|A - CMC^T - S\|_F^2 + \alpha \cdot \|S\|_p + \beta \cdot Tr(C^T LC), \tag{5}$$

$$s.t. \ C \ge 0, M \ge 0, C^T C = I,$$

where $Tr(V)$ denotes the trace of the matrix V. L is the graph Laplacian and defined as $L = D - A$ where $D_{ii} = \sum_l A_{il}$. This framework is flexible to incorporate different types of sparse regularization and constraints. It is worth noting that:

- In this work we only consider undirected networks where the adjacency matrix A is symmetric. This framework can be extended to directed networks straightforwardly by changing the first term to $\|A - CMB^T - S\|_F^2$.
- Empirically we find that adding the interaction matrix M and orthogonal constraint would achieve worse performance so in the following discussion we will simplify the objective function as:

$$\min_{C,M,S} \|A - CC^T - S\|_F^2 + \alpha \cdot \|S\|_p + \beta \cdot Tr(C^T LC), \tag{6}$$

$$s.t. \ C \ge 0, M \ge 0,$$

- In this work, we exploit three different norms, L_0, L_1 and $L_{2,1}$, to capture the reconstruction error, i.e., q is set to be 0, 1 and 2,1, respectively.

4.1 Optimization

The objective function in Eq. (6) is not convex for all parameters simultaneously. We use the multiplicative update rules to solve this optimization problem due to its good compromise between speed and ease of implementation [11]. We will optimize the objective with respect to one variable while fixing the other variables.

The update rules for matrices C and M (since we will not consider M we can simply set $M = I$) are similar to the standard NMF, which is defined as:

$$C \leftarrow C \circ \frac{(A - S)CM + \beta AC}{CMC^T CM + \beta DC} \tag{7}$$

$$M \leftarrow M \circ \frac{C(A - S)C}{C^T CMCC^T} \tag{8}$$

where $\circ$ denotes the element-wise product.

The update rules for the residual matrix S is different from different norms.

L_0 **Regularized Residual.** For L_0 norm, a hard-thresholding update rule has been used in [18]. Formally, it is define as:

$$S_{ij} = \begin{cases} 0 & \text{if } |R_{ij}| \leq \alpha \\ R_{ij} & \text{otherwise} \end{cases} \tag{9}$$

where $R_{ij} = (A - CC^T)_{ij}$.

L_1 **Regularized Residual.** For L_1 norm, a soft-thresholding update rule has been used in [7]. Formally, it is define as:

$$S_{ij} = \begin{cases} 0 & \text{if } |R_{ij}| \leq \frac{\alpha}{2} \\ R_{ij} - \frac{\alpha}{2} sign(R_{ij}) & \text{otherwise} \end{cases} \tag{10}$$

where $R_{ij} = (A - CC^T)_{ij}$, $sign(R_{ij})$ is the sign function: if $R_{ij} > 0$, $sign(R_{ij}) = 1$; if $R_{ij} < 0$, $sign(R_{ij}) = -1$; otherwise $sign(R_{ij}) = 0$.

$L_{2,1}$ **Regularized Residual.** $L_{2,1}$ norm is superior to L_0 and L_1 norm because: (1) it can model the sparsity effectively, and (2) it has a simple and efficient solution to solve the optimization problem [14]. Based on [14], we have the update rule for $L_{2,1}$ regularized S as:

$$S \leftarrow S \circ \frac{A - CC^T}{S + \alpha DS}, \tag{11}$$

where D is the diagonal matrix with the j-th diagonal element which is defined as:

$$D_{jj} = \frac{1}{\|S_j\|_2} \tag{12}$$

Algorithm 1. Optimization Algorithm

Input: Adjacency matrix A, number of communities c, trade-off parameter α and β, regularization type q
Output: Community membership matrix C, community interaction matrix M and residual matrix S
1: Initialize C and S
2: **while** not converge **do**
3:     Calculate graph Laplacian L
4:     Update C according to Eq. (7)
5:     Update M according to Eq. (8)
6:     **if** $q = 0$ **then**
7:         Update S according to Eq. (9)
8:     **else if** $q = 1$ **then**
9:         Update S according to Eq. (10)
10:     **else if** $q = 2, 1$ **then**
11:         Update S according to Eq. (11)
12:     **end if**
13: **end while**

4.2   Computational Complexity

Computational complexity. For simplicity, given two matrices $M_{n \times r}$ and $N_{r \times f}$, the computational complexity of the multiplication of M and N is $O(nrf)$. The complexity of updating rules in Algorithm 1 (Line 3–5) is $O(n^2c + nc^2 + n^3)$. To update S (Line 6–12), the complexity for L_0 and L_1 is $O(n^2 + nc^2)$ and for $L_{2,1}$ is $O(nc^2 + n^3)$. By taking the number of iteration i into consideration, the complexity is $O(i(n^2c + nc^2 + n^3))$.

5   Experiments

To validate the effectiveness of *NRMF* in community detection, we conduct experiments on two types of real-world networks from different domains, i.e., networks with and without known communities. To further exploit the influence of different regularization terms in sparsity modeling, we compare *NRMF* with L_0, L_1 and $L_{2,1}$ regularization.

5.1   Datasets and Evaluation Metrics

We conduct experiments on two types of data: networks with and without known communities. For networks with known communities, we select Football, Email, and Wiki networks. There are 5, 42 and 19 communities in Football, Email and Wiki networks[1], respectively. For networks without known communities, we select Jazz, Email-Univ and Hamsterster networks. The optimal number of

---

[1] These datasets are from http://www-personal.umich.edu/~mejn/netdata/, http://snap.stanford.edu/data/index.html and https://linqs.soe.ucsc.edu/data.

communities for each network is inferred using Louvain algorithm [1]. A brief summary of these datasets is shown in Table 2. We employ purity and normalized mutual information (NMI) as the evaluation metrics for networks with known communities. Purity measures the extent to which each cluster contained data points from primarily one class. The purity of a clustering is obtained by the weighted sum of individual cluster purity values which defined as:

$$Purity = \frac{1}{N} \sum_{i=1}^{k} max_j |c_i \cap t_j|, \tag{13}$$

where N is number of objects, k is number of clusters, c_i is a cluster in C, and t_j is the classification which has the max count for cluster c_i. NMI evaluates the clustering quality based on information theory, and is defined by normalization on the mutual information between the cluster assignments and the pre-existing input labeling of the classes:

$$NMI(\mathcal{C}, \mathcal{D}) = \frac{2 * \mathcal{I}(\mathcal{C}, \mathcal{D})}{\mathcal{H}(\mathcal{C}) + \mathcal{H}(\mathcal{D})}, \tag{14}$$

where obtained cluster $\mathcal{C}$ and ground-truth cluster $\mathcal{D}$. The mutual information $\mathcal{I}(\mathcal{C}, \mathcal{D})$ is defined as $\mathcal{I}(\mathcal{C}, \mathcal{D}) = \mathcal{H}(\mathcal{C}) - \mathcal{H}(\mathcal{C}|\mathcal{D})$ and $\mathcal{H}(\cdot)$ is the entropy.

For networks without known communities we use modularity as the evaluation metric. Formally, modularity is defined as:

$$Q = \frac{1}{2m} \sum_{ij} \left(A_{ij} - \frac{k_i k_j}{2m} \right) \delta(c_i, c_j) \tag{15}$$

where m is the number of edges, A is the adjacency matrix of the input graph, k_i is the degree of node i and $\delta(c_i, c_j)$ is 1 if node i and node j are in the same community and 0 otherwise.

Table 2. Summary of data sets used in the experiments where n, e and c denote the number of nodes, edges, and communities, respectively.

Data		n	e	c
Known communities	Football	115	613	5
	Email	1005	25571	42
	Wiki	2405	17981	19
Unknown communities	Jazz	198	2742	4
	Email-Univ	1133	5451	11
	Hamsterster	1858	12534	32

5.2   Baseline Methods

To demonstrate our model detects communities more effectively than other NMF based community detection approaches, we select several representative NMF based methods as our baselines:

Table 3. Community detection performance w.r.t. Purity. Best performance is in bold font.

	Football	Email	Wiki
NMF	0.9074	0.5834	0.2025
ONMF	0.8754	0.5734	0.3177
PNMF	0.9153	0.6002	0.2531
BNMF	0.9067	0.5587	0.3497
GNMF	0.9143	0.5675	0.2561
BigClam	0.9209	0.6358	0.2995
NSED	0.8994	0.6072	0.2773
$NRMF\text{-}L_0$	0.8914	0.5801	0.2880
$NRMF\text{-}L_1$	**0.9335**	0.6843	0.3774
$NRMF\text{-}L_{2,1}$	**0.9335**	**0.7021**	**0.4154**

- NMF [22]: NMF is the basic matrix factorization framework. To make a fair comparison, the objective function for NMF is $\|A - CC^T\|_F^2$.
- Orthogonal NMF (ONMF) [4]: ONMF is a variant of NMF by enforcing orthogonal constraints on the community membership matrix C, i.e., $C^T C = I$. In particular, since we add contraint to C, the objective is $\|A - CMC^T\|_F^2$.
- Projected NMF (PNMF) [27]: PNMF directly projects the original network to a subspace by minimizing $\|A - CC^T A\|_F^2$.
- Bayesian NMF (BNMF) [20]: BNMF is a bayesian NMF model. It models the matrix factorization in a Bayesian fashion.
- Graph regularized NMF (GNMF) [2]: GNMF incorporates an affinity graph which is constructed to encode the geometrical information to NMF, and then seeks a matrix factorization which respects the graph structure.
- BigClam [24]: BigClam is a cluster affiliation model. It relaxes the graph fitting problem into a continuous optimization problem to find overlapping communities.
- Nonnegative Symmetric Encoder-dedcoder (NSED) [21]: NSED is a nonnegative symmetric encoder-decoder approach proposed for community detection. It extracts the community membership by integrating a decoder and an encoder into a unified loss function.

For our framework, we compare three variants: L_0 regularized NRMF ($NRMF\text{-}L_0$), L_1 regularized NRMF ($NRMF\text{-}L_1$) and $L_{2,1}$ regularized NRMF ($NRMF\text{-}L_{2,1}$).

5.3   Experimental Results

The experimental results on networks with known communities are shown in Table 3 based on Purity and Table 4 based on NMI. Based on these results, it can be observed that:

- $NRMF$-$L_{2,1}$ achieves the best performance in community detection based on both the purity and NMI metrics. It indicates $L_{2,1}$ regularization can better capture the sparsity of the residual matrix compared to other types of regularization such as L_0 and L_1.
- $NRMF$ methods with L_1 and $L_{2,1}$ sparse regularization outperform other NMF based approaches. It demonstrates that our proposed framework can effectively identify the communities because it explicitly takes noise into consideration when factorizing the input adjacency matrix.
- $NRMF$ methods with L_0 performs worse than some other NMF based methods. It shows that L_0 norm is not a good choice for sparsity modeling. This may because L_0 norm only considers the number of nonzero elements but ignore the values of these elements.
- It is interesting to observe that ONMF and GNMF achieve worse performance than other NMF based methods including the standard NMF. This may result from that real-world network data is very sparse and constraints on the latent representations may be difficult to achieve.

The experimental results on networks without known communities are shown in Table 5 based on Modularity. From these results, we can draw the same conclusions to that on the networks with known communities. Since modularity is a measure only based on the network structure (ground-truth community labels may be related not only to structures but also semantics), good performance on modularity further demonstrate the effectiveness of our framework from the structural perspective.

Table 4. Community detection performance w.r.t. NMI. Best performance is in bold font.

	Football	Email	Wiki
NMF	0.8973	0.6234	0.1876
ONMF	0.8843	0.6451	0.2108
PNMF	0.9035	0.6211	0.2404
BNMF	0.9142	0.6021	0.2411
GNMF	0.9038	0.5893	0.1976
BigClam	0.8963	0.5796	0.2722
NSED	0.9096	0.6845	0.2659
$NRMF$-L_0	0.8903	0.6746	0.2235
$NRMF$-L_1	0.9214	0.6746	0.2626
$NRMF$-$L_{2,1}$	**0.9233**	**0.7032**	**0.2737**

5.4   Parameter Sensitivity

The *NRMF* involves two parameters shown in Eq. (5): α controls the trade-off of redisual modeling and β controls the graph regularization. In this section, we examine how the different choices of parameters affect the performance of *NRMF* in community detection. Specifically, we measure the NMI and Purity on Email data and Modularity on Email-Univ data. The results are shown in Figs. 1, 2 and 3. Note that only *NRMF* with $L_{2,1}$ is evaluated because it achieves the best performance in the experiments above.

It can be observed from these results that:

- Large α is preferred in order to achieve better NMI and Purity in detecting communities. In specific, when α is 0.6–0.7, the performance is the best.
- In contrast, smaller β brings better community detection performance. In the experiments, $\beta = 0.1$ is the best choice w.r.t. both NMI and Purity.

Table 5. Community detection performance w.r.t. Modularity. Best performance is in bold font.

	Jazz	Email-Univ	Hamsterster
NMF	0.4136	0.2876	0.2347
ONMF	0.4221	0.2751	0.2378
PNMF	0.4201	0.4015	0.1313
BNMF	0.4161	0.3963	0.1226
GNMF	0.1852	0.2747	0.2060
BigClam	0.4005	0.4024	0.2796
NSED	0.4231	0.3987	0.2553
NRMF-L_0	0.3843	0.3369	0.1892
NRMF-L_1	0.4355	0.4448	0.2856
NRMF-$L_{2,1}$	**0.4463**	**0.5011**	**0.2903**

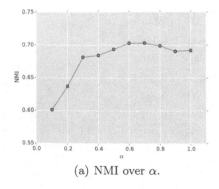
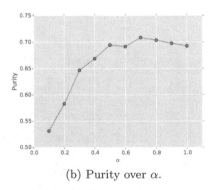

(a) NMI over α.   (b) Purity over α.

Fig. 1. Parameter sensitivity of α on Email data.

– For the networks without ground-truth labels, the choices of α and β are different. In Email-Univ, smaller α and larger β give better performance. In this specific experiment, $\alpha = 0.1$ and $\beta = 0.8$ would be the best parameters.

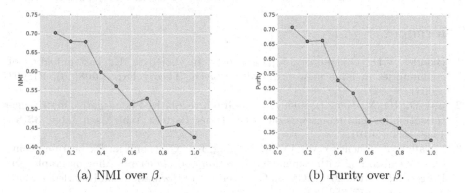

(a) NMI over β.   (b) Purity over β.

Fig. 2. Parameter sensitivity of β on Email data.

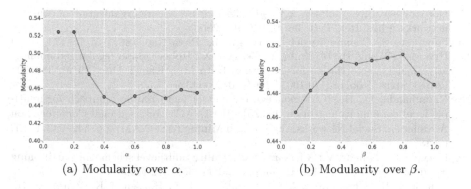

(a) Modularity over α.   (b) Modularity over β.

Fig. 3. Parameter sensitivity results on Email-Univ data.

6   Conclusion

We proposed *NRMF*, a novel framework using nonnegative residual matrix factorization for community detection, that addresses the limitation of existing NMF based community detection approaches: being sensitive to noise and outliers. In *NRMF*, a residual matrix has been introduced explicitly to capture the noise. To model the sparsity of the *NRMF*, we exploit three types of regularization, i.e., L_0, L_1 and $L_{2,1}$. To optimize the objective w.r.t different regularization terms, different updating rules have been employed. Our experimental study demonstrated that *NRMF* effectively preserves community structures and captures noisy information, outperforming state-of-the-art NMF based methods in community detection.

On the basis of *NRMF*, several new research lines can be pursued. For example, it is interesting to exploit more advanced optimization method, e.g., ADMM, or extend it to attributed networks. We leave these extensions for future work.

References

1. Blondel, V.D., Guillaume, J.L., Lambiotte, R., Lefebvre, E.: Fast unfolding of communities in large networks. J. Stat. Mech. Theory Exp. **2008**(10), P10008 (2008)
2. Cai, D., He, X., Han, J., Huang, T.S.: Graph regularized nonnegative matrix factorization for data representation. IEEE Trans. Pattern Anal. Mach. Intell. **33**(8), 1548–1560 (2010)
3. Cavallari, S., Zheng, V.W., Cai, H., Chang, K.C.C., Cambria, E.: Learning community embedding with community detection and node embedding on graphs. In: Proceedings of the 2017 ACM on Conference on Information and Knowledge Management, pp. 377–386 (2017)
4. Ding, C., Li, T., Peng, W. and Park, H.: Orthogonal nonnegative matrix t-factorizations for clustering. In: Proceedings of the 12th ACM SIGKDD International Conference on Knowledge Discovery and Data Mining, pp. 126–135. ACM (2006)
5. Dorogovtsev, S.N., Goltsev, A.V., Mendes, J.F.F.: K-core organization of complex networks. Phys. Rev. Lett. **96**(4), 040601 (2006)
6. Fortunato, S.: Community detection in graphs. Phys. Rep. **486**(3–5), 75–174 (2010)
7. Huang, S., Wang, H., Li, T., Li, T., Zenglin, X.: Robust graph regularized nonnegative matrix factorization for clustering. Data Min. Knowl. Disc. **32**(2), 483–503 (2018). https://doi.org/10.1007/s10618-017-0543-9
8. Kamuhanda, D., He, K.: A nonnegative matrix factorization approach for multiple local community detection. In: 2018 IEEE/ACM International Conference on Advances in Social Networks Analysis and Mining (ASONAM), pp. 642–649. IEEE (2018)
9. Karypis, G., Kumar, V.: A fast and high quality multilevel scheme for partitioning irregular graphs. SIAM J. Sci. Comput. **20**(1), 359–392 (1998)
10. Laurberg, H., Christensen, M.G., Plumbley, M.D., Hansen, L.K., Jensen, S.H.: Theorems on positive data: on the uniqueness of NMF. Comput. Intell. Neurosci. **2008**, (2008)
11. Lee, D.D., Seung, H.S.: Algorithms for non-negative matrix factorization. In: Advances in Neural Information Processing Systems, pp. 556–562 (2001)
12. Ma, X., Dong, D., Wang, Q.: Community detection in multi-layer networks using joint nonnegative matrix factorization. IEEE Trans. Knowl. Data Eng. **31**(2), 273–286 (2018)
13. Newman, M.E.J.: Modularity and community structure in networks. Proc. Int. Acad. Sci. **103**(23), 8577–8582 (2006)
14. Nie, F., Huang, H., Cai, X. and Ding, C.H.: Efficient and robust feature selection via joint $\ell2, 1$-norms minimization. In: Advances in neural information processing systems, pp. 1813–1821 (2010)
15. Pei, Y., Chakraborty, N., Sycara, K.: Nonnegative matrix tri-factorization with graph regularization for community detection in social networks. In: Twenty-Fourth International Joint Conference on Artificial Intelligence (2015)

16. Pei, Y., Du, X., Fletcher, G., Pechenizkiy, M.: Dynamic network representation learning via gaussian embedding. In: NeurIPS 2019 Workshop on Graph Representation Learning (2019)
17. Pei, Y., Fletcher, G., Pechenizkiy, M.: Joint role and community detection in networks via l 2, 1 norm regularized nonnegative matrix tri-factorization. In: Proceedings of the 2019 IEEE/ACM International Conference on Advances in Social Networks Analysis and Mining, pp. 168–175 (2019)
18. Peng, C., Kang, Z., Yunhong, H., Cheng, J., Cheng, Q.: Robust graph regularized nonnegative matrix factorization for clustering. ACM Trans. Knowl. Discov. Data (TKDD) 11(3), 33 (2017)
19. Psorakis, I., Roberts, S., Ebden, M., Sheldon, B.: Overlapping community detection using Bayesian non-negative matrix factorization. Phys. Rev. E 83(6), 066114 (2011)
20. Schmidt, M.N., Winther, O., Hansen, L.K.: Bayesian non-negative matrix factorization. In: Adali, T., Jutten, C., Romano, J.M.T., Barros, A.K. (eds.) ICA 2009. LNCS, vol. 5441, pp. 540–547. Springer, Heidelberg (2009). https://doi.org/10.1007/978-3-642-00599-2_68
21. Sun, B.J., Shen, H., Gao, J., Ouyang, W. and Cheng, X.: A non-negative symmetric encoder-decoder approach for community detection. In: Proceedings of the 2017 ACM on Conference on Information and Knowledge Management, pp. 597–606 (2017)
22. Wang, F., Li, T., Wang, X., Zhu, S., Ding, C.: Community discovery using nonnegative matrix factorization. Data Min. Knowl. Discov. 22(3), 493–521 (2011)
23. Wang, X., Cui, P., Wang, J., Pei, J., Zhu, W., Yang, S.: Community preserving network embedding. In: Thirty-first AAAI Conference on Artificial Intelligence (2017)
24. Yang, J., Leskovec, J.: Overlapping community detection at scale: a nonnegative matrix factorization approach. In: Proceedings of the Sixth ACM International Conference on Web Search and Data Mining, pp. 587–596. ACM (2013)
25. Yang, L.: Autonomous semantic community detection via adaptively weighted low-rank approximation. ACM Trans. Multimedia Comput. Commun. Appl. (TOMM) 15(3s), 1–22 (2019)
26. Ye, F., Chen, C., Zheng, Z.: Deep autoencoder-like nonnegative matrix factorization for community detection. In: Proceedings of the 27th ACM International Conference on Information and Knowledge Management, pp. 1393–1402 (2018)
27. Yuan, Z., Oja, E.: Projective nonnegative matrix factorization for image compression and feature extraction. In: Kalviainen, H., Parkkinen, J., Kaarna, A. (eds.) SCIA 2005. LNCS, vol. 3540, pp. 333–342. Springer, Heidelberg (2005). https://doi.org/10.1007/11499145_35
28. Zhang, Y., Yeung, D.Y.: Overlapping community detection via bounded nonnegative matrix tri-factorization. In: Proceedings of the 18th ACM SIGKDD International Conference on Knowledge Discovery and Data Mining, pp. 606–614. ACM (2012)

Graph Query

Fast Algorithm for Distance Dynamics-Based Community Detection

Maram Alsahafy and Lijun Chang$^{(\boxtimes)}$

University of Sydney, Sydney, Australia
mals4485@uni.sydney.edu.au, lijun.chang@sydney.edu.au

Abstract. Community detection is a fundamental problem in graph-based data analytics. Among many models, the distance dynamics model proposed recently is shown to be able to faithfully capture natural communities that are of different sizes. However, the state-of-the-art algorithm Attractor for distance dynamics does not scale to graphs with large maximum vertex degrees, which is the case for large real graphs. In this paper, we aim to scale distance dynamics to large graphs. To achieve that, we propose a fast distance dynamics algorithm FDD. We show that FDD has a worst-case time complexity of $\mathcal{O}(T \cdot \gamma \cdot m)$, where T is the number of iterations until convergence, γ is a small constant, and m is the number of edges in the input graph. Thus, the time complexity of FDD does not depend on the maximum vertex degree. Moreover, we also propose optimization techniques to alleviate the dependency on T. We conduct extensive empirical studies on large real graphs and demonstrate the efficiency and effectiveness of FDD.

Keywords: Community detection · Distance dynamics · Power-law graphs

1 Introduction

Graph representation has been playing an important role in modelling and analyzing the data from real applications such as social networks, communication networks, and information networks. In the graph representation of these applications, community structures naturally exist [8], where entities/vertices in the same community are densely connected and entities/vertices from different communities are sparsely connected. For example, in social networks, users in the same community share similar characteristics or interests.

In view of the importance of community structures, many approaches have been proposed in the literature for identifying communities, *e.g.*, based on betweenness centrality [9], normalized cut [17], or modularity [12]. The betweenness centrality-based approach [9] divides a graph by iteratively removing from the graph the edge that has the largest betweenness centrality (and thus likely to be cross-community edges). The result is a dendrogram that compactly encodes

© Springer Nature Switzerland AG 2020
Z. Huang et al. (Eds.): WISE 2020, LNCS 12342, pp. 213–226, 2020.
https://doi.org/10.1007/978-3-030-62005-9_16

the division process, which is then post-processed based on the modularity measure [13] to identify the best division point. In addition, heuristic algorithms, *e.g.*, the Greedy algorithm [12] and the Louvain algorithm [3], have also been developed to directly optimize the modularity measure which results in an NP-hard optimization problem [4]. In particular, the Louvain algorithm has gained attention recently due to its low time complexity and fast running time. However, it is known that optimizing the modularity measure suffers from the resolution limit [8], *i.e.*, small communities cannot be identified as they are forced to join other communities to maximize the modularity.

The distance dynamics model was recently proposed in [16] to resolve the resolution limit, which does not optimize any specific qualitative measure. Instead, it envisions a graph as an adaptive dynamic system and simulates the interaction among vertices over time, which leads to the discovery of qualified communities. Specifically, each vertex interacts with its neighbors (*i.e.*, adjacent vertices) such that the distances among vertices in the same community tend to decrease while those in different communities increase; thus, the interaction information among vertices transfers through the graph. Finally, the distances will converge to either 0 or 1, and the graph naturally splits into communities by simply removing all edges with distance 1. It is shown in [16] that the distance dynamics model is able to extract large communities as well as small communities. However, the state-of-the-art algorithm Attractor [16] does not scale to graph with large maximum vertex degrees, while large real graphs are usually power-law graphs with large maximum vertex degrees [2]. Note that, Attractor is claimed in [16] to run in approximately $\mathcal{O}(m + k \cdot m + T \cdot m)$ time, where m is the number of edges in the input graph, T is the number of iterations until convergence, and k is the average number of exclusive neighbors for all pairs of adjacent vertices. However, we show in Sect. 3 that this is inaccurate, and in fact the time complexity of Attractor highly depends on deg_{max}—the maximum vertex degree of the input graph—which prevents Attractor from scaling to graphs with large maximum vertex degrees.

In this paper, we design a fast distance dynamics algorithm FDD to scale distance dynamics to large graphs. We first propose efficient techniques to compute the initial distances for all relevant vertex pairs in $\mathcal{O}(deg_{max} \cdot m)$ time, which is worst-case optimal. We then propose two optimization techniques to improve the efficiency of distance updating by observing that (1) converged vertex pairs can be excluded from the computation and (2) not the distances of all relevant vertex pairs need to be computed. Finally, we reduce the total time complexity to $\mathcal{O}(T \cdot \gamma \cdot m)$, where γ is a small constant. As a result, FDD can efficiently process large real graphs that have large maximum vertex degrees. Note that, our optimization techniques also alleviate the dependency on T such that FDD is more likely to run in $\mathcal{O}(\gamma \cdot m)$ time in practice.

Contributions. We summarize our main contributions as follows.

- We analyze the time complexity of the state-of-the-art distance dynamics algorithm Attractor, and show that it highly depends on deg_{max}. (Section 3)

– We design a fast distance dynamics algorithm FDD by proposing efficient initialization, optimization, and time complexity reduction techniques. (Section 4)
– We conduct extensive empirical studies on large real graphs and demonstrate the efficiency and effectiveness of FDD. (Section 5)

Related Works. Besides the methods mentioned above, there are many other community detection methods such as graph partitioning [10,19], hierarchical clustering [15,20], spectral algorithms [6,7], and clique percolation for overlapping communities [14]. A comprehensive survey about community detection can be found in [8].

2   Preliminaries

In this paper, we focus on undirected and unweighted graphs $G = (V, E)$, where V represents the set of vertices and E represents the set of edges. We use n and m to denote the sizes of V and E, respectively. The edge between vertices u and v is denoted by $(u, v) \in E$. The set of neighbors of u is $N(u) = \{v \in V \mid (u, v) \in E\}$, and the degree of u is $deg(u) = |N(u)|$. The closed neighborhood $N[u]$ of u is the union of $\{u\}$ and its neighbors, $i.e.$, $N[u] = \{u\} \cup N(u)$; note that, $deg(u) = |N[u]| - 1$. In the remaining of the paper, we simply refer to closed neighborhood as neighborhood.

2.1   Distance Dynamics

The distance dynamics model is recently proposed in [16]. It consists of two stages: initial distance computation, and distance updating.

State-I: Initial Distance Computation. For each edge $(u, v) \in E$, the initial distance $d^{(0)}(u, v)$ between u and v is computed as

$$d^{(0)}(u, v) = 1 - s^{(0)}(u, v) \tag{1}$$

where $s^{(0)}(u, v)$ is the initial similarity between u and v which is measured in [16] by the Jaccard similarity between their neighborhoods, $i.e.$,

$$s^{(0)}(u, v) = \frac{|N[u] \cap N[v]|}{|N[u] \cup N[v]|} \tag{2}$$

Intuitively, the more common neighbors u and v have, the more similar they are and the smaller distance they have. It is easy to see that all the distance and similarity values range between 0 and 1.

Stage-II: Distance Updating. By Eq. (1), if the initial distance $d^{(0)}(u, v)$ is close to 0, then u and v are likely to belong to the same community; if $d^{(0)}(u, v)$ is close to 1, then u and v are likely to belong to different communities. However, if the initial distance is neither close to 0 nor close to 1, then it is hard to tell at the moment whether u and v should belong to the same community or not. To resolve this issue, distance dynamics [16] proposes to iteratively update the distance values for edges (*i.e.*, adjacent vertex pairs) based on the neighborhoods of the two end-points, until convergence (*i.e.*, become 0 or 1). Updating the distance/similarity for edge (u, v) is achieved by considering three forces: *influence from direct link* $DI(u, v)$, *influence from common neighbors* $CI(u, v)$, and *influence from exclusive neighbors* $EI(u, v)$. Specifically,

$$s^{(t+1)}(u, v) = s^{(t)}(u, v) + DI^{(t+1)}(u, v) + CI^{(t+1)}(u, v) + EI^{(t+1)}(u, v) \quad (3)$$

and

$$d^{(t+1)}(u, v) = 1 - s^{(t+1)}(u, v) \quad (4)$$

where superscript $^{(t)}$ is used to refer to the corresponding values in iteration t.

1) Influence from Direct Link. Firstly, the similarity $s(u, v)$ will increase as a result of the influence of the direct edge between u and v. That is, $DI^{(t+1)}(u, v)$ is computed as

$$DI^{(t+1)}(u, v) = \frac{f\big(s^{(t)}(u,v)\big)}{deg(u)} + \frac{f\big(s^{(t)}(u,v)\big)}{deg(v)} \quad (5)$$

where $f(\cdot)$ is a coupling function and $\sin(\cdot)$ is used in [16].

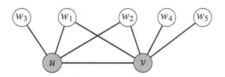

Fig. 1. Distance dynamics

2) Influence from Common Neighbors. Secondly, the similarity $s(u, v)$ will also increase as a result of the influence of the common neighbors of u and v. Let $CN(u, v) = N(u) \cap N(v)$ be the set of common neighbors of u and v; for example, $CN(u, v) = \{w_1, w_2\}$ in Fig. 1. Then, $CI^{(t+1)}(u, v)$ is computed as

$$CI^{(t+1)}(u, v) = \sum_{w \in CN(u,v)} \left(\frac{f\big(s^{(t)}(w,u)\big)}{deg(u)} \times s^{(t)}(w, v) + \frac{f\big(s^{(t)}(w,v)\big)}{deg(v)} \times s^{(t)}(w, u) \right)$$
$$(6)$$

3) Influence from Exclusive Neighbors. Thirdly, the similarity $s(u, v)$ is also affected by the influence of the **exclusive neighbors** $EN_v(u)$ of u with respect to v, and the exclusive neighbors $EN_u(v)$ of v with respect to u, where $EN_v(u) = N(u) \backslash CN(u, v)$. For example, $EN_v(u) = \{w_3\}$ and $EN_u(v) = \{w_4, w_5\}$ in Fig. 1. However, whether the influence of an exclusive neighbor w on $s(u, v)$ is positive or negative will depend on the similarity between w and the other vertex [16]. Specifically, $EI^{(t+1)}(u, v)$ is computed as

$$EI^{(t+1)}(u, v) = \sum_{w \in EN_v(u)} \left(\frac{f\big(s^{(t)}(w,u)\big)}{deg(u)} \times \rho(w, v) \right) + \sum_{w \in EN_u(v)} \left(\frac{f\big(s^{(t)}(w,v)\big)}{deg(v)} \times \rho(w, u) \right)$$
$$(7)$$

where $\rho(w, u) = s^{(0)}(w, u)$ if $s^{(0)}(w, u) \geq \lambda$, and $\rho(w, u) = s^{(0)}(w, u) - \lambda$ otherwise; note that, in the latter case, $\rho(w, u) < 0$. Here, λ is a parameter and is recommended in [16] to be in the range $[0.4, 0.6]$.

3   Time Complexity of Attractor

The Attractor algorithm, along with the distance dynamics model, is proposed in [16] for computing the distance dynamics. The pseudocode of Attractor is shown in Algorithm 1, which is self-explanatory and directly follows from the distance dynamics model in Sect. 2. It is worth pointing out that, although only the distances for edges (*i.e.*, adjacent vertex pairs) are required in the community detection and in the initial distance computation, Attractor also computes and stores $\rho(w, u)$ for vertex pairs w and u that are not directly connected but have common neighbors (see Line 2). This is because these $\rho(w, u)$ values will be used later by Eq. (7) for updating the distances.

Algorithm 1: Attractor(G) [16]

1 Compute $d^{(0)}(u, v)$ for every edge $(u, v) \in E$;
2 Compute $\rho(w, u)$ for every pair of vertices that are not directly connected but have common neighbors;
3 **while** *not converge* **do**
4   **for** *each edge* $(u, v) \in E$ **do**
5   **if** $0 < d^{(t)}(u, v) < 1$ **then**
6   Compute $d^{(t+1)}(u, v)$ from $d^{(t)}(u, v)$ by using Equations (3) – (7);

Attractor is claimed in [16] to approximately run in $\mathcal{O}(m + k \cdot m + T \cdot m)$ time, where k is the average number of exclusive neighbors for all pairs of adjacent vertices, and T is the total number of iterations (*i.e.*, t). Specifically, it is claimed that Line 1 runs in $\mathcal{O}(m)$ time, Line 2 runs in $\mathcal{O}(k \cdot m)$ time, and Lines 3–6 run in $\mathcal{O}(T \cdot m)$ time. However, we find that this analysis is not accurate, which is also evidenced by its poor performance on graphs with large maximum vertex degrees (see our experiments in Sect. 5).

Issue-1. Firstly, Line 1 computes $d^{(0)}(u, v)$ for all m edges in the graph, which cannot be conduct in $\mathcal{O}(m)$ time by assuming the triangle detection conjecture in [1]. This is because the number of triangles in a graph can be directly obtained from the values $d^{(0)}(u, v)$ of all edges in the graph in linear time (see Sect. 4.1).

Issue-2. Secondly, Line 2 computes $\rho(u, v)$ for $k \cdot m$ non-adjacent pairs of vertices. It is unlikely that this can be conducted in $\mathcal{O}(k \cdot m)$ total time.

Issue-3. Thirdly, an iteration of distance updating (Lines 4–6) may update m edges in the worst case. It is unlikely that this can be achieved in $\mathcal{O}(m)$ time for an iteration.

We illustrate Issue-3 by analyzing the time complexity of an iteration of distance updating (*i.e.*, Lines 4–6). From Eqs. (5), (6), and (7), we can see that updating $d(u, v)$ for a specific edge (u, v) takes $2|CN(u, v)| + |N_v(u)| + |N_u(v)| = deg(u) + deg(v)$ time, by assuming that the values of $\rho(w, u)$ are precomputed and each $\rho(w, u)$ value can be retrieved in constant time. Thus, the time complexity of an iteration of distance updating is $\mathcal{O}\left(\sum_{(u,v) \in E} \left(deg(u) + deg(v)\right)\right) = \mathcal{O}\left(\sum_{v \in V} \left(deg(v)\right)^2\right) = \mathcal{O}(deg_{max} \cdot m)$, where deg_{max} is the maximum vertex degree in G. Note that, $\mathcal{O}(deg_{max} \cdot m)$ cannot be replaced by $\mathcal{O}(deg_{ave} \cdot m)$, where deg_{ave} is the average vertex degree in G. In fact, $\mathcal{O}(deg_{max} \cdot m)$ can be n times larger than $\mathcal{O}(deg_{ave} \cdot m)$ in extreme cases; for example, in a star graph we have $deg_{max} = m = n - 1$ and $\sum_{v \in V}(deg(v))^2 = (n - 1) \cdot n$, while $deg_{ave} < 2$.

Moreover, as we will show in Sect. 4.1, computing $\rho(u, v)$ naively at Line 2 has an even higher time complexity than $\mathcal{O}(deg_{max} \cdot m)$. As a result, Attractor will not be able to process large real graphs as it is well known that most of the large real graphs, although have small average degrees, are usually power-law graphs with a few vertices of extremely high degrees [2]. This is also confirmed by our empirical studies in Sect. 5.

4   Fast Distance Dynamics

In this section, we design a fast distance dynamics algorithm FDD. We first propose techniques to compute $d^{(0)}(u, v)$ and $\rho(u, v)$ for all relevant vertex pairs in $\mathcal{O}(deg_{max} \cdot m)$ total time in Sect. 4.1, and then develop optimization techniques to improve the efficiency of distance updating in Sect. 4.2. Finally, we reduce the total time complexity of FDD to $\mathcal{O}(T \cdot \gamma \cdot m)$ for a small constant γ in Sect. 4.3.

4.1   Efficient Initialization

A straightforward approach for initialization (*i.e.*, computing $d^{(0)}(u, v)$ and $\rho(u, v)$ at Lines 1–2 of Algorithm 1) is to compute the values independently for all relevant vertex pairs by conducting an intersection of $N[u]$ and $N[v]$ in $deg(u) + deg(v)$ time. Then, the total time complexity of computing $d^{(0)}(u, v)$ for all edges (*i.e.*, all adjacent vertex pairs) is $\mathcal{O}\left(\sum_{(u,v) \in E} \left(deg(u) + deg(v)\right)\right) = \mathcal{O}(deg_{max} \cdot m)$. Let E_2 be the set of vertex pairs that are not directly connected but share common neighbors (*i.e.*, E_2 is the set of vertex pairs whose $\rho(u, v)$ values need to be computed), and $deg_2(u)$ be the number of vertex pairs in E_2 containing u (equivalently, the number of 2-hop neighbors of u in the graph G). Then, the total time complexity of computing $\rho(u, v)$ for all $(u, v) \in E_2$ will be $\mathcal{O}\left(\sum_{(u,v) \in E_2} \left(deg(u) + deg(v)\right)\right) = \mathcal{O}\left(\sum_{v \in V} \left(deg(v) \cdot deg_2(v)\right)\right)$. This can be much larger than $\mathcal{O}\left(\sum_{v \in V} deg(v) \cdot deg(v)\right) = \mathcal{O}(deg_{max} \cdot m)$; consider a tree where every vertex except leafs has exactly x neighbors, we have $deg_2(v) \approx deg(v) \cdot x$.

Algorithm 2: Initialization of FDD

---

1 Initialize an empty hash table $\mathcal{C}$ for storing common neighbor counts;
2 **for** *each vertex* $u \in V$ **do**
3   **for** *each pair of vertices* $\{v, w\} \subseteq N(u)$ *with* $v < w$ **do**
4   **if** $(v, w) \notin \mathcal{C}$ **then** $\mathcal{C}(v, w) \leftarrow 1$;
5   **else** $\mathcal{C}(v, w) \leftarrow \mathcal{C}(v, w) + 1$;

6 **for** *each edge* $(u, v) \in E$ **do**
7 $s^{(0)}(u, v) \leftarrow \frac{\mathcal{C}(u,v)+2}{deg(u)+deg(v)-\mathcal{C}(u,v)}$;
8 $d^{(0)}(u, v) \leftarrow 1 - s^{(0)}(u, v)$;

9 **for** *each vertex pair* $\{u, v\} \in \mathcal{C} \backslash E$ **do**
10 $s^{(0)}(u, v) \leftarrow \frac{\mathcal{C}(u,v)}{deg(u)+deg(v)-\mathcal{C}(u,v)+2}$;

In this subsection, we propose efficient techniques to compute $d^{(0)}(u, v)$ and $\rho(u, v)$ for all relevant vertex pairs (equivalently, compute $s^{(0)}(u, v)$ for all vertex pairs in $E \cup E_2$) in $\mathcal{O}(deg_{max} \cdot m)$ total time. The general idea is to incrementally count the number of common neighbors for all vertex pairs of $E \cup E_2$ simultaneously. Let $c(u, v)$ be the number of common neighbors of u and v (*i.e.*, $c(u, v) = |N(u) \cap N(v)|$). It is easy to see that we have

$$
s^{(0)}(u, v) = \begin{cases} \dfrac{c(u,v) + 2}{deg(u) + deg(v) - c(u,v)} & \text{if } (u, v) \in E \\[2ex] \dfrac{c(u,v)}{deg(u) + deg(v) - c(u,v) + 2} & \text{if } (u, v) \notin E \end{cases} \tag{8}
$$

Thus, after computing $c(u, v)$, each $d^{(0)}(u, v)$ and $\rho(u, v)$ can be calculated in constant time. In order to efficiently compute $c(u, v)$, we enumerate all wedges in the graph, where a wedge is a triple (v, u, w) such that $(u, v) \in E$ and $(u, w) \in E$. The set of all wedges can be obtained by enumerating all vertex pairs in the neighborhood of each vertex, and for each wedge (v, u, w), we increase $c(v, w)$ by 1. The pseudocode is shown in Algorithm 2, where a hash table $\mathcal{C}$ is used for efficiently accessing the counts $c(u, v)$.

It is easy to see that the time complexity of Algorithm 2 is $\mathcal{O}\left(\sum_{u \in V} deg^2(u)\right) = \mathcal{O}(deg_{max} \cdot m)$, by assuming that each access to the hash table $\mathcal{C}$ takes constant time. Note that, this time complexity (for computing $s^{(0)}(u, v)$ for all vertex pairs in $E \cup E_2$) is worst-case optimal. For example, consider a star graph with n vertices, the time complexity is $\mathcal{O}(n^2)$, and $E \cup E_2$ is the set of all vertex pairs (*i.e.*, $|E \cup E_2| = \binom{n}{2} = \frac{n(n-1)}{2}$).

From the above discussions, it can be verified that $\sum_{(u,v) \in E} c(u, v)$ equals the total number of triangles in the graph. That is, the total number of triangles can be obtained from the $s^{(0)}(u, v)$ values of all $(u, v) \in E$ in $\mathcal{O}(m)$ time. As a result, by assuming the triangle detection conjecture in [1], computing $s^{(0)}(u, v)$ (or $d^{(0)}(u, v)$) for all pairs of directed connected vertices cannot be conducted in $\mathcal{O}(m)$ time, which invalidates the claim in [16] regarding the time complexity of Line 1 of Algorithm 1.

Algorithm 3: Distance Updating of FDD

---

1 **for** *each edge* $(u, v) \in E$ **do**
2 $\quad$ **if** $0 < d^{(0)}(u, v) < 1$ **then**
3 $\quad\quad$ Push (u, v) into a queue $\mathcal{Q}^{(0)}$;

4 Initialize $t \leftarrow 0$, and a disjoint-set data structure S for V;
5 **while** $\mathcal{Q}^{(t)} \neq \emptyset$ **do**
6 $\quad$ **while** $\mathcal{Q}^{(t)} \neq \emptyset$ **do**
7 $\quad\quad$ $(u, v) \leftarrow$ pop an edge from $\mathcal{Q}^{(t)}$;
8 $\quad\quad$ **if** *u and v are in different sets in* S **then**
9 $\quad\quad\quad$ Compute $d^{(t+1)}(u, v)$ from $d^{(t)}(u, v)$ by using Equations (3) – (7);
10 $\quad\quad\quad$ **if** $0 < d^{(t+1)}(u, v) < 1$ **then** Push (u, v) into $\mathcal{Q}^{(t+1)}$;
11 $\quad\quad\quad$ **if** $d^{(t+1)}(u, v) > 1$ **then** $d^{(t+1)}(u, v) \leftarrow 1$;
12 $\quad\quad\quad$ **if** $d^{(t+1)}(u, v) < 0$ **then** Union u and v in S, and $d^{(t+1)}(u, v) \leftarrow 0$;
13 $\quad\quad$ **else** $d^{(t+1)}(u, v) \leftarrow 0$;

14 $\quad$ $t \leftarrow t + 1$;

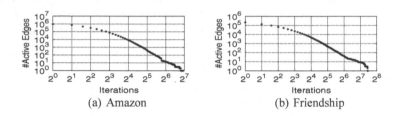

(a) Amazon $\qquad\qquad\qquad$ (b) Friendship

Fig. 2. Number of active edges in iterations

4.2 Optimizing Distance Updating

For distance updating, **Attractor** blindly tests each edge to update its distance if it is not converged (see Line 4 in Algorithm 1), which is inefficient. We propose two optimization techniques to improve the efficiency of distance updating.

Active-Edge Queue Optimization. We observe that although the total number of iterations (*i.e.*, T) that is needed for all edges to converge could be large (*e.g.*, can be up-to a hundred), most of the edges actually converge after the first few iterations. For example, Fig. 2 demonstrates the number of active (*i.e.*, not converged) edges in the iterations for graphs "Amazon" and "Friendship"; please refer to Sect. 5 for the descriptions of these graphs. Thus, it is a waste of time to check the converged (*i.e.*, non-active) edges. Motivated by this, we propose to maintain a queue for all the active edges, and only check and update the active edges in each iteration. The pseudocode of using queue to maintain active edges is shown in Lines 1–3, 5–7, and 10 of Algorithm 3.

Transitivity Optimization. Our next optimization is based on the following observation. In the distance dynamics model [16], after the distances of all edges converge, the communities are formed by the connected components of the graph after removing all edges of distance 1. Thus, during the iterations of distance dynamics, if the distance between u and v does not yet converge but there is a path between u and v consisting only of edges of distance 0, then u and v must be in the same community in the final result. As a result, we can directly set the distance between u and v to be 0 without following the updating equations. We call this transitivity optimization. To efficiently test whether there is a path between u and v consisting only of edges of distance 0 regarding the current distance values of the graph, we use a disjoint-set data structure [5] to maintain the connected components formed by edges of distance 0. Whenever the distance of an edge (u, v) converges to 0, we union the two corresponding connected components of u and v in the disjoint-set data structure. Note that, according to the distance dynamics model [16], once the distance of an edge becomes 0, it will never increase again. The pseudocode of transitivity optimization is shown in Lines 4, 8, and 12 of Algorithm 3.

4.3 Reducing Time Complexity

Following Sects. 3 and 4.1, the (worst-case) time complexity of FDD is $\mathcal{O}(T \cdot deg_{max} \cdot m)$, where T is the number of iterations of distance updating until convergence. Note that, the two optimization techniques in Sect. 4.2 do not increase nor decrease the time complexity. Specifically, the two operations of the disjoint-set data structure, Find (Line 8 of Algorithm 3) and Union (Line 12 of Algorithm 3), have amortized time complexity of $\mathcal{O}(\alpha(n))$, where $\alpha(\cdot)$ is a extremely slow-growing function that is at most 4 for all practical values of n [5]; thus, we consider them as constant operations in the analysis.

As observed in [2] and also demonstrated in Table 1 in Sect. 5, most real graphs are power-law graphs. That is, although the average degree is very small, the maximum degree deg_{max} could be very large or even in the same order of magnitude as n. Thus, the time complexity $\mathcal{O}(T \cdot deg_{max} \cdot m)$ is still too high for large real graphs that have large deg_{max}. In order to scale FDD to large real graphs, we propose to firstly remove from the graphs all vertices whose degrees are larger than a threshold γ. Consequently, FDD is only run on the resulting graph whose deg_{max} is at most γ, and the time complexity becomes $\mathcal{O}(T \cdot \gamma \cdot m)$. Moreover, as a result of the two optimizations proposed in Sect. 4.2, the term T in the time complexity is also largely alleviated. This is because later iterations take an insignificant amount of time due to the small number of active edges (see Fig. 2). Thus, FDD is more likely to run in $\mathcal{O}(\gamma \cdot m)$ time in practice.

5 Experiments

In this section, we conduct empirical studies to demonstrate the effectiveness and efficiency of our techniques. We evaluate our fast distance dynamics algorithm FDD against two existing algorithms: the existing distance dynamics algorithm Attractor [16], and the popular modularity-based community detection algorithm

Table 1. Statistics of real graphs, where deg_{max} is the maximum degree and #*Community* is the number of provided ground-truth communities.

Graphs	n	m	deg_{max}	#*Community*	Source
Karate	34	78	17	2	Network Repository
Dolphins	62	159	12	2	Network Repository
Polbooks	105	441	25	3	Network Repository
Football	115	613	12	12	Network Repository
Email-Eu	986	16,064	345	42	SNAP
Polblogs	1,224	16,715	351	2	KONECT
AS	23,752	58,416	2,778	176	KONECT
Cora	23,166	89,157	377	70	KONECT
Amazon	334,863	925,872	549	5,000	SNAP
Youtube	1,134,890	2,987,624	28,754	5,130	SNAP
Collaboration	9,875	25,973	65	-	SNAP
Friendship	58,228	214,078	1,134	-	SNAP
RoadNet	1,088,092	1,541,898	9	-	SNAP
Live-Journal	3,997,962	34,681,189	14,815	-	SNAP

Table 2. Running time of FDD against Attractor and Louvain (in seconds)

Graphs	FDD	Attractor	Louvain	Graphs	FDD	Attractor	Louvain
Karate	0.001	0.003	0.001	Cora	1.706	14.949	0.235
Dolphins	0.001	0.004	0.002	Amazon	14.988	93.256	5.508
Polbooks	0.006	0.014	0.002	Youtube	22.667	–	>2hrs
Football	0.006	0.012	0.002	Collaboration	0.414	1.628	0.107
Email-Eu	0.113	5.572	0.012	Friendship	2.738	119.988	0.624
Polblogs	0.110	7.681	0.013	RoadNet	11.975	27.399	18.623
AS	0.346	222.784	0.202	Live-Journal	271.555	–	546.271

Louvain [3]. For FDD and Attractor, we set $\lambda = 0.5$ as suggested in [16]. In addition, we choose $\gamma = 50$ by default for FDD. All algorithms are implemented in C++, and all experiments are conducted on a machine with an Intel Core 2.6 GHz CPU and 8 GB memory.

We evaluate the algorithms on 14 real graphs that are widely used in the existing studies. The graphs are downloaded from Network Repository, Stanford Network Analysis Platform (SNAP), and the Koblenz Network Collection (KONECT). Statistics of these graphs are shown in Table 1. The first ten graphs come with ground-truth communities, while the last four graphs do not have ground-truth communities and are mainly used for efficiency testings. In addition to these real graphs, we also generated synthetic graphs based on the LFR benchmark [11] for evaluating the sensitivity of Attractor and FDD to the maximum degree deg_{max}.

Eval-I: Evaluate the Efficiency of FDD **Against** Attractor **and** Louvain.
The results are shown in Table 2. We can see that the running time of Attractor
generally correlates to the maximum degree deg_{max} and it does not scale to
graphs with large deg_{max}, while FDD is not affected by deg_{max} and is much
faster than Attractor. For example, Attractor runs out-of-memory on Youtube
and Live-Journal where $deg_{max} > 10^4$. On the other two graphs AS and Friendship that have $10^3 < deg_{max} < 10^4$, FDD is 643x and 43.8x faster than Attractor,
respectively. When compared with Louvain, the running time of FDD is generally similar to that of Louvain on these graphs, and is smaller on Youtube and
Live-Journal; FDD computes the communities for Youtube in 22 s while Louvain
does not finish within 2 h. This demonstrates the efficiency-superiority of FDD
over existing algorithms Attractor and Louvain.

Table 3. NMI and purity (abbreviated as Pur) of FDD against Attractor and Louvain

Graphs	FDD		Attractor		Louvain		FDD-P	
	NMI	Pur	NMI	Pur	NMI	Pur	NMI	Pur
Karate	**0.782**	**1**	**0.782**	**1**	0.602	0.971	**0.782**	**1**
Dolphins	0.201	0.677	0.201	0.677	**0.516**	**0.968**	0.201	0.677
Polbooks	0.520	**0.886**	0.520	**0.886**	**0.537**	0.848	0.520	**0.886**
Football	**0.931**	**0.939**	**0.931**	**0.939**	0.856	0.800	**0.931**	**0.939**
Email-Eu	**0.726**	**0.824**	0.564	0.585	0.309	0.329	0.707	0.788
Polblogs	0.195	**0.963**	0.201	**0.963**	**0.647**	0.952	0.200	**0.963**
AS	**0.412**	**0.821**	0.362	0.651	0.480	0.509	0.411	0.818
Cora	**0.555**	**0.701**	0.547	0.667	0.459	0.314	0.554	0.700
Amazon	**0.873**	**0.050**	**0.873**	**0.050**	0.794	0.023	**0.873**	**0.050**
Youtube	**0.828**	0.030	–	–	–	–	0.794	**0.031**

Table 4. Running time (in seconds) of our algorithms

Graphs	Attractor	FDD_1	FDD_2	FDD_3	FDD
AS	222.784	33.368	31.093	30.477	0.346
Cora	14.949	4.190	3.568	3.355	1.706
Amazon	93.256	25.717	21.918	19.422	14.988
Youtube	–	–	–	–	22.667
Collaboration	1.628	0.561	0.480	0.442	0.414
Friendship	119.988	24.870	22.889	20.664	2.738
RoadNet	27.399	14.988	12.844	12.242	11.975
Live-Journal	–	–	–	–	271.555

Eval-II: Evaluate the Effectiveness of FDD **Against** Attractor **and** Louvain.
We run the algorithms on the ten graphs that come with ground-truth communities. We measure two well-known metrics, Normalize Mutual Information (NMI) and Purity [18], for the communities extracted by the algorithms with respect to the ground-truth communities. For both metrics, the larger the value, the better the quality. The results are shown in Table 3, where best results are highlighted by bold font. We can see that the communities extracted by FDD in general has the highest quality regarding both NMI and purity. This confirms the effectiveness of the distance dynamics model [16].

Recall that FDD removes from the graph the vertices that have degrees larger than γ and does not assign these vertices into communities. Nevertheless, the extracted communities actually has a no worse quality than Attractor, as shown in Table 3. One possible reason could be that the high-degree vertices may mislead the community extraction algorithms due to connecting to too many other vertices. In Table 3, we also include the algorithm FDD-P which is FDD with post-processing that assigns the removed vertices to communities; specifically, each removed vertex is assigned to the community that contains most of its neighbors. We can see that the community quality is similar to that of FDD. This suggests that it may be a good idea to first ignore high-degree vertices.

Eval-III: Evaluate Our Techniques. We implemented another three variants of FDD, FDD_1, FDD_2, FDD_3, by adding the techniques one-by-one. FDD_1 is improved from Attractor by adding the efficient initialization proposed in Sect. 4.1. FDD_2 is improved from FDD_1 by adding the active-edge queue optimization. FDD_3 is improved from FDD_2 by adding the transitivity optimization. Note that, FDD then is the improvement from FDD_3 by adding the technique in Sect. 4.3. The results are shown in Table 4. We can see that each of our techniques contributed to the efficiency of FDD.

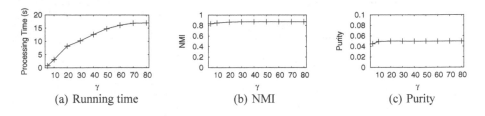

(a) Running time   (b) NMI   (c) Purity

Fig. 3. Varying γ on Amazon

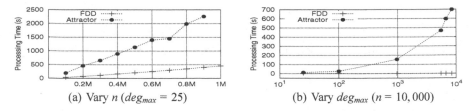

(a) Vary n ($deg_{max} = 25$)   (b) Vary deg_{max} ($n = 10,000$)

Fig. 4. Running time of FDD and Attractor on LFR graphs

Eval-IV: Evaluate the Effect of γ. We in this testing run FDD on Amazon by varying γ from 5 to 80. The results are shown in Fig. 3. Firstly, when γ increases, the running time of FDD also increases; this conforms with our theoretical analysis of FDD, *i.e.*, its time complexity is $\mathcal{O}(T \cdot \gamma \cdot m)$. Secondly, NMI increases slightly along with the increasing of γ. Thirdly, the purity also increases slightly with respect to γ as more vertices are assigned to communities. Based on these results, we need to make a trade-off between the running time and the quality of extracted communities; we observe that $\gamma = 50$ works well in practice.

Eval-V: Evaluate the Effect of deg_{max}. In order to evaluate the performance of FDD and Attractor more thoroughly on graphs with different maximum degrees (deg_{max}), we also generated synthetic graphs based on the LFR benchmark [11]. We fix the average degree to be 20, and vary either the number of vertices n or the value of deg_{max}. The results of varying n from 0.2×10^6 to 10^6 is shown in Fig. 4(a), where deg_{max} is fixed at 25. We can see that the running time of both FDD and Attractor increases as expected. However, the running time of Attractor increases much faster than that of FDD. The results of varying deg_{max} from 25 to 10^4 is shown in Fig. 4, where n is fixed at 10^4. We can see that, when the maximum degree increases, the running time of Attractor also increases significantly. On the other hand, the running time of our algorithm FDD remains almost the same. This confirms our theoretical analysis in Sects. 3 and 4 that the time complexity of Attractor depends on deg_{max} while the time complexity of FDD is not related to deg_{max}. As large real graphs usually have large maximum degrees, FDD is more suitable than Attractor for processing large real graphs.

6   Conclusion

In this paper, we first showed that the time complexity of the state-of-the-art distance dynamics algorithm Attractor highly depends on the maximum vertex degree, and then developed a fast distance dynamics algorithm FDD to scale distance dynamics to large real graphs that have large maximum vertex degrees. In FDD, we proposed efficient techniques to compute the initial distance for all relevant vertex pairs in $\mathcal{O}(deg_{max} \cdot m)$ time, as well as optimization techniques to improve the practical efficiency of distance updating. Moreover, we also reduced the time complexity to $\mathcal{O}(T \cdot \gamma \cdot m)$ for a small constant γ. Experimental results on large real graphs demonstrated the efficiency and effectiveness of FDD.

References

1. Abboud, A., Williams, V.V.: Popular conjectures imply strong lower bounds for dynamic problems. In: Proceeding of FOCS 2014, pp. 434–443 (2014)
2. Barabasi, A.L., Albert, R.: Emergence of scaling in random networks. Science **286**, 509–512 (1999)
3. Blondel, V.D., Guillaume, J.L., Lambiotte, R., Lefebvre, E.: Fast unfolding of communities in large networks. J. Stat. Mech. Theory Exp. **2008**(10), 10008 (2008)

4. Brandes, U., et al.: On Modularity-NP-Completeness and Beyond. Universität Fak. für Informatik, Bibliothek (2006)
5. Cormen, T.H., Leiserson, C.E., Rivest, R.L., Stein, C.: Introduction to algorithms. MIT press, Cambridge (2009)
6. Donetti, L., Munoz, M.A.: Detecting network communities: a new systematic and efficient algorithm. J. Stat. Mech. Theory Exp. **2004**(10), 10012 (2004)
7. Eriksen, K.A., Simonsen, I., Maslov, S., Sneppen, K.: Modularity and extreme edges of the internet. Phys. Rev. Lett. **90**(14), 148701 (2003)
8. Fortunato, S.: Community detection in graphs. Phys. Rep. **486**(3–5), 75–174 (2010)
9. Girvan, M., Newman, M.E.: Community structure in social and biological networks. Proc. Int. Acad. Sci. **99**(12), 7821–7826 (2002)
10. Kernighan, B.W., Lin, S.: An efficient heuristic procedure for partitioning graphs. Bell Syst. Tech. J. **49**(2), 291–307 (1970)
11. Lancichinetti, A., Fortunato, S., Radicchi, F.: Benchmark graphs for testing community detection algorithms. Phys. Rev. E: Stat., Nonlin, Soft Matter Phys. **78**(4), 046110 (2008)
12. Newman, M.E.: Fast algorithm for detecting community structure in networks. Phys. Rev. E **69**(6), 066133 (2004)
13. Newman, M.E., Girvan, M.: Finding and evaluating community structure in networks. Phys. Rev. E **69**(2), 026113 (2004)
14. Palla, G., Derényi, I., Farkas, I., Vicsek, T.: Uncovering the overlapping community structure of complex networks in nature and society. Nature **435**(7043), 814 (2005)
15. Ravasz, E., Barabási, A.L.: Hierarchical organization in complex networks. Phys. Rev. E **67**(2), 026112 (2003)
16. Shao, J., Han, Z., Yang, Q., Zhou, T.: Community detection based on distance dynamics. In: Proceedings of SIGKDD 2015, pp. 1075–1084 (2015)
17. Shi, J., Malik, J.: Normalized cuts and image segmentation. IEEE Trans. Pattern Anal. Mach. Intell. **22**(8), 888–905 (2000)
18. Strehl, A., Ghosh, J., Mooney, R.: Impact of similarity measures on web-page clustering. In: Workshop on Artificial Intelligence for Web Search (AAAI 2000), vol. 58, p. 64 (2000)
19. Suaris, P.R., Kedem, G.: An algorithm for quadrisection and its application to standard cell placement. IEEE Trans. Circuits Syst. **35**(3), 294–303 (1988)
20. Wilkinson, D.M., Huberman, B.A.: A method for finding communities of related genes. Proc. Nat. Acad. Sci. **101**(1), 5241–5248 (2004)

A Hybrid Index for Distance Queries

Junhu Wang[1]([✉]), Shikha Anirban[1], Toshiyuki Amagasa[2], Hiroaki Shiokawa[2], Zhiguo Gong[3], and Md. Saiful Islam[1]

[1] Griffith University, Queensland, Australia
j.wang@griffith.edu.au
[2] University of Tsukuba, Tsukuba, Japan
[3] University of Macau, Zhuhai, China

Abstract. Shortest distance queries not only find important applications in real-word systems, it is also the foundation for numerous other graph analysis problems. State-of-the-art techniques for such queries use 2-hop labeling, in particular, the Pruned Landmark Labeling (PLL) index is among the best performing for small world networks. However, PLL suffers from large label size and index computation time when the graph is large. In this paper, we propose two techniques to address the problem. The first technique is to limit the landmarks to vertices in a minimum vertex cover, and the second is to use a hybrid index by decomposing the graph into a *core* and a forest, and combining the PLL index for the core and a simpler distance labeling for trees. Extensive experiments with real-world graphs verified the effectiveness of these techniques.

1 Introduction

Given a graph G and a pair of vertices s and t, the distance query asks for the distance from s to t, that is, the length of the shortest path from s to t. Distance queries find numerous applications such as in route planning and spatial databases [7,12], community search and influence maximization in location-based social networks [25]. They are also the foundation for several other problems such as graph homomorphism search [8] and distributed subgraph enumeration[21]. Answering such queries instantly is crucial for many of these applications.

Distance queries can be answered using one of two naive methods. The first method is to do a breath-first search[1] from s until we reach t, and the second method is to pre-compute the distances between all vertex pairs and store them as an index. With large graphs nowadays both methods are impractical: the former is too slow, and the latter takes too much space and pre-computation time. Therefore, many indexing techniques have been proposed that tried to strike a balance between index size/index construction time and online query

---

[1] Or to use Dijkstra's algorithm for weighted graphs.

© Springer Nature Switzerland AG 2020
Z. Huang et al. (Eds.): WISE 2020, LNCS 12342, pp. 227–241, 2020.
https://doi.org/10.1007/978-3-030-62005-9_17

processing time. As noted in several recent studies [16,17], for complex networks the state-of-the-art technique is *pruned landmark labelling* (PLL) [3]. PLL is a type of 2-hop labeling [6], which constructs a label for each vertex u, denoted *label*(u), consisting of some vertex-distance pairs (v, d) and stores them as index (v is called a *hub* of u if $(v, d) \in$ *label*(u)). At query time, the distance between s and t can be directly answered by examining the common hubs of s and t, and adding the distance from s to the hub and the distance from the hub to t together. Although PLL works very well compared with other techniques, for large graphs the index can be too large, which will not only take up memory but also make the queries slower. Several subsequent works tried to address the problem by better vertex ordering [2,17], graph compression, and elimination of local minimum nodes [16], however, these methods still generate label sets that are unnecessarily large.

In this paper, we propose two novel techniques to tackle the problem. Specifically, (1) we propose to use only vertices in a minimum vertex cover as landmarks; (2) we propose to decompose the graph into a core and a set of trees, and combine two types of labels, one for the core and one for the trees, as the distance index; We conduct extensive experiments with real-world graphs, which demonstrate that the proposed techniques can significantly reduce the index size and index time, and make the queries faster.

Organization. We present our techniques for undirected, unweighted graphs, but most of them can be easily adapted to directed and/or weighted graphs. Section 2 provides the preliminaries, Sect. 3 discusses computing labels using minimum vertex cover. Section 4 presents hybrid labeling by decomposing the graph into core and trees, and our experiments are reported in Sect. 5. Section 6 discusses related work. We conclude the paper in Sect. 7.

2 Preliminaries

A graph $G \equiv (V, E)$ consists of a set V of vertices and a set $E \subseteq V \times V$ of edges. In this paper we implicitly assume all graphs are *simple* graphs, that is, there is at most one edge from one vertex to another. We will focus on undirected and unweighted graphs where the edges have no direction and every edge has a length of 1.

If (u, v) is an edge in an undirected graph, we say u is neighbor of v and v is a neighbor of u, and we use $N(v)$ to denote the set of all neighbors of v. Let u, v be vertices in graph G. A path from u to v is a sequence of neighboring edges $(u, v_1), (v_1, v_2), \ldots, (v_k, v)$, which is also commonly represented as the sequence of vertices $u, v_1, \ldots, v_k, v$. The number of edges in the path is called its *length*. Given a pair of vertices u and v in the graph, a *shortest path* from u to v is a path from u to v with the smallest length, and we call this length the *distance* from u to v, denoted $d_G(u, v)$ or simply $d(u, v)$ when G is clear from the context. Note that for undirected graphs $d(u, v) = d(v, u)$.

Pruned Landmark Labeling (PLL). The pruned landmark labeling is a type of 2-hop labeling scheme [6] designed for the efficient processing of distance queries. The general idea of a 2-hop labeling scheme is as follows: Suppose G is an undirected graph. For each vertex $u \in V$, we compute a label $L(u)$ which is a set of $(vertex, distance)$ pairs $(v, d(u, v))$, where v is a vertex in V, and $d(u, v)$ is the distance from u to v. A 2-hop labeling scheme is *complete* for G if it can be used to answer all distance queries in G as follows: let $H(u) = \{v | (v, d(u, v)) \in L(u)\}$. For every pair of vertices s and t in the graph:

$$d(s, t) = \min_{v \in H(s) \cap H(t)} d(s, v) + d(v, t) \tag{1}$$

A naive complete 2-hop label $L(u)$ contains $(v, d(u, v))$ for every vertex $v \in V$. However, it is unnecessary and can be too large and too expensive to compute. Using PLL we can compute a much smaller label set for each u using a pruning strategy which also makes the computation much faster.

Let $v_1, v_2, \ldots, v_n$ be the vertices in V, the PLL algorithm in [3] computes $L(u)$ as follows: First, we choose a fixed order of the nodes. Without loss of generality, we assume the order is $v_1, v_2, \ldots, v_n$. Then, we do BFS from v_1, then from v_2, ..., and finally from v_n, each time we traverse from v_j to u, we will decide whether $(v_j, d(u, v_j))$ should be added into $L(u)$.

1. Initially for every $u \in V$, $L_0(u) = \emptyset$;
2. When we reach u from v_j, we create $L_j(u)$ from $L_{j-1}(u)$ as follows: if there does not exist $i < j$ such that $d(v_j, u) \geq d(v_j, v_i) + d(v_i, u)$,

$$L_j(u) = L_{j-1}(u) \cup \{(v_j, d(v_j, u))\}$$

   Otherwise $L_j(u) = L_{j-1}(u)$, and we stop going further from u.
3. Finally, $L(u) = L_n(u)$.

The index L consists of the labels of all vertices, that is, $L = \{L(u) | u \in V\}$.

Observe that the condition in Step 2 above means that if some shortest path from v_j to u goes through one of the vertices v_i $(i < j)$, then $(v_j, d(v_j, u))$ will not be added into $L(u)$, and otherwise it will.

Due to the smart pruning strategy, PLL usually generates a much smaller label for each vertex u without losing completeness. That is, for only some vertices $v \in V$, the pair $(v, d(u, v))$ is in $L(u)$. For easy explanation and to be consistent with the terminology in [16], we will call v a *hub* of u if $(v, d(u, v)) \in L(u)$. We will also say v_i is ranked higher than v_j if v_i precedes v_j in the vertex ordering when computing PLL labels and denote it by $rank(v_i) > rank(v_j)$.

3   Restricting Landmarks to a Minimum Vertex Cover

Intuitively a landmark is a vertex that many shortest paths pass through. If there is a subset S of vertices that every shortest path in G passes through, then we can use S as landmarks when computing the 2-hop labels. Our key insight here is that the original PLL essentially treats every vertex in the graph as a potential landmark, i.e., every vertex can be a hub of some other vertices. However, this is usually not necessary. We observe that using vertices in a *vertex cover* C is sufficient (A *vertex cover* is a set C of vertices in G such that every edge in G has at least one end in C). This is because every edge is incident on some vertex in C. Therefore, every non-trivial path, i.e., path of length ≥ 1, passes through a vertex in C. In other words, for any path between s and t ($s \neq t$), there is a vertex $v \in C$ such that $d(s,t) = d(s,v) + d(v,t)$. Therefore, our first intuition is to find a minimum vertex cover C, and limit the BFS in the PLL computation to vertices in C. For convenience, we denote the PLL labels computed using a set S of vertices by $PLL(S)$.

Example 1. Consider the graph G_1 shown in Fig. 1. $C = \{v_0, v_2, v_4, v_6\}$ is a minimum vertex cover. The computation of PLL labels using the vertex cover and not using the vertex cover are shown in Fig. 2. Using the vertex cover, we create a label set that contains a total of 21 labels, and using (an arbitrary ordering $v_0, v_1, v_2, v_3, v_4, v_5, v_6, v_7$ of) all vertices, we obtain a label set of 30 labels.

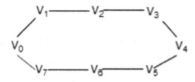

Fig. 1. Graph G_1

(a) $PLL(C)$

	v_0	v_2	v_4	v_6
v_0	0			
v_1	1	1		
v_2	2	0		
v_3	3	1	1	
v_4	4	2	0	
v_5	3	3	1	1
v_6	2		2	0
v_7	1		3	1

(b) $PLL(V)$

	v_0	v_1	v_2	v_3	v_4	v_5	v_6	v_7
v_0	0							
v_1	1	0						
v_2	2	1	0					
v_3	3	2	1	0				
v_4	4	3	2	1	0			
v_5	3		3	2	1	0		
v_6	2			3	2	1	0	
v_7	1				3	2	1	0

Fig. 2. PLL label computation using vertex cover $S \equiv \{v_0, v_2, v_4, v_6\}$ and not using the vertex cover for graph G_1. The labels for each vertex are contained in a row. For instance, in $PLL(C)$, $L(v_3) = \{(v_0, 3), (v_2, 1), (v_4, 1)\}$.

One might wonder whether there are smaller landmark sets than the minimum vertex cover that can make the labels complete. The following proposition answers the question.

Proposition 1. *Let S be a subset of vertices in G. If S is not a vertex cover of G, then there exist a pair of vertices u, v ($u \neq v$) such that $d(u, v)$ can not be computed using the PLL labels computed from S.*

Proof. Since S is not a vertex cover, there must be an edge (u, v) such that neither u nor v is in S. Clearly the edge (u, v) is a shortest path from u to v, while any path from u to v that passes through a vertex in S is of length at least 2. Therefore, $d(u, v)$ can not be computed using the PLL labels computed from S.

It was observed that in the original PLL algorithm, v is a hub of u if and only if v has the highest rank on all shortest paths from u to v.

Lemma 1 ([16]). *For any pair of vertices $u, v \in V$, v is a hub of u if and only if v is ranked highest on all shortest paths from u to v.*

Let C be a vertex cover of G. Without loss of generality, let us assume $v_1, \ldots, v_k$ ($k < n$) are the vertices in C, and $v_{k+1}, \ldots, v_n$ are vertices not in C. Using Lemma 1, we can show

Theorem 1. *If we use the original PLL algorithm to compute the vertex labels in the order $v_1, v_2, \ldots, v_n$, then for every $v \in V - C$, v is the hub of only itself. That is, for any $u \neq v$, v is not a hub of u.*

Proof. Let $v \in V - C$ be any vertex outside the vertex cover C. For every vertex $u \in C$, $rank(u) > rank(v)$, therefore, according to Lemma 1, v is not a hub of u. For every vertex $v' \notin C$ such that $v' \neq v$, any path from v' to v must pass through some vertex $u \in C$ and u is ranked higher than v. Therefore, v is not ranked the highest on any path from v' to v. Thus v is not a hub of v' according to Lemma 1.

Observe that, if we order the vertices in C before those in $V - C$, then all the vertices in $V - C$ are *local minimum vertices* [16] which are vertices whose ranks are lower than that of their neighbours. Hence by Lemma 4.12 of [16], they are hubs of only themselves. This serves as an alternative proof of Theorem 1.

Vertices that are hubs of only themselves are not really helpful for distance queries, hence they can be removed from the label set L. Theorem 1 indicates that if the vertices in V are appropriately ordered, we can find a 2-hop label which is equivalent to the labels we find using a vertex cover. However, our result is still useful since finding a good vertex ordering is not easy, and the approach in [16] is the same as that in [3], that is, to order the vertices by degree and vertex ID, and then find the local minimum set L. In comparison, our method is to order the vertices in the minimum vertex cover before those not in the cover, and by doing this we are likely to get a larger local minimum set.

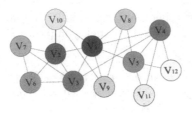

Fig. 3. Example graph in [16]

This is verified in our experiments where the average number of labels of each vertex using our approach is significantly smaller then using the degree-based ordering. To further illustrate this point, we provide the following example.

Example 2. Consider the graph in Fig. 3, which is the graph in the running example of [16]. If the vertices are ordered using degree and vertex ID, then $\{v_7, v_{10}, v_{11}, v_{12}\}$ is the local minimum set (Note that v_8 is not local-minimum). So in computing the PLL labels only the other 8 vertices need to be used as landmarks. For the same graph, we can see that the vertices $v_1, v_2, v_3, v_4, v_5, v_6, v_9$ form a vertex cover. If we rank these vertices before others, then v_8 will be local minimum as well. In our approach we only need to use these 7 vertices as landmarks when computing the PLL labels.

Finding a Minimum Vertex Cover. It is well known that finding a minimum vertex cover is an NP-hard problem. However, we can use the following greedy heuristic algorithm to find an approximate minimum vertex cover.

Algorithm 1: FindMVC

   Input: Graph G
   Output: A vertex cover C of G, initially $C = \emptyset$
1 **while** *there is edge in G* **do**
2     $u \leftarrow$ a vertex in G with largest degree
3     $C \leftarrow C \cup \{u\}$
4   Remove u and all edges incident on u from G
5 Return C

Suppose we have found a minimum vertex cover C. When computing the PLL labels using C, we arrange the vertices in C according to non-increasing degree.

Removing 0-Distance Labels. To further reduce the number of labels, we note that distance 0 labels can be eliminated if we slightly modify the algorithm for computing the distance. Specifically, when computing the distance $d(s, t)$, we first check whether t is a hub of s or s is a hub of t, if yes, then the distance

$d(s,t)$ is recorded in $L(s)$ or $L(t)$ already, so we can directly retrieve it; otherwise we use formula (1).

For example, we can remove all labels of the form $(v,0)$ from the labels in Example 1. To compute $d(v_4, v_7)$, we find that v_4 is a hub of v_7, that is, $(v_4, 3) \in L(v_7)$, therefore, we know $d(v_4, v_7) = 3$. However, for $d(v_3, v_6)$, neither v_3 is a hub of v_6, nor v_6 is a hub of v_3. Therefore, we use Eq. (1), and find $d(v_3, v_6) = d(v_3, v_4) + d(v_4, v_6) = 1 + 2 = 3$.

Although the above idea is simple, for large graphs with millions of vertices, the amount of savings by eliminating all 0-distance labels can be significant.

4   Combining Two Types of Labels - a Hybrid Index

Consider a graph that consists of a single path v_0, v_1, v_2, v_3, v_4. Observe that using PLL, even with a minimum vertex cover $\{v_1, v_3\}$, we will create a label where v_1 is a hub of every other vertex, and v_3 is a hub of v_2, v_3, v_4. However, we can regard v_0 as a landmark, and create a label L' for every other vertex as follows: $L'(v_i) = i$ (note $i = d(v_i, v_0)$) for $i \in [0, 4]$. To find the distance between v_i and v_j, we can use $d(v_i, v_j) = |d(v_j, v_0) - d(v_i, v_0)|$. Here v_0 is a landmark that can be used differently than the 2-hop labels discussed in Sect. 2. We can also use any other vertex as the landmark. For example, if we use v_3 as the landmark, we can compute the distances as follows: for v_i and v_j, if the two vertices lie on the same side of v_3, we use difference, for instance $d(v_0, v_2) = d(v_0, v_3) - d(v_2, v_3)$; if the two vertices lie on different sides of v_3, we use addition, for instance, $d(v_2, v_4) = d(v_2, v_3) + d(v_3, v_4)$. In other words, we can use either addition or difference to compute the difference between two vertices.

More generally, consider a tree T rooted at vertex r. For each vertex v in T, we only need to record its distance $d(v, r)$ from the root r. Given any two vertices v_i, v_j in T, we can compute their distance as follows: (1) find the lowest common ancestor of u, v, denoted $lca(u, v)$, (2) $d(u, v) = d(u, lca(u, v)) + d(v, lca(u, v))$ where $d(u, lca(u, v)) = d(u, r) - d(r, lca(u, v))$ and $d(v, lca(u, v)) = d(v, r) - d(r, lca(u, v))$. In other words,

$$d(u, v) = d(u, r) + d(v, r) - 2d(r, lca(u, v)) \tag{2}$$

Thus for a tree, it is sufficient to use its root as the landmark, and record the distances between each vertex from the root. Note that if u is an ancestor of v, then $lca(u, v) = u$.

Now consider a general graph G. We note that G can be decomposed into a 2-core and a set of trees, where the root of each tree is a vertex in the 2-core [4]. Let V_c be the vertex set of the 2-core, $G(V_c)$ be the 2-core, and $T_1, \ldots, T_k$ be the distinct trees, and r_i be the root of tree T_i. Note that for any pair of vertices within $G(V_c)$, the shortest path will not pass through any non-root vertex in the trees. Based on this observation, we design a hybrid label index for G as follows:

1. Construct the PLL label $L(u)$ for each vertex $u \in V_c$ for the graph $G(V_c)$ as described in Sect. 3, ignoring the vertices not in V_c.

2. Construct a label for each $u \in T_i$, denoted as $L'(u)$, such that $L'(u) = d(r_i, u)$ for $i \in [1, k]$.

When it comes to query processing, we can use the process described in Algorithm 2. The algorithm is self-explanatory. We will use the following following example to further explain it.

Algorithm 2: Computing distance using hybrid landmark labeling

---

Input: Graph G, label L for $G(V_c)$ and L' for the tree vertices, and two vertices $s, t \in V$

Output: $d(s, t)$

1 **if** *both s and t are in V_c* **then**
2 $\quad$ Compute $d(s, t)$ using $L(s)$, $L(t)$ and equation (1)
3 **else if** *$s \in V_c$ and $t \in V(T_i) - \{r_i\}$* **then**
4 $\quad$ $d(s, t) = d(s, r_i) + L'(t)$ where $d(s, r_i)$ is computed using equation (1)
5 **else if** *$s, t \in V(T_i)$* **then**
6 $\quad$ Compute $d(s, t)$ using equation (2)
7 **else if** *$s \in T_i$, $t \in T_j$, and $i \neq j$* **then**
8 $\quad$ $d(s, t) = L'(s) + d(r_i, r_j) + L'(t)$ where $d(r_i, r_j)$ is computed using equation (1)
9 **Return** $d(s, t)$

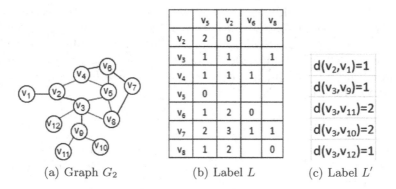

	v_5	v_2	v_6	v_8
v_2	2	0		
v_3	1	1		1
v_4	1	1	1	
v_5	0			
v_6	1	2	0	
v_7	2	3	1	1
v_8	1	2		0

$d(v_2, v_1) = 1$
$d(v_3, v_9) = 1$
$d(v_3, v_{11}) = 2$
$d(v_3, v_{10}) = 2$
$d(v_3, v_{12}) = 1$

(a) Graph G_2 $\qquad$ (b) Label L $\qquad$ (c) Label L'

Fig. 4. Graph G_2 and its hybrid labels

Example 3. Consider the graph G_2 in Fig. 4 (a). The graph can be decomposed into a core, denoted $G_2[V_c]$, consisting of the vertices v_2 to v_8 and two trees, denoted T_1 and T_2, rooted at v_2 and v_3 respectively. For the core $G_2[V_c]$, we find a minimum cover $\{v_5, v_2, v_6, v_8\}$, and compute the PLL labels as shown in Fig. 4 (b). The tree vertex label L' is as shown in Fig. 4 (c).

Consider the distance between vertices v_6 and v_{11}. We have $d(v_{11}, v_6) = d(v_1 1, v_3) + d(v_3, v_6) = 2 + (1+1) = 4$. Similarly $d(v_1, v_{10}) = d(v_2, v_1) + d(v_2, v_3) + d(v_3, v_{10}) = 1 + 1 + 2 = 4$

It is observed that many real-world graphs are of a core-periphery structure [22], where the core consists of a central, densely connected set of vertices, and the periphery is the set of sparsely connected vertices that are also linked to the core. The periphery part of such graphs is likely to contain many tree vertices. For those graphs that do contain a large forest, our hybrid label index can significantly reduce the index size. For example, for the DBpedia, Email-enron, and DBLP data sets, using core-forest decomposition and our hybrid index can reduce the average number of labels per vertex by 45.5%, 22% and 14% respectively, as shown in Table 2 of our experiments.

Finding a Core-Forest Decomposition. To find the core, we start with a vertex of degree one, remove it and all incident edges, and repeat this process until all vertices have a degree of 2 or more. This will obtain the core. We can then identify the root vertices of the forest, which are vertices in the core that are neighbors of at least one vertex not in the core. The forest (trees) can then be obtained using BFS traversal over vertices not in the core.

Complexity. Finding the core-forest Sdecomposition can be done in time $O(|V| + |E|)$. Building the label L' for a tree T can be done by a single BFS from the root of the tree, hence it takes $O(|T|)$ where $|T|$ represents the tree size.

5   Experiments

In this section, we report our experiments with the proposed methods (i.e., the vertex cover as landmarks and the hybrid-index with core-forest decomposition). We compare the label set size and query time of the original PLL where the vertices are ordered according to decreasing degree, with the following methods:

PLL$^+$. PLL using landmarks in an approximate minimal vertex cover.
HLL. The hybrid landmark labeling with the index for core computed using an approximate minimal vertex cover of the core.

Datasets. We use 16 real-world graphs as listed in Table 1. The size of each data set is also shown in the table. Wiki-croc, wiki-charm and wiki-squi are Wikipedia page-page networks on specific topics in December 2018. Github is the GitHub author-follower network collected in June 2019. Ro and Hu are friendship networks of users from 2 European countries collected from the music streaming service Deezer in November 2017. DBLP is an co-authorship network where two authors are connected if and only if they publish at least one paper together. Email-enron is an email communication network of Enron. DBpedia is a snapshot of the DBpedia knowledge graph downloaded from http://pan.baidu.com/s/1c00Jq5E. The remaining datasets are the blue verified Facebook page networks of different categories collected in November 2017. All data sets, except **DBpedia**, are downloaded from the Stanford Large Network Data Collection (https://snap.stanford.edu/data/). We would like to stress that, although most of these data sets are not large, they can still provide a good indication of the ratio of index size/time reduction by using our techniques.

Implementation Environment. All algorithms are implemented in C++ and compiled with gcc 7.4.0 compiler. The experiments are done in a machine with Intel Core i7-7700 with 3.60 GHz CPU, 32 GB memory and Linux (Ubuntu 18.04.3 LTS) operating system.

Table 1 also lists the number of vertices in the approximate minimum vertex cover ($|C_G|$), the number of vertices in the core $|V_{core}|$, the percentage of tree vertices, and the average tree size (number of vertices in the tree), as well as the number of vertices in the approximate minimum vertex cover of the core $|C_{core}|$. It can be seen that the percentage of minimal vertex cover vertices ranges from 0.26% to 0.75%. The percentages of tree vertices is about 35%, 31.1%, and 19% for **DBpedia**, **Email-enron**, **Company** respectively, while they are less than 10% for several other data sets. The average sizes of the tree, shown as $|t|$, are all small except for email-enron wiki-croc, facebook, and DBpedia.

Table 1. Datasets

Dataset	$|V|$	$|E|$	$|C|$	Tree node ratio	Avg. tree size	$|V_{core}|$	$|C_{core}|$
Wiki-croc	11631	180020	3037	0.044	5.72	11116	3033
Wiki-cham	2277	36101	810	0.041	3.2	2184	806
Wiki-squi	5201	217073	2147	0.027	2.42	5063	2143
Github	37700	289003	15511	0.139	1.76	32464	14775
RO	41773	125826	22726	0.153	1.33	35402	20831
HU	47538	222887	29535	0.06	1.13	44669	28902
Politician	5908	41729	3015	0.109	1.68	5262	2917
Athletes	13866	86858	6889	0.092	1.53	12590	6764
Company	14113	52310	7078	0.188	1.68	11457	6515
TVshow	3892	17262	1994	0.179	1.73	3194	1849
New-sites	27917	206259	15862	0.082	1.42	25626	15460
Government	7057	89455	4271	0.053	1.33	6681	4210
Facebook	4039	88234	3038	0.019	7.5	3964	3038
DBLP	317080	1049866	165229	0.143	1.77	271646	160824
Email-enron	36692	183831	14480	0.311	16.18	25286	13504
DBpedia	3365623	7989191	700756	0.35	4	2188839	457547

Table 2. Index size and index time

Dataset	PLL			PLL$^+$			HLL		
	Index size(KB)	Avg. $|L(v)|$	Index time(s)	Index size(KB)	Avg. $|L(v)|$	Index time(s)	Index size(KB)	Avg. $|L(v)|$	Index time(s)
Wiki-croc	1739.8	51.1	11	1274.9	37.4	8	1244	36.5	8
Wiki-cham	225.3	33.8	<1	177.4	26.6	<1	172.84	25.9	<1
Wiki-squi	1351.8	88.7	18	1032.8	67.8	11	1014.45	66.6	11
Github	6166.8	55.8	18	5117.5	46.3	76	4536.9	41	71
RO	72830	595.1	3722	49320.6	403	2592	41722.3	340.9	2255
HU	174318	1252	25262	110914	796.4	15515	103714.5	744.7	14808
Politician	1695.3	97.94	11	903.9	52.2	5	828.5	47.9	5
Athlete	5959.6	146.7	100	3987.8	98.2	66	3666.9	90.3	64
Company	5253.8	127.1	49	3557.5	86	35	2946.9	71.3	31
TVshow	951.4	83.4	3	548.2	48.1	1	456	40	1
New-sites	18699.9	228.6	555	11987	146.6	354	11085.5	135.5	343
Government	2544.4	123.1	36	1819	88	26	1720.1	83.2	25
Facebook	330.41	27.9	1	292	24.7	1	291.5	24.6	1
DBLP	672804.1	724.3	34795	469598.4	505.5	25973	402638.7	433.4	24826
Email-enron	4490.1	62.16	65	3255.1	44.9	51	2547.6	35.1	40
DBpeida	69568	10.1	526	45255	6.4	524.6	260.8	3.5	217.8

5.1   Index Size and Index Time

Table 2 shows the index size in both space consumption (in KB) and average number of labels per vertex (avg. $|L(v)|$). The index time, which includes the time for finding the approximate minimum vertex cover and the core-forest decomposition, is also shown in the table[2]. As can be seen, using the minimal vertex cover as landmarks alone can significantly reduce the index size, with the *Politician* and *TVshow* datasets achieving 46.7% and 42.4% reduction in the index size respectively. Using core-forest decomposition further reduces the index sizes for all of the data sets, with the DBpedia and Email-enron datasets achieving 45.5% and 21.8% reduction respectively from using the vertex cover alone. Note that the DBpedia and the Email-enron datasets have relatively large percentage of tree vertices. The index size of HLL is significantly smaller than that of PLL for all data sets, for instance, the index size of HLL is 37.4% of that of PLL for the DBpedia dataset. The index time of HLL is also much shorter than that of PLL for most datasets (for some of the small datasets, the index time of all methods are similar).

5.2   Query Time

To test the query time, we randomly generated 1 million vertex pairs for each data set. The average query time for each data set is shown in Table 3. As can bee seen from the table, the average query time for PLL$^+$ is significantly shorter than that of the original PLL for all data sets. However, the query time for PLL$^+$ and HLL are similar. This could be because of the extra query processing steps used in HLL offset the gain obtained by the smaller index size.

---

[2] For most datasets, the time spent on finding the minimum vertex cover and core-forest decomposition is small compared with that of PLL. One exception in our experiments is the DBpedia, for which computing the vertex cover too significant amount of time.

Table 3. Query time

Dataset	PLL	PLL$^+$	HLL
Wiki-croc	154	109.11	113.87
Wiki-cham	11.43	9.18	9.52
Wiki-squi	92.29	67.13	68.93
Github	545.74	453.78	464.92
RO	7740.05	5434.42	5271.85
HU	21678.9	14090.3	12547.2
Politician	140.58	76.21	78.38
Athletes	553.63	374.85	374.69
Company	493.42	331.45	331.21
TVshow	75.63	43.8	46.26
New-sites	1823.71	1161.12	1142.11
Government	200.57	150.33	151.6
Facebook	18.11	14.21	15.08
DBLP	87079.9	59721.2	58832.6
Email-enron	696.3	505	498.6
DBPedia	6978	5890.2	5875.4

6   Related Work

The problem of finding point-to-point shortest distances in graphs is one of the most basic, and most studied, problems in graph databases. Algorithms for the problem can be divided into exact (e.g., [3,5,9,10,15,16,26]) and approximate (e.g., [11,18,19,23,24]), or grouped by the targeted graph types such as road-networks [1,2,28,29], small-world networks [3,16] and general graphs [5,26]. The exact algorithms can be further divided into tree-decomposition based [26,27], 2-hop index based [3] or multi-hop index based [5,26]. There are also works that focus on dynamic graphs and incremental maintenance of indexes [13,14,20]. As noted in recent studies [16,17], overall the pruned landmark labeling (PLL) [3] and its variants (such as [2] and [17], to be discussed in some detail below) are the best-performing.

It was noted in [3] that vertex ordering plays an important role in the label size. However, other than the heuristic degree-based ordering, no specific vertex ordering method was given. The degree-based labeling heuristic some-times generates labels that are too large, especially for graphs that have "high-way" structures that resemble those in real-world road networks. Based on this observation [2] proposed *pruned highway labeling* which first partitions the vertices into $\{P_1, P_2, \ldots, P_N\}$ where each P_i is a shortest path $p_{i,1}, p_{i,2}, \ldots, p_{i,l_i}$ between $p_{i,1}$ and p_{i,l_i} (intuitively these shortest paths represent the "highways"), and then the labels of node v, $label(v)$, are changed to triples of the form

$(i, d(p_{i,1}, p_{i,j}), d(v, p_{i,j}))$, and the way to compute the distance between s and t is also slightly changed accordingly. For each vertex v, it computes $label(v)$ using BFS from the vertices $p_{i,j}$ one by one, with similar prune strategy as that in [3]. To get a good ordering of vertices in $\{P_1, P_2, \ldots, P_N\}$, it first computes a shortest path tree rooted at a randomly selected vertex, then starting from the root, the next vertex is selected as the child with the largest number of descendants. If v and its child w are both selected, and the number of descendants of v and w are similar, then it will *skip* v and only do BFS from w. The authors of [17] proposes an idea similar to [2] for ordering the vertices in PLL, specifically it uses the multiplication of vertex degree and difference in descendant size to rank the vertices. Experiments conducted in [17] demonstrate the vertex ordering strategies in [2] and [17] can lead to good improvement in label size and query time over degree-based labeling. [9] proposes a highway cover labelling where a set H of landmarks (called highway nodes) is chosen, the pairwise distance between the vertices in H are pre-computed, and given any two vertices s and t outside H, the distance of the shortest path between s and t that pass through some vertex in S can be computed directly using the labels of s, and this distance is used as an upper bound of $d(s, t)$. Although the index size is much smaller than PLL, the actual shortest distance needs to be computed using a two-way BFS over the graph obtained from $G - H$, guided by the upper bound. More recently, [16] proposes a way to parallelize the PLL label computation by doing BFS from multiple vertices simultaneously, based on the observation that PLL label computation can be ordered by distance instead of vertices. To make the label size smaller, it proposes to skip local-minimum nodes (which themselves are based on ranking the vertices in non-increasing degrees). [15] proposes a 2-hop labeling technique that is I/O efficient.

To the best of our knowledge, none of the previous works has considered using a minimum vertex cover as potential landmarks, and none of them has used core-forest decomposition and a combination of two types of labels. The vertex ordering techniques proposed in [2] and [17] may be combined with our techniques in computing the PLL labels in the core (by ordering the vertices in the vertex cover using the techniques in [17]) in order to further reduce the index size.

7   Conclusion

In this paper, we first proposed to use the vertex cover as the landmarks, which can significantly reduce the index size of PLL. This method can be viewed as a method to order the vertices in the computation of PLL labels, that is, to order the vertices in a minimum vertex cover before those that are not in the cover. Query time is also significantly shorter due to the smaller label size. We then proposed a hybrid index based on core-forest decomposition, where vertices in the core use the PLL labels while the vertices in the forest use a simpler labeling scheme. For datasets where there are large percentages of forest vertices, the hybrid index can further significantly reduce index size, without compromising query performance. These techniques not only works for undirected and unweighted graphs, they can also be applied to directed graphs and weighted graphs.

Acknowledgement. This work is partly supported by The Science and Technology Development Fund, Macau SAR, with the code FDCT-SKL-IOTSC-2018–2020 and FDCT/0045/2019/A1.

References

1. Abraham, I., Delling, D., Goldberg, A.V., Werneck, R.F.: A Hub-based labeling algorithm for shortest paths in road networks. In: Pardalos, P.M., Rebennack, S. (eds.) SEA 2011. LNCS, vol. 6630, pp. 230–241. Springer, Heidelberg (2011). https://doi.org/10.1007/978-3-642-20662-7_20
2. Akiba, T., Iwata, Y., Kawarabayashi, K., Kawata, Y.: Fast shortest-path distance queries on road networks by pruned highway labeling. In: 2014 Proceedings of the Sixteenth Workshop on Algorithm Engineering and Experiments, ALENEX 2014, Portland, Oregon, USA, 5 January 2014, pp. 147–154 (2014)
3. Akiba, T., Iwata, Y., Yoshida, Y.: Fast exact shortest-path distance queries on large networks by pruned landmark labeling. Proc. ACM SIGMOD Int. Conf. Manag. Data **2013**, 349–360 (2013)
4. Bi, F., Chang, L., Lin, X., Qin, L., Zhang, W.: Efficient subgraph matching by postponing cartesian products. In: Proceedings of the 2016 International Conference on Management of Data, SIGMOD Conference 2016, San Francisco, CA, USA, June 26–July 01 2016, pp. 1199–1214 (2016)
5. Chang, L., Yu, J.X., Qin, L., Cheng, H., Qiao, M.: The exact distance to destination in undirected world. VLDB J. **21**(6), 869–888 (2012). https://doi.org/10.1007/s00778-012-0274-x
6. Cohen, E., Halperin, E., Kaplan, H., Zwick, U.: Reachability and distance queries via 2-hop labels. SIAM J. Comput. **32**(5), 1338–1355 (2003)
7. Delling, D.: Route planning in transportation networks: from research to practice. In: Proceedings of the 26th ACM SIGSPATIAL International Conference on Advances in Geographic Information Systems, SIGSPATIAL 2018, Seattle, WA, USA, 06–09 November 2018, p. 2. ACM (2018)
8. Fan, W., Li, J., Ma, S., Wang, H., Wu, Y.: Graph homomorphism revisited for graph matching. PVLDB **3**(1), 1161–1172 (2010)
9. Farhan, M., Wang, Q., Lin, Y., McKay, B.D.: A highly scalable labelling approach for exact distance queries in complex networks. In: Advances in Database Technology - 22nd International Conference on Extending Database Technology, EDBT 2019, Lisbon, Portugal, 26–29 March 2019, pp. 13–24 (2019)
10. Fu, A.W., Wu, H., Cheng, J., Wong, R.C.: IS-LABEL: an independent-set based labeling scheme for point-to-point distance querying. PVLDB **6**(6), 457–468 (2013)
11. Gubichev, A., Bedathur, S.J., Seufert, S., Weikum, G.: Fast and accurate estimation of shortest paths in large graphs. In: Proceedings of the 19th ACM Conference on Information and Knowledge Management, CIKM 2010, pp. 499–508 (2010)
12. Haryanto, A.A., Islam, M.S., Taniar, D., Cheema, M.A.: Ig-tree: an efficient spatial keyword index for planning best path queries on road networks. World Wide Web **22**(4), 1359–1399 (2019)

13. Hassan, M.S., Aref, W.G., Aly, A.M.: Graph indexing for shortest-path finding over dynamic sub-graphs. In: Özcan, F., Koutrika, G., Madden, S. (eds.) Proceedings of the 2016 International Conference on Management of Data, SIGMOD Conference 2016, San Francisco, June 26 - July 01 June 2016, pp. 1183–1197. ACM (2016)
14. Hayashi, T., Akiba, T., Kawarabayashi, K.: Fully dynamic shortest-path distance query acceleration on massive networks. In: Proceedings of the 25th ACM International Conference on Information and Knowledge Management, CIKM 2016, Indianapolis, IN, USA, 24–28 October 2016, pp. 1533–1542. ACM (2016)
15. Jiang, M., Fu, A.W., Wong, R.C., Xu, Y.: Hop doubling label indexing for point-to-point distance querying on scale-free networks. PVLDB **7**(12), 1203–1214 (2014)
16. Li, W., Qiao, M., Qin, L., Zhang, Y., Chang, L., Lin, X.: Scaling distance labeling on small-world networks. In: Proceedings of the 2019 International Conference on Management of Data, SIGMOD Conference 2019, Amsterdam, The Netherlands, 30 June - 5 July 2019, pp. 1060–1077 (2019)
17. Li, Y., U, L.H., Yiu, M.L., Kou, N.M.: An experimental study on hub labeling based shortest path algorithms. PVLDB **11**(4), 445–457 (2017)
18. Potamias, M., Bonchi, F., Castillo, C., Gionis, A.: Fast shortest path distance estimation in large networks. In: Proceedings of the 18th ACM Conference on Information and Knowledge Management, CIKM 2009, Hong Kong, China, 2–6 November 2009, pp. 867–876. ACM (2009)
19. Qi, Z., Xiao, Y., Shao, B., Wang, H.: Toward a distance oracle for billion-node graphs. PVLDB **7**(1), 61–72 (2013)
20. Qin, Y., Sheng, Q.Z., Falkner, N.J.G., Yao, L., Parkinson, S.: Efficient computation of distance labeling for decremental updates in large dynamic graphs. World Wide Web **20**(5), 915–937 (2016). https://doi.org/10.1007/s11280-016-0421-1
21. Ren, X., Wang, J., Han, W., Yu, J.X.: Fast and robust distributed subgraph enumeration. PVLDB **12**(11), 1344–1356 (2019)
22. Rombach, P., Porter, A.A., Fowler, J.H., Mucha, P.J.: Core-periphery structure in networks (revisited). SIAM review **59**(3), 619–646 (2014)
23. Sadri, A., Salim, F.D., Ren, Y., Zameni, M., Chan, J., Sellis, T.: Shrink: Distance preserving graph compression. Inf. Syst. **69**, 180–193 (2017)
24. Sarma, A.D., Gollapudi, S., Najork, M., Panigrahy, R.: A sketch-based distance oracle for web-scale graphs. In: Proceedings of the Third International Conference on Web Search and Web Data Mining, WSDM 2010, pp. 401–410. ACM (2010)
25. Wang, X., Zhang, Y., Zhang, W., Lin, X.: Efficient distance-aware influence maximization in geo-social networks. IEEE Trans. Knowl. Data Eng. **29**(3), 599–612 (2017)
26. Wei, Fang: Efficient graph reachability query answering using tree decomposition. In: Kučera, Antonín, Potapov, Igor (eds.) RP 2010. LNCS, vol. 6227, pp. 183–197. Springer, Heidelberg (2010). https://doi.org/10.1007/978-3-642-15349-5_13
27. Wei-Kleiner, F.: Tree decomposition-based indexing for efficient shortest path and nearest neighbors query answering on graphs. J. Comput. Syst. Sci. **82**(1), 23–44 (2016)
28. Wu, L., Xiao, X., Deng, D., Cong, G., Zhu, A.D., Zhou, S.: Shortest path and distance queries on road networks: an experimental evaluation. PVLDB **5**(5), 406–417 (2012)
29. Zhu, A.D., Ma, H., Xiao, X., Luo, S., Tang, Y., Zhou, S.: Shortest path and distance queries on road networks: towards bridging theory and practice. In: Proceedings of the ACM SIGMOD International Conference on Management of Data, SIGMOD 2013, New York, NY, USA, 22–27 June 2013, pp. 857–868. ACM (2013)

DySky: Dynamic Skyline Queries on Uncertain Graphs

Suman Banerjee[1(✉)], Bithika Pal[2], and Mamata Jenamani[2]

[1] Indian Institute of Technology, Gandhinagar, India
suman.b@iitgn.ac.in
[2] Indian Institute of Technology, Kharagpur, India
bithikapal@iitkgp.ac.in, mj@iem.iitkgp.ac.in

Abstract. Given a graph, and a set of query vertices (subset of the vertices), the dynamic skyline query problem returns a subset of data vertices (other than query vertices) which are not dominated by other data vertices based on certain distance measure. In this paper, we study the dynamic skyline query problem on uncertain graphs (DySky). The input to this problem is an uncertain graph, a subset of its nodes as query vertices, and the goal here is to return all the data vertices which are not dominated by others. We employ two distance measures in the context of uncertain graphs, namely, *Majority Distance*, and *Expected Distance*. Our approach is broadly divided into three steps: *Pruning*, *Distance Computation*, and *Skyline Vertex Set Generation*. We implement the proposed methodology with three publicly available datasets and observe that it can find out skyline vertex set without taking much time even for million sized graphs if expected distance is concerned. Particularly, the pruning strategy reduces the computational time significantly.

Keywords: Uncertain graph · Reliability · Skyline query

1 Introduction

'Skyline' has emerged as an effective multi-criteria decision making operator and hence an extensively researched topic in data management community for almost two decades [4]. Borzsony et al. [2] fist introduced this operator. Given a set of data points D, the skyline operator in it returns the subset of them that are not dominated by other data points present in the dataset. For any two points d_1 and d_2, we say that d_1 dominates d_2, if with respect to each dimension d_1 is not worse than d_2, however, strictly better in at least one dimension. Without loss of generality, in this study, we assume that lower value means better in all dimensions. This problem has been studied in the context of *graph data* as well [20]. In real-world scenarios, the relationship among agents are uncertain in nature and this uncertainty is caused due to several reasons, like noisy

---

The work of the first author is supported by the institute post-doctoral fellowship grant by IIT-Gandhinagar. (Project Number: MIS/IITGN/PD-SCH/201415/006).

Z. Huang et al. (Eds.): WISE 2020, LNCS 12342, pp. 242–254, 2020.
https://doi.org/10.1007/978-3-030-62005-9_18

measurements, unknown values, explicit manipulations, etc. Hence, this kind of situations are modeled as an uncertain graph, where edges are marked with existence probabilities. In case of social networks, these probabilities signify the influence probability between two users, in case of computer networks these signify the successful packet transfer probability between two systems etc.

After introduced by Borzsony et al. [2], skyline queries have been studied on different kinds of data, for different purposes, with different system architectures, such as on road networks [14], on uncertain data [21], on spatial data, finding perspective customers [18], resisting outliers, in distributed environment [21], map-reduce framework [15], and so on. Keeping the topic of this paper in our mind, here we briefly elaborate the skyline query processing on probabilistic and uncertain data. He et al. [5] studied the skyline query on uncertain time series data and developed a two step methodology to answer this probabilistically. Park et al. [15] studied the skyline query processing on uncertain data and proposed parallel algorithms for computing the same using map reduce framework. Zhou et al. [21] studied the skyline query processing over uncertain data in distributed environments. Le et al. [12] studied the skyline queries on uncertain data to return the user specific relevant results without enumerating all possible worlds. Though there are several studies in this direction, skyline query has not been studied yet in the context of uncertain graphs, to the best of our knowledge.

Due to different practical applications, in recent times analysis of uncertain graphs have emerged as an important research topic [11]. Several problems have been studied such as embedding [6], subgraph search [7], structural pattern findings [1] and so an. ke et al. [9] recently studied the $s-t$ reliability problem which asks with how much probability a target node t is reachable from a source node s in a given uncertain graph. Chen et al. [3] studied the frequent pattern finding in uncertain graphs and enumeration-evaluation algorithm for this problem. Look into [8] for survey. Motivated by the scenario that most of the real world networks are uncertain in nature, in this paper, we introduce the problem of skyline queries on uncertain graphs. Particularly, it has the following contributions:

- We propose the noble problem *"Dynamic Skyline Queries on Uncertain Graph Problem"* (**DySky**). Given an uncertain graph with a subset of vertices as query vertices, the goal of this problem is to obtain the subset of the data vertices that are not dominated by the other data vertices with respect to some distance measure from the query vertices.
- We propose a solution approach for this problem, which is divided into three steps: pruning, distance computation, and skyline vertex set generation.
- The proposed methodology has been implemented with three datasets, for different query size, selection strategies, and distance measures.

Rest of the paper is organized as follows: Sect. 2 describes required preliminaries and defines the problem formally. Section 3, contains the proposed methodology, followed by the experimental evaluations in Sect. 4. Finally, Sect. 5 draws the conclusions.

2   Preliminaries and Problem Definition

In this section, we present required preliminary concepts and then define the problem formally. We denote an *uncertain graph* by $\mathcal{G}(\mathcal{V}, \mathcal{E}, \mathcal{W}, \mathcal{P})$, where $\mathcal{V}(\mathcal{G}) = \{v_1, v_2, \ldots, v_n\}$ is the set of n vertices, $\mathcal{E}(\mathcal{G}) \subseteq \mathcal{V}(\mathcal{G}) \times \mathcal{V}(\mathcal{G})$ is the set of m edges, $\mathcal{W}$ is the distance function that assigns each edge to a positive real number, i.e., $\mathcal{W} : \mathcal{E}(\mathcal{G}) \longrightarrow \mathbb{R}^+$, and $\mathcal{P}$ is the existence function that assigns each edge to a probability value, i.e., $\mathcal{P} : \mathcal{E}(\mathcal{G}) \longrightarrow (0, 1]$. In our study, we consider only simple, finite, undirected, and weighted graphs. The number of nodes and edges of the graph $\mathcal{G}$ is denoted by n and m, respectively. For an edge $e \in \mathcal{E}(\mathcal{G})$ its weight and existence probability is denoted by $\mathcal{W}(e)$ and $\mathcal{P}(e)$, respectively. In the literature, an uncertain graph $\mathcal{G}(\mathcal{V}, \mathcal{E}, \mathcal{W}, \mathcal{P})$ can be conceptualized as the probability distribution over a set of deterministic graphs, which is called as the *possible world of the uncertain graph*, and denoted as $\mathcal{L}(\mathcal{G})$. Each $G(V, E, W) \in \mathcal{L}(\mathcal{G})$ is obtained from $\mathcal{G}$ by keeping all its vertices, keeping its edges with existing probability, and if an edge of $\mathcal{G}$ is also there in G, then $\mathcal{W}(e) = W(e)$. Now, the probability that the deterministic garph G will be generated can be computed by the Eq. 1.

$$\mathcal{P}_{G \sqsubseteq \mathcal{G}} = \prod_{e \in E(G)} \mathcal{P}(e) \prod_{e \in E(\mathcal{G}) \setminus E(G)} (1 - \mathcal{P}(e)) \tag{1}$$

In any deterministic graph G, two vertices v_i and v_j are said to be reachable if there exist a path between v_i and v_j. However, in case of uncertain graphs, the reachability between any two given vertices can be defined in probabilistic way, which we call *reliability*. The term reliability between any two vertices v_i and v_j in the uncertain graph $\mathcal{G}$ is defined as the probability that the vertices v_i and v_j are reachable from each other. We define the reliability ($\mathcal{R}^{\mathcal{G}}_{(v_i v_j)}$) between two vertices v_i and v_j as,

$$\mathcal{R}^{\mathcal{G}}_{(v_i v_j)} = \sum_{G \in \mathcal{L}(\mathcal{G})} I^{G}_{(v_i v_j)} \mathcal{P}_{G \sqsubseteq \mathcal{G}}. \tag{2}$$

where, $I^{G}_{(v_i v_j)}$ is the boolean variable, whose value is 1 if v_i and v_j are connected in G and 0 otherwise. In case of a deterministic weighted graph, distance between any two vertices is defined as the sum of individual edge weights constituting shortest path. However, in case of uncertain graphs distance between any two vertices can be defined in many ways. Here, we quote two of them that we use in our study.

Definition 1 (Majority Distance). *[16] Given an uncertain graph $\mathcal{G}$ and its two vertices $v_i, v_j \in V(\mathcal{G})$, its majority distance is denoted by $dist_{md}(v_i, v_j)$ and defined as the most probable shortest path distance. Mathematically, it can be represented by Eq. 3, where $p_{v_i v_j}$ (Eq. 4) is the shortest path distribution between the vertices v_i and v_j that gives probability value for every distance d.*

$$dist_{md}(v_i, v_j) = \underset{d}{argmax} \; p_{v_i v_j}(d) \tag{3}$$

$$p_{v_i v_j}(d) = \sum_{G \mid d_G(v_i, v_j) = d} \mathcal{P}_{G \sqsubseteq \mathcal{G}} \qquad (4)$$

Definition 2 (Expected Distance). *Given an uncertain graph $\mathcal{G}$ and its two vertices $v_i, v_j \in V(\mathcal{G})$, let $P^l_{(v_i v_j)}$ denotes the set of paths upto length l. For each path $p_k \in P^l_{(v_i v_j)}$, the path probability is defined by $\mathbb{P}(p_k)$ in Eq. 5, and, Eq. 6 defines the expected distance between v_i and v_j.*

$$\mathbb{P}(p_k) = \frac{\prod_{e \in p_k} \mathcal{P}(e)}{\sum_{p_j \in P^l_{(v_i v_j)}} \prod_{e \in p_j} \mathcal{P}(e)} \qquad (5)$$

$$dist_E(v_i, v_j) = \sum_{p_k \in P^l_{(v_i v_j)}} dist(p_k).\mathbb{P}(p_k) \qquad (6)$$

For any $p \in \mathbb{Z}^+$, $[p]$ denotes the set $\{1, 2, \ldots, p\}$. Given a set of 2 or more dimensional data points $\mathcal{D}$, the problem of *skyline query computation* asks to find out the data points that are not dominated by any other data points in $\mathcal{D}$, which is formally defined in Definition 3.

Definition 3 (Skyline Query). *Given a set of p dimensional data points $\mathcal{D} = \{d_1, d_2, \ldots, d_{|D|}\}$, we say that d_i dominates d_j, if for all $k \in [p]$, $d_i(k) \leq d_j(k)$ and there exist atleast one $k \in [p]$ such that $d_i(k) < d_j(k)$. Skyline of the dataset $\mathcal{D}$ is the subset of the data points that are not dominated by any of the data points in $\mathcal{D}$.*

Since past one decade or so, skyline queries have been studied extensively [10] in graphs as well, which we define next.

Definition 4 (Skyline Query in Graphs). *Given a graph $G(V, E)$, and a subset of vertices $\mathcal{Q}$ (called query vertices), for any two data vertices (vertices that are not query vertices, i.e., $V(G) \setminus \mathcal{Q}$) v_i and v_j, we say v_i dominates v_j, if $\forall w \in \mathcal{Q}$, $dist(w, v_i) \leq dist(w, v_j)$ and $\exists x \in \mathcal{S}$, such that $dist(x, v_i) < dist(x, v_j)$. The skyline query asks to return data vertices that are not dominated by other data vertices.*

Though, the skyline query problem has been studied in the context of probabilistic data [19], to the best of our knowledge this problem has not been studied in the context of uncertain graphs. In this paper, we introduce the problem of finding the dynamic skyline queries on uncertain graphs (DySky) as follows.

Definition 5 (Dynamic Skyline Queries on Uncertain Graphs). *Given an uncertain graph $\mathcal{G}$, a subset of vertices $\mathcal{Q}$ (called query vertices), and a distance measure (i.e., expected distance, majority distance etc.) the problem of dynamic skyline queries on uncertain graphs asks to find out the subset of the data vertices such that none of them are dominated by the other data vertices.*

Just to highlight, here the term 'Dynamic' does not mean the graph is time varying; rather it means that the query vertices are not fixed and can be varied. Figure 1 shows a toy example of an uncertain graph with its majority distance, expected distance, and shortest path distance (for deterministic version) tables, where the skyline vertices are marked in orange color. It is important to observe as the distance measure changes, the skyline vertex set is also getting changed. This motivates us to study the DySky Problem under two different distance measures.

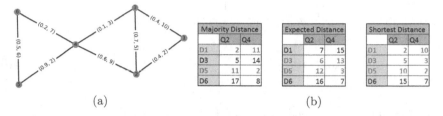

(a)   (b)

Fig. 1. (a) An uncertain graph with 6 vertices and 8 edges. The vertices 2 and 4 are the Query Vertices (denoted as $Q2$ and $Q4$) and remaining are data vertices (i.e., $D1$, $D3$, $D5$, and $D6$). (b) Different distance tables (Majority Distance, Expected Distance, Shortest Path Distance in deterministic version) between the query and data vertices. Skyline vertices in each cases are marked in Orange. (Color figure online)

3   Proposed Methodology

3.1   Overview

The proposed methodology is broadly divided into three steps:

- **Step 1 (Pruning):** In this step, a subset of the data vertices are returned as the candidate skyline vertices. This step comprises of two subsets. First, pruning is done by performing *Breadth First Search* (henceforth mentioned as B.F.S.) from the query vertices and subsequently, pruning is done based on path length computation.
- **Step 2 (Distance Computation):** In this step, distance computation is done between the candidate skyline vertices and the query vertices. As mentioned previously, in our study we have used majority distance and expected distance.
- **Step 3 (Skyline Vertex Set Generation):** Based on the previously computed distance, any existing skyline finding algorithm can be used to find out the actual skyline vertices. In our study, we have used the *Block Nested Loop (BNL)* Algorithm proposed by Borzsonyi et al. [2].

3.2   The Algorithm

Algorithm 1, 2, and 3 together constitute the proposed methodology for the DySky Problem. We describe the entire procedure in two subsections.

The Pruning Step. Algorithm 1 describes the B.F.S. and distance based pruning strategies. It takes the uncertain graph, the set of query vertices, and distance threshold as inputs and outputs the candidate skyline vertices. In B.F.S. pruning, from each of the query vertices, B.F.S. trees are constructed to check the connectivity. First, we create the dictionary $\mathcal{D}$. If a query vertex and data vertex is connected and the data vertex has the entry in the dictionary $\mathcal{D}$, the query vertex is included as a value corresponding to this key. Otherwise, a key corresponding to the data vertex is created and the query vertex is added as a value to this 'key'. Now, the data vertices that are reachable from all the query vertices are kept as the candidate skyline vertices. Here, the B.F.S. pruning ends.

In reality, even if two vertices are connected by a path of large distance (i.e., more than certain threshold), reachability becomes costlier. Hence, to eliminate such vertices, we perform the path length-based pruning. For this purpose, shortest path length between candidate skyline vertex and query vertex is computed. For a candidate skyline vertex, if there exist atleast one query vertex for which the computed length value is more than the user defined threshold, the candidate skyline vertex set is updated by removing the candidate skyline vertex.

Any pruning strategy to work correctly should guarantee that it does not remove any skyline vertices. In Lemma 1, we state the proposed pruning strategy is correct. Due to space limitation, we are not able to give the proof, though it is easy to follow.

Algorithm 1: Step 1 (B.F.S and Distance based pruning)

Data: Uncertain Graph $\mathcal{G}(\mathcal{V}, \mathcal{E}, \mathcal{W}, \mathcal{P})$, the
Set of Query Vertices $\mathcal{Q} \subseteq \mathcal{V}(\mathcal{G})$,
Distance Threshold T
Result: Candidate Skyline Vertices
$\mathcal{CS} \subseteq \mathcal{V}(\mathcal{G}) \setminus \mathcal{Q}$

1 . Create Dictionary $\mathcal{D}$
2 **for** $All\ u \in \mathcal{Q}$ **do**
3 $\quad$ **for** $All\ v \in V(\mathcal{G}) \setminus \mathcal{Q}$ **do**
4 $\quad\quad$ **if** $Isconnected(uv)$ **then**
5 $\quad\quad\quad$ **if** $v \in \mathcal{D}.Keys()$ **then**
6 $\quad\quad\quad\quad$ $\mathcal{D}[v].values() =$
$\quad\quad\quad\quad\quad$ $\mathcal{D}[v].values() \cup \{u\}$
7 $\quad\quad\quad$ **else**
8 $\quad\quad\quad\quad$ $\mathcal{D}.Create\_Key(v)$
9 $\quad\quad\quad\quad$ $\mathcal{D}[v].Add\_Value(u)$

10 $\mathcal{CS} = \emptyset$
11 **for** $All\ u \in \mathcal{D}.Keys()$ **do**
12 $\quad$ **if** $\mathcal{D}[u].Values() == \mathcal{Q}$ **then**
13 $\quad\quad$ $\mathcal{CS} = \mathcal{CS} \cup \{u\}$

14 **for** $All\ v \in \mathcal{CS}$ **do**
15 $\quad$ **for** $All\ u \in \mathcal{Q}$ **do**
16 $\quad\quad$ **if** $distance(uv) > T$ **then**
17 $\quad\quad\quad$ $\mathcal{CS} = \mathcal{CS} \setminus \{v\}$

Lemma 1 *The proposed pruning strategy (Algorithm 1) is correct.*

Now, we do an analysis for time and space requirement of Algorithm 1. Let q be the number of query vertices, i.e., $|\mathcal{Q}| = q$. For creating the B.F.S. trees rooted at the query vertices requires $\mathcal{O}(q(m + n))$ time. The maximum number of values associated with a 'key' in the dictionary $\mathcal{D}$ is of $\mathcal{O}(q)$. Execution time from Line No. 2 to 9 and 11 to 17 requires $\mathcal{O}(q(n-q)^2)$ and $\mathcal{O}((n-q)q^2)$. Now, in distance-based pruning, the number of distance computations is $\mathcal{O}(q(n-q))$. Computing shortest path between two vertices in a weighted graph with positive edge weights requires $\mathcal{O}(m+n\log n)$ time. Hence, time requirement for distance-based pruning requires $\mathcal{O}(q(n-q)(m+n\log n))$ time. Total time requirement for Algorithm 1 is of $\mathcal{O}(q(m+n)+nq(n-q)+q(n-q)(m+n\log n)) = \mathcal{O}(q(n-q)(m+n\log n))$. Extra space requirement of Algorithm 1 is to store the dictionary $\mathcal{D}$, which is of $\mathcal{O}(q(n-q))$, to store the candidate skyline vertices, which is of $\mathcal{O}(n-q)$, and to perform the B.F.S., which is of $\mathcal{O}(n)$. Hence, total space requirement of Algorithm 1 is of $\mathcal{O}(q(n-q))$. Lemma 2 describes the formal statement.

Lemma 2 *Time and space requirement of Algorithm 1 is of* $\mathcal{O}(q(n-q)(m+n\log n))$ *and* $\mathcal{O}(q(n-q))$, *respectively.*

Distance Computation and Skyline Vertex Set Generation. Now, we describe Step 2 and 3 of our proposed methodology. It is important to observe that depending upon which distance measure is used (i.e., majority distance or expected distance) Step 2 will be different. Algorithm 2 and 3 describes the last two steps for the majority distance and expected distance, respectively.

Algorithm 2: Step 2 and 3 (Distance Computation and Skyline Vertex Set Generation) for Majority Distance

Data: Candidate Skyline Vertices $\mathcal{CS}$
Result: The Skyline Vertex Set $\mathcal{S}$
1   Generate $|\mathcal{R}|$ number of sample graphs
2   Store the graph probabilities in $\mathcal{P}_G[1\ldots|\mathcal{R}|]$
3   Create_Matrix $\mathcal{M} \in \mathbb{R}^{|\mathcal{CS}|\times|\mathcal{Q}|}$
4   for *All* $u \in \mathcal{CS}$ do
5   for *All* $v \in \mathcal{Q}$ do
6   Create dictionary $Temp$
7   for *All* $r \in \mathcal{R}$ do
8         $d =$ shortest distance between u and v in r
9         $Temp[d] = Temp[d] + \mathcal{P}_G[r]$
10      $\mathcal{M}[u][v] = \underset{d}{argmax}\ Temp[d]$
11   $\mathcal{S} =$ Apply BNL on $\mathcal{M}$
12   return $\mathcal{S}$

Algorithm 3: Step 2 and 3 (Distance Computation and Skyline Vertex Set Generation) for Expected Distance

Data: Candidate Skyline Vertices $\mathcal{CS}$
Result: The Skyline Vertex Set $\mathcal{S}$
1   . Create_Matrix $\mathcal{M} \in \mathbb{R}^{|\mathcal{CS}|\times|\mathcal{Q}|}$
2   for *All* $u \in \mathcal{CS}$ do
3   for *All* $v \in \mathcal{Q}$ do
4       $path =$ Compute all paths from u to q upto length l
5       $prob[1\ldots|path|] = 0$; $dist[1\ldots|path|] = 0$
6   for *All* $t \in path$ do
7   for *All* $e \in \mathcal{E}(t)$ do
8           $prob[t] = prob[t] + \mathcal{P}(e)$
9           $dist[t] = dist[t] + \mathcal{P}(e) * \mathcal{W}(e)$
10      $\mathcal{M}[u][v] = \sum dist / \sum prob$
11   $\mathcal{S} =$ Apply BNL on $\mathcal{M}$
12   return $\mathcal{S}$

For the majority distance case, first we generate $|\mathcal{R}|$ number of subgraphs, as mentioned, while defining possible world semantics, and the corresponding generation probabilities are stored in the array $\mathcal{P}_G$. Next, the majority distance is computed between a candidate skyline vertex and a query vertex. Finally, the BNL Algorithm is applied on the distance matrix $\mathcal{M}$ to obtain the skyline vertex set.

Now, we analyze Algorithm 2 for time and space requirement. As mentioned in the definition of possible world semantics, generation of $|\mathcal{R}|$ number of subgraphs require $\mathcal{O}(m|\mathcal{R}|)$ time. Using Dijkstra's algorithm computing the shortest path between a pair of vertices requires $\mathcal{O}(m+n\log n)$ time. Hence, execution time from Line 4 to 10 requires $\mathcal{O}(q(n-q)|\mathcal{R}|(m+n\log n))$. Now, BNL algorithm requires $\mathcal{O}((n-q)^2)$ time. Extra space consumed by Algorithm 2 is to store the array $\mathcal{P}_G$, $Temp$, and the matrix $\mathcal{M}$ which requires $\mathcal{O}(|\mathcal{R}|)$, $\mathcal{O}(|\mathcal{R}|)$, and $\mathcal{O}(q(n-q))$ space, respectively. The formal statement is given in Lemma 3.

Lemma 3 *Time and space requirement of Algorithm 2 is of $\mathcal{O}(q(n-q)|\mathcal{R}|(m+n\log n)+(n-q)^2)$ and $\mathcal{O}(q(n-q)+|\mathcal{R}|)$, respectively.*

Lemma 2 and 3 together imply the statement mentioned in Theorem 1.

Theorem 1 *If majority distance is concerned, the proposed methodology returns the skyline vertex set in $\mathcal{O}(q(n-q)|\mathcal{R}|(m+n\log n)+(n-q)^2)$ time and $\mathcal{O}(q(n-q)+|\mathcal{R}|)$ space.*

It is trivial to observe that Algorithm 3 just implements the expected distance, and hence, without explanation we move to analyze the algorithm. Assume that maximum degree of the input uncertain graph is d_{max}. Hence, the maximum number paths upto length l between any pair of vertices is of $\mathcal{O}(d_{max}^l)$. Hence, running time from Line 2 to 10 is of $\mathcal{O}(q(n-q)ld_{max}^l)$. Hence, total running time of Algorithm 3 is of $\mathcal{O}(q(n-q)ld_{max}^l+(n-q)^2)$. Extra space consumed by the Algorithm 3 is to store the matrix $\mathcal{M}$, array $Path$, $Prob$ and $dist$ which requires $\mathcal{O}(q(n-q)+ld_{max}^l)$. Hence, Lemma 4 holds.

Lemma 4 *The running time and space requirement of Algorithm 3 is of $\mathcal{O}(q(n-q)ld_{max}^l+(n-q)^2)$ and $\mathcal{O}(q(n-q)+ld_{max}^l)$, respectively.*

Lemma 2 and 4 together imply the statement mentioned in Theorem 2.

Theorem 2 *If expected distance is concerned, the proposed methodology returns the skyline vertex set in $\mathcal{O}(q(n-q)(m+n\log n+ld_{max}^l)+(n-q)^2)$ time and $\mathcal{O}(q(n-q)+|\mathcal{R}|)$ space.*

4   Experimental Evaluations

In this section, we describe the experimental validations of our proposed approach. We have used three different **datasets**, appeared in two different contexts, which are described below.

- **Minnesota Road Network (MRN)** [17]: This is a road network dataset of the Minnesota city, with $n = 2642$, $m = 3300$, and avg. degree $= 2$. Here, the junctions are represented by the nodes. If two junctions are connected by a road, then the corresponding two vertices are connected by an edge.
- **P2P Network** [13]: This dataset contains a sequence of snapshots of the Gnutella peer-to-peer file sharing network from August 2002, with $n = 8114$, $m = 26013$, and avg. degree $= 6.41$. There are total of 9 snapshots of Gnutella network collected in August 2002. Nodes represent hosts in the Gnutella network topology and edges represent connections between the Gnutella hosts.

- **USA Road Network (URN)** [17]: This dataset describes a road network from the United States, with $n = 129164$, $m = 165435$, and avg. degree $= 2.56$. Here, vertices represent junctions, and an edge between signifies that the corresponding junctions are are connected by road.

All the datasets are undirected and unweighted. Probability of existence and weight of each edge is chosen from the intervals $(0, 1]$ and $[10, 100]$ uniformly at random. In our experiments, we consider l as 4. In our study the following query vertex selection strategies have been adopted, for the **experimental setup**.

- **RAND**: By this method, to select k query vertices first one is chosen randomly and remaining $(k - 1)$ query vertices are chosen from the two hop neighbors of the initially selected vertex uniformly at random.
- **HDEG**: By this method, to select k query vertices first the subset of the nodes whose degree is more than a threshold value are marked and a node is chosen uniformly at random as a query vertex. Remaining $(k - 1)$ are chosen from the two hop neighbors of the initially selected vertices uniformly at random.
- **HCLUS**: This method is exactly the same as HDEG, except the case that, for choosing the first query vertex the subset of vertices are chosen based on the clustering coefficient of nodes.

Based on the selection strategy, we choose the query size from the set $\{2, 3, 5, 8, 10, 15, 20\}$. The experiments are repeated for 10 times. All the algorithms have been implemented with Python 3.5 + NetworkX 2.1 environment on a HPC Cluster with 5 nodes each of them having 64 cores and 160 GB of memory and the implementations are available at https://github.com/BITHIKA1992/Skyline_Uncertain_Graph/. Now, the **goals of the experiments** are defined below.

- *Efficiency of the Pruning Strategies*: As the number of query vertices increases, what is the fraction of data vertices removed before distance computation?
- *Query Size Vs. Skyline Vertices*: Under different query vertex selection strategies how the cardinality of the skyline vertex set changes with respect to the query size?
- *Distance Metric Vs. Skyline Vertices*: For a fixed query selection strategy and query size, how the cardinality of the skyline vertices changes with respect to the distance metric?
- *Query Selection Strategy Vs. Skyline Vertices*: For a fixed query size and distance metric, how the cardinality of the skyline vertices changes with respect to query selection strategy?
- *Query Size Vs. Computational Time*: For a fixed query size and distance metric, how computational time grows with respect to the query size?

Now, we discuss the **experimental results** of the proposed approach.

Efficiency of the Pruning Strategies. It is easy to observe that the B.F.S pruning will return the vertices from the component in which the query vertices belong. As we have selected the query vertices from the largest component, the BFS pruning returns the vertices from the largest component only. For distance-based pruning, we have taken the threshold value as 400, considering 4-hop path with the maximum edge weight 100. In Fig. 2, we show the box plot for the candidate size with respect to each query size and the query selection strategy. It can be observed that the candidate size for RAND selection strategy is less than the other two, in all the datasets, which is easy to convince. For P2P network, the interquartile range is very high compared to other datasets. This is due to the fact that the network is of high average degree. Also, for RAND selection strategy, this range is the highest for small query size. This is due to the existence of various small size components in the network. Both the road networks are very sparse and for the large query sizes like 15, 20, the candidate size becomes very small and the variance also reduces. With this sparsity for small road network MRN, it is impossible to find the connected vertices from all the query vertices within the distance of 400. So, we remove the results for query size of 15 and 20 for the MRN dataset.

Query Size Vs. Skyline Vertices. In Fig. 3, we show the plot for query size Vs. skyline size, with two distance metrics and three query selection strategies. In this part, we describe the comparison of sizes. From all the 10 executions, here we report the mean values for the skyline size. With the increase in query

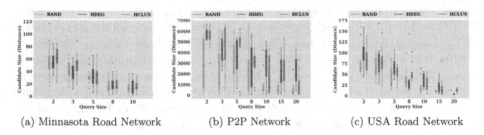

(a) Minnesota Road Network   (b) P2P Network   (c) USA Road Network

Fig. 2. Box plot for the candidate skyline size with respect to the query size for the Minnesota Road Network, P2P Network, and USA Road network datasets.

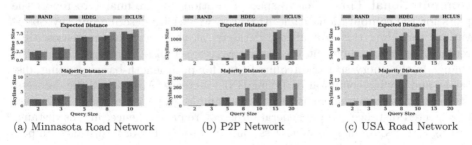

(a) Minnesota Road Network   (b) P2P Network   (c) USA Road Network

Fig. 3. Query size Vs. Skyline size plot for the Minnesota Road Network, P2P Network, and USA Road network datasets.

size, the skyline size increases. However, for the URN dataset in Fig. 3(c), the skyline size decreases for large value of query size. The reason is due to the small size of the candidate skyline, which can be verified from Fig. 2(c). Also, for both the road network datasets the maximum skyline size reaches approximately 15, whereas for the P2P network it reaches around 1500. This is due to its candidate size. For, both the cases, at a large value of query size, the ratio of candidate to skyline size is very small. As the number of query vertices increases, the possibility of domination decreases.

Distance Metric Vs. Skyline Vertices. In this part, referring to Fig. 3, we describe the behavior of skyline size with respect to different distance metrics. For the road networks in Fig. 3(a) and (c), the skyline size is similar in both the datasets. However, for the P2P network in Fig. 3(b), the skyline size in the expected distance ($\approx$max 1500) is much more than the majority distance ($\approx$max 300). The reason lies in the networks' high average degree value and density. As the number of paths increases between a query vertex to a data vertex, the expected distance value is unable to dominate other data vertices. This results in the large size of the skyline vertex set. This can be verified from Fig. 3(b), by looking into HDEG and HCLUS selection strategies, where it differs from the expected distance results. However, for RAND, the size is similar in both the distances. From the experiments, we also observe that for a particular query vertex set the skyline vertices may not be the same from both the distances.

Query Selection Strategy Vs. Skyline Vertices. In this part, referring to Fig. 3, we describe the behavior of skyline size with respect to two different query selection strategies. First, we describe the threshold value selected for HDEG and HCLUS for different datasets. As the P2P network dataset consists of high degree nodes, we select the high degree threshold value as 15, and it returns 440 nodes. In case of both the road networks, the maximum degree is around 5. Hence, for MRN and URN datasets, this threshold value is considered as 2 and 3, respectively. The clustering coefficient threshold is taken as 0 as the clustering coefficient for all the networks are very less. From Fig. 3, the main observation is that for all the selection strategies the skyline size does not vary much for smaller query size. Whereas, for the large value of query size, HCLUS gives the maximum skyline vertices.

Computational Time. Due to space limitation, we are unable to report the plots for the computational time. However, we briefly describe our key observations regarding this. For all the datasets, as the query size increases, time requirement for finding out the skyline vertex set also increases. Due to the change in the query size, the required time for path length-based pruning, distance, and skyline computation (using BNL) increase. Also, for all the datasets, the main time requirement is due to the sample graph generation. As in case of expected distance sample generation is not required, hence, in this distance setting time requirement is much less compared to the majority distance. In particular, for query size 2, the ratio between the computational time requirement for majority distance to expected distance for MRN, P2P, and URN are 47, 28, and 2556, respectively. Now, for the P2P Network dataset, when the query size increases beyond 10, there is a sharp increase in the skyline computation time. This is because for higher query sizes the candidate and skyline size are more compared to the previous query sizes and observed in 2(b) and 3(b).

5 Conclusion

In this paper, we introduce the problem of dynamic skyline queries on uncertain graphs for two different distance measures, namely, majority distance and expected distance. For this problem, we have proposed a methodology having three main steps: pruning, distance computation, and skyline vertex set generation. The proposed methodology has been analyzed to understand its time and space requirements. The experimental results demonstrate that it can find out the skyline vertex set with reasonable computation time.

References

1. Bonchi, F., Gullo, F., Kaltenbrunner, A., Volkovich, Y.: Core decomposition of uncertain graphs. In: Proceedings of the 20th ACM SIGKDD International Conference on Knowledge Discovery and Data Mining, pp. 1316–1325. ACM (2014)
2. Borzsony, S., Kossmann, D., Stocker, K.: The skyline operator. In: Proceedings 17th International Conference on Data Engineering, pp. 421 430. IEEE (2001)
3. Chen, Y., Zhao, X., Lin, X., Wang, Y., Guo, D.: Efficient mining of frequent patterns on uncertain graphs. IEEE Trans. Knowl. Data Eng. **31**(2), 287–300 (2018)
4. Chomicki, J., Ciaccia, P., Meneghetti, N.: Skyline queries, front and back. ACM SIGMOD Rec. **42**(3), 6–18 (2013)
5. He, G., Chen, L., Zeng, C., Zheng, Q., Zhou, G.: Probabilistic skyline queries on uncertain time series. Neurocomputing **191**, 224–237 (2016)
6. Hu, J., Cheng, R., Huang, Z., Fang, Y., Luo, S.: On embedding uncertain graphs. In: Proceedings of the 2017 ACM on Conference on Information and Knowledge Management, pp. 157–166. ACM (2017)
7. Jin, R., Liu, L., Aggarwal, C.C.: Discovering highly reliable subgraphs in uncertain graphs. In: Proceedings of the 17th ACM SIGKDD International Conference on Knowledge Discovery and Data Mining, pp. 992–1000. ACM (2011)
8. Kassiano, V., Gounaris, A., Papadopoulos, A.N., Tsichlas, K.: Mining uncertain graphs: an overview. In: Sellis, T., Oikonomou, K. (eds.) ALGOCLOUD 2016. LNCS, vol. 10230, pp. 87–116. Springer, Cham (2017). https://doi.org/10.1007/978-3-319-57045-7_6
9. Ke, X., Khan, A., Quan, L.L.H.: An in-depth comparison of st reliability algorithms over uncertain graphs. Proc. VLDB Endowment **12**(8), 864–876 (2019)
10. Khan, A., Singh, V., Wu, J.: Finding skyline nodes in large networks. In: 28th International Conference on Data Engineering Workshops, pp. 198–204. IEEE (2012)
11. Khan, A., Ye, Y., Chen, L.: On uncertain graphs. Synth. Lect. Data Manage. **10**(1), 1–94 (2018)
12. Le, T.M.N., Cao, J., He, Z.: Answering skyline queries on probabilistic data using the dominance of probabilistic skyline tuples. Inf. Sci. **340**, 58–85 (2016)
13. Leskovec, J.: Gnutella peer-to-peer network, august 4 2002 (2002). https://snap.stanford.edu/data/p2p-Gnutella04.html

14. Miao, X., Gao, Y., Guo, S., Chen, G.: On efficiently answering why-not range-based skyline queries in road networks. IEEE Trans. Knowl. Data Eng. **30**(9), 1697–1711 (2018)
15. Park, Y., Min, J.K., Shim, K.: Processing of probabilistic skyline queries using mapreduce. Proc. VLDB Endowment **8**(12), 1406–1417 (2015)
16. Potamias, M., Bonchi, F., Gionis, A., Kollios, G.: Nearest-neighbor queries in probabilistic graphs. Technical report, Boston University Comp. Science Department (2009)
17. Rossi, R.A., Ahmed, N.K.: The network data repository with interactive graph analytics and visualization. In: AAAI (2015). http://networkrepository.com
18. Yin, B., Gu, K., Wei, X., Zhou, S., Liu, Y.: A cost-efficient framework for finding prospective customers based on reverse skyline queries. Knowl. Based Syst. **152**, 117–135 (2018)
19. Zhang, K., Gao, H., Han, X., Cai, Z., Li, J.: Modeling and computing probabilistic skyline on incomplete data. IEEE Trans. Knowl. Data Eng. **32**, 1405–1418 (2019)
20. Zheng, W., Zou, L., Lian, X., Hong, L., Zhao, D.: Efficient subgraph skyline search over large graphs. In: Proceedings of the 23rd ACM International Conference on Information and Knowledge Management, pp. 1529–1538. ACM (2014)
21. Zhou, X., Li, K., Zhou, Y., Li, K.: Adaptive processing for distributed skyline queries over uncertain data. IEEE Trans. Knowl. Data Eng. **28**(2), 371–384 (2015)

Leveraging Double Simulation to Efficiently Evaluate Hybrid Patterns on Data Graphs

Xiaoying Wu[1], Dimitri Theodoratos[2(✉)], Dimitrios Skoutas[3], and Michael Lan[2]

[1] Computer School of Wuhan University, Wuhan, China
xiaoying.wu@whu.edu.cn
[2] New Jersey Institute of Technology, Newark, USA
{dth,ml122}@njit.edu
[3] Athena R.C., Marousi, Greece
dskoutas@athenarc.gr

Abstract. Labeled graphs are used to represent entities and their relationships in a plethora of Web applications. Graph pattern matching is a fundamental operation for the analysis and exploration of data graphs. In this paper, we address the problem of efficiently finding homomorphic matches for hybrid patterns, where each edge may be mapped either to an edge or to a path, thus allowing for higher expressiveness and flexibility in query formulation. We design a novel holistic graph simulation-based algorithm, called *GraphMatch-Sim*, which leverages simulation to precisely identify, in advance, all the graph nodes that participate in the pattern matches returned. *GraphMatch-Sim* can flexibly employ any reachability index as a plug-in component. Unlike existing methods, it produces no redundant intermediate results, thus achieving worst-case optimality. An extensive experimental evaluation on both real and synthetic datasets shows that our method evaluates hybrid patterns orders of magnitude faster than existing algorithms and has much better scalability.

1 Introduction

Labeled graphs can model complex relationships among entities and are used to represent data in a plethora of Web applications. Graph matching, which refers to identifying the matches of a query pattern in a graph, is a fundamental operation for graph analysis and exploration, and is becoming increasingly important due to the massive growth in the volume of graph data being produced.

Existing approaches can be characterized by: (a) the type of *edges* the patterns have, and (b) the type of *morphism* used to map the patterns to the data structure. An edge in a pattern can be either a *child* edge, which is mapped to an *edge* in the data graph, or a *descendant* edge, which is mapped to a *path* in

The research of the first author was supported by the National Natural Science Foundation of China under Grant No. 61872276.

© Springer Nature Switzerland AG 2020
Z. Huang et al. (Eds.): WISE 2020, LNCS 12342, pp. 255–269, 2020.
https://doi.org/10.1007/978-3-030-62005-9_19

the data graph. The morphism determines how a pattern is mapped to the data graph. Over the years, research has advanced from considering *isomorphisms* for matching *patterns with child edges* [12] towards considering *homomorphisms* for matching *patterns with descendant edges* [3,4,7]. By allowing *edge-to-path* mappings, homomorphisms can extract matches "hidden" deep within large graphs, which might be missed by isomorphisms. This is an important feature for numerous applications, especially when users explore a large and heterogeneous graph without being very familiar with its schema, which is often with Web data.

In this paper, we focus on finding homomorphic matches of tree patterns on large graphs. Tree patterns are the basic building blocks of general graph patterns. Given a graph pattern Q, a common approach taken by many graph pattern matching methods [1,2,11,14] is to decompose or transform Q into tree patterns and use them as the basic processing unit. Thus, efficiently finding tree pattern matches on graphs is crucial in graph processing.

Existing algorithms for homomorphically matching tree patterns to graphs face two problems. First, they often produce a large amount of *intermediate results* which do not participate in the final query answer, hence their performance does not scale when the size of the graph increases. Second, they cannot efficiently handle *hybrid patterns* that allow both descendant and child edges. Often, an explicit parent-child relationship in the data graph is important. For instance, when child edges represent causal effects, patterns with child edges extract specific information that is missed by those with descendant edges. Hybrid patterns generalize both types, enabling more flexibility in query formulation and more fine-grained results.

Contribution. We address the problem of efficiently evaluating hybrid tree patterns using homomorphisms over a large graph. We introduce a novel simulation-based technique for identifying and excluding nodes of the data graph which do not participate in the final solutions. Simulation is a very recently proposed powerful technique for node pruning in graph matching but has never been used before on edge-to-path matching or in hybrid patterns. Our main contributions are as follows:

- To increase efficiency, our approach computes a *binary match simulation relation* to identify, in advance, exactly the graph nodes occurring in the answer, instead of directly computing homomorphic matches.
- We design a *holistic algorithm*, called *GraphMatch-Sim*, that integrates pruning and graph pattern matching, by matching the query pattern against the input data as a whole. It processes the child edges of hybrid patterns directly without resorting to a post-processing step. We also present *tuning strategies* to further improve the performance of our algorithm. Contrary to previous algorithms, *GraphMatch-Sim* produces no redundant intermediate results thus achieving worst-case optimality.
- Unlike most existing graph matching algorithms that allow edge-to-path mapping [4,7], *GraphMatch-Sim* is not tied to a specific *reachability indexing scheme*, thus being more generic and flexible. It can employ any reachabil-

ity index as a plug-in component. In our implementation, we use the recent efficient reachability scheme *Bloom Filter Labeling* (BFL) [10].

- We run extensive experiments to evaluate the performance and scalability of our algorithm on real and synthetic datasets. We also compare with the best known previous graph matching algorithms and their suggested pre-filtering techniques [13,14]. The results show that our simulation-based technique has greater pruning power than other pre-filtering techniques and that *GraphMatch-Sim* significantly outperforms previous algorithms even when they are modified to use BFL.

2   Related Work

We next discuss the major issues of existing homomorphism-based graph pattern evaluation methods in the literature.

Existing algorithms spend a considerable amount of time on intermediate results (candidate results in construction that do not participate in the final answer). For example, decomposition-based algorithms such as R-Join [4], Twig-StackD [14], and EmptyHeaded [1] decompose the query pattern into several subpatterns, produce the candidates for each subpattern, and join them to form the final answer. These algorithms decompose the input query into a set of binary (reachability) relationships between pairs of nodes (R-Join [4]), or into a set of root-to-leaf paths (TwigStackD [14]), or into a tree of subqueries (Empty-Headed [1]). The decomposition-based approach suffers from a large number of intermediate candidates, thus spending a prohibitive amount of time on examining false positives. Follow-up work [7] has shown the limited scalability of the decomposition-based approach.

To efficiently check the reachability relationship between two given nodes in a graph, every graph pattern matching algorithm that allows edge-to-path mapping uses some kind of reachability indexing scheme. Most schemes associate with every node a label which is an entry in the index for the data graph. However, most existing graph matching algorithms are tied to a specific reachability indexing scheme, such as 2HOP labeling in R-Join [4], and 3HOP labeling in TPQ-3Hop [7]. This tight-coupling of the matching algorithm with a specific labeling scheme compromises generality and flexibility.

Finally, the above mentioned algorithms were designed either for child-only [1] or descendant edge-only query patterns [4,7,14]. They cannot efficiently handle hybrid patterns that allow both descendant and child edges. TPQ-3Hop [7] suggests a post-processing strategy for handling possible child edges. Our experimental results in Sect. 6 show that this post-processing is expensive, as there exists a large amount of intermediate results which do not participate in the query answer. A recent algorithm, called *GPM* [13], evaluates hybrid patterns in a bottom-up way. It incrementally evaluates the pattern query from the results of the sub-patterns rooted at the children of the current pattern node. However, it does not avoid generating redundant intermediate results.

The *GraphMatch-Sim* algorithm we develop in this paper overcomes the aforementioned shortcomings. It can evaluate hybrid patterns on data graphs in

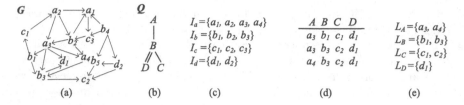

Fig. 1. (a) Data graph G, (b) Query Q, (c) Inverted lists of G, (d) Answer of Q on G, (e) Occurrence lists of Q on G.

a holistic way, computing exactly the nodes occurring in the final query answer without generating any redundant nodes. Moreover, it can flexibly employ any reachability index as a plug-in component.

3   Problem Definition

Data graph. Assume a directed graph $G = (V, E)$ where V and E denote the sets of nodes and edges, respectively. Each node $v \in V$ is associated with a label $label(v) \in \mathcal{L}$ from a finite set of labels $\mathcal{L}$. For each label $x \in \mathcal{L}$, the inverted list I_x contains the nodes in G with label x. A node u is said to *reach* node v, denoted $u \prec v$, if there exists an edge or a path from u to v. Abusing tree notation, we call v a *child* of u (and u a *parent* of v) if $(u, v) \in E$, while v a *descendant* of u (and u an *ancestor* of v) if $u \prec v$. Figure 1(a) shows a data graph G, and Fig. 1(c) shows the corresponding inverted lists.

Query. We focus on queries that are tree-patterns. Every node x in a pattern Q has a label $label(x) \in \mathcal{L}$. There can be two types of edges in Q. A *child* (resp. *descendant*) edge denotes a child (resp. descendant) relationship between the respective two nodes. Figure 1(b) shows a query. Single line edges denote child edges while double line edges represent descendant edges.

Answer. Given a tree pattern Q and a data graph G, a *homomorphism* from Q to G is a function m mapping the nodes of Q to nodes of G, such that: (1) for any node $x \in V_Q$, $label(x) = label(m(x))$; and (2) for any edge $(x, y) \in E_Q$, if (x, y) is a child edge, $(m(x), m(y))$ is an edge of G, while if (x, y) is a descendant edge, $m(x) \prec m(y)$ in G.

Let (q_i, q_j) be a query edge in Q, and v_i and v_j be two nodes in G, such that $label(q_i) = label(v_i)$ and $label(q_j) = label(v_j)$. The pair (v_i, v_j) is called an *occurrence* of the query edge (q_i, q_j) if: (a) (q_i, q_j) is a child edge in Q and (v_i, v_j) is an edge in G, or (b) (q_i, q_j) is a descendant edge in Q and $v_i \prec v_j$ in G. For instance, in the data graph G and query Q of Fig. 1, the edges (b_4, d_2) and (b_4, d_1) in G are two occurrences of the query edge (B, D).

An *occurrence* of a tree pattern Q on a data graph G is a tuple indexed by the nodes of Q whose values are the images of the nodes in Q under a homomorphism from Q to G. The *answer* of Q on G is a relation whose schema is the set of nodes of Q, and whose instance is the set of occurrences of Q under all possible homomorphisms from Q to G. Fig. 1(d) shows the answer (relation of occurrences) of Q on G. There are three occurrences of Q in the answer. If x is a node in Q labeled by a, the *occurrence list of x on G* is a sublist L_x of the inverted list I_a containing only those nodes that occur in the answer of Q on G for x (that is, nodes that occur in the column x of the answer). Figure 1(e) shows the occurrence lists of Q on G.

Given a large directed graph G and a tree pattern query Q, our goal is to efficiently find the answer of Q on G.

4   The Graph Simulation Relation

Simulations have been implemented in different graph database tasks [5,6,8,9]. As opposed to a homomorphism, which is a function, a simulation is a binary relation on the node sets of two directed graphs. It provides one possible notion of structural equivalence between the nodes of two graphs.

Double Simulation. Since the structure of a node is determined by its incoming and outgoing paths, we define a type of simulation called *double simulation*, which handles both the incoming and the outgoing paths of the graph nodes. Double simulation is an extension of dual simulation [8] to allow edge-to-path mappings. In this paper, we consider the double simulation of hybrid tree pattern queries on graph data.

Definition 1 (Double Simulation). *The double simulation of a query $Q = (V_Q, E_Q)$ by a directed data graph $G = (V_G, E_G)$ is the largest binary relation $S \subseteq V_Q \times V_G$ such that, whenever $(q, v) \in S$, the following conditions hold:*

1. *$label(q) = label(v)$.*
2. *For each $(q, q') \in E_Q$, there exists $v' \in V_G$ such that $(q', v') \in S$ and (v, v') is an occurrence of the edge (q, q').*
3. *If $q \neq root(Q)$, let q' be the parent of q in Q; then there exists $v' \in V_G$ such that $(q', v') \in S$, and (v', v) is an occurrence of the edge (q', q) in E_Q.*

The double simulation of Q by G is unique, since there is exactly one largest binary relation S satisfying the above three conditions. This can be proved by the fact that, whenever we have two binary relations S_1 and S_2 satisfying the three conditions between Q and G, their union $S_1 \cup S_2$ also satisfies those conditions.

We call the largest binary relation that satisfies conditions 1 and 2 *forward simulation* of Q by G, and we call the largest binary relation that satisfies conditions 1 and 3 *backward simulation*. We denote the forward, backward, and double simulation by $\mathcal{F}$, $\mathcal{B}$, and $\mathcal{FB}$, respectively. For $q \in V_Q$, $\mathcal{F}(q)$, $\mathcal{B}(q)$, and $\mathcal{FB}(q)$ denote the set of all nodes of V_G that forward, backward, and double simulate q, respectively. $\mathcal{FB}$ preserves both incoming and outgoing edge types (child or descendant) between Q and G, whereas $\mathcal{F}$ and $\mathcal{B}$ preserve only outgoing and incoming edge types, respectively. Table 1 shows the simulation relations $\mathcal{F}$, $\mathcal{B}$, and $\mathcal{FB}$ of the query Q on the graph G of Fig. 1.

The following theorem shows the significance of double simulation in graph pattern matching. All graph data nodes captured by the double simulation participate in the query's final answer. The proof is omitted in the interest of space.

Theorem 1 *Given a tree pattern query Q and a data graph G, there exists a homomorphism from Q to G that maps node $q \in V_Q$ to node $v \in V_G$ if and only if $v \in \mathcal{FB}(q)$.*

Table 1. Forward ($\mathcal{F}$), backward ($\mathcal{B}$), and double ($\mathcal{FB}$) simulation relations of the query Q on the graph G of Fig. 1.

Query node q	$\mathcal{F}(q)$	$\mathcal{B}(q)$	$\mathcal{FB}(q)$
A	$\{a_3, a_4\}$	$\{a_1, a_2, a_3, a_4\}$	$\{a_3, a_4\}$
B	$\{b_1, b_3\}$	$\{b_1, b_2, b_3, b_4\}$	$\{b_1, b_3\}$
C	$\{c_1, c_2, c_3\}$	$\{c_1, c_2, c_3\}$	$\{c_1, c_2\}$
D	$\{d_1, d_2\}$	$\{d_1, d_2\}$	$\{d_1\}$

Computing Double Simulation. We develop a 2-pass algorithm called *FB-Sim* to compute $\mathcal{FB}$ for a query Q and a data graph G by traversing the nodes of Q two times. *FBSim* leverages the acyclic nature of the rooted tree pattern. It first computes the forward simulation $\mathcal{F}$ (which considers outgoing query edges), and then refines $\mathcal{F}$ by computing a subset of the backward simulation $\mathcal{B}$ (which considers the incoming query edge).

Algorithm, 1 shows the pseudocode of *FBSim*. *FBSim* first invokes procedure *getFSim()* to compute $\mathcal{F}$ by traversing nodes of Q in a bottom-up way (line 2). When $q \in V_Q$ is a leaf node in Q, every node in I_q which is the inverted list of G having the same label with node q, is in $\mathcal{F}(q)$. Otherwise, a node v is removed from $\mathcal{F}(q)$ if there exist no nodes $v_1, \ldots, v_k \in V_G$ for the corresponding child query nodes $q_1, \ldots q_k$ of q such that: for every $i \in [1, k]$, $v_i \in \mathcal{F}(q_i)$, and (v, v_i) is an occurrence of the query edge $(q, q_i) \in E_Q$ in G. Due to the bottom-up traversal, $\mathcal{F}(q_i)$ of each child node q_i of $q \in V_Q$ are available from the previous iteration.

Based on the forward simulation relation $\mathcal{F}$, *FBSim* proceeds to compute a subset B of the forward simulation relation $\mathcal{F}$ using procedure *getBSim()* (line 3). The procedure computes B by traversing nodes of Q in a top-down manner. When $q \in V_Q$ is the root node, $B(q)$ is set to be equal to $\mathcal{F}(q)$. Otherwise, let q' be the parent of q in Q. A node $v \in \mathcal{F}(q)$ is in $B(q)$ if there exists node $v' \in B(q')$, such that (v', v) is an occurrence of the edge $(q', q) \in Q$ in G.

After procedure *getBSim()* terminates, for every node $q \in V_Q$, we have $B(q) = \mathcal{FB}(q) \subseteq \mathcal{F}(q)$. Since, based on Theorem 1, $\mathcal{FB}(q) = L_q$, $B = L_q$. The correctness of the algorithm is shown by the next theorem.

Theorem 2 *Algorithm* FBSim *correctly computes the double simulation $\mathcal{FB}$ of Q by G.*

Complexity. The worst case time complexity for computing both $\mathcal{F}$ and B is $O(|V_Q| \times |I_{max}|^2 \times R)$, where $|V_Q|$ is the cardinality of V_Q, $|I_{max}|$ is the size of the largest inverted list of G, and R is the time for checking if a pair of nodes in G is a query edge occurrence. Time R is bound by the time for checking reachability for a pair of nodes in the data graph.

Algorithm 1. Algorithm *FBSim* for computing double simulation.

Input: Data graph G, pattern query Q
Output: Double simulation $\mathcal{FB}$ of Q by G

1. $\mathcal{F}(q) := I_q$ and $\mathcal{B}(q) := \emptyset$, for every node q in V_Q;
2. getFSim();
3. getBSim();
4. $\mathcal{FB} := \mathcal{B}$;

Procedure getFSim()

1. **for** (every node q in V_Q in a bottom-up order) **do**
2.   **for** ($v_q \in \mathcal{F}(q)$ and all children q_i of q in Q) **do**
3.   **if** ($\nexists v_{q_i} \in \mathcal{F}(q_i)$ s.t. (v_q, v_{q_i}) is an occurrence of (q, q_i)) **then**
4. $\mathcal{F}(q) := \mathcal{F}(q) - \{v_q\}$;

Procedure getBSim()

1. $\mathcal{B}(root(Q)) := \mathcal{F}((root(Q)))$;
2. **for** (every non-root node q_i in V_Q in a top-down order) **do**
3. $q := parent(q_i)$;
4.   **for** ($v_{q_i} \in \mathcal{F}(q_i)$) **do**
5.   **if** ($\exists v_q \in \mathcal{B}(q)$ s.t. (v_q, v_{q_i}) is an occurrence of $(q, q_i) \in E_Q$) **then**
6. $\mathcal{B}(q_i) := \mathcal{B}(q_i) \cup \{v_{q_i}\}$;

---

5   The Pattern Matching Algorithm

We now present our holistic pattern matching algorithm *GraphMatch-Sim*. The algorithm uses a data structure called *answer graph* [13] to compactly encode all possible homomorphisms of a pattern to a data graph. The outline of *GraphMatch-Sim* is presented in Algorithm 2. It takes as input a data graph G and a pattern query Q. It constructs the answer graph G_A of Q on G in two phases: (a) the *node selection* phase, and (b) the *node linking* phase.

The node selection phase computes the double simulation relation of Q by G using algorithm *FBSim* of Algorithm 1 (line 1). As shown in the previous section, after relation $\mathcal{FB}$ is computed, the occurrence list L_q for each node q of Q is available (line 2). The node linking phase traverses Q in a top-down manner and links nodes in occurrence lists L_q with edges to produce the answer graph G_A (lines 3–5). This is implemented by procedure *expand*. Let q be the current query node under consideration. For each node $v_q \in L_q$, it expands G_A by adding incident edges to v_q. More concretely, it iterates over every child q_i of q (line 1). For each node $v_{q_i} \in L_{q_i}$ of q_i, it determines whether (v_q, v_{q_i}) is an occurrence of the query edge (q, q_i) (line 3). If so, it adds the edge (v_q, v_{q_i}) to G_A (line 4).

Complexity. As shown in Sect. 4, the time complexity of the first phase is $O(|V_Q| \times |I_{max}|^2 \times R)$. The time complexity of the second phase is $O(|V_Q| \times |L_{max}|^2 \times R)$, where $|L_{max}|$ is the size of the largest occurrence list of Q on G. As $L_{max} \subseteq I_{max}$, the time complexity of *GraphMatch-Sim* is bound by $O(|V_Q| \times |I_{max}|^2 \times R)$. The memory consumption is determined by the size of the answer graph which is bounded by $|V_Q| \times |I_{max}|^2$.

Algorithm 2. Algorithm *GraphMatch-Sim*.

Input: Data graph G, pattern query Q
Output: Answer graph G_A of Q on G

1. Use Algorithm *FBSim* to compute $\mathcal{FB}$ of Q by G;
2. Initialize G_A as a k-partite graph without edges having one data node which contains the occurrence list $L_q := \mathcal{FB}(q)$ for every node $q \in V_Q$;
3. **for** ($q \in V_Q$ in a top-down order) **do**
4.   **for** $(v_q \in L_q)$ **do**
5.   expand(q, v_q);

Procedure expand(q, v_q)

1. **for** $(q_i \in children(q))$ **do**
2.   **for** $(v_{q_i} \in L_{q_i})$ **do**
3.   **if** $((v_q, v_{q_i})$ is an occurrence of edge $(q, q_i) \in E_Q)$ **then**
4.   Add the edge (v_q, v_{q_i}) to G_A;

5.1   Optimizations

In the following, we discuss techniques to further improve the performance of *GraphMatch-Sim* on large graphs.

Leveraging Bitwise Operations to Check Parent-child Relationships. As aforementioned, using binary search to find parent-child relationships can be costly as it repeatedly incurs random memory accesses. Below, we present a more efficient method for finding all v_{q_i} in L_{q_i} that have a parent-child relationship with v_q in one step.

We convert the parent-child relationship checking into set containment testing. More concretely, let A_{v_q} denote the adjacency list of v_q, and L_{q_i} be the occurrence list of q_i; every element $v_{q_i} \in A_{v_q} \cap L_{q_i}$ is a child of v_q in G. We store A_{v_q} and L_{q_i} as bit vectors, and implement the intersection using a bitwise AND operation. Thus, we can obtain all nodes in L_{q_i} satisfying a child relationship with v_q in one batch, avoiding the random memory access problem.

Cardinality-Based Checking Order. When computing $\mathcal{F}$ with algorithm *FBSim*, in order to decide whether a node $v_q \in I_q$ is in $\mathcal{F}_q$, we need to check, for every child query node q_i of q, the existence of $v_{q'} \in \mathcal{F}_{q'}$, such that $(v_q, v_{q'})$

is an occurrence of query edge (q, q'). Usually, it is beneficial to examine the $\mathcal{F}_{q'}$ lists in order of increasing cardinality. This way, a node v_q that does not satisfy the simulation condition will be detected early on.

6   Experimental Evaluation

In this section we present experimental results to show the efficiency of our graph pattern matching approach for evaluating hybrid patterns. For the performance evaluation, we compare our algorithm *GraphMatch-Sim* with *GPM* [13] and *TwigStackD* [14], since the latter two are the only known approaches that can, or can be adapted to, directly evaluate hybrid pattern queries, and they are not coupled with a specific reachability index scheme.

We compare also our simulation-based pruning technique with the node pre-filtering technique described in [14]. The pre-filtering technique conducts two traversals on the data graph and maintains data structures that record, for each data node, whether it has ancestors or descendants matching a particular query node. Designed for descendant edge-only patterns, the pre-filtering technique is unable to filter out nodes violating the parent/child structural constraints in queries. Also, it selects nodes only guaranteed to appear in the final answer of a subpattern of the input query. In contrast, our simulation-based technique prunes *all* the graph nodes that do not participate in the final query answer.

We implemented our algorithm *GraphMatch-Sim* with the optimizations described in Sect. 5.1 (denoted as *SIM*). We implemented also a version of *SIM* which applies the node pre-filtering [14] as a preprocessing step (denoted as *SIM-flt*).

We implemented *GPM* as described in [13], as well as a version that uses the node pre-filtering as a preprocessing step (denoted as *GPM-flt*).

TwigStackD [14] was designed to evaluate descendant edge-only patterns. In order to evaluate hybrid patterns, we implemented an algorithm called *TS-post*, which first uses *TwigStackD* to finds all the solutions for the input pattern regarded as a descendant-only pattern, and then filters out solutions violating the child edge constraints (which is the postprocessing step suggested in [7]). We also extended *TwigStackD* to evaluate hybrid patterns directly, and produce answer graphs instead of tuples (denoted as *TS-ag*). A version of *TS-ag* called *TS-ag-flt*, which includes the node pre-filtering as a preprocessing step, was also implemented.

All the above graph matching algorithms were impemented used a recent efficient reachability scheme called *Bloom Filter Labeling* (BFL) which was shown to greatly outperform most existing schemes such as 2HOP and 3HOP [10].

Our implementation was coded in Java. All the experiments reported here were performed on a workstation running Ubuntu 16.04.

Datasets. We ran experiments on two graph datasets with different structural properties. Their main characteristics are summarized in Table 2.

citation[1] models citations extracted from the DBLP, ACM, and MAG (Microsoft Academic Graph) publications and consists of a directed graph with approximately 1,397K nodes and 3,021K edges. Nodes represent papers while edges represent citations.

Table 2. Dataset statistics. $|V|$, $|E|$ and $|L|$ are the number of nodes, edges and distinct labels, respectively. *maxout* and *maxin* are the maximum out-degree and in-degree of the graph. d_{avg} denotes the average degree of a graph.

dataset	$|V|$	$|E|$	$|L|$	*maxout*	*maxin*	d_{avg}
Citation	1,397,240	3,016,539	16,442	717	4,090	4.32
Cite-lb10000	6,540,401	15,011,260	8,343	181,247	203,695	4.59
Cite-lb8000	6,540,401	15,011,260	6,124	181,247	203,695	4.59
Cite-lb7000	6,540,401	15,011,260	5,662	181,247	203,695	4.59
Cite-lb6000	6,540,401	15,011,260	4,969	181,247	203,695	4.59
Cite-lb5000	6,540,401	15,011,260	3,970	181,247	203,695	4.59
Cite-lb4000	6,540,401	15,011,260	3,227	181,247	203,695	4.59
Cite-lb3000	6,540,401	15,011,260	2,434	181,247	203,695	4.59
Cite-lb2000	6,540,401	15,011,260	1,674	181,247	203,695	4.59
Cite-lb1000	6,540,401	15,011,260	885	181,247	203,695	4.59

Table 3. Parameters for Query Generation.

Parameters	Range	Description
Q	300 to 3300	Number of queries
D	6 to 16	Maximum depth of queries
DS	0 to 1	Probability of setting an edge to be a descendant edge ('//')
NP	1 to 3	Number of branches per query node

citeseerx[2] represents a directed graph consisting of 6.3M publications (nodes) and 14.3M citations between them (edges). The original graph does not have labels. We wrote a label assignment program which randomly adds a specified number of distinct labels to graph nodes. Using this program, we generated nine labeled citeseerx graphs whose number of labels ranges from 1,000 to 10,000. Each graph is named as cite-lb$x$, where x is the number of labels in the graph. Table 2 shows statistics for these datasets.

Queries. Queries were randomly generated. We implemented a query generator that creates a set of tree pattern queries based on the parameters listed in Table 3. Random queries are generated according to the input data graph and these parameters. For each data graph, we first generated a number of queries (in the range of

---

[1] www.aminer.cn/citation.

[2] www.citeseerx.ist.psu.edu.

300 to 3300) using different value combinations of the parameters listed in Table 3, and then formed a query set by randomly selecting 10 among them.

6.1   Performance Evaluation on *Citation* Graph

We measured the performance of the different versions of *SIM*, *GPM* and *TS* for evaluating the ten queries over *citation*. The characteristics of the queries can been found in Table 4.

In Table 5, we report the normalized query time of the algorithms (the timing of *SIM* has been normalized to 1 for each query). The average query evaluation time and memory usage of the algorithms in comparison is shown in Table 6. In Table 4, we show for each query the percentage of the number of inverted list nodes accessed by the algorithms using simulation-based technique (SIM%) and

Table 4. Query statistics over *citation*. '//' denotes the descendant pattern edge. SIM% and FLT% are the percentage of the number of inverted list nodes retained by the simulation-based filtering and the node pre-filtering, respectively. #TUP denotes the query solution tuples. AG denotes the answer graph.

query	|V|	height	% of '//'	maxout	SIM%	FLT%	FLT/SIM	#TUP	size of AG
0	8	2	37.50	3	0.007	3.90	557.51	10	25
1	7	2	28.57	3	0.09	3.82	42.40	930	857
2	17	6	35.29	3	0.007	0.02	2.93	108	55
3	16	6	50.00	2	0.33	0.75	2.31	**3.52548E + 11**	6,863
4	11	3	36.36	3	1.22	2.33	1.91	255,816	14,255
5	22	7	22.73	3	1.23	3.23	2.61	**6.88239E + 12**	20,509
6	13	4	53.85	4	2.14	2.57	1.20	88,051	71,234
7	13	4	38.46	3	5.02	6.23	1.24	11,784,410,550	20,355
8	13	5	46.15	2	2.92	3.13	1.07	**4.96216E + 11**	63,058
9	18	5	50.00	3	3.22	5.99	1.86	**1.35026E + 17**	339,276

Table 5. Normalized query evaluation time on *citation*.

query	SIM	SIM-flt	GPM	GPM-flt	TS-ag	TS-ag-flt	TS-post
0	1	8.17	9.09	16.48	5.96	7.74	**2210.22**
1	1	4.11	17.75	8.23	238.84	4.25	**1260.34**
2	1	1.45	742.06	2.93	822.73	1.34	**983.15**
3	1	0.90	257.48	2.52	**862.73**	7.69	na
4	1	0.76	1.24	0.39	17.20	0.85	**56.80**
5	1	0.74	**426.77**	5.40	289.94	24.33	na
6	1	0.87	**81.10**	1.60	**81.18**	1.50	**80.03**
7	1	0.91	1.07	1.10	**26.56**	0.99	na
8	1	0.68	12.97	0.50	**34.28**	0.98	na
9	1	0.88	59.13	6.44	75.10	27.09	na

Table 6. Average query evaluation time and memory usage on *citation*.

average	*SIM*	*SIM-flt*	*GPM*	*GPM-flt*	*TS-ag*	*TS-ag-flt*	*TS-post*
time (sec.)	23.991	20.229	1573.778	90.824	1962.846	352.864	na
memory (MB)	2586	3024	3093	2843	1832	2317	na

the node pre-filtering technique (FLT%) during the matching process. The three algorithms *GPM*, *TS-ag*, and *TS-post* access all the nodes (100%) of the inverted lists during the matching process since they do not have a filtering phase. We next present our findings.

Performance Comparison. *SIM* has the best time performance overall among all the algorithms in comparison. Except from Q_4 and Q_7, *GPM* is up to $742 \times$ slower than *SIM* on the ten queries, and it consumes 20% more memory than *SIM* (Table 6), due to the redundant intermediate results produced and recorded by *GPM* during the matching process. The overall performance of *GPM* is better than both *TS-post* and *TS-ag*: it is up to $243 \times$ faster than *TS-post* for the queries that *TS-post* can run to finish, and it outperforms *TS-ag* by up to $24\times$.

   TS-post has the worst performance among all the algorithms. This is mainly due to the postprocessing strategy in *TS-post* for handling the child constraints in hybrid queries, which results in a large amount of intermediate results that do not participate in the query answer. For queries having a large number of solution tuples, *TS-post* was aborted with out-of-memory error, before returning solution tuples. Unlike *TS-post*, the other algorithms in comparison encode solutions in answer graphs. As we can see from Table 4, the size of each query's answer graph is much smaller than the number of solution tuples.

Simulation vs. Pre-filtering. As we can see from Table 4, both the simulation-based pruning (abbreviated as simulation) and the pre-filtering (abbreviated as pre-filtering) drastically reduce the number of accessed graph nodes. However, simulation has a stronger pruning power than pre-filtering, since it retains only nodes in the final query answer. Also, unlike simulation, pre-filtering needs data structures which record, for each data graph node, whether it has ancestors or descendants matching a particular query node. This increases the memory footprint of the algorithms. As shown in Table 6, the memory usage increase of *SIM-flt* over *SIM*, *TS-ag-flt* over *TS-ag*, and *GPM-flt* over *GPM* is 14%, 26% and 7%, respectively.

   Our experimental results show that pre-filtering can significantly improve the time performance of *TS* and *GPM*. For example, *TS-ag-flt* has a max and an average speedup of $612 \times$ and $5 \times$ over *TS-ag*, respectively (Table 6), since *TS-ag-flt* accesses a small percentage of the graph nodes accessed by *TS-ag* (Table 4). Nevertheless, the improvement is only marginal for *SIM*. On Q_0 and Q_1, *SIM-flt*, in fact, increases the evaluation time of *SIM* by over $7 \times$ and $3 \times$, respectively. This is because the overhead cost associated with node pre-filtering offsets the reduction in the query execution cost.

6.2 Scalability Comparison on *Citeseerx*

In this experiment, we examine the impact of the total number of distinct graph labels on the performance of the algorithms in comparison. We used the aforementioned nine labeled *citeseerx* graphs *cite-lbx* (Table 2), where the number of labels x increases from 1 K to 10 K. For each *cite-lbx* graph, we generated a query set *cite-lbx.qry* with 10 distinct queries. The statistics of the query sets are summarized in Table 7. Figure 2 reports the execution time and memory usage of the seven algorithms.

As with the experiment on *citation*, *SIM* has the best overall performance, while *TS-post* again exhibits the worst time and memory performance. From Fig. 2(a), we observe that the execution time of the algorithms tends to increase while decreasing the total number of graph labels. This is expected since the average cardinality $(=|V|/|L|)$ of the input inverted list per node label in a graph increases when the number of distinct labels in the graph decreases. This observation confirms the complexity results which show dependency of the execution time on the input size.

Table 7. Query set statistics on *citeseerx*. '//' denotes the descendant pattern edge. SIM% and FLT% are the percentage of the number of inverted list nodes retained by the simulation-based filtering and the traversal-based filtering (pre-filtering), respectively. #TUP denotes the query solution tuples. AG denotes the answer graph.

query set	#queries	$|V|$	height	% of '//'	maxout	SIM%	FLT%	FLT/SIM	avg. #TUP	avg. size of AG
Cite-lb10000.qry	10	3.5	1.4	65.71	1.9	1.08	3.97	3.67	219.9	145.3
Cite-lb8000.qry	10	3.2	1.2	65.63	2	0.97	5.18	5.33	93.4	166.3
Cite-lb7000.qry	10	3.1	1.1	67.74	2	1.57	5.29	3.36	1,859.1	494.9
Cite-lb6000.qry	10	3.3	1.2	63.64	2	1.07	5.00	4.65	760.9	335.2
Cite-lb5000.qry	10	3.2	1.1	65.63	2	1.50	5.87	3.91	582.9	788.7
Cite-lb4000.qry	10	3.2	1.1	65.62	2	1.25	5.69	4.57	4531.8	943
Cite-lb3000.qry	10	3.4	1.4	67.64	2	1.57	5.24	3.35	17,543.3	1,959.2
Cite-lb2000.qry	10	3.1	1.1	67.74	2	0.18	2.02	11.10	67.7	125.2
Cite-lb1000.qry	10	3.3	1.2	63.64	2	1.46	6.57	4.50	40,064.1	37,966.6

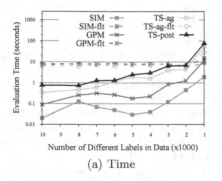
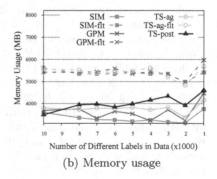

(a) Time   (b) Memory usage

Fig. 2. Scalability comparison on the citeseer graph.

In contrast to the results on *citation*, in *citeseerx* we observe a degenerate time performance of the algorithms when the pre-filtering technique is applied to graphs with a large (>1000) number of labels. The reason is that when the number of graph labels is large, the cardinality of the inverted lists is relatively small. In this case, the potential benefit of reducing intermediate results during graph matching can be offset by the overhead of node pre-filtering.

The memory performance of the algorithms on *citeseerx* (Fig. 2(b)) is consistent with the results on *citation*.

7   Conclusion

We have addressed the problem of evaluating hybrid tree pattern queries over a large graph like those encountered in web-based applications. Hybrid patterns allow the extraction of interesting information which cannot be extracted with patterns that involve only child or descendant structural relationships between nodes. We have designed a novel graph simulation-based node-filtering technique to prune nodes that do not contribute to the final query answer. This technique is integrated into a graph pattern matching holistic algorithm. In contrast to existing approaches, our algorithm does not produce any redundant intermediate results, it is not tied to a specific reachability indexing scheme, and it handles child constraints in the given pattern directly instead of resorting to postprocessing. An extensive experimental evaluation verified the efficiency and scalability of our approach, showing that it outperforms the best-known approach by orders of magnitude.

We are currently working on efficiently matching graph patterns to large data graphs in a dynamic graph setting where the matches of the graph pattern are computed incrementally.

References

1. Aberger, C.R., Tu, S., Olukotun, K., Ré, C.: Emptyheaded: a relational engine for graph processing. In: SIGMOD, pp. 431–446 (2016)
2. Bi, F., Chang, L., Lin, X., Qin, L., Zhang, W.: Efficient subgraph matching by postponing cartesian products. In: SIGMOD, pp. 1199–1214 (2016)
3. Chen, L., Gupta, A., Kurul, M.E.: Stack-based algorithms for pattern matching on dags. In: VLDB, pp. 493–504 (2005)
4. Cheng, J., Yu, J.X., Yu, P.S.: Graph pattern matching: a join/semijoin approach. IEEE Trans. Knowl. Data Eng. **23**(7), 1006–1021 (2011)
5. Fan, W., Li, J., Ma, S., Tang, N., Wu, Y., Wu, Y.: Graph pattern matching: from intractable to polynomial time. Proc. VLDB Endowment **3**(1), 264–275 (2010)
6. Kaushik, R., Bohannon, P., Naughton, J.F., Korth, J.F.: Covering indexes for branching path queries. In: SIGMOD, pp. 133–144 (2002)
7. Liang, R., Zhuge, H., Jiang, X., Zeng, Q., He, X.: Scaling hop-based reachability indexing for fast graph pattern query processing. IEEE Trans. Knowl. Data Eng. **26**(11), 2803–2817 (2014)

8. Ma, S., Cao, Y., Fan, W., Huai, J., Wo, T.: Strong simulation: capturing topology in graph pattern matching. ACM Trans. Database Syst. **39**(1), 1–46 (2014)
9. Mennicke, S., Kalo, J., Nagel, D., Kroll, H., Balke, W.: Fast dual simulation processing of graph database queries. In: ICDE, pp. 244–255 (2019)
10. Su, J., Zhu, Q., Wei, H., Yu, J.X.: Reachability querying: can it be even faster? IEEE Trans. Knowl. Data Eng. **29**(3), 683–697 (2017)
11. Sun, Z., Wang, H., Wang, H., Shao, B., Li, J.: Efficient subgraph matching on billion node graphs. Proc. VLDB Endowment **5**(9), 788–799 (2012)
12. Ullmann, J.R.: An algorithm for subgraph isomorphism. J. ACM **23**(1), 31–42 (1976)
13. Wu, X., Theodoratos, D., Skoutas, D., Lan, M.: Efficiently computing homomorphic matches of hybrid pattern queries on large graphs. In: Ordonez, C., Song, I.-Y., Anderst-Kotsis, G., Tjoa, A.M., Khalil, I. (eds.) DaWaK 2019. LNCS, vol. 11708, pp. 279–295. Springer, Cham (2019). https://doi.org/10.1007/978-3-030-27520-4_20
14. Zeng, Q., Zhuge, H.: Comments on stack-based algorithms for pattern matching on dags. Proc. VLDB Endowment **5**(7), 668–679 (2012)

Knowledge Graph and Entity Linkage

Knowledge-Infused Pre-trained Models
for KG Completion

Han Yu, Rong Jiang, Bin Zhou, and Aiping Li[⊠]

College of Computer, National University of Defense Technology, Changsha, China
{yuhan17,jiangrong,liaiping}@nudt.edu.cn

Abstract. Knowledge graphs (KG) are the basis for many artificial intelligence applications but still suffer from incompleteness. In this paper, we introduce a novel method for KG completion task by knowledge-infused pre-trained language models. We represent each triple in the KG as textual sequences and transform the KG completion task into a sentence classification task that fits the input of the language model. Our KG completion framework based on the knowledge-infused pre-trained language model which can capture both linguistic information and factual knowledge to compute the plausible of the triples. Experiments show that our method achieves better results than previous state-of-the-art on multiple benchmark datasets.

Keywords: Knowledge graph completion · Link prediction · Relation prediction · Pre-trained language model

1 Introduction

Knowledge graphs (KG) are structured knowledge bases, where facts are represented in the form of entities and relations. The entities are the nodes of the knowledge graph, and the relations are the edge between entities. Each edge and connected nodes form a triple $(head\ entity, relation, tail\ entity)$, indicating the relationship between entities, e.g.., $(Mark\_Twain, is\_a, writer)$. KG can be the basis for many applications: semantic search, recommendation, question answering, and data integration, etc. [11]. However, even large knowledge graphs such as FreeBase [2], YAGO [28], and WordNet [17], are still far from being complete, that is, missing relations or entities in the graphs [36]. This problem prompt the KG completion task which mainly includes link prediction and relation prediction to be proposed.

Many research efforts are devoted to KG completion, among them, knowledge graph embedding is an effective approach in which entities and edges are

The work described in this paper is partially supported by the National Key Research and Development Program of China (No. 2017YFB0802204, 2016QY03D0603, 2016QY03D0601, 2017YFB0803301, 2019QY1406), the Key R&D Program of Guangdong Province (No. 2019B010136003), and the National Natural Science Foundation of China (No. 61732004, 61732022, 61672020).

Z. Huang et al. (Eds.): WISE 2020, LNCS 12342, pp. 273–285, 2020.
https://doi.org/10.1007/978-3-030-62005-9_20

represented by embedding vectors. The embedding methods that use only knowledge graph structure information are often suffer from the sparsity of KG [14]. Therefore, Some recent studies incorporate extra text information to enrich knowledge representation [27,38,40]. These methods encode extra information as a unified word embeddings representation and cannot express the contextual information of the words in different contexts. For instance, in the two triples $(Mark\_Twain, is\_a, writer)$ and $(Mark\_Twain, born\_in, America)$, the same words in the description of $Mark\_Twain$ should have different importance weights related to the two relations is_a and $born\_in$. Besides, sufficient semantic and syntactic information cannot be learned by the small text of these methods such as the entity description.

Recently, BERT [6] and its various variants XLNet [42], RoBERTa [16], and ALBERT [12] have achieved great success in the field of natural language processing (NLP). These methods pre-trained with a large amount of unlabeled corpus and achieve state-of-the-art performance on several downstream NLP tasks by simply fine-tuning all pre-trained parameters. BERT can capture rich linguistic knowledge in pre-trained. BERT-based models have already effectively applied to various applications of NLP, such as question answering, reading comprehension, relationship extraction, dialogue generation, and it is also used in the KG completion [43]. However, some inference tasks require not only linguistic knowledge but also factual knowledge. To alleviate this problem, ERNIE [45], KnowBERT [21], K-BERT [15] and K-Adapter [34] inject knowledge into the language model.

For KG completion task, factual knowledge is particularly important for inferring relations between entities. Consider our previous example $(Mark\_Twain, born\_in, America)$. Given the head entity $Mark\_Twain$ and the relation $born\_in$. [22] suggest that the pre-trained language models may infer the tail entity $America$ by the surface form of entity name, because $Mark\_Twain$ to be a common American name. But when a person with an Italian name was born in the America, we need to use factual knowledge to reason the tail entity. In this study, we propose a novel method for KG completion using knowledge-infused pre-trained language models. For each triple, we span the entities and relation into text sequences and convert the completion of the knowledge graph into sequence classification problems. Then, we fine-tune the knowledge-infused BERT on these sequences to predict the plausibility of triples. The contributions of our paper are as follows:

- We propose a novel method for KG completion using knowledge-infused pre-trained language models. And to the best of our knowledge, this is the first study to use a knowledge-infused pre-trained language model for KG completion.
- Evaluating results on several benchmark datasets show that our method can achieve state-of-the-art performance in KG completion tasks.

2   Related Work

KG Embedding KG embedding methods can be classified into translational distance models and semantic matching models based on different scoring functions [33]. The representative translational distance models are TransE [3] and its extensions include TransH [35], TransD [9], etc. These models use distance-based scoring functions to evaluate the plausibility of a triple. The semantic matching models employ similarity-based scoring functions, and the typical models are RESCAL [20] and DistMult [44]. In addition, the convolutional neural networks (CNN) based methods ConvKB [18], ConvE [5], R-GCN [24] show promising results for KG completion.

The above methods only use structure information for KG completion, while some methods introduce external information to improve the performance [33]. NTN [27] represents the entities by word embeddings that are learned from the external corpus. DKRL [40] encodes the entity descriptions and learn embeddings with both triples and descriptions. SSP [37] learns the topic and KG embeddings together by characterizing the correlation between fact triples and text descriptions. Through external information, the effectiveness of these models can be improved, but these methods use the same word embedding weights to represent the entities and relations in different triples which would have different meanings.

To alleviate the above problems, TEKE [24] assigns different word embeddings to the relation in different triples. AATE [1] enhances representations by exploiting the entity descriptions and triple specific relation mention, then uses the mutual attention mechanism to learn more accurate textual representations. These methods can handle the semantic variety of entities and relations in distinct triples, but the ability of textual representation is limited by the small corpus such as entity descriptions. Compared with these methods, KG-BERT [43] can capture rich linguistic information via pre-trained language models. But they lack the factual knowledge information to grasp the relationship between entities which is important for KG completion tasks. Our method uses knowledge-infused language models to solve this problem.

Pre-trained Language Model Pre-trained language representation models can be divided into feature-based and fine-tuning methods. Feature-based methods only pre-trained word embedding parameters while fine-tuning methods learn the parameters of the pre-trained model architecture. Through fine-tuning, the pre-trained model can be applied in downstream tasks with few parameters need to be learned scratch. The representative fine-tuning method BERT achieves state-of-the-art results for various NLP tasks. Currently, BERT-based models are explored in fields such as question answering [8,41], reading comprehension [46], relation extraction [26], text classification [23], etc. And it also used in KG completion [43]. Though BERT can capture rich semantic information, but ignore the incorporation of knowledge information. Therefore, some works [21,34,45] injecting extra knowledge information into pre-trained language representation. In this study, we take a knowledge-infused pre-trained language model as the framework and fine-tune on the KG completion task.

3   Methodology

3.1   Knowledge-Infused BERT

BERT is a pre-trained language model built on the multi-layer bidirectional Transformer encoder. And it applied to downstream tasks through two steps of pre-training and fine-tuning. For pre-training, BERT is trained in a self-supervised way, and it trained with large corpus data (3,300 M words from BooksCorpus and English Wikipedia). For fine-tuning, BERT is initialized by the pre-trained parameters, and use labeled data from downstream tasks (such as sentence classification, question answering, etc.) to fine-tune all parameters. BERT can obtain rich contextual semantic and syntactic information through pre-training. Knowledge embedding methods (such as TransE [3]) which vectorize the structured KG can learn the knowledge information of entities and relations. To take full advantage of the contextual language representation of the pre-trained BERT and the factual knowledge of entities and relations in the KG, we apply Knowledge-infused BERT for KG completion.

We use ERNIE [45] as the knowledge-infused language model which consists of two encoders, T-Encoder and K-Encoder, to construct our framework. T-Encoder is responsible to capture basic lexical and syntactic information from the input tokens, and K-Encoder is to integrate extra factual knowledge information into textual information. The structure of T-Encoder is the same as BERT, which consists of multi-layer self-attention Transformer. Given the token sequence $\{Tok_1, ..., Tok_n\}$, T-Encoder generates semantic and syntactic embedding as follows,

$$\{T_1, ..., T_n\} = T{-}Encoder(\{Tok_1, ..., Tok_n\}). \tag{1}$$

Where $\{T_1, ..., T_n\}$ is the output embedding. K-Encoder is composed of stacked aggregators, which is similar to Transformer in structure that consists of multi-head self-attentions and infusion layer. The input of the K-Encoder is the output embedding of T-Encoder $\{T_1, ..., T_n\}$ and the entity embeddings $\{Ent_1, ..., Ent_n\}$ which is pre-trained by KG embedding methods. Then the K-Encoder injects the knowledge into language representation,

$$\{E_1, ..., E_n\} = K{-}Encoder(\{T_1, ..., T_n\}, \{Ent_1, ..., Ent_n\}). \tag{2}$$

Where $\{E_1, ..., E_n\}$ is the final output embedding. With K-Encoder, the heterogeneous information of semantic and syntactic information and factual knowledge can be integrated into a unified vector space. As T-Encoder and K-Encoder is identical to its implementation in ERNIE, readers can refer [45] for a more detailed description of the model.

3.2   KG Completion Framework

KG consists of structured entities and relationships. We define h as the head entity, r as the relationship, t as the tail entity, and (h, r, t) as the triple. We fine-tune pre-trained knowledge-infused BERT for KG completion. When pre-training original knowledge-infused BERT, the input of the T-Encoder is continuous text or word sequence. Therefore, we turn the entity and relation which are their names or descriptions, into a sequence form as the input of the T-Encoder. And take the pre-trained KG embedding of the entity as the input of the K-Encoder.

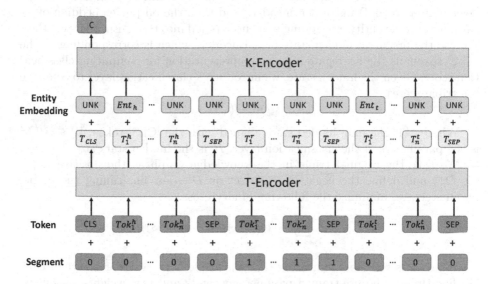

Fig. 1. The architecture of our framework. The input embeddings of K-Encoder are the sum of the token, segment and the default position embeddings of BERT. The input of K-Encoder are the sum of the output embeddings of T-Encoder and entity embedding sequence.

Link Prediction. The architecture of our framework for predicting the plausibility of a triple is shown in Fig. 1. For triple classification and link prediction task, the first input token of the T-Encoder is a special classification token [CLS]. The head entity is represented as a sequence of tokens $\{Tok_1^h, ..., Tok_n^h\}$, the relation is represent as $\{Tok_1^r, ..., Tok_n^r\}$ and the tail entity is represent as $\{Tok_1^t, ..., Tok_n^t\}$. The sequence of entity tokens and relation tokens are separated by a special token [SEP]. We then concatenate three token sequence to construct the input token sequence. For a given input token, its input representation of T-Encoder is constructed by summing the corresponding token, segment, and position embeddings. We set head entity and tail entity to the same segment embedding, and relation to another different segment embedding. We use the default position embedding of BERT that all token sequences in same location has the same position embedding. For K-Encoder, the input is the output embeddings of T-Encoder sums the head and tail entity embeddings $\{Ent_h, Ent_t\}$. We set the special token [UNK] as the first token of the entity embedding sequence. If the entity is represented by name, the second token is the pre-trained knowledge embedding of the entity, and the remaining positions are filled with [UNK] to the same length as the T-Encoder input. If the entity is represented by a entity description, we first mark the entities in the description, then set the first token of the marked entity with knowledge embedding, and fill the rest with [UNK].

Firstly, the input representations are fed into the T-Encoder which is a multi-layer bidirectional Transformer encoder. And then the output embedding of T-Encoder and the entity embedding sequence are fed into the K-Encoder together. We use the first final hidden state of K-Encoder, which is corresponding to the [CLS] token, as the aggregate sequence representation for computing classification score. Given the hidden state, we introduce a classification layer to compute the triple scores,

$$s = sigmoid(CW). \tag{3}$$

Where C is the final hidden state aligned with the [CLS] token, $W \in \mathbb{R}^{H \times 2}$ is the parameters of the classification layer, H is the hidden state size.

We take the original triple in the knowledge graph as the positive triple set D^+, and define the negative triple set as D^-. For fine-tuning the model parameters, we minimize the following binary cross,

$$L = - \sum_{D^+ \cup D^-} (ylog(s) + (1 - y)log(s)). \tag{4}$$

Where y is the label indicating that the triple is negative or positive. During fine-tuning, the pre-trained parameter weights and new weights W can be updated via gradient descent.

Relation Prediction. The framework of relation prediction is roughly the same as link prediction, except that there are no relation tokens in the token sequence. The architecture of relation prediction task is shown in Fig 2. We construct the token sequence composed by head entity tokens and tail entity tokens, without relation tokens. The head entity and tail entity tokens have different segment embeddings. We set the special token [CLS] as the first input token, and separate the head entity and tail entity with [SEP]. We also use the final hidden state C corresponding to [CLS] as the representation of the two entities. The scoring function for predicting relation is:

$$s' = sigmoid(CW'). \tag{5}$$

Where $W' \in \mathbb{R}^{H \times R}$ is the parameters of the classification layer for predicting relation, R is the number of relations in a knowledge graph. We minimize the cross-entropy loss to fine-tuning the model:

$$L' = - \sum_{D^+} \sum_{i=1}^{R} y'_i log(s'_i), \tag{6}$$

where y_i is the relation indicator for the triple.

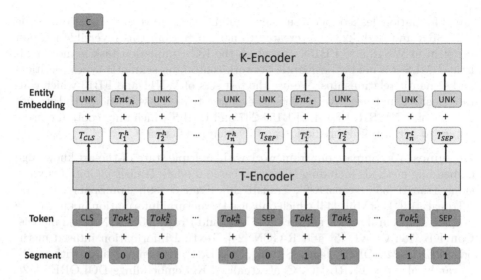

Fig. 2. The framework of relation prediction. The input embeddings of K-Encoder are the sum of the token, segment and the default position embeddings of BERT. The input of K-Encoder are the sum of the output embeddings of T-Encoder and entity embedding sequence.

Table 1. The statistics of datasets. Number of entities, relations, and observed triples in each split for benchmarks.

Dataset	Entities	Relations	Train	Dev	Test
WN11	38,696	11	112,581	2,609	10,544
FB13	75,043	13	316,232	5,908	23,733
WN18RR	40,943	11	86,835	3,034	3,134
FB15K	14,951	1,345	483,142	50,000	59,071
FB15k-237	14,541	237	272,115	17,535	20,466
UMLS	135	46	5,216	652	661

4 Experiments

In this section we evaluate our KG completion framework on three experimental tasks, triple classification, link prediction and relation prediction.

Datasets. We evaluate our experiments on six widely used benchmark KG datasets: WN11, FB13 [27], FB15K [3], WN18RR, FB15k-237 and UMLS [5]. Table 1 provides statistics of all datasets used in our experiment. WN11 and WN18RR are the subsets of WordNet which is a large lexical knowledge graph of English. FB15K and FB15k-237 are the subsets of Freebase which is a large knowledge graph about general facts. As noted by [30], WN18 and FB15k are information leaked because they contain many reversible relation, while WN18RR and FB15k237 are created to not suffer from this reversible relation problem in WN18 and FB15k, for which the KG completion task is more realistic. UMLS is a medical semantic network containing semantic types (entities) and semantic relationships. We use the test sets of WN11 and FB13 which contain positive and negative triplets to evaluate triple classification. And we use test set of WN18RR, FB15K, FB15k-237 and UMLS which only contain correct triples to perform link prediction and relation prediction.

Baselines. We compare our framework with multiple state-of-the-art knowledge embedding methods including transition-based models TransE [3] and its extensions TransH [35], TransD [9], TransR [14], TransG [39], TranSparse [10] and PTransE [13], DistMult [44] which only used structural information in knowledge graphs. The neural tensor network NTN [27] and ProjE [25]. CNN-based models: ConvKB [18], ConvE [5] and R-GCN [24]. Textual information infused methods: TEKE [24], DKRL [40], SSP [37], AATE [1]. KG embeddings with entity hierarchical types TKRL [40]. Contextualized KG embeddings DOLORES [32]. Complex-valued KG embeddings ComplEx [31] and RotatE [29]. Adversarial learning framework KBGAN [4]. BERT-based framework KG-BERT [43].

Settings. We use pre-trained ERNIE [45] model with 6 layers of T-Encoder and 6 layers of K-Encoder. We denote the hidden dimension of token embeddings is $H_w = 768$, and hidden dimension of entity embeddings is $H_e = 100$. We set 12 self-attention heads for token embeddings and 4 self-attention heads for entity embeddings. In our framework, we set the hyper-parameters of batch size to 32, learning rate to 5e-5, and the dropout rate to 0.1. We fine-tune our framework with Adam implemented in BERT. We tuned 3 epochs for triple classification, 5 for link prediction and 20 for relation prediction. For triple classification training, we sample 1 negative triple for a positive triple. For link prediction training, we sample 5 negative triples for a positive triple. For relation prediction training, we only use positive triple.

To capture the relationships among entities in the pre-trained language model, we add relation classification task as the pre-training processes. We use a subset of T-REx [7] which is a large scale alignment dataset to pre-train the model to classify relation labels of given entity pairs based on context. For the UMLS dataset, since there lacks sufficient medical knowledge in the pre-trained knowledge-infused language model, we pre-train the knowledge-infuse framework with PubMed abstracts, PubMed Central full-text papers and the entity embeddings of UMLS.

Triple Classification. Triple classification is to infer whether a triple is the correct triple or not. Table 2 shows the results of various models performing triple classification on WN11 and FB13 datasets. We ran our models 3 times and average the accuracy of each time as the final result. We can see that the BERT-based methods have a large improvement over the results of other baseline models. Our framework performance better than KG-BERT, proving the effectiveness of our

Table 2. Results on triple classification for different embedding methods.

Method	WN11	FB13	avg
NTN	86.2	90.0	88.1
TransE	75.9	81.5	78.7
TransH	78.8	83.3	81.1
TransR	85.9	82.5	84.2
TransD	86.4	89.1	87.8
TEKE	86.1	84.2	85.2
TransG	87.4	87.3	87.4
TranSparse-S	86.4	88.2	87.3
DistMult	87.1	86.2	86.7
DistMult-HRS	88.9	89.0	89.0
AATE	88.0	87.2	87.6
ConvKB	87.6	88.8	88.2
DOLORES	87.5	89.3	88.4
KG-BERT	93.5	90.4	91.9
Ours	**93.5**	**90.5**	**92.0**

method. Analysis of the results, the effectiveness of the BERT-based methods have two main fold: First, the baseline models no matter uses structural information or extra text information not the utilization of rich language patterns, and the BERT-based methods can obtain rich linguistic patterns information from a large amount of corpus through pre-training. Second, entities connected by different relationships have different meanings, and words have different semantics in the corpus according to different contexts. BERT-based methods can make full use of the learned contextual information in the triple classification process. In addition, the BERT-based methods have achieved a greater improvement on WN11, that may because WordNet is a linguistic knowledge graph, which is closer to the language model. Our model obtained more improvements in the FB13 dataset. The reason for the improvement is that the original BERT model learns more about the semantic association between tokens than the knowledge between entities. The knowledge-infused BERT can use the extra factual knowledge between entities by injecting knowledge into the language model. Note that, if we do not use the entity embeddings as the input of K-Encoder (that is, replace all *Ent* with [UNK]), the performance of our framework will decline compared with KG-BERT. This is because knowledge-infused language model forgets part of the linguistic information during the pre-training process.

Link Prediction. Link prediction task predicts the head entity given the relation and tail entity, or predicts the tail entity given the relation and head entity. We following the protocol of [19] that only report results under the filtered setting [3] which removes all corrupted triples appeared in training data and testing data before getting the ranking lists. We use two common metric Mean Rank (MR) and Hit@10 to evaluate the performance of models. A lower MR is better while a higher Hits@10 is better.

Table 3 represents link prediction performance of various models. We take the results of baseline models from the original papers. We observe that BERT-based methods get lower MR than other baseline models. It because pre-trained BERT can capture the semantic relatedness of entity and relation sentences to avoid very high ranks. BERT has not learned the knowledge graph structural information between entities, so BERT-based methods not achieve higher Hit@10 than some state-of-the-art models. The knowledge-infused BERT injects entity

Table 3. Link prediction results on WN18RR, FB15k-237 and UMLS datasets.

Method	WN18RR		FB15k-237		UMLS	
	MR	Hit@10	MR	Hit@10	MR	Hit@10
TransE	2365	50.5	223	47.4	1.84	98.9
TransH	2524	50.3	255	48.6	1.80	**99.5**
TransR	3166	50.7	237	51.1	1.81	99.4
TransD	2768	50.7	246	48.4	1.71	99.3
DistMult	3704	47.7	411	41.9	5.52	84.6
ComplEx	3921	48.3	508	43.4	2.59	96.7
ConvE	5277	48	246	49.1	–	–
ConvKB	2554	52.5	257	51.7	–	–
R-GCN	–	–	–	41.7	–	–
KBGAN	–	48.1	–	45.8	–	–
RotatE	3340	**57.1**	177	**53.3**	–	–
KG-BERT	97	52.4	153	42.0	1.47	99.0
Ours	**96**	52.7	**149**	43.1	**1.45**	99.3

Table 4. Relation prediction results on FB15k dataset.

Method	Mean rank	Hit@1
TransE	2.5	84.3
TransR	2.1	91.6
DKRL	2.0	90.8
TKRL	1.7	92.8
PTransE	1.2	93.6
SSP	1.2	–
ProjE	1.2	95.7
KG-BERT	1.2	96.0
Ours	**1.2**	**96.2**

knowledge into the language model to better learn the relationship between entities during the pre-training, therefore, our framework get higher Hit@10 than KG-BERT. Without the relation classification task added in the pre-training, our framework has lower performance than KG-BERT due to the lack of learned relationship between entities and the loss of some semantic information.

Relation Prediction. Relation prediction task is to predict relation given the head and the tail entities. The procedure is similar to link prediction and we evaluate the models using MR and Hits@1 with filtered setting. Table 4 shows the results of relation prediction task on FB15K. We note that our framework also shows promising results and achieves the highest Hits@1 so far. The relation prediction task is analogous to sentence pair classification in BERT fine-tuning and can also benefit from BERT pre-training. Knowledge-infused BERT not only learns the semantic representation of entities, but also the knowledge representation between entities, so we get better results than KG-BERT.

5   Conclusion

In this paper, we presented a novel method for KG completion, and outperforming existing methods on multiple benchmark datasets. Our method use knowledge-infused pre-trained language model and turn the triples in the KG into token sequences. Then transform the KG completion task into sentence classification. The experiments demonstrate that our method outperforms state-of-the-art results on several benchmark datasets. In the future, it is promising to exploit incorporate more KG structured information, and inject factual knowledge to compute the plausible of the triples without re-training.

References

1. An, B., Chen, B., Han, X., Sun, L.: Accurate text-enhanced knowledge graph representation learning. In: Proceedings of the 2018 Conference of the North American Chapter of the Association for Computational Linguistics: Human Language Technologies, Volume 1 (Long Papers), pp. 745–755 (2018)
2. Bollacker, K., Evans, C., Paritosh, P., Sturge, T., Taylor, J.: Freebase: a collaboratively created graph database for structuring human knowledge, pp. 1247–1250 (2008)
3. Bordes, A., Usunier, N., Garciaduran, A., Weston, J., Yakhnenko, O.: Translating embeddings for modeling multi-relational data, pp. 2787–2795 (2013)
4. Cai, L., Wang, W.Y.: Kbgan: adversarial learning for knowledge graph embeddings. arXiv preprint arXiv:1711.04071 (2017)
5. Dettmers, T., Minervini, P., Stenetorp, P., Riedel, S.: Convolutional 2D knowledge graph embeddings. In: Thirty-Second AAAI Conference on Artificial Intelligence (2018)
6. Devlin, J., Chang, M., Lee, K., Toutanova, K.: Bert: Pre-training of deep bidirectional transformers for language understanding. arXiv: Computation and Language (2018)

7. Elsahar, H., et al.: A large scale alignment of natural language with knowledge base triples, T-rex (2018)
8. Godbole, A., Kavarthapu, D., Das, R., Gong, Z., McCallum, A.: Multi-step entity-centric information retrieval for multi-hop question answering
9. Ji, G., He, S., Xu, L., Liu, K., Zhao, J.: Knowledge graph embedding via dynamic mapping matrix. In: Proceedings of the 53rd Annual Meeting of the Association for Computational Linguistics and the 7th International Joint Conference on Natural Language Processing (Volume 1: Long Papers), pp. 687–696 (2015)
10. Ji, G., Liu, K., He, S., Zhao, J.: Knowledge graph completion with adaptive sparse transfer matrix. In: Thirtieth AAAI Conference on Artificial Intelligence (2016)
11. Jin, J., Luo, J., Khemmarat, S., Dong, F., Gao, L.: GSTAR: an efficient framework for answering top-k star queries on billion-node knowledge graphs. World Wide Web **22**(4), 1611–1638 (2019)
12. Lan, Z., Chen, M., Goodman, S., Gimpel, K., Sharma, P., Soricut, R.: Albert: a lite bert for self-supervised learning of language representations
13. Lin, Y., Liu, Z., Luan, H., Sun, M., Rao, S., Liu, S.: Modeling relation paths for representation learning of knowledge bases. arXiv preprint arXiv:1506.00379 (2015)
14. Lin, Y., Liu, Z., Sun, M., Liu, Y., Zhu, X.: Learning entity and relation embeddings for knowledge graph completion, pp. 2181–2187 (2015)
15. Liu, W., et al.: K-Bert: enabling language representation with knowledge graph. arXiv preprint arXiv:1909.07606 (2019)
16. Liu, Y., et al.: Roberta: A robustly optimized Bert pretraining approach
17. Miller, G.A.: Wordnet: a lexical database for English. Commun. ACM **38**(11), 39–41 (1995)
18. Nguyen, D.Q., Nguyen, D.Q., Nguyen, T.D., Phung, D.: A convolutional neural network-based model for knowledge base completion and its application to search personalization. Semant. Web **10**(5), 947–960 (2019)
19. Nguyen, D.Q., Nguyen, T.D., Nguyen, D.Q., Phung, D.: A novel embedding model for knowledge base completion based on 22 convolutional neural network
20. Nickel, M., Tresp, V., Kriegel, H.-P.: A three-way model for collective learning on multi-relational data. ICML **11**, 809–816 (2011)
21. Peters, M.E., et al.: Knowledge enhanced contextual word representations, pp. 43–54 (2019)
22. Poerner, N., Waltinger, U., Schütze, H.: Bert is not a knowledge base (yet): Factual knowledge vs. name-based reasoning in unsupervised QA
23. Reimers, N., Schiller, B., Beck, T., Daxenberger, J., Gurevych, I.: Classification and clustering of arguments with contextualized word embeddings. In: Proceedings of the 57th Annual Meeting of the Association for Computational Linguistics (2019)
24. Schlichtkrull, M., Kipf, T.N., Bloem, P., van den Berg, R., Titov, I., Welling, M.: Modeling relational data with graph convolutional networks. In: Gangemi, A., et al. (eds.) ESWC 2018. LNCS, vol. 10843, pp. 593–607. Springer, Cham (2018). https://doi.org/10.1007/978-3-319-93417-4_38
25. Shi, B., Weninger, T.: Proje: embedding projection for knowledge graph completion. In: Thirty-First AAAI Conference on Artificial Intelligence (2017)
26. Soares, L.B., FitzGerald, N., Ling, J., Kwiatkowski, T.: Matching the blanks: distributional similarity for relation learning
27. Socher, R., Chen, D., Manning, C.D., Ng, A.Y.: Reasoning with neural tensor networks for knowledge base completion, pp. 926–934 (2013)
28. Suchanek, F.M., Kasneci, G., Weikum, G.: Yago: a core of semantic knowledge, pp. 697–706 (2007)

29. Sun, Z., Deng, Z.-H., Nie, J.-Y., Tang, J.: Rotate: knowledge graph embedding by relational rotation in complex space. arXiv preprint arXiv:1902.10197 (2019)
30. Toutanova, K., Chen, D.: Observed versus latent features for knowledge base and text inference. In: Proceedings of the 3rd Workshop on Continuous Vector Space Models and their Compositionality, pp. 57–66 (2015)
31. Trouillon, T., Welbl, J., Riedel, S., Gaussier, É., Bouchard, G.: Complex embeddings for simple link prediction (2016)
32. Wang, H., Kulkarni, V., Wang, W.Y.: Dolores: deep contextualized knowledge graph embeddings. arXiv preprint arXiv:1811.00147 (2018)
33. Wang, Q., Mao, Z., Wang, B., Guo, L.: Knowledge graph embedding: a survey of approaches and applications. IEEE Trans. Knowl. Data Eng. **29**(12), 2724–2743 (2017)
34. Wang, R., et al.: K-adapter: infusing knowledge into pre-trained models with adapters. arXiv: Computation and Language (2020)
35. Wang, Z., Zhang, J., Feng, J., Chen, Z.: Knowledge graph embedding by translating on hyperplanes. In: Twenty-Eighth AAAI Conference on Artificial Intelligence (2014)
36. Wu, T., et al.: Knowledge graph construction from multiple online encyclopedias. World Wide Web, pp. 1–28 (2019)
37. Xiao, H., Huang, M., Meng, L., Zhu, X.: SSP: semantic space projection for knowledge graph embedding with text descriptions. In: Thirty-First AAAI Conference on Artificial Intelligence (2017)
38. Xiao, H., Huang, M., Zhu, X.: SSP: semantic space projection for knowledge graph embedding with text descriptions. arXiv: Computation and Language (2016)
39. Xiao, H., Huang, M., Zhu, X.: Transg: a generative model for knowledge graph embedding. In: Proceedings of the 54th Annual Meeting of the Association for Computational Linguistics (Volume 1: Long Papers), pp. 2316–2325 (2016)
40. Xie, R., Liu, Z., Sun, M.: Representation learning of knowledge graphs with hierarchical types, pp. 2965–2971 (2016)
41. Yang, W., et al.: End-to-end open-domain question answering with Bertserini
42. Yang, Z., Dai, Z., Yang, Y., Carbonell, J., Salakhutdinov, R., Le, Q.V.: XlNet: generalized autoregressive pretraining for language understanding
43. Yao, L., Mao, C., Luo, Y.: KG-Bert: Bert for knowledge graph completion
44. Zhang, Z., Zhuang, F., Qu, M., Lin, F., He, Q.: Knowledge graph embedding with hierarchical relation structure. In: Proceedings of the 2018 Conference on Empirical Methods in Natural Language Processing, pp. 3198–3207 (2018)
45. Zhang, Z., Han, X., Liu, Z., Jiang, X., Sun, M., Liu, Q.: Ernie: enhanced language representation with informative entities, pp. 1441–1451 (2019)
46. Zhu, H., Dong, L., Wei, F., Wang, W., Qin, B., Liu, T.: Learning to ask unanswerable questions for machine reading comprehension

TransMVG: Knowledge Graph Embedding Based on Multiple-Valued Gates

Xiaobo Guo[1,2], Neng Gao[1], Jun Yuan[3], Xin Wang[1(✉)], Lei Wang[1], and Di Kang[4]

[1] Institute of Information Engineering, Chinese Academy of Sciences, Beijing, China
{guoxiaobo,gaoneng,wangxin,wanglei}@iie.ac.cn
[2] School of Cyber Security, University of Chinese Academy of Sciences, Beijing, China
[3] College of Traffic Engineering, Hunan University of Technology, Zhuzhou, China
yuanjun@iie.ac.cn
[4] National Secrecy Science and Technology Evaluation Center, Beijing, China
nsstec_kangd@163.com

Abstract. The essence of knowledge representation learning is to embed the knowledge graph into a low-dimensional vector space to make knowledge computable and inferable. Semantic discriminate models greatly improve the performance of knowledge embedding through increasingly complex feature engineering. For example, the projection calculation based on matrixes can achieve more detailed semantic interactions and higher accuracies. However, complex feature engineering results in high time complexity and discriminate parameters pressure, which make them difficult to effectively applied to large-scale knowledge graphs. TransGate is proposed to relieve the pressure of the huge number of parameters in semantic discriminate models and obtains better performance with much fewer parameters. We find that the gate filtering vector obtained by the traditional gate used by TransGate would rapidly fall in the state of a nearly boundary binary-valued distribution (most values are near 0 or near 1) after only a few hundred rounds of training. This means that most filtering gate values either allow the information element to pass completely or not at all, which can be called extreme filtering. We argue that this filtering pattern ignore the interaction between information elements. In this paper, TransMVG model is proposed to improve the traditional boundary binary-valued gate to a multiple-valued gate on the premise of ensuring the randomness. The experiments results show that TransMVG outperforms the state-of art baselines. This means it is feasible and necessary to multivalue the filter gate vectors in the process of knowledge representation learning based-on the gate structure.

Keywords: Knowledge representation learning · Boundary binary-valued gate · Multiple-valued gate.

© Springer Nature Switzerland AG 2020
Z. Huang et al. (Eds.): WISE 2020, LNCS 12342, pp. 286–301, 2020.
https://doi.org/10.1007/978-3-030-62005-9_21

1   Introduction

Nowadays, knowledge graph has become an important resource to support AI related applications, including relation extraction, question answering, semantic search and so on. Generally, a knowledge graph is a set of facts, usually represented as a triplet $(head, relation, tail)$, denoted as (h, r, t). Although the current knowledge graph, such as WordNet [14] and Freebase [1], has a large amount of data, it is far from perfect. For example, according to Google, 71% of the people in Freebase lack birthplace records and 75% lack nationality records. The lack of completeness of knowledge graph seriously affects the downstream applications.

Semantic indiscriminate models assume that the vector representation of entities and relations in any condition should be the same, regardless of the importance of semantic environments. Semantic discriminate models are proposed to distinguish multiple semantics. TransH [22] embeds entities in the relation hyperplane to distinguish relation-specific information. TransR [13] learns a mapping matrix for each relation and map each entity into the relation space respectively. TranSparse [11] replaces the mapping matrix in the TransR with two sparse matrices for each relation. TransD [10] dynamically constructs two mapping matrices for each triplet by setting projection vectors for each entity and relation. TransG [23] models the different semantics of the same relation into different Gaussian distributions and assumes that all semantics of a relation contribute to the fractional function of the fact triples. These models have greatly improved accuracies through increasingly complex feature engineering, but the problem of large number of parameters and high computation complexity also come. Despite their high accuracies, it is difficult to apply these models to large-scale real knowledge graphs.

The primary cause of the large number of parameters in the semantic discriminated models is that they do not pay attention to the intrinsic correlation between relations, and by default they assume relations are independent. As a result, these models have to learn a set of parameters for each relation for a relation-specific semantic discrimination. TransGate [25] is proposed to relieve the pressure of the huge number of parameters in semantic discriminate models with two fixed-size shared parameter gates based on the traditional Long Short Term Memory (LSTM) [26] gate by utilizing the inherent correlation between relations. It obtains better performance with much fewer parameters. Unlike relation-specific matrices in other models, the two learned global shared parameter gates of TransGate [25] will not grow with the expansion of the data set and makes it much easier to be applied to the large-scale real knowledge graphs.

We run TransGate [25] on FB15K [2] multiple times with different parameter configurations and find that the gate filtering vector obtained by the traditional gate rapidly falls in the state of a nearly boundary binary-valued distribution (most values are near 0 or near 1) after only a few hundred rounds of training. It means that most filtering gate values either allow the information element to pass completely or not at all. This extreme filtering pattern is shown in Fig. 1 (The parameter configuration is the same with Fig. 4). We argue that it ignores the interaction between information elements. We believe that the gate filtering vector should choose how much information elements to pass through rather than whether to pass through. In different semantic environments, the proportion combinations of information elements allowed to pass by the gate filtering vector are different. This is the fundamental reason why the same vector can express multiple semantics. Inspired by [12], TransMVG proposed in this paper not only inherits the advantages of global parameter sharing of TransGate [25], but also makes the gate filtering vector have stronger information selection ability.

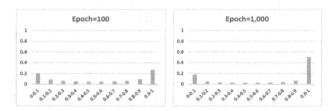

Fig. 1. The extreme filtering of TransGate

Our main contributions are as the following:

Contribution I: We find that extreme filtering phenomenon exists in the gate filtering vectors obtained by the traditional shared gate method. We propose that the gate filtering vector should choose how much information elements to pass through instead of whether to pass through, so as to obtain richer semantic interaction and more refined semantic expression.

Contribution II: The proposed TransMVG model improves the traditional boundary binary-valued gate to the multiple-valued gate, which means all values between 0 and 1 can be randomly selected as a filtering gate value, not mostly 0 and 1. In this way, more different proportion combinations of information elements allowed to pass by the gate filtering vector make the learned semantics more precise and clear.

Contribution III: Experiments show that TransMVG obtains some significant improvement compared to most state-of-art baselines with fewer parameters. This indicates that it is feasible and necessary to filter information elements precisely by using the multiple-valued gate.

2   Related Work

2.1   Semantic Indiscriminate Models

Indiscriminate models usually focus more on the scalability on real-world knowledge graphs. They assume that the vector representations of entities and relations are consistent in any semantic environment. As a result, they often have low accuracies:

TransE [2] represents a relation as a translation vector r indicating the semantic translation from the head entity h to the tail entity t, so that the pair of embedded entities in a triplet (h, r, t) can be connected by r with low error. The score function is $f_r(\mathbf{h}, \mathbf{t}) = \|\mathbf{h} + \mathbf{h} - \mathbf{t}\|_2^2$. It is very efficient, but only suitable for $1 - to - 1$ relations, and has flaws in dealing with $1 - to - N$, $N - to - 1$ and $N - to - N$ relations.

DistMult [24] has the same time and space complexity as TransE. It uses weighted element-wise dot product to define the score function $f_r(h, t) = \sum h_k r_k t_k$. Although DistMult has better overall performance than TransE, but it is unable to model asymmetric relations.

ComplEx [21] makes use of complex valued embeddings and Hermitian dot product to address the antisymmetric problem in DistMult. However, TransE and DistMult perform better on symmetry relations than ComplEx.

CombinE [19] considers triplets features from two aspects: relation $\mathbf{r}_p \approx \mathbf{h}_p + \mathbf{t}_p$ and entity $\mathbf{r}_m \approx \mathbf{h}_m - \mathbf{t}_m$. The score function is $f_r(h, t) = \|\mathbf{h}_p + \mathbf{t}_p - \mathbf{r}_p\|_{L1/L2}^2 + \|\mathbf{h}_m - \mathbf{t}_m - \mathbf{r}_m\|_{L1/L2}^2$. CombinE doubles the parameter size of TransE, but does not yield significant boost in performance.

2.2 Semantic Discriminate Models

Discriminate models focus more on precision. They assume that the vector representations should depend on the specific semantic environment. They usually contain two stages: relation-specific information discrimination and score computation.

TransH [22] is proposed to enable an entity to have distinct distributed representations when involved in different relations. TransH projects the entity embeddings to the hyperplane with a certain norm vector. By projecting entity embeddings into relation hyperplanes, it allows entities playing different roles for different relations.

TransR/CTransR [10] is proposed based on the idea that entities and relations should be considered in two different vector spaces. TransR set a mapping matrix for each relation to map entity embedding into a relation vector space. CTransR is a clustering-based extension of TransR, where diverse head-tail entity pairs are clustered into different groups and each group has only one relation vector for all pairs in the group.

TranSparse [11] consider the heterogeneity (some relations link many entity pairs and others do not) and the imbalance (the number of head entities and that of tail entities in a relation could be different) of knowledge graphs. It uses adaptive sparse matrices to replace transfer matrices, in which the sparse degrees are determined by the number of entity pairs linked by relations.

KG2E [10] uses Gaussian embedding to explicitly model the certainty of entities and relations. Each entity or relation is represented by a Gaussian distribution, where the mean denotes its position and the covariance can properly represent its certainty. It performs well on $1 - to - N$ and $N - to - 1$ relations.

TransG [23] can discover the latent semantics of a relation automatically through Chinese Restaurant Process, and leverages a mixture of multiple relation-specific components for translating entity pair to address new issues.

TransGate [25] is proposed to relieve the pressure of the huge number of parameters in semantic discriminate models. Across the whole knowledge graph, it establishes two fixed-size shared parameter gates based on the traditional LSTM [26] gate by utilizing the inherent correlation between relations. The shared parameter gate is used to filter entity vectors according to certain semantic environment, and the filtered entity vectors only represent the semantic in the current semantic environment.

2.3   Other Models

Many researches attempt to introduce some novel techniques of deep learning into knowledge graph embedding. KBGAN [4] introduces GAN (Generative Adversarial Networks) to boost several embedding models. ProjE [17] uses a learnable combination operator to combine embeddings and feeds combined embeddings in a multi-class classifier to handle complex relations. R-GCN [16] and ConvE [5] introduces multi-layer convolution network in knowledge graph embedding. ConvKB [15] employs a convolutional neural network to capture global relations and transitional characteristics between entities and relations .

TransMVG does not use many sets of semantic parameters specific to each relationship, nor does it use deep learning frameworks. It only uses a smaller set of global shared parameters based on two multiple-valued gates and a shallow learning framework to get a better performance on link prediction task and triplet classification task.

3   Embedding Based on Multiple-Valued Gate

In this paper, we propose the TransMVG model. Its gate filtering vectors of all dimensions can take the values between 0 and 1 with nearly the same proportions, ensuring that the values of each dimension obey different *Bernoulli* distribution respectively at the same time. In this section, we first introduce some background knowledge and then introduce the TransMVG model and its training methods. Finally, we perform a complexity analysis, comparing TransMVG with other baselines.

3.1   Background Knowledge

LSTM. LSTM [26] is a kind of recurrent neural network which overcomes long-term dependence. It consists of a memory cell and three gates, namely input gate, forget gate and output gate. The input gate selects which information elements are related to the existing state and can be added to the memory cell. The forget gate determines which information elements can be filtered out. The output gate determines which inputs can be entered to the next step based on the existing state.

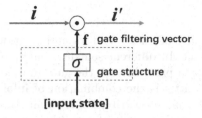

Fig. 2. Traditional gate

Gate is the core mechanism of LSTM, and its function is to let information pass selectively. A gate consists of a full connection layer and a sigmoid activation function. The gate vector and the vector to be filtered perform the Hadamard product operation to finish the information filtering. The feedforward form of gate is:

$$\mathbf{f} = \sigma\left(\mathbf{W}\left[\boldsymbol{input}, \boldsymbol{state}\right] + \mathbf{b}\right) \tag{1}$$

$$\boldsymbol{i}^{'} = \boldsymbol{i} \bigodot \mathbf{f} \tag{2}$$

$\forall x \in \mathbb{R}$, $\sigma\left(x\right) = \frac{1}{1+exp(-x)}$, $0 \leq \sigma\left(x\right) \leq 1$. That is, all the values of gate filtering vector $\mathbf{f}$ is between 0 and 1. As shown in Figure 2, *input* represents new information of the current gate, and *state* represents the memory sum of all previous input information. Based on the current *state* and the new *input*, a gate filtering vector $\mathbf{f}$ is generated. Each value of $\mathbf{f}$ indicates how much of each information element in a vector should be allowed passing and how much should be forgotten. $\boldsymbol{i}$ is the vector to be filtered, $\boldsymbol{i}^{'}$ is the vector filtered through the gate structure, $\bigodot$ is the counterpoint multiplication.

Noise Theorem. The following theorem has already been proved in [12]. It can be used to extend one Bernoulli distribution to many independent Bernoulli distributions by adding random noise. The value of $G\left(\alpha, \tau\right)$ can be controlled by controlling the temperature parameter τ . τ is a parameter greater than zero, which controls the soft degree of *sigmoid*. The higher the temperature, the smoother the generated distribution. The lower the temperature, the closer the generated distribution is to the discrete one-hot distribution.

Theorem 1. *Assume $\sigma\left(\cdot\right)$ is the sigmoid function. Given $\alpha \in \mathbb{R}$ and temperature $\tau > 0$, we random variable $D_\alpha \sim B\left(\sigma\left(\alpha\right)\right)$ where $B\left(\sigma\left(\alpha\right)\right)$ is the Bernoulli distribution with parameter $\sigma\left(\alpha\right)$, and define*

$$G\left(\alpha, \tau\right) = \sigma\left(\frac{\alpha + logU - log\left(1 - U\right)}{\tau}\right) \tag{3}$$

where $U \sim Uniform\left(0, 1\right)$. Then the distribution of $G\left(\alpha, \tau\right)$ can be considered as an approximation of Bernoulli distribution $B\left(\sigma\left(\alpha\right)\right)$.

3.2   TransMVG

A value of each dimension of entity vectors and relation vectors corresponds to an information element. In different semantic environment, the information elements of each dimension pass through the gate in different proportions. The essence of multiple semantics is the combination of information elements with different proportions. Our main contribution is eliminating the extreme filtering of traditional gates and making the information elements get more accurate combinations.

Model. Shown in Fig. 3, TransMVG uses three real number vectors $h, r, t \in \mathbb{R}^m$ in a same vector space to represent each triplet in a knowledge graph. Two full connection layers $W_h \cdot [h, r] + b_h$ and $W_t \cdot [t, r] + b_t$ are set for learning the global shared parameter gates of head entity vectors and tail entity vectors respectively. Gate filtering vectors $\mathbf{f}_h$ and $\mathbf{f}_t$ generated by full connection layers after a *sigmoid* operation will carry on Hadamard multiplication with head entity vectors and tail entity vectors respectively to select how much the information element of each dimension should pass.

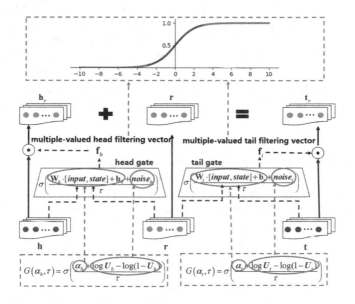

Fig. 3. TransMVG

The relation vector (semantic environment) in a triplet can be regarded as the *state* in the aforementioned LSTM gate. Taking the head gate as an example, the working object of the full connection weight in the traditional LSTM gate is the concatenation of new input and the current state. The new input can be understood as the entity vector to be filtered h , and the current state can be regarded as the semantic environment, namely the relation vector r. Therefore, the object that the full connection layer will work on is the concatenating of the head entity vector h and the relation vector r. The weight matrix of full connection layer is $W_h \in \mathbb{R}^{m \times 2m}$, the bias is $b_h \in \mathbb{R}^m$. The output of the head entity gate is the filtering vector $\mathbf{f}_h$ of the head entity vector h. A Hadamard multiplication is performed between $\mathbf{f}_h$ and h to realize the specific semantic representation of entity h in the sematic environment r. In a similar way, the specific semantic representation of entity t in the sematic environment r can be obtained.

The values gate filtering vectors generated by traditional gates are mostly in a boundary binary-valued state(0 or 1). We use the **Theorem** 1 to change the extreme filtering pattern to a general one. α in Eq. 3 can be regarded as any value of the vector generated by the full connection layer, $(noise_h)_i$ and $(noise_t)_i$ can be regarded as the noise for each dimension. According to Eq. 3, we can get Eqs. 6 and 7 as following:

$$(W_h \cdot [h, r] + b_h)_i = (\alpha_h)_i \tag{4}$$

$$(W_t \cdot [t, r] + b_t)_i = (\alpha_t)_i \tag{5}$$

$$(noise_h)_i = log\,(u_h)_i - log(1 - (u_h)_i) \tag{6}$$

$$(noise_t)_i = log\,(u_t)_i - log(1 - (u_t)_i) \tag{7}$$

$i = 1, ..., k, k$ is the number of the vector dimension. Then, according to **Theorem** 1, values of the i-th dimension of all head filtering gates and all tail filtering gates nearly obey Bernoulli distribution respectively, as Eqs. 8 and 9.

$$\sigma\left(\frac{(W_h \cdot [h, r] + b_h)_i + (\text{noise}_h)_i}{\tau}\right) \sim Bernoulli \tag{8}$$

$$\sigma\left(\frac{(W_t \cdot [t, r] + b_t)_i + (\text{noise}_t)_i}{\tau}\right) \sim Bernoulli \tag{9}$$

For the whole vector generated by the full connection layer, the above operation is performed for each dimension. Then we can get general vectors by adding noise to each dimension respectively, as Eqs. 10-13:

$$\mathbf{f}_h = \sigma\left(\frac{W_h \cdot [h, r] + b_h + \text{noise}_h}{\tau}\right) \tag{10}$$

$$\mathbf{f}_t = \sigma\left(\frac{W_t \cdot [t, r] + b_t + \text{noise}_t}{\tau}\right) \tag{11}$$

$$noise_h = logU_h - log\,(1 - U_h) \tag{12}$$

$$noise_t = logU_t - log\,(1 - U_t) \tag{13}$$

The temperature τ is a hyper-parameter and the gate value of the gate filtering vectors can be approximated to the none-flat region (The blue part of the *sigmoid* function in Fig. 3) of the *sigmoid* function by adjusting τ. The elements in U_h and U_t are independent with each other and are sampled from a uniform distribution (0,1) respectively. Since the noise added for the values of each dimension is independent with each other, the gate values obtained for each dimension will conform to the Bernoulli distribution with different independent distributions respectively, that is, all values between 0 and 1 can be possible to be taken with similar possibilities.

After generating the gate filtering vector with precise filtering function, the entity can achieve the specific semantic representation in a certain semantic environment as shown in Eqs. 14 and 15.

$$\boldsymbol{h}_r = \boldsymbol{h} \odot \mathbf{f}_h \tag{14}$$

$$\boldsymbol{t}_r = \boldsymbol{t} \odot \mathbf{f}_t \tag{15}$$

After the semantic representation filtering, we make a translation in a specific semantic environment to obtain the distance function:

$$d_r(h,t) = \|\boldsymbol{h}_r + \boldsymbol{r} - \boldsymbol{t}_r\|_{L1/L2} \tag{16}$$

The smaller the distance, the better for the correct triplets, and the larger the distance, the better for the wrong triplets.

Training. In the training process, the maximum interval method is used to optimize the objective function to enhance the distinguishing ability of knowledge representation. For each (h, r, t) and its negative sample (h', r, t'), TransMVG aims to minimize the hinge-loss as following:

$$L = max\left(\gamma + d_r(h, t) - d_r(h', t'), 0\right) \tag{17}$$

where γ is a margin hyper-parameter and (h', r, t') is a negative sample from the negative sampling set. A negative sample can be obtained by randomly replacing the head or the tail of a correct triplet with an entity from the entity list. The loss function 17 is used to encourage discrimination between training triplets and corrupted triplets by favoring lower scores for correct triplets than for corrupted ones. The training process of TransMVG is carried out using Adam optimizer with constant learning rate.

Complexity Analyisis. As shown in Table 1, we compare TransMVG with famous semantic discriminate models and some classical indiscriminate models. The statistics of parameters of these models are from [25]. N_e is the number of entities, m is the dimension of entities, N_r is the number of relations, n is the dimension of relations. k is the number of hidden nodes of a neural network and s is the number of slice of a tensor. $\widehat{\Theta}$ denotes the average sparse degree of all transfer matrices.

From the table, we can see TransMVG enables semantic discrimination with fewer parameters and lower complexity, and requires no pretraining. Compared with TransGate, TransMVG adds only one hyper parameter τ to get noise and temperature control for all values of the fully connected layers.

4   Experiments

In this section, we empirically evaluate TransMVG on two key tasks: link prediction and triplet classification. We demonstrate that TransMVG outperforms most state-of-art baselines on multiple benchmark datasets.

Table 1. Complexity analyisis

Models	Embedding parameters	Discriminate parameters	Hyper parameters	Time complexity	Pre-training
TransE	$O(N_e m + N_r n), (m = n)$	None	2	$O(m)$	None
TransH	$O(N_e m + N_r n), (m = n)$	$O(N_r n)$	4	$O(m)$	TransE
DistMult	$O(N_e m + N_r n), (m = n)$	None	2	$O(m)$	None
TransR	$O(N_e m + N_r n)$	$O(N_r mn)$	3	$O(mn)$	TransE
CTransR	$O(N_e m + N_r n)$	$O(N_r mn)$	4	$O(mn)$	TransR
TransD	$O(N_e m + N_r n)$	$O(N_e m + N_r n)$	3	$O(m)$	TransE
TranSparse	$O(N_e m + N_r n)$	$O(2N_r(1 - \hat{\theta})mn), (0 \le \hat{\theta} \le 1)$	5	$O(2(1 - \hat{\theta})mn), (0 \le \hat{\theta} \le 1)$	TransE
ComplEx	$O(2N_e m + 2N_r n), (m = n)$	None	2	$O(m)$	None
CombinE	$O(2N_e m + 2N_r n), (m = n)$	None	2	$O(2m)$	None
ProjE	$O(N_e m + N_r n + 5m), (m = n)$	None	2	$O(N_e m + 2m)$	None
TransGate	$O(N_e m + N_r m + 5m)$	$O(4m^2 + 2m)$	2	$O(m^2)$	None
TransMVG	$O(N_e m + N_r n + 5m)$	$O(4m^2 + 2m)$	3	$O(m^2)$	None

4.1   Datasets

Link prediction and triplets classification are implemented on two large-scale knowledge bases: WordNet [14] and Freebase [1]. Twodata sets are employed from WordNet. Among them, WN11 [2] is for link prediction, and WN18RR [5] is for triplet classification. Three datasets from Freebase are used. Among them, FB15K [2] and FB15K-237 [20] are for link prediction, and FB13 [18] is for triplet classification. The details of these datasets are in Table 2.

4.2   Link Prediction

Link prediction aims to predict the missing h or t for a triplet (h, r, t). i.e., predict t given (h, r) or predict h given (r, t). Instead of giving one best answer,

Table 2. Statistics of datasets

Datasets	Rel	Ent	Train	Valid	Test
WN11 [2]	11	38,696	112,581	2,609	10,544
WN18RR [5]	11	40,943	86,835	3,034	3,134
FB13 [18]	13	75,043	316,232	5,908	23,733
FB15K [2]	1,345	14,951	483,142	50,000	59,071
FB15K-237 [20]	237	14,541	272,115	17,535	20,466

this task ranks a set of candidate entities from the knowledge graph. For each testing triplet (h, r, t), we corrupt it by replacing the tail t with every entity e in the knowledge graph and calculate all distance scores. Then we rank the scores in ascending order, and get the rank of the original. In fact, a corrupted triplet may also exist in the knowledge graph, which should be also considered as correct.

We filter out the correct triplets from corrupted triplets that have already existed in the knowledge graph to get the filtered results. With use of these filtered ranks, we get three commonly used metrics for evaluation: the average rank of all correct entities (Mean Rank), the mean reciprocal rank of all correct entities (MRR), and the proportion of correct entities ranked in top k(Hits@k). A good link prediction result expects a lower Mean Rank, a higher MRR, and a higher (Hits@k).

In this task, we use three datasets: WN18RR [5], FB15K [2] and FB15K-237 [20]. For three datasets, we search the learning rate α for Adam among $\{0.001, 0.01, 0.1\}$, the temperature τ among $\{100, 200, 500\}$ the margin γ among $\{2, 4, 6, 8, 10\}$, the embedding dimension m among $\{50, 100, 200\}$, and the batch size B among $\{1440, 2880, 5760\}$. The optimal configurations are as follow: on WN18RR, $\gamma = 8$, $\alpha = 0.1$, $\tau = 200$, $m = 200$, $B = 2880$ and taking $L1$ distance; on FB15K, $\gamma = 4$, $\alpha = 0.1, \tau = 200$, $m = 200$, $B = 5760$ and taking $L1$ distance; on FB15K-237, $\gamma = 4$, $\alpha = 0.1$, $\tau = 200$, $m = 200$, $B = 5760$ and taking $L1$ distance.

In Table 3, the best scores are in bold, while the second best scores are in underline. In the table, we observe that: (1) On FB15K-237, TransMVG outperforms all baselines at MRR, Hits@10 and Hits@1 metrics, improved by 1.4%, 6.0% and 20.7% respectively compared with the underlined second ranks. (2) On FB15K, TransMVG outperforms all baselines at Mean Rand metric and get a second good rank at Hits@10 metric. (3) On WN18RR, TransMVG outperforms all baselines at Hits@10 metric, improved by 2.7% compared with the underlined second ranks. This indicates the great ability of TransMVG on precise link prediction.

As the reversing relations have been removed in WN18RR, the semantic hierarchy of the relations in the database is no longer complete. As a result, the TransMVG model does not have a complete corpus to adequately learn

Table 3. Evaluation results on link prediction

Datasets	WN18RR				FB15K				FB15K-237			
Metrics	MRR	MR	Hits@10	Hits@1	MRR	MR	Hits@10	Hits@1	MRR	MR	Hits@10	Hits@1
TransE [2]	0.226	3384	50.1	-	0.220	125	47.1	23.1	0.294	347	46.5	14.7
DistMult [24]	0.43	5110	49.0	39	0.654	97	82.4	54.6	0.241	254	41.9	15.5
TransD [10]	-	-	42.8	-	0.252	67	77.3	23.4	-	-	45.3	-
CombineE [19]	-	-	-	-	0.283	-	85.2	55.4	-	-	-	-
ComplEX [21]	0.44	5261	51.0	41.0	0.692	-	84.0	59.9	0.247	339	42.8	15.8
KB-LRN [6]	-	-	-	-	0.794	44	87.5	74.8	0.309	209	49.3	21.9
NLFeat [20]	-	-	-	-	0.822	-	87.0	-	0.249	-	41.7	-
RUGE [7]	-	-	-	-	0.768	-	86.5	70.3	-	-	-	-
KBGAN [4]	0.213	-	48.1	-	-	-	-	-	0.278	-	45.8	-
R-GCN [16]	-	-	-	-	0.696	-	84.2	60.1	0.248	-	41.7	15.3
TransG [23]	-	-	-	-	0.657	51	83.1	55.8	-	-	-	-
ConvE [5]	0.43	4187	52.0	40.0	0.657	51	83.1	55.8	0.325	244	50.1	23.7
ConvKB[15]	0.248	2554	52.5	-	0.768	-	-	-	0.396	257	51.7	-
TransGate [25]	0.409	3420	51.0	39.1	0.832	33	91.4	75.5	0.404	177	58.1	25.1
TransMVG	0.253	4391	53.9	3.9	0.630	31	88	46.5	0.410	223	61.2	30.3

the information element interactions of various relations. Thus, TransMVG only perform best on one metirc on WN18RR.

FB15K contains a number of redundant relations. This may inhibit fine semantic recognition ability of TransMVG to some extent. Therefore, on FB15K, TransMVG only performed best in one metric. Fb15K-237 is obtained by removing the redundant relations in FB15K. TransMVG has a good performance almost beyond various baselines at all metrics. The result also shows that the multi-valued gate in TransMVG have more powerful multi-semantic learning ability than the boundary binary-valued gate in TransGate.

4.3 Triplet Classification

Triplet classification aims to judge whether a given triplet (h, r, t) is correct or not. It is first used in [18] to evaluate knowledge graph embeddings learned by NTN model. In this paper we use WN11 [2] and FB13 [18] as the benchmark datasets for this task. These two datasets contain positive and negative triplets. For each triplet (h, r, t), if the value calculated by the distance score Eq. 16 is above a relation-specific threshold Δ, then the triplet will be classified as positive, otherwise it will be classified as negative.

For WN11 and FB13, we compare TransMVG with baselines reported in [25]. In training, for the two datasets, we search the learning rate α among $\{0.001, 0.01, 0.1\}$, the temperature τ among $\{100, 200, 500\}$, the margin γ among $\{2, 4, 6, 8, 10\}$, the embedding dimension m among $\{50, 100, 200\}$, the batch size B among $\{1440, 2880, 5760\}$. The optimal configurations are as follow: On WN11, $\gamma = 10$, $\alpha = 0.01$, $\tau = 100$, $m = 100$, $B = 2880$ and taking L1 distance. On FB13, the best configurations are: $\gamma = 6$, $\alpha = 0.001$, $\tau = 100$, $m = 100$,

Table 4. Evaluation results on triplet classification

Datasets	WN11	FB13
SE [3]	53.0	75.2
SME [3]	70.0	63.7
LFM [9]	73.8	84.4
SLM [18]	69.9	85.3
NTN [18]	70.4	87.1
TransE [2]	75.9	70.9
TransH [22]	77.7	76.5
Trans R[13]	85.5	74.7
CTransR [13]	85.7	-
KG2E [8]	85.4	85.3
TransD [10]	85.6	**89.1**
TransSparse [11]	86.8	86.5
TransG [23]	<u>87.4</u>	87.3
TransGate [25]	87.3	<u>88.8</u>
TransMVG	**89.5**	84

$B = 2880$ and taking L1 distance. Table 4 shows the detailed evaluation results of triplets classification. From the table, we observe that: (1) On WN11, our method obtains the accuracy of 89.5% and outperforms all baseline models. (2) On FB13, the accuracy of our method is only in the middle of the rank list.

The WN11 dataset includes $1-1$, $1-N$, $N-1$ three relation types. There are 113,000 triples in its training set. The FB13 dataset includes $N-1$, $N-N$ two relation types. There are 316,000 triplets in its training set. This indicates that FB13 is more dense than WN11 both in relation nature and the number of entity pairs connected by each relation. So we guess if the dimension of TransMVG can go beyond the current maximum set 200, it will achieve better performance under richer proportion combination of more information elements.

4.4 Distribution Visualization

As shown in Fig. 5, we run TransGate and TransMVG on FB15K with the same settings, the initial values of the two models conform to both nearly binary-valued distributions. TransGate has an obvious trend of boundary binary-valued trend when it is run 100 times. TransMVG, on the other hand, obtains multiple-valued gates in a gentle manner. In fact, in both the TransGate and TransMVG models, the value of any dimension of the gate filtering vector conforms to a binary distribution, and each dimension distribution is independent with each other. The binary distribution of the gate values of each dimension in TransGate tends to be a boundary binary-valued distribution, while that of the gate values in TransMVG tends to be many independent different distributions due to the addition of noise, resulting in a mutiple-valued gate.

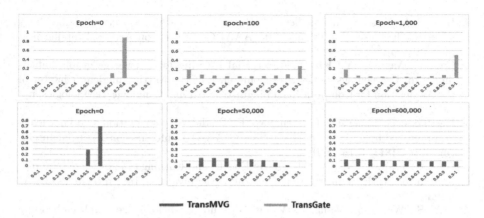

Fig. 4. The value distributions of the gate filtering vectors in TransGate and Trans-MVG

5 Conclusion

In this paper, we focus on embedding the knowledge graph into a low-dimension vector space for knowledge graph completion. We find that extreme filtering problem exsits in the traditional method based on shared parameter gate, and the main reason is the boundary binary-valued distribution of its gate filter values. We propose an information element interaction mechanism to explain the multi-semantic representation of the same vector in different semantic environments. Our TransMVG model refined the interaction of information elements by adding independently distributed noise to the full connection layer of the shared parameter gate and pushing the gate values to be multi-valued. We have conduct a number of experiments on link prediction task and triplet classification task. The experiments results show that TransMVG almost outperforms state-of-the-art baselines. This means it is feasible and necessary to multivalue the filter gate vectors in the process of knowledge representation learning.

In TransMVG, only the multi-semantics of entities have been taken into consideration. In the future, we will try to deal with the multi-semantics of both entities and relations at the same time.

References

1. Bollacker, K., Evans, C., Paritosh, P., Sturge, T., Taylor, J.: Freebase: a collaboratively created graph database for structuring human knowledge. In: Proceedings of the 2008 ACM SIGMOD International Conference on Management of data, pp. 1247–1250 (2008)
2. Bordes, A., Usunier, N., Garcia-Duran, A., Weston, J., Yakhnenko, O.: Translating embeddings for modeling multi-relational data. In: Advances in Neural Information Processing Systems, pp. 2787–2795 (2013)

3. Bordes, A., Weston, J., Collobert, R., Bengio, Y.: Learning structured embeddings of knowledge bases. In: Twenty-Fifth AAAI Conference on Artificial Intelligence, pp. 301–306 (2011)
4. Cai, L., Wang, W.Y.: Kbgan: Adversarial learning for knowledge graph embeddings. In: Proceedings of the 2018 Conference of the North American Chapter of the Association for Computational Linguistics: Human Language Technologies, pp. 1470–1480 (2018)
5. Dettmers, T., Minervini, P., Stenetorp, P., Riedel, S.: Convolutional 2d knowledge graph embeddings. In: The Thirty-Second AAAI Conference on Artificial Intelligence, pp. 1811–1818 (2018)
6. Garcia-Duran, A., Niepert, M.: Kblrn: End-to-end learning of knowledge base representations with latent, relational, and numerical features. In: Proceedings of UAI (2017)
7. Guo, S., Wang, Q., Wang, L., Wang, B., Guo, L.: Knowledge graph embedding with iterative guidance from soft rules. In: Thirty-Second AAAI Conference on Artificial Intelligence, pp. 4816–4823 (2018)
8. He, S., Liu, K., Ji, G., Zhao, J.: Learning to represent knowledge graphs with gaussian embedding. In: Proceedings of the 24th ACM International on Conference on Information and Knowledge Management, pp. 623–632 (2015)
9. Jenatton, R., Roux, N.L., Bordes, A., Obozinski, G.: A latent factor model for highly multi-relational data. Adv. Neural Inf. Process. Syst. **4**, 3167–3175 (2012)
10. Ji, G., He, S., Xu, L., Liu, K., Zhao, J.: Knowledge graph embedding via dynamic mapping matrix. In: Proceedings of the 53rd Annual Meeting of the Association for Computational Linguistics and the 7th International Joint Conference on Natural Language Processing, vol. 1, pp. 687–696 (2015)
11. Ji, G., Liu, K., He, S., Zhao, J.: Knowledge graph completion with adaptive sparse transfer matrix. In: Thirtieth AAAI Conference on Artificial Intelligence, pp. 985–991 (2016)
12. Li, Z., et al.: Towards binary-valued gates for robust LSTM training. In: Proceedings of the 35th International Conference on Machine Learning (2018)
13. Lin, Y., Liu, Z., Sun, M., Liu, Y., Zhu, X.: Learning entity and relation embeddings for knowledge graph completion. In: Twenty-ninth AAAI Conference on Artificial Intelligence, pp. 2181–2187 (2015)
14. Miller, G.: Wordnet: a lexical database for English. Commun. ACM. **38**, 39–41 (1995)
15. Nguyen, D.Q., Nguyen, T.D., Nguyen, D.Q., Phung, D.: A novel embedding model for knowledge base completion based on convolutional neural network. In: Proceedings of the 2018 Conference of the North American Chapter of the Association for Computational Linguistics: Human Language Technologies, pp. 327–333 (2017)
16. Schlichtkrull, M., Kipf, T.N., Bloem, P., Van Den Berg, R., Titov, I., Welling, M.: Modeling relational data with graph convolutional networks. In: European Semantic Web Conference, pp. 593–607 (2018)
17. Shi, B., Weninger, T.: Proje: Embedding projection for knowledge graph completion. In: Proceedings of the Thirty-First AAAI Conference on Artificial Intelligence, pp. 1236–1242 (2017)
18. Socher, R., Chen, D., Manning, C.D., Ng, A.Y.: Reasoning with neural tensor networks for knowledge base completion. In: Advances in Neural Information Processing Systems, pp. 926–934 (2013)
19. Tan, Z., Zhao, X., Wang, W.: Representation learning of large-scale knowledge graphs via entity feature combinations. In: Proceedings of the 2017 ACM on Conference on Information and Knowledge Management, pp. 1777–1786 (2017)

20. Toutanova, K., Chen, D.: Observed versus latent features for knowledge base and text inference. In: Proceedings of the 3rd Workshop on Continuous Vector Space Models and their Compositionality, pp. 57–66 (2015)
21. Trouillon, T., Welbl, J., Riedel, S., Gaussier, E., Bouchard, G.: Complex embeddings for simple link prediction. In: Proceedings of the 33rd International Conference on Machine Learning, pp. 2071–2080 (2016)
22. Wang, Z., Zhang, J., Feng, J., Chen, Z.: Knowledge graph embedding by translating on hyperplanes. In: Twenty-Eighth AAAI Conference on Artificial Intelligence, pp. 1112–1119 (2014)
23. Xiao, H., Huang, M., Zhu, X.: Transg: A generative model for knowledge graph embedding. In: Proceedings of the 54th Annual Meeting of the Association for Computational Linguistics, vol. 1, pp. 2316–2325 (2016)
24. Yang, B., Yih, W., He, X., Gao, J., Deng, L.: Embedding entities and relations for learning and inference in knowledge bases. In: Proceedings of the International Conference on Learning Representations (2014)
25. Yuan, J., Gao, N., Xiang, J.: TransGate: Knowledge graph embedding with shared gate structure. Proc. AAAI Conf. Artif. Intell. **33**, 3100–3107 (2019)
26. Hochreiter, S., Schmidhuber, J.: Long short-term memory. Neural Comput. **9**, 1735–1780 (1997)

ADKT: Adaptive Deep Knowledge Tracing

Liangliang He[1], Jintao Tang[1(✉)], Xiao Li[2], and Ting Wang[1]

[1] College of Computer, National University of Defense Technology,
No. 137, Yanwachi Street, Changsha 410073, Hunan, People's Republic of China
{heliangliang19,tangjintao,tingwang}@nudt.edu.cn
[2] Information Center, National University of Defense Technology,
No. 137, Yanwachi Street, Changsha 410073, Hunan, People's Republic of China
xiaoli@nudt.edu.cn

Abstract. Deep Learning based Knowledge Tracing (DLKT) has been shown to outperform other methods due to its strong representational ability. However, DLKT models usually exist a common flaw that all learners share the same model with identical network parameters and hyper-parameters. The drawback of doing so is that the learned knowledge state for each learner is only affected by the specific learning sequence, but less reflect the personalized learning style for each learner. To tackle this problem, we proposes a novel framework, called Adaptive Deep Knowledge Tracing (ADKT), to directly introduce personalization into DLKT. The ADKT framework tries to retrain an adaptive model for each learner based on a pre-trained DLKT model and trace the knowledge states individually for each learner. To verify the effectiveness of ADKT, we further develop the ADKVMN (Adaptive Dynamic Key-Value Memory Network) model by combining the ADKT and the classic DKVMN, which has been widely referred as a state-of-the-art DLKT model. With extensive experiments on two popular benchmark datasets, including the ASSISTments2009 and ASSISTments2015 datasets, we empirically show that ADKVMN has superior predictive performance than DKVMN.

Keywords: Knowledge Tracing (KT) · Deep learning · Dynamic Key-Value Memory Network (DKVMN) · Personalized learning

1 Introduction

Knowledge Tracing (KT) [2,4,11,12,14] is an important task in E-learning. The goal of KT is to model the knowledge state of learners, i.e., the mastery level of knowledge components (KCs) (e.g., skills, concepts and so on) [15]. Generally, in a KT task, given a learner's historical interaction sequence $\mathbf{X}_t = (\mathbf{x}_1, \mathbf{x}_2 \ldots \mathbf{x}_t)$

Liangliang He and Jintao Tang are co-first authors of this article.

© Springer Nature Switzerland AG 2020
Z. Huang et al. (Eds.): WISE 2020, LNCS 12342, pp. 302–314, 2020.
https://doi.org/10.1007/978-3-030-62005-9_22

up to the timestamp t on a specific learning scenario, the KT model tries to predict the probability that the learner will correctly perform a learning action (e.g., answering an exercise) in the next timestamp, i.e., $p(a_{t+1} = 1|q_{t+1}, \mathbf{X}_t)$ where $\mathbf{x}_t = (q_t, a_t)$ is an input tuple containing the exercise tag q_t at the timestamp t and the learner's answer a_t to q_t [6,15,17].

The classic method used for the KT task is Bayesian Knowledge Tracing (BKT) [4]. In BKT, the knowledge state of a learner consists of the known status of each predefined KC. During the learning process, BKT updates the knowledge status of each KC based on the Hidden Markov Model (HMM). However, BKT is unable to capture the relationship between different concepts and lacks the ability to extract undefined KCs [17].

Recent years, Deep Learning based Knowledge Tracing (DLKT) methods [11,15,17] have been shown to outperform other methods. The framework of DLKT is shown in Fig. 1. The learning process usually consists of two stages. First, all learners' complete sequences are input to the Deep Neural Network (DNN) to learn a universal DLKT model. Second, the knowledge state of each user is traced by the universal model from the first timestamp. The DNN method is the core component in DLKT. Representative DNN methods in DLKT are Deep Knowledge Tracing (DKT) [11] and Dynamic Key-Value Memory Network (DKVMN) [17]. DKT uses Recurrent Neural Network (RNN) and Long Short-Term Memory (LSTM) network [5] to capture the sequential dependency between each KC. DKVMN models the relationship between KCs with memory network and traces the learner's knowledge state about each underlying KC.

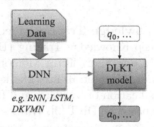

Fig. 1. Deep Learning based Knowledge Tracing framework

However, The DLKT framework exists an obvious defect that all learners share the same network parameters as well as hyper-parameters. This means that such a knowledge tracing pattern does not reflect the personalized learning behavior of learner. To address this problem, we propose a new deep knowledge tracing framework, named ADKT (Adaptive Deep Knowledge Tracing) which is an extension of DLKT. The proposed framework aims to train an ADKT model for each learner on her existing sequence, and trace the future knowledge state of each learner more accurately. To verify the effectiveness of the ADKT framework, we apply the proposed framework to DKVMN which is the state-of-the-art deep learning based KT model in recent years, and propose the Adaptive Dynamic

Key-Value Memory Network (ADKVMN) model. With large-scale experiments on two benchmark datasets including ASSISTments2009 and ASSISTments2015, we empirically demonstrate that ADKVMN has superior predictive performance over DKVMN.

To summarize, the contributions of this paper are summarized as follows:

(1) We propose a new deep learning based knowledge tracing framework named Adaptive Deep Knowledge Tracing (ADKT) by introducing the idea of personalized learning, aiming at tracing the knowledge state of each learner individually.
(2) We apply the adaptive framework to the Dynamic Key-Value Memory Network (DKVMN) and propose the Adaptive Dynamic Key-Value Memory Network (ADKVMN) model.
(3) Experiments on two benchmark datasets show that ADKVMN has superior predictive performance than DKVMN.

The rest of this paper is organized as follows. Section 2 reviews the previous works. Section 3 proposes the ADKT framework. ADKVMN is proposed in Sect. 4. Section 5 conducts experiments and result analysis. The last section concludes this paper and presents the future work.

2   Related Works

In this section, we make a brief review of previous works in the field of deep learning based knowledge tracing and personalization in KT.

Deep Learning Based Knowledge Tracing. Piech et al. pioneer the work of DLKT by introducing Deep Knowledge Tracing (DKT) [11] with Recurrent Neural Network (RNN) and Long Short-Term Memory (LSTM) network [5] to the KT application. After that, Minn et al. uses the k-means clustering method [7] to obtain a dynamic classification of learners, and traces the knowledge state of learners in each clustering based on DKT [9]. Wang et al. further extend binary answer embedding in DKT to open multi-step answer embedding [13]. Recently, the most widely concerned and applied DLKT model is the Dynamic Key-Value Memory Network (DKVMN) [17] which can automatically learn the correlation between the input exercise and the underlying KCs, so as to trace the knowledge state of learner about each KC. Chaudhry et al. extend DKVMN by modeling learners' hint data [3]. Minn et al. [8] propose Dynamic Student Classification on Memory Networks (DSCMN) that combines the advantages of DKVMNN and DKT-DSC [9]. Yeung [15] extracts the parameters which is required in IRT model [14] from the trained DKVMN to provide a interpretability for the DLKT.

Personalized Exploration of KT. Bayesian Knowledge Tracing (BKT) [4] contains four parameters: $P(L_0)$, $P(T)$, $P(G)$ and $P(S)$, which denote the initial mastery probability, the learning rate, the guessing probability and the slip probability, respectively. These parameters, which are updated based on the Hidden

Markov Model (HMM), are suitable for all learners and exercises. Many works have been proposed to explore personalization in BKT. For example, Corbett et al. personalize $P(G)$ and $P(S)$ in BKT, which assumes that different learners are individualized in the two aspects of guess and slip [1]. Pardos et al. personalize $P(L_0)$ and $P(T)$ to reflect the differences between learners [10]. In addition, Pardos et al. show that $P(G)$ and $P(S)$ can reflect the difficulty of different exercises [16]. Unlike BKT, less works have explored personalization in DLKT. However, Wang et al.'s works [8,9] on DLKT illustrate that the performances of the DLKT model shared by learners with similar learning histories are better than that shared by all learners.

3   Adaptive Deep Knowledge Tracing

3.1   Motivations for Adaptive Deep Knowledge Tracing

Motivation 1. In fact, the parameters in DLKT model which is trained on all learners' sequences can be seen as a condensed representation of all learners' *global* learning patterns (e.g., learning rate, forgetting, guess, slip, etc.). However, there are obvious *local* differences in the learning patterns of different learners. Therefore, a one-size-fit-all model parameters cannot meet the needs of different learners for adaptive learning. It is imperative to mine the personalized learning pattern from each sequence itself.

Motivation 2. When DLKT is used for the KT task, as the model is trained on all sequences, the parameters from DLKT model may provide basic information for an unknown exercise, such as the related underlying KCs and difficulty. In addition, DLKT model simulates the whole learning process of all learners. Thus, DLKT can provide basic answer prediction for each exercise. If each ADKT model is trained based on the DLKT model, there will be no large prediction bias caused by inadequate information especially for learners with short sequences.

3.2   Proposed Adaptive Deep Knowledge Tracing Framework

Based on the above motivations, this paper proposes the Adaptive Deep Knowledge Tracing (ADKT) framework. The framework is illustrated in Fig. 2. The framework firstly trains a DLKT model on all learners' sequences. Then, it retrains the DLKT model on a new learner's sequence and results in the ADKT model of the learner. Finally, the framework can trace individual learner' future knowledge state based on the ADKT model.

The most important component in ADKT is the DNN. We formalize the ADKT framework as $ADKT(DNN)$. The DNN (e.g. RNN, LSTM and DKVMN) is required to perform the KT task based on sequences which include Q&A interaction and hints usage, etc. The whole adaptive knowledge tracking process is divided into **DLKT model training** and **ADKT model training**.

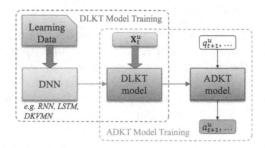

Fig. 2. Adaptive Deep Knowledge Tracing framework.

DLKT Model Training. The DNN is trained on all learners' sequences, and the trained DNN forms the DLKT model. In fact, the DLKT model simulates the learning patterns of all learners. Therefore, the parameters of the model cannot show differences for each learner.

ADKT Model Training. Based on the sequence $\mathbf{X}_t^u$ of the new learner u up to the current timestamp t. The DLKT model is individually trained as an ADKT model that is especially tuned for tracing the future knowledge state of u, i.e., predicting the answers $a_{t+1}^u, \ldots$ to the exercises $q_{t+1}^u, \ldots$ that user u will begin at the next timestamp $t + 1$.

Algorithm 1. Adaptive deep knowledge tracing algorithm.

Input:

Deep neural network: DNN; The training and validation set of the DLKT model: $\mathbf{s}^{train}$ and $\mathbf{s}^{valid}$; the training and validation set of the ADKT model of user u: $\mathbf{s}_u^{train}$ and $\mathbf{s}_u^{train}$.

Output:

Learner u's ADKT model: $ADKT_u$.

1: # DLKT model training.
2: Initialize $random\_initialize(DNN)$.
3: **while** the predicted accuracy of DNN is optimal in $\mathcal{S}^{valid}$ **do**
4:     $train(DNN, \mathbf{s}^{train})$
5:     $test(DNN, \mathbf{s}^{valid})$
6: **end while**
7: $DLKT \leftarrow DNN$
8: # ADKT model training.
9: Initialize $optimal\_initialize(DLKT)$
10: **while** the predicted accuracy of $DLKT$ is optimal in $\mathbf{s}_u^{valid}$ **do**
11:     $train(DLKT, \mathbf{s}_u^{train})$
12:     $test(DLKT, \mathbf{s}_u^{valid})$
13: **end while**
14: $ADKT_u \leftarrow DLKT$
15: **return** $ADKT_u$

The algorithm for the ADKT framework is described in Algorithm 1. The adaptive deep knowledge tracing can be implemented based on a DNN, only if the function $optimal\_initialize(\cdot)$ is determined.

4 Improving DKVMN with ADKT

In this section, we make an instance of ADKT to illustrate how to personalize a DLKT model. We use Dynamic Key-Value Memory Network (DKVMN) [17] as the DNN component, marked as $\boldsymbol{ADKT}(DKVMN)$ or ADKVMN.

4.1 DKVMN

Assume there are Q non-repeating exercises and N underlying KCs. These underlying KCs are stored in the *key* matrix $\mathbf{M}^k$ (of size $N \times d_k$), where d_k is the dimension of each *key* memory slot. And the learner's mastery levels of each KC are stored in the *value* matrix $\mathbf{M}^v$ (of size $N \times d_v$), where d_v is the dimension of each *value* memory slot. The knowledge tracing process of DKVMN is shown in Fig. 3.

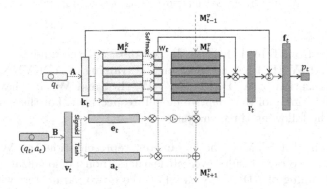

Fig. 3. Architecture for Dynamic Key-Value Memory Network (DKVMN).

In this figure, $\mathbf{A}$ (of size $Q \times d_k$) denotes an embedding matrix used to transform the one-hot vector of the input exercise q_t into a low-dimensional continuous vector $\mathbf{k}_t$ (of dimension d_k); $\mathbf{B}$ (of size $2Q \times d_v$) denotes an embedding matrix used to transform the one-hot vector of the input question-answer interaction (q_t, a_t) into a low-dimensional continuous vector $\mathbf{v}_t$ (of dimension d_v); $\mathbf{r}_t$ is a read vector which is treated as a summary of a learner's mastery level to q_t; $\mathbf{f}_t$ denotes a summary vector contained both the learner's mastery level and the difficulty of q_t; p_t is a scalar that denotes the probability of answering q_t correctly; $\mathbf{e}_t$ and $\mathbf{a}_t$ are respectively an erase vector and an add vector transformed from $\mathbf{v}_t$.

At the timestamp t, an exercise q_t is input firstly, and then the probability of correct answer q_t is predicted, and finally (q_t, a_t) is used to update $\mathbf{M}^v$.

4.2   ADKVMN

Parameter Initialization. For the DKVMN model, the parameters which can be used to initialize ADKVMN models of all learners are: $\mathbf{A}$, $\mathbf{B}$, $\mathbf{M}^k$, $\mathbf{M}^v$, $\mathbf{W}^f(\mathbf{b}^f)$, $\mathbf{W}^p(\mathbf{b}^p)$, $\mathbf{W}^e(\mathbf{b}^e)$ and $\mathbf{W}^a(\mathbf{b}^a)$. In order to determine the optimal parameter initialization scheme, 8 schemes are explored in this paper. To illustrate the situation better, all schemes are tested on two data subsets, and the results of the test AUC are shown in Table 1. Where AR, BR, MKR, MVR, WER, WAR, WFR and WPR denote that $\mathbf{A}$, $\mathbf{B}$, $\mathbf{M}^k$, $\mathbf{M}^v$, $\mathbf{W}^f(\mathbf{b}^f)$ and $\mathbf{W}^p(\mathbf{b}^p)$ are respectively initialized in ADKVMN, and the residual randomly; NR denotes no parameter in ADKVMN is randomly initialized; $[400, 500)$ and $[160, 200)$ are respectively sub-sets from ASSISTments2009 and ASSISTments2015 (see Sect. 5.2).

Table 1. Comparison of different initialization schemes for ADKVMN model.

Sub-sets	The initialized parameters in ADKVMN								
	AR	BR	MKR	MVR	WER	WAR	WFR	WPR	NR
$[400, 500)$	0.66201	0.68069	0.71850	0.84306	0.84235	0.84235	0.82456	0.76282	0.84235
$[160, 200)$	0.66242	0.64501	0.72554	0.74780	0.75302	0.75302	0.72910	0.68024	0.75302

We can see from Table 1 that: for the two datasets, the random initialization of $\mathbf{W}^e(\mathbf{b}^e)$ and $\mathbf{W}^a(\mathbf{b}^a)$ has no effect on the training of ADKVMN, while the random initialization of $\mathbf{A}$, $\mathbf{B}$, $\mathbf{M}^k$, $\mathbf{M}^v$, $\mathbf{W}^f(\mathbf{b}^f)$ and $\mathbf{W}^p(\mathbf{b}^p)$ have different effects on the training of the adaptive model, respectively. For these results, this paper gives the following three analysis:

- In the trained DKVMN, the parameter representations of $\mathbf{W}^e(\mathbf{b}^e)$ and $\mathbf{W}^a(\mathbf{b}^a)$ are average results over all learners (non personalization). Therefore, the training of ADKVMN based on these two parameters with random initialization is equivalent to the training of ADKVMN without random initialization.
- The parameter representations of $\mathbf{A}$, $\mathbf{B}$ and $\mathbf{M}^k$ are trained based on the sequence data of all learners, so they contains information about all exercises ($\mathbf{A}$ and $\mathbf{B}$) and underlying KCs ($\mathbf{M}^v$). However, the information cannot be obtained based on a learner's sequence in the training process of ADKVMN.
- When a learner has enough data, ADKVMN may get more complete $\mathbf{M}^v$ information about underlying KCs. In this case, $\mathbf{M}^v$ in the trained DKVMN will have a bad impact on the training results of ADKVMN. Therefore, ADKVMN based on random initialization of $\mathbf{M}^v$ can achieve better results than nonrandom initialization. However, for cold-start, the results are just the opposite. Since a relatively complete $\mathbf{M}^v$ can not be obtained based on the learning data of a learner, DKVMN parameter reflecting the average ability state of all learners may provide help for the prediction process of ADKVMN.

5   Experiments and Analysis

5.1   Datasets and Metric

Datasets. To evaluate performance, two benchmark datasets: ASSISTments2009 and ASSISTments2015 [17] are used in this paper. The statistical information of these two datasets can be found in Table 2. Considering the need of the ADKT framework, short sequences are abandoned in dataset, i.e., the sequence length is not less than 20 for ASSISTments2009, and 10 for ASSISTments2015. And the 3-fold cross validation is used to tune hyper-parameters of all compared models in this paper.

Table 2. Statistical information of ASSISTments2009 and ASSISTments2015. Where MaxSeqlen, MinSeqlen and MeanSeqlen denote maximum, minimum and mean sequence length, respectively.

Datasets	Overview					
	Learners	Exercise tags	Records	MaxSeqlen	MinSeqlen	MeanSeqlen
ASSISTments2009	4,151	110	325,637	1261	1	78
ASSISTments2015	19,840	110	683,801	618	1	34

Metric. In the field of KT, a widely used metric is the Area Under the Curve (AUC). The role of AUC is to evaluate the prediction accuracy of a model on specific datasets: AUC of 0.5 is equivalent to the predicted results obtained by random guessing; the higher the AUC value, the better the prediction performance of a model [17].

5.2   Experimental Configuration

Test Scheme. Unlike DLKT, The sequence of testing ADKT model does not begin with the first timestamp. Therefore, this paper proposes a new test scheme, as shown in Fig. 4. Firstly, *test-seq* and *train-seq* are concatenated into *target-seq* in order. Secondly, the position of *test-seq* in *target-seq* is marked by a flag vector (*flag-vec*) that is not the only. Thirdly, *target-seq* is used to test the ADKT model which is trained based on *train-seq*, and get a corresponding predicted sequence (*predict-seq*). Finally, *flag-vec* is used to retrieve the predicted results of *test-seq* (*test-predict-seq*) from *predict-seq*.

Hyper-parameter. It is seriously time-consuming to adjust the hyperparameters for each model, and these hyper-parameters are not suitable to the whole learning process due to the longer and longer sequence. Moreover, it is found that the adjustment of hyper-parameters is more sensitive to the length of a sequence and less sensitive to the content during the experiment. Therefore, this paper suggests that all the sequence in the same specified range of length

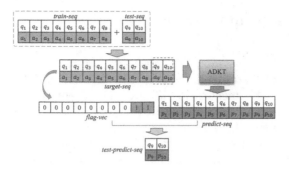

Fig. 4. The testing process of the ADKT model.

share a set of hyper-parameters. In this way, an ADKT model is initialized by retrieving the range of the sequence length. The range division scheme in the experiment is that: [20, 50), [50, 100), [100, 200), [200, 300), [300, 400), [400, 500), [500, 600) for ASSISTments2009, and [10, 20), [20, 40), [40, 80), [80, 120), [120, 160), [160, 200), [200, 300) for ASSISTments2015. Where longer sequences are cut into sub-sequences belonging all shorter ranges.

Data Partitioning. The whole dataset is partitioned into three parts: training set (*train_set*), validating set (*valid_set*) and testing set (*test_set*), which are used to get the DLKT model. Then, each sequence in *test_set* is cut into different sub-sets (*sub_test_sets*) of the specified length range (in Sect. 5.2). Finally, each sequence in each *sub_test_set* is cut into three parts: an adaptive training sequence (*a_train_seq*), an adaptive validating sequence (*a_valid_seq*) and an adaptive testing sequence (*a_test_seq*), which are used to train, validate and test the ADKT model of the corresponding learner. For the DLKT model, 30% of the sequences in datasets are held out as a *test_set*, and 20% of the *train_set* is split to form a *valid_set*. For each ADKT model, 80% of each sequence in each *sub_test_set* is held out as *a_train_seq*, and *a_valid_seq* and *a_test_seq* are all 10%. All the *a_train_seqs*, *a_valid_seqs* and *a_test_seqs* are respectively marked as *a_train_set*, *a_test_set* and *u_test_set* in this paper.

Compared Methods. This paper compares the performance of the following four models on two datasets. The relationship among these models is shown in Table 3:

- **DKVMN** [17] is widely concerned and studied in the field of KT in recent years. The DKVMN model trained based on *train_set* is directly used to test *a_test_set*.
- **DKVMN++** is proposed in this paper, Is a derivative model of the ADKT framework, i.e., the above trained DKVMN is retrained on *a_train_set* as a whole and the retrained model is used to test *a_test_set*.

- **ADKVMN** is the product of the ADKT framework applied to DKVMN. Te above trained DKVMN is individually retrained based on each sequence in *a_train_set* and the retrained models (ADKVMN) are used to test the corresponding sequence in *a_test_set*, respectively.
- **ADKVMN^R** is as the same as the above ADKVMN. However, the DKVMN is only initialized randomly before individually retraining. In other words, DKVMN is individually trained based on each sequence in *a_train_set* and the trained models are used to test the corresponding sequence in *a_test_set*, respectively.

Table 3. The relationship during DKVMN, DKVMN++, DKVMN++ and ADKVMN^R.

Models	Datasets		Training mode	
	train_set	*a_train_set*	Whole	Individually
DKVMN	√	×	√	×
DKVMN++	√	√	√	×
ADKVMN	√	√	√	√
ADKVMN^R	×	√	×	√

5.3 Result and Analysis

The AUC results of testing DKVMN, DKVMN++, ADKVMN and ADKVMN^R on *a_test_set* of ASSISTments2009 and ASSISTments2015 are respectively shown in Table 4 and Table 5. It can be seen from these two tables that: ADKVMN which is initialized based on the trained DKVMN achieves the best performance overall; For ASSISTments2009, the best performance is achieved by ADKVMN on all the ranges except for the slight underperformance on the range of [500, 600) compared to DKVMN. For ASSISTments2015, although DKVMN++ performs better than other three models on ranges of [160, 200) and [300, 500), its performance is worse than DKVMN and ADKVMN when the sequence of learners is shorter (on 10–160), which is especially obvious on the range of [10, 20). On the contrary, the performance of ADKVMN which perform better than DKVMN on the range of [20, 300) is relatively stable. Moreover, ADKVMN^R based on random initialization showed the worst performance.

The fact that DKVMN is worse than DKVMN++ and ADKVMN but better than ADKVMN^R shows the importance of data. Since ADKVMN^R uses *a-train-set* data, DKVMN users *train-set* data, and DKVMN++ and ADKVMN use both *train-set* and *a-train-set* data, DKVMN++ and ADKVMN achieve the best performance on the whole. However, personalization makes that ADKVMN performs better than DKVMN++ on the whole.

Table 4. Results of the test AUC on ASSISTments2009.

Models	The range of #exercise						
	[20, 50)	[50, 100)	[100, 200)	[200, 300)	[300, 400)	[400, 500)	[500, 600)
DKVMN	0.81256	0.80814	0.82565	0.81170	0.80731	0.81304	0.80071
DKVMN++	0.81258	0.80939	0.82605	0.81318	0.81105	0.81691	**0.80861**
ADKVMN	**0.81340**	**0.81144**	**0.82977**	**0.81438**	**0.81803**	**0.82144**	0.80318
ADKVMN^R	0.68147	0.67913	0.69123	0.66041	0.68562	0.62857	0.58465

Table 5. Results of the test AUC on ASSISTments2015.

Models	The range of #exercise						
	[10, 20)	[20, 40)	[40, 80)	[80, 120)	[120, 160)	[160, 200)	[200, 300)
DKVMN	**0.75255**	0.74834	0.72846	0.72205	0.72857	0.71886	0.71329
DKVMN++	0.72174	0.73442	0.72146	0.71841	0.72727	**0.73225**	0.72628
ADKVMN	0.75055	**0.74848**	**0.72953**	**0.72354**	**0.73156**	0.72091	**0.72910**
ADKVMN^R	0.62468	0.60706	0.61133	0.61422	0.61989	0.60633	0.66181

Although ADKVMN^R shows the worst performance, with the increasing number of exercises, the gap between ADKVMN^R and other models is narrowing. In addition, the performance of DKVMN++ is often poor when the learner's exercise sequence is relatively short. A possible reason is that: unlike the ADKVMN model, DKVMN++ concatenates all short sequences into longer sequences, which has a great impact on the trained DKVMN. When the learner's exercise sequence is relatively long, although the performance of DKVMN++ is significantly improved, the results are questionable, because the number of learners evaluated in this case is limited.

6   Conclusion and Future Work

This paper introduces the idea of personalized learning to improve DLKT, and proposes the Adaptive Deep Knowledge Tracing (ADKT) framework. The proposed framework aims to train a ADKT model for each learner on her (or his) existing sequence, and trace the future knowledge state of each learner more accurately. In addition, to verify the effectiveness of the ADKT framework, this paper applies the proposed framework to DKVMN which is a widely used DLKT model recently, and proposes ADKVMN model based on the framework. Finally, experiments on two popular benchmark datasets, including ASSISTments2009 and ASSISTments2015, show that the effectiveness of introducing the personalized learning into DLKT and the importance of rich available data to the performance of the knowledge tracing model. In our future work, as we only study the performance of ADKT on the DKVMN model, we plan to apply ADKT in more DLKT models, such as DKT.

Acknowledgment. We would like to thank the anonymous reviewers for their helpful comments. The research is supported by the National Key Research and Development Program of China (2018YFB1004502) and the National Natural Science Foundation of China (61532001, 61702532, 61690203).

References

1. d Baker, R.S.J., Corbett, A.T., Aleven, V.: More accurate student modeling through contextual estimation of slip and guess probabilities in Bayesian knowledge tracing. In: Woolf, B.P., Aïmeur, E., Nkambou, R., Lajoie, S. (eds.) ITS 2008. LNCS, vol. 5091, pp. 406–415. Springer, Heidelberg (2008). https://doi.org/10.1007/978-3-540-69132-7_44
2. Cen, H., Koedinger, K., Junker, B.: Learning factors analysis – a general method for cognitive model evaluation and improvement. In: Ikeda, M., Ashley, K.D., Chan, T.W. (eds.) ITS 2006. LNCS, vol. 4053, pp. 164–175. Springer, Heidelberg (2006). https://doi.org/10.1007/11774303_17
3. Chaudhry, R., Singh, H., Dogga, P., Saini, S.K.: Modeling hint-taking behavior and knowledge state of students with multi-task learning. International Educational Data Mining Society (2018)
4. Corbett, A.T., Anderson, J.R.: Knowledge tracing: modeling the acquisition of procedural knowledge. User Model. User-Adap. Interact. **4**(4), 253–278 (1994)
5. Hochreiter, S., Schmidhuber, J.: Long short-term memory. Neural Comput. **9**(8), 1735–1780 (1997)
6. Khajah, M., Lindsey, R.V., Mozer, M.C.: How deep is knowledge tracing? arXiv preprint arXiv:1604.02416 (2016)
7. MacQueen, J.B.: Some methods for classification and analysis of multivariate observations. In: Proceedings of Berkeley Symposium on Mathematical Statistics & Probability (1967)
8. Minn, S., Desmarais, M.C., Zhu, F., Xiao, J., Wang, J.: Dynamic student classification on memory networks for knowledge tracing. In: Yang, Q., Zhou, Z.-H., Gong, Z., Zhang, M.-L., Huang, S.-J. (eds.) PAKDD 2019. LNCS (LNAI), vol. 11440, pp. 163–174. Springer, Cham (2019). https://doi.org/10.1007/978-3-030-16145-3_13
9. Minn, S., Yu, Y., Desmarais, M.C., Zhu, F., Vie, J.J.: Deep knowledge tracing and dynamic student classification for knowledge tracing. In: 2018 IEEE International Conference on Data Mining (ICDM), pp. 1182–1187. IEEE (2018)
10. Pardos, Z.A., Heffernan, N.T.: Modeling individualization in a Bayesian networks implementation of knowledge tracing. In: De Bra, P., Kobsa, A., Chin, D. (eds.) UMAP 2010. LNCS, vol. 6075, pp. 255–266. Springer, Heidelberg (2010). https://doi.org/10.1007/978-3-642-13470-8_24
11. Piech, C., et al.: Deep knowledge tracing. In: Advances in Neural Information Processing Systems, pp. 505–513 (2015)
12. Pliakos, K., Joo, S.H., Park, J.Y., Cornillie, F., Vens, C., Van den Noortgate, W.: Integrating machine learning into item response theory for addressing the cold start problem in adaptive learning systems. Comput. Educ. **137**, 91–103 (2019)
13. Wang, L., Sy, A., Liu, L., Piech, C.: Learning to represent student knowledge on programming exercises using deep learning (2017)
14. Wilson, K.H., Karklin, Y., Han, B., Ekanadham, C.: Back to the basics: Bayesian extensions of IRT outperform neural networks for proficiency estimation. arXiv preprint arXiv:1604.02336 (2016)
15. Yeung, C.K.: Deep-IRT: make deep learning based knowledge tracing explainable using item response theory. arXiv preprint arXiv:1904.11738 (2019)

16. Yudelson M.V., Koedinger K.R., Gordon G.J.: Individualized Bayesian knowledge tracing models. In: Lane, H.C., Yacef, K., Mostow, J., Pavlik, P. (eds.) AIED 2013. LNCS, vol. 7926, pp. 171–180 (2013). Springer, Heidelberg. https://doi.org/10.1007/978-3-642-39112-5_18
17. Zhang, J., Shi, X., King, I., Yeung, D.Y.: Dynamic key-value memory networks for knowledge tracing. In: Proceedings of the 26th International Conference on World Wide Web, pp. 765–774 (2017)

MULCE: Multi-level Canonicalization with Embeddings of Open Knowledge Bases

Tien-Hsuan Wu$^{(\boxtimes)}$, Ben Kao, Zhiyong Wu, Xiyang Feng, Qianli Song, and Cheng Chen

Department of Computer Science, The University of Hong Kong, Hong Kong, China
{thwu,kao,zywu}@cs.hku.hk, {andyfeng,u3502802,susume}@conenct.hku.hk

Abstract. An open knowledge base (OKB) is a repository of facts, which are typically represented in the form of ⟨*subject; relation; object*⟩ triples. The problem of *canonicalizing* OKB triples is to map different names mentioned in the triples that refer to the same entity into a basic canonical form. We propose the algorithm Multi-Level Canonicalization with Embeddings (MULCE) to perform canonicalization. MULCE executes in two steps. The first step performs *word-level canonicalization* to coarsely group subject names based on their GloVe vectors into semantically similar clusters. The second step performs *sentence-level canonicalization* to refine the clusters by employing BERT embedding to model relation and object information. Our experimental results show that MULCE outperforms state-of-the-art methods.

Keywords: Open Knowledge Base · Entity resolution · Canonicalization · Word embedding

1 Introduction

A *knowledge base* (KB) is a collection of facts about the world. This information is often stored as structured records with which inference and search engines are run to answer user questions. KBs can be generally classified into *curated* KBs and *open* KBs. A curated KB is one that is manually and collaboratively created. Examples include Freebase [5], DBpedia [1], Wikidata [28] and YAGO [26]. A curated KB models knowledge as entities and the relations among them. For example, the fact "Apple was founded by Steve Jobs, Steve Wozniak, and Ronald Wayne" is represented by 4 entities "Apple Incoprorated', "Steven Paul Jobs", "Stephen Gary Wozniak", and "Ronald Wayne"; and the relation "founded-by" that connects the company to the three founders. Because curated KBs are manually created, they are accurate and unambiguous. In particular, each entity is given a unique id that helps distinguish different entities that share the same name (e.g., *Apple* (company) vs. *Apple* (fruit)). The drawbacks of curated KBs, however, are inefficient update and limited scopes. An open KB (OKB) [11,31]

© Springer Nature Switzerland AG 2020
Z. Huang et al. (Eds.): WISE 2020, LNCS 12342, pp. 315–327, 2020.
https://doi.org/10.1007/978-3-030-62005-9_23

is constructed by collecting vast amounts of web documents and applying *open information extraction* (OIE) on the documents to extract *assertions*. Some OIE systems include TextRunner [2], ReVerb [10], OLLIE [22] and ClausIE [7]. Assertions are extracted from document sentences and are represented in the form of ⟨*subject*; *relation*; *object*⟩ triples. For example,

A_1 : ⟨Mumbai; is the largest city of; India⟩
A_2 : ⟨Bombay; is the capital of; Maharashtra⟩
A_3 : ⟨Bombay; is the economic hub of; India⟩

are three assertions extracted by ReVerb from the ClueWeb09[1] corpus which consists of around 500 million English Web pages. An *entity name* is a string that describes an entity, e.g., "Mumbai". An *entity mention* is an occurrence of an entity name in an assertion (e.g., there are two entity mentions of the entity name "Bombay" in assertions A_2 and A_3). OKBs can be automatically constructed, which gives them the advantages of wider scopes and being up-to-date.

An important issue of OKBs are ambiguity among entities and relations. For example, "Mumbai" and "Bombay" are two different entity names, but they refer to the same (city) entity; "Apple", on the other hand, can refer to a company or a fruit. In order to properly answer queries using OKBs, we need to perform *entity resolution* (ER), which is the process of determining the physical entity that a given entity mention refers to [3,4].

One approach to solving the ER problem is *entity linking* (EL) [9,12,14,16, 21,23,24,29]. Given an entity mention with a name x, EL identifies an entity e in a curated KB that is most likely the one that x refers to. All entity mentions that are linked to e are then treated as referring to the same entity. For EL to work, an entity mention in an OKB has to have an equivalence in a curated KB. This obviously limits EL's applicability. For example, the dataset provided by [27] consists of 45,031 assertions extracted by ReVerb from the ClueWeb09 corpus. In this dataset, about 23.7% of the entity mentions cannot be linked to any Wikipedia entities.

In this paper, we study canonicalization as another approach to solving ER. In [13], *canonicalization* is done by applying *hierarchical agglomerative clustering* (HAC) to cluster assertions (and thus the entity mentions of the assertions). The idea is to group assertions with similar entity names and context into clusters. Entity mentions of assertions that are grouped in the same cluster are then considered to refer to the same entity. A canonical form is then given to represent the subject names of these mentions. In [27], Vashishth, et al. propose the CESI algorithm that uses assertion embedding [18] as feature for clustering. It is shown that CESI generally outperforms the algorithms given in [13].

The above methods follow a similar clustering framework in which a certain similarity measure is applied. In practice, it is tricky to control the similarity threshold based on which clusters are formed. For example, "Barack Obama" and "Barack Hussein Obama" are highly similar in terms of word overlap, but "Mumbai" and "Bombay" are totally different words. In this paper we propose

---

[1] http://www.lemurproject.org/clueweb09.php/.

a new approach called **Multi-Level Canonicalization with Embeddings (MULCE)**. MULCE utilizes the state-of-the-art language models BERT [8] and GloVe to construct assertion embeddings. The key difference between MULCE and existing methods is that MULCE splits the clustering process into two steps. The first step clusters assertions at a coarse-granularity level, followed by a fine tuning step. Our experiments show that MULCE outperforms existing methods and produces high-quality canonicalization results.

The rest of the paper is organized as follow. Section 2 summarizes some related works. Section 3 presents MULCE. Section 4 presents experiment results. Finally, Sect. 5 concludes the paper.

2 Related Works

In this section we briefly describe some related works. We focus on three topics, namely, *noun phrase clustering*, *entity linking*, and *word embedding*.

[Noun Phrase Clustering]. Noun phrase clustering solves the entity resolution problem by clustering entity mentions. Example works include ConceptResolver [15], Resolver [30], Galarraga et al. [13] and CESI [27].

ConceptResolver is designed to process noun phrases extracted by the OIE system NELL [6]. It operates in two phases. The first phase performs disambiguation under the *one-sense-per-category* assumption. For example, *apple* can be a company or a fruit, but there cannot be two companies called *Apple*. Entity names are type-augmented (e.g., *apple* becomes *apple:company* or *apple:fruit*). The second phase uses HAC to cluster entity mentions under each category.

Resolver clusters entity mentions derived from TextRunner's Open IE triples. The similarity of two given entity mentions is based on the string similarity of their names, as well as the similarity of the relation phrases and objects of the assertions that contain the mentions. To improve HAC efficiency, some pruning techniques are applied.

Galarraga et al. [13] proposed the *blocking-clustering-merging* framework. Given an OKB, a *canopy* is created for each word found in some entity name of the assertions. For example, given the assertion A : ⟨Barack Obama; is the president of; the US⟩. The entity name "Barack Obama" induces two canopies: P_{Barack} and P_{Obama}. A canopy P_w is a set of assertions such that an assertion $A_i \in P_w$ if the entity name of A_i contains the word w. For example, the assertion A above is a member of both P_{Barack} and P_{Obama}. The process of constructing canopies is called *token blocking*. HAC clustering is performed for each canopy. In [13], it is shown that *IDF Token Overlap* is the most effective similarity measure evaluated. Specifically, the similarity of two entity names is given by the number of overlapping words between them, with a weighting given to each word according to the inverse frequency of that word. In the following, we call this algorithm G-IDF.

CESI [27] adopts a comprehensive word-embedding model, which is mainly based on GloVe. Furthermore, CESI utilizes some side information such as entity linking results and a paraphrase database. CESI canonicalizes an OKB in three steps: (1) acquire side information for each assertion; (2) learn assertion embeddings using the side information; (3) perform clustering to canonicalize the OKB. With word embeddings and side information, CESI has a high comprehensiveness, allowing it to outperform existing methods. However, as the side information requires entities to exist in a curated KB before canonicalization, one potential limitation is that CESI cannot handle emerging entities effectively.

[Entity Linking (EL)]. Entity Linking is another approach to solve the ER problem. EL links entity mentions with their corresponding entities in a curated KB. Readers are refer to survey papers [14,23,29] on entity linking. Here, we describe some representative works [9,21,24].

In [9,21], entity mentions are linked to corresponding Wikipedia pages, which is also known as *Wikification*. Both methods adopt a two-phase approach: *candidate selection* followed by *ranking*, with some differences in their selection and ranking strategies. For a given entity name, some candidate entity Wikipedia pages are first selected. These pages are then ranked according to their similarity with the given entity name in the ranking phase. The entity name will be linked to the highest-ranked Wikipedia page. In [9], candidate selection is based on measuring string similarity between the entity name and the title of a Wikipedia page, with special considerations given to acronyms and aliases. During the ranking phase, candidates are ranked based on some Wikipedia features such as the rank of the page in Google search, and other advanced string similarity features such as character n-grams shared by the entity names and titles of pages. On the other hand, the method proposed in [21] uses the *anchor text* of hyperlinks (the displayed text of a clickable link) for candidate selection. Given an entity name, if the name appears frequently as the displayed text of a hyperlink that links to a Wikipedia page, then the Wikipedia page is considered a candidate for the given entity name. The candidates are then ranked according to some *local features* or some *global features*. Given an entity name e, the document d from which e is extracted, and a candidate Wikipedia page p, local features measure the similarity between p and e, and the similarity between p and d. Global features model the relatedness among all the candidate pages such that the entity linking results of the names in a document are coherent.

The method proposed in [24] links entity mentions from unstructured Web texts to an entity heterogeneous information network (HIN) such as DBLP and IMDb networks. Entity linking is based on some probabilistic models. The *entity popularity model* computes the popularity of an entity in an HIN based on PageRank [19]. Given an entity mention m in a document d and an entity e in the HIN, the *entity object model* computes the similarity between e and m based on the texts in d and the neighbors of e in the HIN. The two models are combined and optimized using expectation-maximization algorithm. Finally, the knowledge population algorithm enriches the HIN using the documents that contain high-confident entity linking results, which can provide more information in the subsequent linking processes.

[Word Embeddings and Sentence Embeddings]. Word embeddings map words to a high dimensional vector space such that two words having similar meanings have a high cosine similarity between their vectors. GloVe [20] and word2vec [17] are two commonly-used models implementing the word embedding methodology. Both models learn embeddings based on the co-occurrence information of words in text documents. In word2vec, a continuous skip-gram model is adopted. GloVe adopts a log-bilinear regression model. It is essentially a count-based model that learns word embeddings by building a word-word co-occurrence matrix and then learning the co-occurrence probability. An extension of word embedding is *sentence embedding*, which generates a vector for every token in a sentence. It also embeds information about token position and co-occurrences with other tokens, etc. Compared with word embedding, sentence embedding can better capture contextual information. BERT [8] is a new deep learning NLP model that produces sentence embeddings effectively. The idea of BERT is to pre-train a general language model with large amounts of training data. The model can then be efficiently fine tuned for different downstream tasks and applications. Our algorithm MULCE uses both GloVe and BERT to provide embeddings for assertions' subject entity mentions. Specifically, we use GloVe as a word embedding technique to capture the semantics of an entity name, and then use BERT to obtain a sentence embedding that refines the meaning of an entity mention by incorporating contextual information of sentences. More details will be given in the next section.

3 MULCE

In this section we describe our algorithm MULCE. We first give some definitions.

Definition 1 (Assertion, subject entity name, entity mention, OKB). *An assertion is a triple of the form $A = \langle s; r; o \rangle$, where s, r, o are the subject, relation, and object fields, respectively. A subject entity name (or simply "subject name", or "subject") is a string s that appears in the subject field of some assertion A. We use $s(A)$ to denote the subject name of assertion A. An entity mention m is a pair of the form $(s(A), A)$ for some assertion A. We use $m(A)$ to denote the mention of assertion A; and $m.s$ to denote the subject name $s(A)$ of the mention. An OKB is a collection of assertions.*

For example, if $A_1 = \langle$Mumbai; is the largest city of; India$\rangle$, then $m(A_1) =$ ("Mumbai", A_1), and $s(A_1) = m(A_1).s =$ "Mumbai". In this paper we focus on resolving subject entity names via canonicalization.

Definition 2 (Canonicalization). *Given an OKB of n assertions $\mathcal{K} = \{A_i\}_{i=1}^n$, the problem of canonicalizing $\mathcal{K}$ is to compute a partitional clustering $\mathcal{C} = \{C_j\}_{j=1}^{|\mathcal{C}|}$. The clustering $\mathcal{C}$ induces a mention-entity mapping ρ. Specifically, $\forall A \in \mathcal{K}$, if $A \in C_j$, then $\rho(m(A)) = e_j$, where e_j denotes a physical entity. e_j is called the canonical form of $m(A)$.*

Given two subject names, we consider the following four scenarios in judging if the two names refer to the same entity.

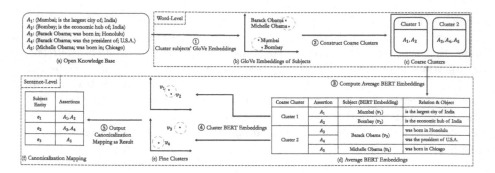

Fig. 1. Workflow of MULCE

(i) **Easy Negatives:** the names are very different strings and they refer to different entities (e.g., Dwayne Johnson vs. Obama);

(ii) **Easy Positives:** the names are very similar strings and they refer to the same entity (e.g., President Obama vs. Barack Obama);

(iii) **Hard Negatives:** the names are similar (either semantically or string-wise) but they refer to different entities (e.g., Johnny Damon vs. Johnny Cash (string-wise) or Hong Kong vs. Macau (semantically));

(iv) **Hard Positives:** the names are very different but refer to the same entity (e.g. Mumbai vs. Bombay).

The challenge lies in designing a canonicalization method that handles an OKB containing instances of different categories as listed above. We propose MULCE, which is a two-stage, coarse-to-fine mechanism to canonicalize an OKB. Figure 1 shows the overall workflow of MULCE. The first stage, word-level canonicalization (Sect. 3.1), produces coarse-grained clusters according to lexical information of subject names. These coarse-grained clusters are further divided into fine-grained clusters in the second stage, sentence-level canonicalization (Sect. 3.2), where contextual information of assertions are considered.

3.1   Word-Level Canonicalization

The first stage of MULCE aims at discovering entity mention pairs that are easy negative. Given such a pair (m_1, m_2), the first stage will attempt to put the mentions m_1 and m_2 into different clusters. This can be achieved by clustering assertions based on the GloVe vectors of their subject names. The power of using GloVe embeddings to extract a coarse taxonomy in the vector space has been previously demonstrated by CESI [27]. Here are the details of word-level canonicalization. Given an OKB containing a set of assertions (Fig. 1(a)), we compute a GloVe embedding for each subject name. For a subject name s, the embedding of s is the average of GloVe embedding vectors[2] of the words in s. Subject names are then clustered with complete-link HAC based on their

---

[2] http://nlp.stanford.edu/data/glove.6B.zip.

embeddings (Fig. 1(b)) using cosine similarity as the distance metric. We adopt complete-link HAC because in [27], it is suggested that small-sized clusters are expected in canonicalization problem. Any two subject names from one cluster should be similar as they both refer to the same entity. After the HAC, assertions are partitioned into *coarse clusters* according to the clustering of their subject names (Fig. 1(c)). The focus of this stage is to construct coarse clustering that correctly splits entity mentions of easy-negative instances while keeping those of positive instances in the same clusters. In particular, we allow some coarse clusters to contain mentions of hard-negative instances. These instances will be further processed in the next stage.

3.2   Sentence-Level Canonicalization

The second stage of MULCE focuses on separating entity mentions of hard-negative instances within each cluster obtained in the first stage, while keeping those of hard-positive instances together in the same cluster. This is achieved by performing sentence-level canonicalization using the other fields of the assertions as context information of the subject entity mentions. MULCE applies BERT to encode the semantic information of an assertion (which includes all three fields of a triple, namely, *subject, relation,* and *object*). We note that directly performing HAC over the sentence embeddings of assertions may not provide satisfactory results. This is because with the simple sentence structures of triples (or assertions), the information encoded into one single sentence embedding may be too limited and biased. Hence, given a subject name s, MULCE uses the mean vector of all sentence embeddings that share the same subject name s as the representation of s.

Based on this idea, we propose the following sentence-level canonicalization. We first compute a BERT embedding for each subject name. For each word in an assertion, we obtain token-level pre-trained BERT vector[3] using bert-as-service[4]. For a subject name s, we collect all assertions $A_i = \langle s_i; p_i; o_i \rangle$ with $s_i = s$ and embed each assertion using BERT. We then extract the token-level BERT vectors v_i corresponding to the subject s_i, which is the final hidden state of the BERT encoder. The BERT embedding vector of the subject name s is computed by averaging all the v_i's. Even though we only extract the embedding vectors corresponding to subjects in this step, we consider this step as *sentence-level* because the information contained in the sentence (i.e., relation and object) is leveraged in the process of generating subject tokens.

After the BERT embeddings of subjects are computed (Fig. 1(d)), we use single-link HAC to cluster the BERT embeddings within each coarse cluster. The resulting clusters are called *fine clusters* (Fig. 1(e)). Finally, we construct an output mapping that resolves assertions' subject names. Specifically, two assertions A_i and A_j are considered to refer to the same physical entity if and only if the BERT embeddings of their subject names, s_i and s_j, co-locate in the same cluster (Fig. 1(f)).

---

[3] Using BERT-Base Uncased Model: https://github.com/google-research/bert.

[4] https://github.com/hanxiao/bert-as-service.

Table 1. Dataset statistics

	All			Test			Validation		
	# assertions	# assertions linked to Wikipedia	# Wikipedia entities	# assertions	# assertions linked to Wikipedia	# Wikipedia entities	# assertions	# assertions linked to Wikipedia	# Wikipedia entities
Ambiguous	34,554	14,000	258	25,364	10,962	227	9,190	3,038	57
ReVerb45K	45,031	34,343	8,630	36,814	28,212	7,060	8,217	6,131	1,886

4   Experiments

We conduct experiments to evaluate the performances of the canonicalization algorithms. First, we provide details of the datasets used in the experiments in Sect. 4.1. This is followed by a summary of the metrics used, presented in Sect. 4.2. Finally, Sect. 4.3 reports experimental results and discussions.

4.1   Datasets

In the experiments, we use two real-world datasets to evaluate MULCE and other state-of-the-art algorithms. Both datasets consist of sampled assertions that are extracted from the ClueWeb09 corpus using ReVerb [10]. We call this collection of assertions *ReVerb OKB*. Table 1 summarizes the statistics of the two datasets. We briefly describe them below.

Ambiguous **Dataset:** The *Ambiguous* dataset was created by Galárraga et al. [13]. Firstly, 150 entities that have at least two different names in the ReVerb OKB were sampled. We call this set *sampled entity set E_s*. An entity e is called a homonym entity if (1) e has the same name as some other entity $e' \in E_s$ and (2) e and e' refer to different physical entities. Assertions whose subjects refer to sampled entities or homonym entities are collected into the dataset *Ambiguous*. Intuitively, entities mentioned in the assertions of the set are ambiguous as the same name can refer to different physical entities. We use the *Ambiguous* dataset released by the authors of CESI [27] as it contains additional information that is necessary to run CESI.

ReVerb45K **Dataset:** The *ReVerb45K* dataset was provided by [27]. Similar to *Ambiguous*, assertions in *ReVerb45K* have subjects referring to entities with at least two different names. *ReVerb45K* is larger (more assertions) and sparser (lower assertion-to-entity ratio) compared with *Ambiguous*.

Both datasets are split into validation sets and test sets by [27]. We use the validation sets to determine the HAC clustering thresholds. For MULCE, as it involves two levels of clustering, we used grid search on validation sets to find optimal thresholds. The ground truth is obtained by linking subject names to Wikipedia pages using Stanford CoreNLP entity linker [25]. Assertions with subject names linked to the same Wikipedia page are considered to have their subjects referring to the same physical entity. We observe that compared with [12], which was adopted in the evaluation of many previous works, Stanford CoreNLP entity linker achieves a higher precision but a lower recall. That is, mostly only high-confidence linking results are produced. If an assertion A in a dataset (*Ambiguous* or *ReVerb45K*) has a subject name that cannot be linked to a Wikipedia page, A will be excluded in the performance evaluation, since its ground truth entity cannot be determined by its Wiki-linkage.

4.2 Evaluation Metrics

We follow [13, 27] and evaluate clustering results with *macro*, *micro*, and *pariwise* scores. Specifically, let C be the clustering produced by a canonicalization algorithm, G be the gold standard clustering (i.e., the set of clusters formed according to the ground truth), and n be the number of assertions.

Macro Analysis: We define macro precision (P_{macro}) as the fraction of clusters in C that have all the assertions with subjects linked to the same wiki entity in the ground truth. Macro recall (R_{recall}) is the fraction of wiki entities that have all the assertions linked to them assigned to the same cluster by a canonicalization method.

Micro Analysis: Micro precision measures the purity of the clusters assuming that the most frequent ground truth entity in a cluster is the correct entity of that cluster. More formally, $P_{micro}(C, G) = \frac{1}{n} \sum_{c \in C} \max_{g \in G} |c \cap g|$. Micro recall is defined symmetrically as $R_{micro}(C, G) = P_{micro}(G, C)$. It measures the fraction of assertions assigned to the correct cluster, assuming that for each entity e, the correct cluster is the one that contain the most assertions with their subjects' ground truth entity being e.

Pairwise Analysis: We say that two assertions in a cluster are a "hit" if their subject refer to the same ground truth entity. Pairwise precision is defined as $P_{pairwise}(C, G) = \frac{\sum_{c \in C} \#hits_c}{\sum_{c \in C} \#pairs_c}$, where $\#pairs_c = \binom{|c|}{2}$ is the number of pairs in a cluster c. Pairwise recall is defined similarly as $R_{pairwise}(C, G) = \frac{\sum_{c \in C} \#hits_c}{\sum_{e \in G} \#pairs_e}$.
For each analysis, we also report the *F1 score*.

4.3 Results and Discussions

We compare MULCE against G-IDF and CESI, which are described in Sect. 2. We also conduct experiments to evaluate the following two ablated versions of MULCE:

- **Word-Level Canonicalization**: Canonicalization is done by clustering subjects' GloVe embeddings.
- **Sentence-Level Canonicalization**: BERT embeddings of subjects are clustered without forming "coarse clusters".

Table 2. Performance comparison using Ambiguous and *ReVerb45K* datasets

Ambiguous Dataset									
	Macro			Micro			Pairwise		
	Precision	Recall	F1	Precision	Recall	F1	Precision	Recall	F1
Galárraga-IDF	0.6688	**0.9692**	0.7914	0.9188	0.9898	0.9530	0.9779	**0.9983**	0.9880
CESI	0.6127	0.9665	0.7500	0.9244	0.8546	0.8881	0.9555	0.6042	0.7403
Word-Level	0.5664	**0.9692**	0.7150	0.9153	**0.9945**	0.9532	0.9801	0.9989	0.9894
Sentence-Level	**0.8889**	0.6211	0.7313	**0.9891**	0.5392	0.6979	**0.9968**	0.1949	0.3261
MULCE	0.8694	0.8811	**0.8752**	0.9797	0.9509	**0.9651**	0.9951	0.9862	**0.9907**
ReVerb45K Dataset									
	Macro			Micro			Pairwise		
	Precision	Recall	F1	Precision	Recall	F1	Precision	Recall	F1
Galárraga-IDF	0.8060	**0.9362**	0.8208	0.9046	0.9146	0.9096	0.9321	0.8173	0.8709
CESI	0.6568	0.9043	0.7609	0.8373	**0.9517**	0.8908	0.8465	**0.9344**	0.8881
Word-Level	0.6400	0.8895	0.7444	0.8200	0.9448	0.8780	0.8152	0.9290	0.8684
Sentence-Level	0.7315	0.6948	0.7127	0.8689	0.7247	0.7903	0.9189	0.3462	0.5029
MULCE	**0.8167**	0.8351	**0.8258**	**0.9123**	0.9160	**0.9141**	**0.9349**	0.8888	**0.9113**

Table 2 presents the results. Overall, MULCE achieves the highest F1 scores across both datasets. We further make the following observations.

G-IDF clusters assertions based on whether their subject names share some uncommon words. Therefore, it requires a corpus that is large enough to provide accurate estimates of words' document frequencies. Another disadvantage of such G-IDF is that it does not utilize the relation or the object fields of assertions in clustering, despite those fields provide valuable contextual information. Moreover, G-IDF only take subject names as strings without considering their semantics. Nevertheless, G-IDF achieves high recall scores. This shows that G-IDF is good at identifying subject names of the same entity, although it lacks the ability to distinguish string-wise similar but semantically different entities.

CESI uses side information (see Sect. 2) to learn assertion embeddings. The embeddings provide semantics information of subject names. CESI has high micro recall and pairwise recall scores for the *ReVerb45K* dataset. This demonstrates its ability to identify literally different subject names that refer to the same physical entity. The shortcoming of CESI is that it does not work well for the hard-negative cases, i.e., subject names that are very similar but in fact refer to different entities. Using MULCE, with the multi-level framework, we can tackle these highly similar cases using sentence-level canonicalization.

Ablation Analysis. We conduct an ablation analysis by applying only word-level canonicalization or applying only sentence-level canonicalization. For both datasets, we see that word-level canonicalization performs better in recall scores, and sentence-level canonicalization performs better in precision scores. The result is expected because it follows our design principle. We identify subjects with similar meanings first in word-level canonicalization (larger and coarser clusters, high recall), and then further split and refine the clusters in sentence-level canonicalization (smaller and finer clusters, high precision). When both word-level and sentence-level canonicalization are employed, i.e., using MULCE, we register the highest F1 scores. This demonstrates that MULCE's two-level clustering method is highly effective.

Case Study. We further illustrate the effectiveness of the algorithms by a case study. The following assertions are extracted from the *ReVerb45K* dataset.

A_1 : ⟨Mumbai; is the largest city of; India⟩
A_2 : ⟨Bombay; is the economic hub of; India⟩
A_3 : ⟨Hong Kong; is a special administrative region of; China⟩
A_4 : ⟨Macau; is a special administrative region of; China⟩

Note that the subjects of A_1 and A_2 refer to the same city and hence the assertions should be put in the same cluster. These two assertions test whether an algorithm can distinguish the same entity with different names (hard positives). Assertions A_3 and A_4 have different entities as subjects (Hong Kong and Macau), but these entities are highly semantically similar (hard negatives).

We observe that G-IDF separates the four assertions into different clusters. This is because the similarity function used, namely, IDF token overlap, is purely word-based and the 4 subject names share no common words. CESI correctly puts A_1 and A_2 in the same cluster. This demonstrates that word embeddings and side information have provided enough clue for the algorithm to infer semantic similarity even though the names have different words. However, CESI incorrectly puts A_3 and A_4 in the same cluster. This shows that the CESI has problems handling the hard-negative cases. Our method, MULCE, can correctly handle all four assertions. In word-level canonicalization, A_1 and A_2 are grouped into one coarse cluster, and A_3 and A_4 are grouped into the other coarse cluster. These clusters are then processed by sentence-level canonicalization, where A_3 and A_4 are separated into two clusters while A_1 and A_2 remain in the same cluster. Through this case study, we demonstratre that MULCE is capable of properly handling hard positive and hard negative cases.

5   Conclusion

In this paper, we studied the problem of OKB canonicalization. We proposed the two-stage canonicalization framework. Using word-level canonicalization, we can get coarse clusters where subject names having similar meanings are grouped together. These coarse clusters are further divided into fine clusters in the sentence-level canonicalization, where BERT embeddings are used to capture the information of the relation and object in an assertion. With experiments, we demonstrated that MULCE outperforms state-of-the-art methods on two datasets. We also conducted an ablation study to show that combining word-level and sentence-level canonicalizations is effective.

References

1. Auer, S., Bizer, C., Kobilarov, G., Lehmann, J., Cyganiak, R., Ives, Z.: DBpedia: a nucleus for a web of open data. In: Aberer, K., et al. (eds.) ASWC/ISWC -2007. LNCS, vol. 4825, pp. 722–735. Springer, Heidelberg (2007). https://doi.org/10.1007/978-3-540-76298-0_52
2. Banko, M., Cafarella, M.J., Soderland, S., Broadhead, M., Etzioni, O.: Open information extraction from the web. IJCAI **7**, 2670–2676 (2007)
3. Benjelloun, O., Garcia-Molina, H., Menestrina, D., Su, Q., Whang, S.E., Widom, J.: Swoosh: a generic approach to entity resolution. VLDB **18**(1), 255–276 (2009)
4. Bhattacharya, I., Getoor, L.: A latent Dirichlet model for unsupervised entity resolution. In: ICDM, pp. 47–58. SIAM (2006)
5. Bollacker, K., et al.: Freebase: a collaboratively created graph database for structuring human knowledge. In: SIGMOD, pp. 1247–1250. ACM (2008)
6. Carlson, A., Betteridge, J., Kisiel, B., Settles, B., Hruschka, E.R., Mitchell, T.M.: Toward an architecture for never-ending language learning. In: AAAI (2010)
7. Corro, L.D., Gemulla, R.: Clausie: clause-based open information extraction. In: WWW, pp. 355–366. ACM (2013)
8. Devlin, J., Chang, M.W., Lee, K., Toutanova, K.: BERT: pre-training of deep bidirectional transformers for language understanding. In: NAACL. ACL (2019)
9. Dredze, M., McNamee, P., Rao, D., Gerber, A., Finin, T.: Entity disambiguation for knowledge base population. In: COLING, pp. 277–285. ACL (2010)
10. Fader, A., Soderland, S., Etzioni, O.: Identifying relations for open information extraction. In: EMNLP, pp. 1535–1545. ACL (2011)
11. Fader, A., Zettlemoyer, L., Etzioni, O.: Open question answering over curated and extracted knowledge bases. In: KDD, pp. 1156–1165. ACM (2014)
12. Gabrilovich, E., Ringgaard, M., Subramanya, A.: FACC1: freebase annotation of clueweb corpora (2013). http://lemurproject.org/clueweb09/FACC1/
13. Galárraga, L.A., Heitz, G., Murphy, K., Suchanek, F.M.: Canonicalizing open knowledge bases. In: CIKM, pp. 1679–1688. ACM (2014)
14. Hachey, B., Radford, W., Nothman, J., Honnibal, M., Curran, J.R.: Evaluating entity linking with Wikipedia. Artif. Intell. **194**, 130–150 (2013)
15. Krishnamurthy, J., Mitchell, T.M.: Which noun phrases denote which concepts? In: HLT, vol. 1, pp. 570–580. ACL (2011)
16. Lin, T., Etzioni, O.: Entity linking at web scale. In: Proceedings of the Joint Workshop on Automatic Knowledge Base Construction and Web-scale Knowledge Extraction, pp. 84–88. ACL (2012)
17. Mikolov, T., Sutskever, I., Chen, K., Corrado, G.S., Dean, J.: Distributed representations of words and phrases and their compositionality. In: NIPS (2013)
18. Nickel, M., Rosasco, L., Poggio, T.: Holographic embeddings of knowledge graphs. In: Thirtieth AAAI Conference on Artificial Intelligence (2016)
19. Page, L., Brin, S., Motwani, R., Winograd, T.: The page rank citation ranking: bringing order to the web. Technical report, Stanford InfoLab (1999)
20. Pennington, J., Socher, R., Manning, C.: Glove: global vectors for word representation. In: EMNLP, pp. 1532–1543 (2014)
21. Ratinov, L., Roth, D., Downey, D., Anderson, M.: Local and global algorithms for disambiguation to Wikipedia. In: HLT, vol. 1, pp. 1375–1384. ACL (2011)
22. Schmitz, M., Bart, R., Soderland, S., Etzioni, O., et al.: Open language learning for information extraction. In: EMNLP-CoNLL, pp. 523–534. ACL, July 2012

23. Shen, W., Wang, J., Han, J.: Entity linking with a knowledge base: issues, techniques, and solutions. TKDE **27**(2), 443–460 (2015)
24. Shen, Wei, et al.: Shine+: A general framework for domain-specific entity linking with heterogeneous information networks. TKDE **30**(2), 353–366 (2018)
25. Spitkovsky, V.I., Chang, A.X.: A cross-lingual dictionary for English Wikipedia concepts. In: Proceedings of the Eighth International Conference on Language Resources and Evaluation (2012)
26. Suchanek, F.M., Kasneci, G., Weikum, G.: YAGO: a core of semantic knowledge. In: WWW, pp. 697–706. ACM (2007)
27. Vashishth, S., Jain, P., Talukdar, P.: CESI: canonicalizing open knowledge bases using embeddings and side information. In: WWW, pp. 1317–1327. IW3C2 (2018)
28. Vrandečić, D., Krötzsch, M.: Wikidata: a free collaborative knowledge base (2014)
29. Wu, G., He, Y., Hu, X.: Entity linking: an issue to extract corresponding entity with knowledge base. IEEE Access **6**, 6220–6231 (2018)
30. Yates, A.P., Etzioni, O.: Unsupervised methods for determining object and relation synonyms on the web. JAIR (2009)
31. Yin, P., Duan, N., Kao, B., Bao, J., Zhou, M.: Answering questions with complex semantic constraints on open knowledge bases. In: CIKM (2015)

Encoding Knowledge Graph Entity Aliases in Attentive Neural Network for Wikidata Entity Linking

Isaiah Onando Mulang'[1]([✉]), Kuldeep Singh[2], Akhilesh Vyas[3],
Saeedeh Shekarpour[4], Maria-Esther Vidal[3], Jens Lehmann[5], and Soren Auer[3]

[1] Fraunhofer IAIS, University of Bonn, and Zerotha Research, Bonn, Germany
isaiah.mulang.onando@iais.fraunhofer.de
[2] Cerence GmbH, and Zerotha Research, Aachen, Germany
kuldeep.singh1@cerence.com
[3] TIB, Hannover, Germany
akhilesh.vyas@tib.eu, maria.vidal@tib.eu
[4] University of Dayton, Dayton, USA
saeedeh@knoesis.org
[5] Fraunhofer IAIS, Sankt Augustin, Germany

Abstract. The collaborative knowledge graphs such as Wikidata excessively rely on the crowd to author the information. Since the crowd is not bound to a standard protocol for assigning entity titles, the knowledge graph is populated by non-standard, noisy, long or even sometimes awkward titles. The issue of long, implicit, and nonstandard entity representations is a challenge in Entity Linking (EL) approaches for gaining high precision and recall. Underlying KG in general is the source of target entities for EL approaches, however, it often contains other relevant information, such as aliases of entities (e.g., Obama and Barack Hussein Obama are aliases for the entity Barack Obama). EL models usually ignore such readily available entity attributes. In this paper, we examine the role of knowledge graph context on an attentive neural network approach for entity linking on Wikidata. Our approach contributes by exploiting the sufficient context from a KG as a source of background knowledge, which is then fed into the neural network. This approach demonstrates merit to address challenges associated with entity titles (multi-word, long, implicit, case-sensitive). Our experimental study shows ≈8% improvements over the baseline approach, and significantly outperform an end to end approach for Wikidata entity linking.

Keywords: Knowledge graph context · Wikidata · Entity linking

1 Introduction

Entity linking (EL) over Web of data often referred as Named Entity Disambiguation (NED) or Entity Disambiguation is a long-standing field of research

The original version of this chapter was revised: an inadvertently forgotten co-author was added. The correction to this chapter is available at https://doi.org/10.1007/978-3-030-62005-9_41

in various research communities such as information retrieval, natural language processing, semantic web, and databases since early approaches in 2003 [2]. EL generally comprises two subtasks: *entity recognition* that is concerned with the identification of entity surface forms in the text, and *entity disambiguation* that aims at linking the surface forms with structures and semi-structured knowledge bases (e.g. Wikipedia), or structured knowledge graphs (e.g. DBpedia [1], Freebase [4] or Wikidata [24]).

Research Objectives, Approach, and Contribution. Uniqueness of Wikidata is that the contents are collaboratively edited. As at April 2020; Wikidata contains 83,151,903 items and a total of over 1.2B edits since the project launch[1]. User-created entities add additional noise and vandalism in Wikidata [13] since users do not follow a strict naming convention nor a standardised approach; for instance, there are 1788134 labels in which each label matches with at least two different URIs. The previous approaches for EL [18,19] on the textual content consider the well-established knowledge bases such as Wikipedia, Freebase, YAGO [22], and particularly DBpedia. Thereby, Wikidata as the core background KG along with its inherent challenges, has not been studied particularly for the task of EL.

Besides the vandalism and noise in underlying data of Wikidata, collaborative editing of its content adds several aliases of the entities and its description as entity properties (attributes). This enables Wikidata as a rich source of additional information which may be useful for EL challenges. Thus, in this work, we analyse the impact of additional context from Wikidata on Attentive Neural Networks (NN) for solving its entity linking challenges. We develop a novel approach called Arjun, first of its kind to recognise entities from the textual content and link them to equivalences from Wikidata KG. An important strength of Arjun is an ability to link non-Wikipedia entities of Wikidata by exploiting unique characteristic of the Wikidata itself (i.e. availability of entity aliases as explained in Sect. 2). Please note, focus of this paper is not to propose a black-box deep learning approach for entity linking using latest deep learning models such as transformers or graph neural networks. In this paper we hypothesise that even though Wikidata is noisy and challenging, but its special property to provide aliases of entities can help an NN better understand the context of the potential entities. Since the concept of informing a neural network using contextual data from a KG is our proposed-solution in this work, we believe that traditional neural networks make it more transparent to understand the impact of KG context. Hence, our approach contributes to model attentive neural networks respecting the contextual content and trained on a sizable dataset. In particular, Arjun is a pipeline of two attentive neural networks coupled as follows:

1. In the first step, Arjun utilises a deep attentive neural network to identify the surface forms of entities within the text.
2. In the second step, Arjun uses a local KG to expand each surface form from previous step to a list of potential Wikidata entity candidates. Unlike [15],

---

[1] https://www.wikidata.org/wiki/Wikidata:Statistics.

Arjun does not use a pre-computed entity candidate list and search entity candidates among all the Wikidata entities.

3. Finally, the surface forms, coupled with potential Wikidata candidates, are fed into the second attentive neural network to disambiguate the Wikidata entities further.

Although simple, our approach is empirically powerful and shows ≈8% improvement over the baseline. We also release the source code and all utilised data for reproducibility and reusability on Github[2]. The remainder of the article is structured as follows: Sect. 2 motivates our work by discussing Wikidata specific entity linking challenges. Section 3 discusses related work. This is followed by the formulation of the problem in Sect. 4. Section 5 describes the approach. In Sect. 6 we discuss the experimental setup and the results of the evaluation. We conclude in Sect. 7.

2 Motivating Examples

We motivate our work by highlighting some challenges associated with linking entities in the text to Wikidata. Wikidata is a community effort to collect and provide an open structured encyclopedic data. The total number of entities described in Wikidata is over 54.1 million [24]. Wikidata entities are represented by unique IDs known as `QID` and `QIDs` are associated with entity labels. Figure 1 shows three sentences extracted from the dataset released by ElSahar et al. [9] which aligns 6.2 million Wikipedia sentences to associated Wikidata triples (<subject, predicate, object>).

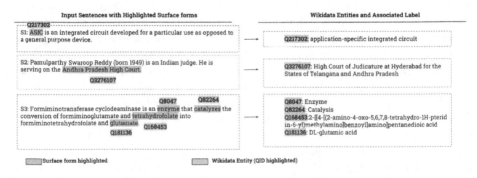

Fig. 1. Wikidata Entity linking Challenges: Besides the challenge of capitalisation of surface forms and implicit nature of entities, Wikidata has several specific challenges, such as very long entity labels and user created entities.

In the first sentence S1, the surface form `ASIC` is linked to a Wikidata entity `wiki:Q217302` and the entity is implicit (i.e. no exact string match between surface form and entity label). However, ASIC is also known as 'Application Specific

---
[2] https://github.com/mulangonando/Arjun.

Integrated Circuit' or Custom Chip. Therefore to disambiguate this entity, background information about the surface form will be useful. Please note, we will use this sentence as a running example "Sentence S1". In the second sentence S2 the surface form Andhra Pradesh High Court is linked to wiki:Q3276107 which has 14 words in the full entity label[3]. It is also important to note here that the surface form Andhra Pradesh High Court also contains two sub-surface forms Andhra Pradesh and High Court which are the entity labels of the two Wikidata entities wiki:Q1159 and wiki:Q671721. An ideal entity linking tool first has to identify Andhra Pradesh High Court as a single surface form then disambiguate the surface form to a long entity label. In Wikidata, entity labels and associated aliases can be long (e.g. wiki:Q1156234, wiki:Q15885502). In addition there are long erroneous entity labels and aliases, such as entity wiki:Q44169790[4] with 62 words in the label and entity wiki:Q12766033 with 129 words in one alias. Presence of long multi-word entity labels is also specific to Wikidata and poses another challenge for entity linking. Furthermore, in sentence S3 illustrated in the Fig. 1, the surface form tetrahydrofolate is linked to wiki:Q168453 which not only has a multi-word entity label and lowercase surface forms but also contains several numeric and special, non-alphanumeric ASCII characters. Such entities are not present in other public KGs. This is because unlike Wikidata other KGs do not allow users to create new entities and the entity extraction process depends on unique IRIs of Wikipedia pages, WordNet taxonomy, and GeoNames. A large number of user-created entities poses specific challenges for entity linking. Therefore, it is evident that in addition to generic entity linking challenges such as the impact of capitalisation of surface forms and the implicit nature of entities which are tackled up to a certain extent by approaches for entity linking over Wikipedia and DBpedia [19], Wikidata adds some specific challenges to the entity linking problem.

3   Related Work

Several comprehensive surveys exist that detail the techniques employed in entity linking (EL) research; see, for example, [2]. An elaborate discussion on NER has been provided by Yadav Bethard [25]. However, the use of Knowledge Graph as background knowledge for EL task is a relatively recent approach. Here, a knowledge graph is not only used for the reference entities but also offers additional signals to enrich both the recognition and the disambiguation processes. For entity linking, FALCON [19] introduces the concept of using knowledge graph context for improving entity linking performance over DBpedia. Falcon creates a local KG fusing information from DBpedia and Wikidata to support entity and predicate linking of questions. We reused the Falcon Background knowledge base and then expand it with all the entities present in the Wikidata (specially non standard entities).

---

[3] High Court of Judicature at Hyderabad for the States of Telangana and Andhra Pradesh.

[4] https://www.wikidata.org/wiki/Q44169790.

The developments in deep learning has introduced a range of models that carry out both NER and NED as a single end to end step using various neural network based models [15]. Kolitsas et al. [15] enforces during testing that gold entity is present in the potential list of candidates, however, Arjun doesn't have such assumption and generates entity candidates on the fly. This is one reason Arjun is not compared with Kolitsas's work in the evaluation section. Please note, irrespective of the model opted for entity linking, the existing EL approaches and their implementations are commonly evaluated over standard datasets (e.g. CoNLL (YAGO) [14]). These datasets contain standard formats of the entities commonly derive from Wikipedia URI label. Recently, researchers have explicitly targeted EL over Wikidata by proposing new neural network-based approach [5]. Contrary to our work, authors assume entities are recognised (i.e. step 1 of Arjun is already done), inputs to their model is a "sentence, one wrong Wikidata Qid, one correct Qid" and using an attention-based model they predict correct Qid in the sentence- more of a classification problem. Hence, 91.6F-score in Cetoli et al.'s work [5] is for linking correct QID to Wikidata, given the particular inputs. Their model is not adaptable for an end to end EL due to input restriction. OpenTapioca [6]) is an end to end EL approach to Wikidata that relies on topic similarities and local entity context, but ignores the Wikidata specific challenges (Sect. 2). Works in [10,20] are other attempts for Wikidata entity linking.

4 Problem Statement

Wikidata is an RDF[5] knowledge graph that contains a set of triples $(s, p, o) \in \mathcal{R} \times \mathcal{P} \times (\mathcal{R} \cup \mathcal{L})$, where $\mathcal{R} = \mathcal{C} \cup \mathcal{E} \cup \mathcal{P}$ is the union of all RDF resources. ($\mathcal{C}, \mathcal{P}, \mathcal{E}$ are respectively a set of classes, properties, and entities), and L is the set of literals ($L \cap R = \emptyset$). An RDF knowledge graph represents a directed graph structure which is formally defined as:

Definition 1 (Knowledge Graph). *A knowledge graph KG is a directed labelled graph $G(V, E)$, where $V = \mathcal{E} \uplus \mathcal{C} \uplus L$ is a disjoint union of entities $\mathcal{E}$, classes $\mathcal{C}$, and literal values L. The set of directed edges is denoted by $E = \mathcal{P}$, where $\mathcal{P}$ are properties connecting vertices. Please note that there is no outgoing edge from literal vertices.*

In this paper, we target **end to end EL task**. The EL for us is defined as recognising the surface forms of entities in the text and then map them to the entities in the background KG. The EL task can be defined as follows:

Definition 2 (Entity Linking). *Assume a given text is represented as a sequence of words $w = (w_1, w_2, ..., w_N)$ and the set of entities of a KG is represented by set $\mathcal{E}$. The EL task maps the text into a subset of entities denoted as $\Theta : w \to \mathcal{E}'$ where $\mathcal{E}' \subset \mathcal{E}$. Herein, the notion of Wididata entity refers to the representation of an entity based on the corresponding label because Wikidata might consider a variety of identifiers (called Q values) for the same label.*

---

[5] https://www.w3.org/RDF/.

The EL task can be divided into two individual sub-tasks. The first sub-task Surface Form Extraction is recognising the surface forms of the entities in the text. This task is similar to Named Entities Recognition (NER). However, it disregards identifying the type of entities (e.g. person, place, date, etc.).

Definition 3 (Surface form Extraction). *Let $w = (w_1, w_2, ..., w_N)$ be a text represented as a sequence of words. The surface form extraction is then a function $\theta_1 : w \rightarrow \mathcal{S}$, where the set of surface forms is denoted by $\mathcal{S} = (s_1, s_2, ..., s_K)$ $(K \leq N)$ and each surface form s_x is a sequence of words from start position i to end position j: $s_x^{(i,j)} = (w_i, w_{i+1}, ..., w_j)$.*

The second sub-task Entity Disambiguation (ED) is mapping each surface form into a set of the most probable entities from the background KG.

Definition 4 (Entity Disambiguation). *Let $\mathcal{S}$ be the set of surface forms and $\mathcal{E}$ the set of entities of the background KG. Entity Disambiguation is a function $\theta_2 : \mathcal{S} \rightarrow \mathbb{P}(\mathcal{E})$, which assigns a set of entities to each surface form.*

Please note that a single surface form potentially might be mapped into multiple potentially suitable entities.

5   Arjun: A Context-Aware Entity Linking Approach

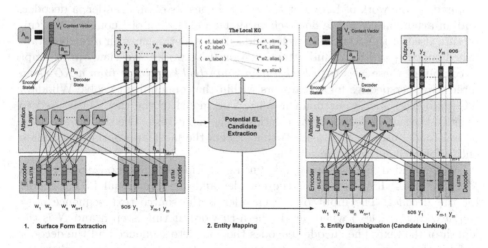

Fig. 2. Proposed Approach Arjun: Arjun consists of three tasks. First task identifies the surface forms using an attentive neural network. Second task induces background knowledge from the Local KG and associate each surface form with potential entity candidates. Third task links the potential entity candidates to the correct entity labels.

Arjun is illustrated in Fig. 2. Arjun performs three sub-tasks:

1. surface form extraction which identifies the surface forms of the entities,
2. entity mapping (or candidate generation) which maps the surface forms to a list of candidate entities from the Local KG,
3. entity disambiguation which selects the most appropriate candidate entity for each surface form.

We devise a context-aware approach based on attentive neural networks for tasks (1) and (3). We initially introduce our derived Local KG. Then we present the details of our approach for the tasks (1), (2) and (3).

Local KG and Refinement Strategies. Arjun relies on Wikidata as the background knowledge graph. Wikidata consists of over 100 million triples in RDF format. Wikidata provides dumps of all the entities and associated aliases[6]. Although Wikidata has specific challenges for EL, its unique characteristic to provide entity aliases can be utilised in developing an approach for entity linking. Since the training dataset is in English, we extracted all 38.6 million Wikidata entities with English labels and 4.8 million associated aliases from the dumps. We use entity labels and aliases as indexed documents in the Local KG and large portion of it is reused from Local KG built by Sakor et al. [19]. For example, the entity described in exemplary "Sentence S1" (cf. Fig. 1), entity wiki:Q217302 with label *application-specific integrated circuit* is enriched in the Local KG with its aliases: ASIC, Custom Chip, and Custom-Chip.

Model Architecture. For task (1) and (3), our attentive neural model is inspired by the work of Luong et al. [16] and consists of an encoder, a decoder, and an attention layer. We don't claim that an extension of Luong's NN architecture used in this work as novelty, indeed we experiment with already established concepts of LSTM and attentive Neural Networks. We view our attempt of combining these NNs with *background contextual knowledge from a KG* as an interesting perspective for researchers within the community to solve Wikidata KG challenges and is our main novelty. The model in the task (1) is used to identify the surface forms of the entities in the input text. The similar attentive neural model used in the task (3) that selects the most appropriate candidate entity for each surface form (cf. Fig. 2).

We extended Luong's model by using a Bidirectional Long Short-Term Memory (Bi-LSTM) model for the encoder and one-directional LSTM model for the decoder. The input of the encoder is the source text sequence $w = (w_1, w_2, ..., w_n, .., w_N)$ where w_n is the n-th word at time step n and N is the length of the text. The encoder encodes the complete sequence and the decoder unfolds this sequence into a target sequence $y = (y_1, y_2, .., y_m, ..., y_M)$ where y_m is the m-th word at time step m and M is the length of the target sequence. In our assumption each target sequence ends with EOS (end of sequence) token. The N and M values can be considered as the last time steps of the source sequence w and the target sequence y respectively.

---

[6] https://dumps.wikimedia.org/wikidatawiki/entities/.

Each word of the source sequence is projected to its vector representation acquired from an embedding model $\mathcal{R}^d$ with dimensionality d. The transformation of the input is represented in the matrix X as:

$$X = [x_1, x_2, ., x_n, ..., x_N] \tag{1}$$

where x_n is a vector with the size d and represents the low dimensional embedding of the word w_n.

The LSTM Layer: In our model, the encoder and the decoder consist of single layer of Bi-LSTM and LSTM respectively. Now we explain the LSTM layer.

We model the first layer of our network using a LSTM layer since it has been successfully applied to various NLP tasks. Each LSTM unit contains three gates (i.e., input i, forget f and output o), a hidden state h and a cell memory vector c. The forget gate is a sigmoid layer applied on the previous state h_{t-1} at time step t-1 and the input x_t at time step t to remember or forget its previous state (Eq. 2).

$$f_t = \sigma(W^f[x_t, h_{t-1}] + b^f) \tag{2}$$

Please note that W is the weight matrix and b is the bias vector. The next step determines the update on the cell state. The input gate which is a sigmoid layer updates the internal value (Eq. 3), and the output gates alter the cell states (Eq. 4).

$$i_t = \sigma(W^i[x_t, h_{t-1}] + b^i) \tag{3}$$

$$o_t = \sigma(W^o[x_t, h_{t-1}] + b^o) \tag{4}$$

The next tanh layer computes the vector of a new candidate for the cell state $\tilde{C}_t$ (Eq. 5). Then the old state C_{t-1} is updated by the new cell state C_t via multiplying the old state with the forget gate and adding the candidate state to the input gate (Eq. 6). The final output is a filtering on the input parts (Eq. 4) and the cell state (Eq. 7).

$$\tilde{C}_t = \tanh(W^C[x_t, h_{t-1}] + b^C) \tag{5}$$

$$C_t = f_t \odot C_{t-1} + i_t \odot \tilde{C}_t \tag{6}$$

$$h_t = o_t \odot \tanh(C_t) \tag{7}$$

where the model learning parameters are weight matrices W^f, W^i, W^o, W^C and bias vectors b^f, b^i, b^o, b^C. The σ denotes the element-wise application of the sigmoid function and $\odot$ denotes the element-wise multiplication of two vectors.

The Bi-LSTM of the encoder consists of two LSTM layers. The first layer takes an input sequence in forward direction (1 to N) and the second layer takes the input in backward direction (N to 1). We employ same Eqs. 2, 3, 4, 5, 6, 7 for each LSTM Layer. The final encoder hidden state is produced by the sum of hidden states from both LSTM layers ($h_n = \overrightarrow{h_n} + \overleftarrow{h_n}$) at timestep n.

Attention and Decoder Layer. The decoder layer takes SOS token (start of the sequence) vector and the encoder final states (h_N and C_N) as the initial inputs to start decoding source text sequence into a target text sequence. Here we differentiate between the encoder hidden state and decoder hidden state by using the notations h_n at time step n and h_m time step m respectively. Below we explain how the decoder generates target text sequence words y_m one by one.

In the attention layer, we define attention weights as $a_m = [a_{m1}, a_{m2}, ..., a_{mN}]$ for a decoder state at time step m which has the size equals to the number of total time steps in the encoder side. The attention weights contain only scalar values which are calculated by comparing all encoder states h_n and decoder state h_m. To calculate the attention weight (a_{mn}) of an encoder state at time step n wrt. A decoder state at time step m, we use the following Eq. (8) [16].

$$a_{mn} = \frac{\exp(h_m \cdot h_n)}{\sum_{n'=1}^{N} \exp(h_m \cdot h_{n'})} \tag{8}$$

Where $(h_m \cdot h_n)$ denotes the dot product. The Eq. 9 computes the context vector V_m as weighted average over all the encoder hidden states (h_n) that captures the relevant encoder side information to help in predicting the current target sequence word y_m at time step m and can be defined as:

$$V_m = \sum_{n=1}^{N} a_{mn} h_n \tag{9}$$

We calculate Attention Vector $(\tilde{h}_m)$ using the concatenation layer on the context vector V_m and decoder hidden state h_m for combining information from both the vectors. The Eq. 10 represent it mathematically (where tanh is an activation function same as describe in [16]).

$$\tilde{h}_m = \tanh(W_v[v_m; h_m]) \tag{10}$$

Finally, we apply softmax layer on the attention vector $\tilde{h}_m$ for predicting a word of a target text sequence from the predefined vocabulary of the complete target text sequences.

$$p(y_m | y_{<m}, x) = \text{softmax}(W_s \tilde{h}_m) \tag{11}$$

Where W_s is weight matrix of softmax layer and p is probability. Please note that the decoder stops producing words once it encounters EOS (end of sequence) token or m is equal to M.

5.1 Entity Mapping Process

The local KG acts as a source of background knowledge. It is an indexed graph (created using the same methodology proposed by Sakor et al. [19] and reusing a large portion of the indexed graph built by the authors) where each entity label is extended with its aliases from Wikidata. Once Task 1 identifies surface forms in the input sentence, the entity mapping step (Task 2) takes each surface form and retrieves all the entities for which entity label(s) in the local KG matches with the surface form. Next, the full list of the entity candidates is then passed into the Step 3 of Arjun as input to predict (disambiguate) the best Wikidata entity labels.

Let us trace our approach for the sentence S1 of Fig. 1 to understand the steps better. The sentence S1 `ASIC is an integrated circuit developed for particular use as opposed to a general-purpose device` is fed to the attentive neural model comprises of an encoder (Bi-LSTM), decoder (LSTM), and an attention layer as an input for the surface form extraction task. Thereby, the term `ASIC` is recognised as a surface form. Then, for the entity mapping task, we populate a Local KG to generate candidate entities associated with this surface form. We employ semantic search (reused from Falcon [19]) to identify entity candidate labels for `ASIC` which returns `Application Specific Integrated Circuit`. The last step of Arjun is entity disambiguation. In this step, the surface form `ASIC` along with `Application Specific Integrated Circuit` is fed into the encoder as the input sequence. Here, we utilise identical attentive neural network used for surface form extraction task. This attentive neural network decides the context of `ASIC` using extra information in the form of associated alias to correctly link to the Wikidata entity `application-specific integrated circuit` (Q217302).

6   Experimental Setup

6.1   Dataset

We rely on the recently released T-REx [9] dataset that contains 4.65 million Wikipedia extracts (documents) with 6.2 million sentences. These sentences are annotated by 11 million Wikidata triples. In total, over 4.6 million surface forms are linked in the text to 938,642 unique entities. T-REx is the only available dataset for Wikidata with such a large number of triple alignment. We are not aware of any other dataset explicitly released for Wikidata entity linking challenges. Please note that the popular entity linking datasets (e.g. CoNLL (YAGO) [14]) have linked entities either to Wikipedia, YAGO, Freebase or DBpedia. Work in [5,6] attempt to develop approaches for EL over Wikidata and simply align (map using owl:sameAs) existing Wikipedia based dataset to Wikidata. However, our focus in this paper is to solve Wikidata specific challenges and these datasets do no embrace Wikidata specific challenges for entity linking. We divide the T-REx dataset into an 80:20 ration for training and testing.

6.2   Baseline

In this work, we pursue the following research question: "How well does the attentive neural network perform for entity linking task leveraging background knowledge particularly for a challenging KG such as Wikidata?" To the best

of our knowledge, it is a pioneering work for the task of entity linking on the
Wikidata knowledge graph where it considers the inherent challenges (noisy
nature, long entity labels, implicit entities). Therefore, we do not compare our
approach to generic entity linking approaches which typically either do not use
any background knowledge or employ the well-established knowledge graphs such
as DBpedia, YAGO, Freebase. Our approach Arjun comprises all three tasks
illustrated in Fig. 2. To elaborate the advantage of inducing additional context
post NER step, we built a "baseline" which is an end to end neural model.
The "baseline" in our case is the attentive neural network employed in Task
1 without any background knowledge (or can be seen as end to end EL using
attentive neural network). In fact, in the task (1) (cf. Fig. 2), the baseline directly
maps the text to a sequence of Wikidata entities without identifying surface form
candidates. Hence, the baseline approach is the modified version of Arjun. With
a given input sentence, the baseline implicitly identifies the surface forms and
directly links them to Wikidata entities. Unlike Arjun, the baseline does not use
any KG context for the expansion of the surface forms. We also compare Arjun
with recently released SOTA for Wikidata entity linking- OpenTapioca [6] which
is an end to end EL approach. We are not aware of any other end to end EL
tool/approach released for Wikidata.

6.3   Training Details

Implementation Details. We implemented all the models using the PyTorch
framework. The local KG and the semantic search is implemented using Apache
Lucene Core[7] and Elastic search [12]. The semantic search returns entity candidates with a score (higher is better). We reuse the implementation of Falcon
local KG [19] for the same. After empirically observing the performance, we
set the threshold score to 0.85 for selecting the potential entity candidates per
surface form (i.e. the parameter is optimised on the test set). We reused pre-
trained word embeddings from Glove [17] for the attention based neural network.
These embeddings have been pre-trained on Wikipedia 2014 and Gigaword 5[8].
We employ 300-dimensional Glove word vectors for the training and testing of
Arjun. The models are trained and tested on two Nvidia GeForce GTX1080 Ti
GPUs with 11 GB size. Due to brevity, detailed description of training details
can be found in our public Github.

Dataset Preparation. We experimented initially with higher text sequence
lengths but resorted to 25 words due to GPU memory limitation. In total, we
processed 983,257 sentences containing 3,133,778 instances of surface forms (not
necessarily unique entities) which are linked to 85,628 individual Wikidata entities. From these 3,133,778 surface forms occurrences, approximately 62% do not
have exact match with a Wikidata entity label.

---

[7] https://lucene.apache.org/core/.
[8] https://nlp.stanford.edu/projects/glove/.

6.4 Results

Table 1 summarises performance of Arjun compared to the baseline model and another NED approach. We observe nearly 8% improvement in the performance over baseline and Arjun significantly outperforms another end to end EL tool OpentTapioca. Arjun and OpenTapioca generate entity candidates on the fly, i.e., out of Millions of Wikidata entities, the task here is to reach to top-1 entity. This contrasts with other end to end entity linking approaches such as [15], which rely on a pre-computed list of 30 entity candidates per surface form. This translates into extra complexity due to the a large search space for generating entity candidates in the case of Arjun. Our solution demonstrates a clear advantage of using KGs as background knowledge in conjunction with a attention neural network model. We now detail some success and failure cases of Arjun.

Table 1. Performance of Arjun compared to the Baseline.

Method	Precision	Recall	F-Score
baseline	0.664	0.662	0.663
OpenTapioca [6]	0.407	0.829	0.579
Arjun	0.714	0.712	0.713

Success Cases of Arjun. Arjun achieves 0.77F-Score for the surface form extraction task. Arjun identifies the correct surface form for our exemplary sentence S1 (i.e. ASIC) and links it to the entity label *Application Specific Integrated Circuit* of wiki:Q217302. The baseline can not achieve the linking for this sentence. In the Local KG, the entity label of wiki:Q217302 is enriched with aliases that also contain ASIC. This allows Arjun to provide the correct linking to the Wikidata entity containing the long label. Background knowledge induced in the attentive neural network also allows us to link several long entities correctly. For example, in the sentence "The treaty of London or London convention or similar may refer to," the gold standard links the surface form London convention with the label *Convention on the Prevention of Marine Pollution by Dumping of Wastes and Other Matter* (c.f. wiki:Q1156234). The entity label has 14 words, and Arjun provides correct linking. OpenTapioca on the other hand have high recall(it has high number of False Positives), however, the precision is relatively quite low. The limited performance of OpenTapioca was due to the fact that it finds limitation in linking non Wikipedia entities which constitute a major portion in the dataset. This demonstrate strength of Arjun in also linking non standard, noisy entities which are not part of Wikipedia.

Failure Cases of Arjun. In spite of the successful empirical demonstration of Arjun, we have a few types of failure cases. For example in the sentence: 'Two vessels have borne the name HMS Heureux, both of them captured from the French' has two gold standard entities (Heureux to *French ship Heureux* (`wiki:Q3134963`) and `French` to *French* (`wiki:Q150`)). Arjun links `Heureux` to *L'Heureux* (`wiki:Q56539239`). This issue is caused by the semantic search over the Local KG while searching for the potential candidates per surface form. In this case, L'Heureux is also returned as one of the potential entity candidates for the surface form `Heureux`. A similar problem has been observed in correctly mapping the surface form `Catalan` to `wiki:Q7026` (*Catalan Language*) where Arjun links `Catalan` to Catalan (`wiki:Q595266`). Another form of failure case is when Arjun identifies and links other entities which are not part of the gold standard. The sentence 'Tom Tailor is a German vertically integrated lifestyle clothing company headquartered in Hamburg' has two gold standard entity mappings: `vertically integrated` to *vertical integration* (`wiki:Q1571520` and `Hamburg` to *Hamburg* (`wiki:Q1055`). Arjun identifies *Tom* (`wiki:Q3354498`) and *Tailor* (`wiki:Q37457972`) as the extra entities and can not link `vertically integrated`. For brevity, a detailed analysis of the failure cases per entity type (very long label, noisy nonstandard entity), performance loss due to semantic search can be found in our Github.

Limitations and Improvements for Arjun. Arjun is the first step towards improving a deep learning model with additional contextual knowledge for EL task. Arjun can be enhanced in various directions considering current limitations. We list some of the immediate future extensions:

1. *Enhancing Neural Network with Multiple layers:* Arjun currently has a Bi-LSTM and a single layer LSTM for the encoder and the decoder respectively. It has been empirically observed in sequence to sequence models for machine translations that the models show significant improvements if stacked with multiple layers [21]. Therefore, with more computing resources, the neural network model used in Arjun can be enhanced with multiple layers.
2. *Alternative Models:* In this article, our focus is to empirically demonstrate how background knowledge can be used to improve an attentive neural network for entity linking. Several recent approaches [7,8,23] enhance the performance NER and can be used in our models for task (1) and task (3).
3. *Improving NER:* there is a room of improvement regarding surface form extraction where Arjun currently achieves an F-score of 0.77. The latest context-aware word embeddings [3] can be re-used in Arjun or completely replacing NER part with latest language models such as BERT [7].
4. *Replacing Semantic Search:* Another possibility of improvement is in the second step of our approach (i.e., inducing background knowledge). Currently, we rely on very trivial semantic search (same as [19]) over the Local KG to extract Wikidata entity candidates per surface form. Ganea et al. [11] developed a novel method to embed entities and words in a common vector space to provide a context in an attention neural network model for entity linking. This approach could potentially replace semantic search. Classification is seen as one of the most reasonable and preferred ways to prevent out of scope entity labels [15]. On the contrary, Sakor et al. [19] illustrated that expanding the surface forms the way we did, works pretty well for short text.

Our hypothesis was that it should also work for Arjun, which is not completely true if we see our empirical results. Hence, in this paper, we do not claim that every step we took was the best, but after our empirical study, we demonstrate that the candidate expansion by Sakor et al. doesn't work well. However, it solves our purpose of inducing context in the NN which is the main focus of the paper. It leads to an interesting discussion: what is the most efficient way to induce KG context in a NN, maybe the classification one?- one need to empirically prove and we leave it for future work.

5. *Coverage restricted to Wikidata:* Effort can be made in the direction to develop common EL approach targeting multiple knowledge graphs with standard and nonstandard entity formats.

7 Conclusion

In this work, we focused on introducing the limitations of EL on Wikidata in general, presented the novel approach Arjun, and outlined deficiencies of Arjun, which in particular will guide future work on this topic. In this work, we empirically illustrate that for a challenging KG like Wikidata, if a model is fused with additional context post-NER step, it improves entity linking performance. However, this work was our first attempt towards a longer research agenda. We plan to extend our contribution particularly in the following directions (i) extending towards joint entity and predicate linking and use latest language models for NER task, (ii) enriching the background KG to several interlinked KG from Linked Open Data (DBpedia, Freebase, YAGO), (iii) extending Arjun for the learning entities across languages (currently limited to English).

Acknowledgments. This work is Co-funded by the European Union's Horizon 2020 research and innovation programme under the QualiChain Project, Grant Agreement No. 822404; and the IASIS project, Grant Agreement No. 727658.

References

1. Auer, S., Bizer, C., Kobilarov, G., Lehmann, J., Cyganiak, R., Ives, Z.: DBpedia: a nucleus for a web of open data. In: Aberer, K., et al. (eds.) ASWC/ISWC - 2007. LNCS, vol. 4825, pp. 722–735. Springer, Heidelberg (2007). https://doi.org/10.1007/978-3-540-76298-0_52

2. Balog, K.: Entity linking. Entity-Oriented Search. TIRS, vol. 39, pp. 147–188. Springer, Cham (2018). https://doi.org/10.1007/978-3-319-93935-3_5

3. Blythe, D., Akbik, A., Vollgraf, R.: Syntax-aware language modeling with recurrent neural networks (2018)

4. Bollacker, Kurt D., Cook, R.P., Tufts, P.: A shared database of structured general human knowledge, Freebase (2007)

5. Cetoli, A., Bragaglia, S., O'Harney, A.D., Sloan, M., Akbari, M.: A neural approach to entity linking on Wikidata. In: Azzopardi, L., Stein, B., Fuhr, N., Mayr, P., Hauff, C., Hiemstra, D. (eds.) ECIR 2019. LNCS, vol. 11438, pp. 78–86. Springer, Cham (2019). https://doi.org/10.1007/978-3-030-15719-7_10

6. Delpeuch, A.: Opentapioca: lightweight entity linking for Wikidata. arXiv preprint arXiv:1904.09131 (2019)
7. Devlin, J., Chang, M.-W., Lee, K., Toutanova, K.: Bert: pre-training of deep bidirectional transformers for language understanding. In: NAACL-HLT, pp. 4171–4186 (2019)
8. Edunov, S., Ott, M., Auli, M., Grangier, D.: Understanding back-translation at scale. In: EMNLP, pp. 489–500 (2018)
9. ElSahar, H., et al.: T-rex: a large scale alignment of natural language with knowledge base triples. In: LREC (2018)
10. Mulang, I.O., et al.: Evaluating the impact of knowledge graph context on entity disambiguation models. In: CIKM 2020 (to appear, 2020)
11. Ganea, O.-E., Hofmann, T.: Deep joint entity disambiguation with local neural attention. In: EMNLP, pp. 2619–2629 (2017)
12. Gormley, C., Tong, Z.: Elasticsearch: The Definitive Guide: A Distributed Real-Time Search and Analytics Engine. O'Reilly Media, Inc. (2015)
13. Heindorf, S., Potthast, M., Stein, B., Engels, G.: Vandalism detection in Wikidata. In: CIKM (2016)
14. Hoffart, J., et al.: Robust disambiguation of named entities in text. In: EMNLP (2011)
15. Kolitsas, N., Ganea, O.-E., Hofmann, T.: End-to-end neural entity linking. In: CoNLL, pp. 519–529 (2018)
16. Luong, T., Pham, H., Manning, C.D.: Effective approaches to attention-based neural machine translation. In: EMNLP, pp. 1412–1421 (2015)
17. Pennington, J., Socher, R., Manning, C.D.: Glove: global vectors for word representation. In: EMNLP, pp. 1532–1543 (2014)
18. Raiman, J., Raiman, O.: Deeptype: multilingual entity linking by neural type system evolution. In: AAAI 2018 (2018)
19. Sakor, A., et al.: Old is gold: linguistic driven approach for entity and relation linking of short text. In: NAACL HLT, pp. 2336–2346 (2019)
20. Sakor, A., Singh, K., Patel, A., Vidal, M.-E.: Falcon 2.0: an entity and relation linking framework over Wikidata. arXiv preprint arXiv:1912.11270 (2019)
21. Shu, R., Miura, A.: Residual stacking of RNNs for neural machine translation. In: WAT@COLING, pp. 223–229 (2016)
22. Suchanek, F.M., Kasneci, G., Weikum, G.: Yago: a core of semantic knowledge. In: WWW, pp. 697–706 (2007)
23. Vaswani, A., et al.: Attention is all you need. In: NeurIPS 2017 (2017)
24. Vrandecic, D.: Wikidata: a new platform for collaborative data collection. In: WWW Companion, pp. 1063–1064 (2012)
25. Yadav, V., Bethard, S.: A survey on recent advances in named entity recognition from deep learning models. In: COLING (2018)

Improving Entity Linking with Graph Networks

Ziheng Deng[1], Zhixu Li[1,2]($\boxtimes$), Qiang Yang[3], Qingsheng Liu[4],
and Zhigang Chen[5]

[1] Institute of Artificial Intelligence, School of Computer Science and Technology,
Soochow University, Suzhou, China
`zhdeng@stu.suda.edu.cn, zhixuli@suda.edu.cn`
[2] iFLYTEK Research, Suzhou, China
[3] King Abdullah University of Science and Technology, Jeddah, Saudi Arabia
`qiang.yang@kaust.edu.sa`
[4] Anhui Toycloud Technology, Hefei, China
`qsliu@iflytek.com`
[5] State Key Laboratory of Cognitive Intelligence, iFLYTEK, Hefei,
People's Republic of China
`zgchen@iflytek.com`

Abstract. Entity linking aims to assign a unique identity to entities mentioned in text given a predefined Knowledge Base. Previous works address this task based on the local or global features or the combination of them. However, they are faced with the following problems: 1) For the local features based models, their decisions tend to choose entities with high external knowledge support due to the unbalanced training data and supporting score combination strategy. 2) For the global features based methods, the collective entity linking methods suffer from high computational complexity while the sequential decision model may ignore the correlation between mentions. To tackle the problem of local models, this paper proposes to leverage graph convolutional network for entity embeddings, which could integrate global semantic information and latent relation between entities. We also utilize multi-hop attention mechanism to strengthen the expression of mention context and balance the contributions of mention context and external knowledge. To tackle the problem of global methods, we put forward a global sequential inference model with graph-based search algorithm to model the coherence between mentions with low computation cost. Extensive experiments show that our model could achieve competitive results on multiple standard datasets.

Keywords: Entity linking · Attention mechanism · Graph convolutional network

1 Introduction

Entity linking is the task of mapping mentions in the text to their corresponding unambiguous entries in a given knowledge base. It is a critical step in text understanding and information extraction. The mainstream entity linking approaches

Z. Huang et al. (Eds.): WISE 2020, LNCS 12342, pp. 343–354, 2020.
https://doi.org/10.1007/978-3-030-62005-9_25

can be seen as a combination of two continuous steps, i.e., local step and global step [5,6,11,13,17,21]. While the local step mainly focuses on modeling the semantic meaning of the context around the mention, the global step seeks to optimize the coherence between all refereed entities in the document. Despite the success of the existing approaches, both local and global models have their problems respectively.

- **Problem 1.** A key prior step to both local and global models is entity embeddings. Traditional entity embeddings usually learn entities and words in the same vector space, thus to capture their semantic information. For instance, Yamada et al. [20] utilize the skip-gram model to predict the entity given its description words. However, due to the intrinsic flaw of their training strategy, it can only preserve local semantic since the context window is limited to certain size [15]. This results in the ignorance of the distant words occurred in the description that carries global word co-occurrence information
- **Problem 2.** After obtaining the entity representation, the preliminary decision made by the entity linking system is to model the contextual information and compare the similarity with the candidate entities. Most work [2,3,6,11,21] use single attention mechanism to perform a soft selection among local contexts independently where the dependencies between words are ignored. Moreover, external knowledge is widely used to improve the performance of local model [1,6,9,10,14,21] such as mention-entity prior. However, previous work integrates similarity score and prior distribution in a linear way where prior distribution could be assigned with high weights since most mentions are pointed to popular entities. This will lead to the bias of these kinds of entities especially for entities with few linked occurrences.
- **Problem 3.** For the global models, some previous works [21] transfer the collective decision problem into a sequential decision problem where the next linking decision is based on the previous linked decisions. Though they are able to reduce the computation cost to a large amount, they prone to fall into the error accumulation problem where previous wrongly linked entities lead the whole linking decision to be wrong.

In this paper, we propose to optimize these models in several ways. Firstly, we novelly apply graph convolutional network in entity embeddings to capture the long-distance semantics for both local and global models. Moreover, it allows latent relation to pass among entities, which is beneficial to global model that optimizes relatedness between linked entities. Secondly, we propose to apply multi-hop attention mechanism to project mention context and external knowledge into multiple representation spaces in the different hops where dependencies between words can be interacted. And the multi-hop mechanism helps the model to focus on different informative parts in each iteration. Thirdly, to alleviate the error accumulation problem, we apply a graph search algorithm in sequential model. In each decision step, we consider all possible linking decision and keep the top K candidates based on the coherence score. In the final step, we choose the linking decisions with highest linking score. By expanding the search space, we are able to reduce the side effect of the error propagation since wrong linked entities will lower the total coherence score.

2   Related Work

2.1   Entity Embeddings

The goal of entity embedding is to compress the information of the entity to a low dimension vector. He et al. [10] are the first to utilize Stacked Denoising Auto-encoders to learn entity representation from entity description. Guo and Barbosa [8] represent the entity into vector space by performing random walks on Knowledge Bases. In their work, they fail to unify the word space and entity space since words and entities did not interact directly which may be inappropriate for local model. Yaghoobzadeh et al. [19] follow the work of word2vec to map words and entities into same vector spaces by extending the skip-gram model where they learn to predict entities given its nearby description words. Yamada et al. [20] jointly optimize the embedding of words and entities simultaneously with extra internal link in knowledge bases to model coherence between entities. To alleviate the entity co-occurrence sparsity problem, Ganea and Hofmann [6] learns entity embeddings from their entity pages and local context of their hyperlink annotations from Wikipedia. Compared with our work, their model fails to capture long distant word semantic information since skip-gram model can only capture local context information. and have trouble in passing latent information between entities which is benefit to global model.

2.2   Local Models

The crucial step in improving the performance of local model is to effectively model the mention context. Ratinov et al. [17] are the first to compute the similarity between mention context and entity title as the local score. Later, these heuristic methods were replaced by representation learning. Francis et al. [4] use convolutional neural network to capture multiple granularities semantic information from entity title and entity description. Ganea and Hofmann [6] adopt attention mechanism to select words that are informative. However, previous methods capture important words by standard attention mechanism, which may neglect the inter-dependencies between words. Furthermore, external knowledges have been proved to be useful in many models, entity prior that provides the probability distribution over candidate entities can achieve 71.9% precision without any sophisticated approach [9]. Mulang et al. [12] exploit background knowledge from external Knowledge Graph to generate more source information. However, these features are integrated by linear combination, which makes model tend to assign high weights to popular entities. Our model can incorporate these external knowledge more efficiently and thus generate more robust mention context representation through multi-hop attention mechanism.

2.3   Global Models

Despite local models achieve promising results, global models can improve the entire model performance better since relations between entity of each mention are considered [5,6,11,16,17,21]. Studies on global model can be categorized into two groups. One group focus on disambiguating all mentions in the same document simultaneously based on the global coherence among the all linking decisions. They normally define a joint probability distribution over all mentions and compute all the pairwise coherence scores between two entities selected arbitrarily from candidate entity sets. Ganea and Hofmann [6] relax the assumption that all mentions are related with each other and utilize a truncated fitting of Loopy Belief Propagation in inference decision procedure. Wu et al. [18] capture the topical coherence among various entity mentions by utilizing GCN model.

Though these models maximize the coherence between linked entities, they often suffer from high computation cost which grows exponentially when the number of candidate entities increases. Another group of studies focus on reducing the computation cost. They [21] turn the global model into a sequential decision problem where each linking decision depends on the previous linked entities. Their model requires only one pass through the document but ignores the global coherence between all mentions. In our model, we combine the advantages of both approaches where we keep efficiency of sequential decision model and take the inter-entity coherence between mentions into consideration by expanding the search space in each linking decision.

3   Our Model

Our model consists of three parts: (1) Learning entity embeddings with graph convolutional network, (2) Utilizing multiple-hop attention mechanism in local model, and (3) Adding a graph-based search algorithm in the global model. We will present the details of each step in the following subsections.

3.1   GCN-based Entity Embeddings

In this section, we use GCN to capture global semantic meaning of entities and transfer latent relations between entities.

In the first step, we get the feature matrix X which is built with words embeddings and entities embeddings [6] by concation. Then, we build a word-entity graph where the weights between entity nodes are calculated as the relatedness between entities:

$$\text{WLM}(e_i, e_j) = 1 - \frac{\log \max(|C_{e_i}|, |C_{e_j}|) - \log(C_{e_i} \cap C_{e_j})}{\log |E| - \log \min(|C_{e_i}|, |C_{e_j}|)} \tag{1}$$

where E is the entity set and C_{e_i} is the set that contains entities linking to entity e_i. To avoid the model only learns the local context semantic information, the weights between entity nodes and word nodes are measured as the normal distribution of word-entity co-occurrence $\hat{p}(w|e) \propto \#(w, e)$ where e is the entity and w is the word sampled from the entire entity description page. The weights between word nodes are calculated as the point-wise mutual information where words with high semantic correlation are assigned with high positive values.

After building the graph, we utilize two graph convolutional network layers to capture semantic meanings and integrate inter-entity information from neighbor nodes.

$$L = \tilde{A}ReLU(\tilde{A}XW_0)W_1 \tag{2}$$

where A is the adjacency matrix, D is its degree matrix where $D_{ii} = \sum_j A_{ij}$ and $\tilde{A} = D^{-\frac{1}{2}}AD^{-\frac{1}{2}}, W_0$ and W_1 are weight matrices. We hope the model can represent similar entities with similar vectors, so we formulate the loss function as:

$$\mathcal{L}_e = \sum_{e_i \in E} \sum_{e_o \in C_{e_i}, e_i \neq e_o} logP(e_o|e_i) \tag{3}$$

where $P(e_o|e_i)$ is the conditional probability of entity vector e_o given entity vector e_i.

$$P(e_o|e_i) = \frac{\exp(L_{e_i}^\intercal L_{e_o})}{\sum_{e \in E} \exp(L_{e_i}^\intercal L_e))} \tag{4}$$

where L_{e_i} denotes the vector of entity e_i in matrix L and E is the set of entities in the training data. We then apply these entity embeddings in our next models.

3.2   External Knowledge-Aware Local Model

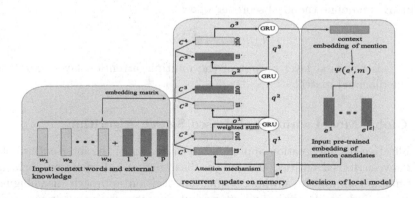

Fig. 1. Neural attention mechanism with multiple hops in local model. Inputs are context word vectors, candidate entity priors, candidate entity type and candidate entity title. Output is the candidate entity score

In order to solve ambiguous mention problem, we first propose our local model by incorporating external knowledge effectively with multi-hop attention. As Fig. 1 shows, we identity the true referent entity of the mention based on its local context and external knowledge. For each mention m and its candidate entity set $\mathcal{E} = \{e^1, ..., e^{|\mathcal{E}|}\}$, we select the context words surrounding the mention $x = \{w_1, ..., w_N\}$ where N is the window size. Then we extract external knowledge entity title l, entity prior p and entity type y from entity description page and get their embedding b. We further concatenate context words and external knowledge $v = [x; b]$ and convert it into memory $U = Av$ where A is the transforming matrix.

Since single attention layer may have trouble in distributing attention over local context and external knowledge, we stack the attention layer to H hops. In each hop h, we compute the attention weights for each memory component i as:

$$p_i^h = Softmax((q^{h^T})c_i^h) \tag{5}$$

where $c_i^h = C^h(u_i)$ and C^h is the embedding matrix that projects memory to different embedding spaces in each hop. q is defined as query vector that searches for the relevant information in memory and the initial query vector q^1 is computed as:

$$q = Qe^i \tag{6}$$

where Q is the embedding matrix and e^i is the candidate entity embeddings. Then, we read out the memory o^h by the weighted sum C_i^{h+1}:

$$o^h = \sum_i p_i^h C_i^{h+1} \tag{7}$$

To integrate different information in each hops, we update the query vector in each hop where $q^{h+1} = GRU(q^h, o^h)$.

Finally, we define the local score as follow:

$$\Psi(m, e^i) = e^{i^T} R o^H \tag{8}$$

where o^H is the final layer output of the multiple attention layer and R is a trainable diagonal matrix.

3.3   Global Model with Graph-Based Search Algorithm

The collective disambiguation approaches usually model the inter-entity coherence between linked entities and jointly disambiguate all mentions, which is very time consuming. Meanwhile, sequential decision approach disambiguates the mention independently in linear time but may ignore the coherence among linked entities and suffer from error accumulation problem. We propose a graph search algorithm by keeping top K scoring linking decisions expanded in the sequential order and choose the best output at last.

Formally, given the list of mentions $M = \{m_1, ..., m_T\}$ where T is the number of mentions and all previously linked entities E_t, At time step t in linking decision, we keep track of K hypotheses and their score $S(E_t^k|M)$. As we move to next $t + 1$ linking decision, it expands each K hypotheses individually and leads to $K * K$ new hypotheses:

$$E_{t+1}^k = \left[E_t^k, e_{t+1}^{k,k'} \right], k' \in [1, K] \tag{9}$$

where $e_{t+1}^{k,k'}$ is one candidate entity of the mention that needs to be linked. We then compute the score of each new generated hypotheses as follows:

$$S(E_t^k, e_{t+1}^{k,k'}|M) = S(E_t^k|M) + \Phi(e_{t+1}^{k,k'}|M, E_t^k) \tag{10}$$

where $\Phi(e_{t+1}^{k,k'}|M, E_t^k)$ defines the coherence score between candidate entity and previously linked entities.

$$\Phi(e_{t+1}^{k,k'}|M, E_t^k) = \Sigma_{e_i^k \in E_t^k} \alpha(e_i) e_{t+1}^{k,k'} Re_i^k \tag{11}$$

where $\alpha(e_i^k) = softmax(u(e_i^k))$ and $u(e_i^k)$ measures the relatedness score between two entities.

$$u(e_i^k) = e_{t+1}^{k,k'\,\mathsf{T}} Be_i^k \tag{12}$$

where B is a learnable diagonal matrix, and $e_i^k \in E_t^k$. We select top K hypothesis based on the score of local model $\Psi(m_t, e_{t+1}^{k,k'})$ and global model $\Phi(e_{t+1}^{k,k'}|E_t^k)$.

By doing so we can alleviate the side effect of error accumulation problem since previous wrong linked entity tends to reduce the coherence between the linked entities. Furthermore, to empower the model with the ability of dealing with unpopular entities, such as mention "World Cup" is more likely to align to the entity "FIFA_WORLD_CUP", but it may actually refer to "FIBA_Basketball_ World_Cup" in some scenarios. Thus, we add an extra term γ that helps in expanding the search space and putting more attentions on various entities.

$$S(E_t^k, e_{t+1}^{k,k'}|M) = S(E_t^k|M) + \Phi(e_{t+1}^{k,k'}|M, E_t^k) - \gamma k' \tag{13}$$

The model now can give more credit to candidate entities that come from different previous linked entity sets. So our global model can successfully modeling the inter-entity coherence in sequential decision model and drop the computation cost by a large margin compared with the collective model.

Finally, the whole model can be seen as a ranking model and we minimize the max-margin loss as follows:

$$\theta^* = \arg \min_{\theta} \sum_{D \in \mathcal{D}} \sum_{m \in D} \sum_{e_t^j \in \mathcal{E}_t} g(e_t^j, m_t) \tag{14}$$

$$g(e_t^j, m_t) = \max(0, \delta - P_\theta(e_t^*) + P_\theta(e_t^j)) \tag{15}$$

where $P_\theta(e_t^j)$ is the output probability from two-layer neural network given the local score and global score. δ is a margin parameter, (m_t, e_t^j) is the mention-candidate pair and e_t^* is the ground-truth entity, θ are the model parameters, $\mathcal{D}$ is a training dataset. We aim to assign higher score to the ground truth entities than other candidate entities.

4 Experiments

In this section, we illustrate our proposed whole model combining the local model and global model on multiple standard datasets and external knowledge-aware local model.

4.1 Datasets and Metrics

To test the performance of our local and global model, we evaluate them on public datasets, and the target knowledge base is Wikipedia. We train our model on AIDA-train which consists of 946 documents for training, 216 documents for validating, and 231 documents for testing. Furthermore, to evaluate the generalization of our model, we also evaluate our model on five cross domain datasets: MSNBC, AQUAINT, ACE2004 created by Guo and Barbosa [9], CWEB and WIKI which were extracted from ClueWeb and Wikipedia corpora automatically.

4.2 Baselines

We compare our model with several state-of-the-art models on public datasets. Following previous work, we use in-KB accuracy and micro F1 metrics to evaluate our results. AIDA-light [13] uses a novel two-stage mapping algorithm where the first stage identifies and links low ambiguity mentions and the second stage establishes reliable contexts to link the rest mentions. WNED [9] applies random walks on disambiguation graphs that combines lexical and statistical features with semantic signatures. Global-RNN [14] develops a framework based on convolutional neural networks and recurrent neural networks that model the local and global features simultaneously. MulFocal-Att [7] applies attention-based approach to collective resolving mentions and models relations between entities in small subset. Deep-ED [6] leverages learned neural representations to construct entity representations, and adopts neural attention mechanism in local model. Ment-Norm [11] models relations between mentions as latent variables without supervision. Yang et al. [21] turn the collective resolving in global model into sequential decision problem where the mentions are resolved in sequential order instead of resolving mentions simultaneously.

4.3 Hyper-parameters Setting

We use Wikipedia corpus to train our entity embeddings. We remove stop words and rare words that appeared fewer than five times to minimize the influence of noise. The embedding size of the graph convolutional layer as 300. We use Adam to train the model for a maximum of 200 epochs with learning rate of 5e-3 and stop training process if the validation loss does not decrease for 10 epochs. For entity linking model, we tune the parameters on AIDA-A dataset, the dimension of word embedding and entity embedding are both 300. The embedding size of the each matrices in local model is equivalent to the word embedding size and we empirically find that the 3 hops in local model can achieve best results, and the window size is set to 50. The dropout rate is set to 0.2, and rank margin δ is set to 0.01. We use Adam with learning rate of 1e-4 until validation accuracy exceeds 92.8%, afterwards we reduce it to 1e-5. We compute the validation accuracy every 5 epochs. The model will stop when the validation accuracy does not increase after 200 epochs.

Table 1. F1 score on AIDA-B test set

Method	F1-score
Local models	
Prior $p(e\|m)$	
Lazic et al. (2015)	86.4
Globerson et al. (2016)	87.9
Yamada et al. (2016)	87.2
Ganea and Hofmann (2017)	88.8
Bert-Entity-Sim (2020)	90.06
Ours	**90.38**
Local & Global models	
Huang, Heck, and Ji (2015)	86.6
Chisholmand Hackey (2015)	88.7
Ganea et al. (2016)	87.6
Guo and Barbosa (2016)	89.0
Globerson et al. (2016)	91.0
Yamada et al. (2016)	91.5
Ganea and Hofmann (2017)	92.22
Le and Titov (2018)	93.07
Bert-Entity-Sim (2020)	93.54
Ours	**93.60**

4.4 Performance Comparison

Following previous work, we test our models using in-KB accuracy for AIDA-B dataset and micro F1 scores for other datasets. First of all, we test our local model using AIDA-B dataset in Table 1. We can see that our local model can achieve the highest in-KB accuracy and outperforms the previous state-of-the-art model by 0.3%. Table 2 shows the results of the model that combined the local model and global model on five cross-domain datasets. As we can see, none of the existing model can consistently outperform all other datasets, and our model achieve the best result on the ACE2004 dataset competitive on other

Table 2. Performance Comparison on Cross-domain Datasets using F1 score

Methods	MSNBC	AQUAINT	ACE2004	CWEB	WIKI	Avg
AIDA	79	56	80	58.6	63	67.32
GLOW	75	83	82	56.2	67.2	72.68
RI	90	**90**	86	67.5	73.4	81.38
WNED	92	87	88	77	**84.5**	85.7
Deep-ED	93.7	88.5	88.5	**77.9**	77.5	85.22
Ment-Norm	93.9	88.3	89.9	77.5	78.0	85.51
Prior(p(e—m))	89.3	83.2	84.4	69.8	64.2	78.18
ETHZ-Attn + DCA-SL	**94.57**	87.38	89.44	73.47	78.16	84.60
ETHZ-Attn + DCA-RL	93.80	88.25	90.14	75.59	78.84	85.32
our model	94.41	89.23	**90.54**	76.64	78.2	**85.84**

datasets. On average, our proposed model outperforms the local and global version of Le and Titov [11] and Yang [21] by an average 0.30 and 0.6 on F1 score. Furthermore, it is worth noting that two heads may not better than one, so the performance of the combination model may not always outperform single model. Thus, our model is competitive in linking accuracy while being comparable in running time.

5   Conclusion and Future Work

In this paper, we propose a novel entity linking model including entity embeddings, local model and global model. The entity embeddings model utilizes graph convolutional network can capture global semantic information and pass relation between entities. The local model uses multi-hop attention mechanism to balance the contributions of mention context and external knowledge while the global model casts our problem into a sequential inference problem with graph-based search algorithm. Extensive experiments show the competitiveness of our approach both in accuracy and efficiency. In the future, we would like to introduce BERT into global model since BERT model has shown promising performance in discovering latent relation between entities.

Acknowledgments. This research is partially supported by National Key R&D Program of China (No. 2018AAA0101900), the Priority Academic Program Development of Jiangsu Higher Education Institutions, National Natural Science Foundation of China (Grant No. 62072323, 61632016, 61836007), Natural Science Foundation of Jiangsu Province (No. BK20191420), Natural Science Research Project of Jiangsu Higher Education Institution (No. 17KJA520003), and the Suda-Toycloud Data Intelligence Joint Laboratory.

References

1. Bunescu, R.C., Pasca, M.: Using encyclopedic knowledge for named entity disambiguation. In: EACL 2006, 11st Conference of the European Chapter of the Association for Computational Linguistics, Proceedings of the Conference, April 3–7, 2006, Trento, Italy (2006). https://www.aclweb.org/anthology/E06-1002/
2. Camacho-Collados, J., Bovi, C.D., Raganato, A., Navigli, R.: Sensedefs: amultilingual corpus of semantically annotated textual definitions -exploiting multiple languages and resources jointly for high-quality wordsense disambiguation and entity linking. Lang. Resour. Eval. **53**(2), 251–278 (2019). https://doi.org/10.1007/s10579-018-9421-3
3. Chen, H., Wei, B., Liu, Y., Li, Y., Yu, J., Zhu, W.: Bilinear joint learning of word and entity embeddings for entity linking. Neurocomputing **294**, 12–18 (2018). https://doi.org/10.1016/j.neucom.2017.11.064
4. Francis-Landau, M., Durrett, G., Klein, D.: Capturing semantic similarity for entity linking with convolutional neural networks. In: NAACL HLT 2016, The 2016 Conference of the North American Chapter of the Association for Computational Linguistics: Human Language Technologies, San Diego California, USA, June 12–17, 2016, pp. 1256–1261 (2016). https://www.aclweb.org/anthology/N16-1150/
5. Ganea, O., Ganea, M., Lucchi, A., Eickhoff, C., Hofmann, T.: Probabilistic bag-of-hyperlinks model for entity linking. In: Proceedings of the 25th International Conference on World Wide Web, WWW 2016, Montreal, Canada, April 11–15, 2016, pp. 927–938 (2016). https://doi.org/10.1145/2872427.2882988
6. Ganea, O., Hofmann, T.: Deep joint entity disambiguation with local neural attention. In: Proceedings of the 2017 Conference on Empirical Methods in Natural Language Processing, EMNLP 2017, Copenhagen, Denmark, September 9–11, 2017, pp. 2619–2629 (2017). https://www.aclweb.org/anthology/D17-1277/
7. Globerson, A., Lazic, N., Chakrabarti, S., Subramanya, A., Ringard, M., Pereira, F.: Collective entity resolution with multi-focal attention. In: Proceedings of the 54th Annual Meeting of the Association for Computational Linguistics, ACL 2016, August 7–12, 2016, Berlin, Germany, Volume 1: Long Papers (2016). https://doi.org/10.18653/v1/p16-1059
8. Guo, Z., Barbosa, D.: Entity linking with a unified semantic representation. In: 23rd International World Wide Web Conference, WWW '14, Seoul, Republic of Korea, April 7–11, 2014, Companion Volume, pp. 1305–1310 (2014). https://doi.org/10.1145/2567948.2579705
9. Guo, Z., Barbosa, D.: Robust named entity disambiguation with random walks. Semantic Web **9**(4), 459–479 (2018). https://doi.org/10.3233/SW-170273
10. He, Z., Liu, S., Li, M., Zhou, M., Zhang, L., Wang, H.: Learning entity representation for entity disambiguation. In: Proceedings of the 51st Annual Meeting of the Association for Computational Linguistics, ACL 2013, 4–9 August 2013, Sofia, Bulgaria, Volume 2: Short Papers, pp. 30–34 (2013). https://www.aclweb.org/anthology/P13-2006/
11. Le, P., Titov, I.: Improving entity linking by modeling latent relations between mentions. In: Proceedings of the 56th Annual Meeting of the Association for Computational Linguistics, ACL 2018, Melbourne, Australia, July 15–20, 2018, Volume 1: Long Papers. pp. 1595–1604 (2018). https://doi.org/10.18653/v1/P18-1148
12. Mulang, I.O., et al.: Context-aware entity linking with attentive neural networks on Wikidata knowledge graph. CoRR abs/1912.06214 (2019). http://arxiv.org/abs/1912.06214

13. Nguyen, D.B., Hoffart, J., Theobald, M., Weikum, G.: Aida-light: high-throughput named-entity disambiguation. In: Proceedings of the Workshop on Linked Data on the Web co-located with the 23rd International World Wide Web Conference (WWW 2014), Seoul, Korea, April 8, 2014 (2014). http://ceur-ws.org/Vol-1184/ldow2014_paper_03.pdf

14. Nguyen, T.H., Fauceglia, N.R., Rodriguez-Muro, M., Hassanzadeh, O., Gliozzo, A.M., Sadoghi, M.: Joint learning of local and global features for entity linking via neural networks. In: COLING 2016, 26th International Conference on Computational Linguistics, Proceedings of the Conference: Technical Papers, December 11–16, 2016, Osaka, Japan, pp. 2310–2320 (2016). https://www.aclweb.org/anthology/C16-1218/

15. Pennington, J., Socher, R., Manning, C.D.: Glove: Global vectors for word representation. In: Proceedings of the 2014 Conference on Empirical Methods in Natural Language Processing, EMNLP 2014, October 25–29, 2014, Doha, Qatar, A meeting of SIGDAT, a Special Interest Group of the ACL. pp. 1532–1543 (2014). https://doi.org/10.3115/v1/d14-1162

16. Phan, M.C., Sun, A., Tay, Y., Han, J., Li, C.: Pair-linking for collectiveentity disambiguation: Two could be better than all. IEEE Trans. Knowl. Data Eng. **31**(7), 1383–1396 (2019). https://doi.org/10.1109/TKDE.2018.2857493

17. Ratinov, L., Roth, D., Downey, D., Anderson, M.: Local and global algorithms for disambiguation to Wikipedia. In: The 49th Annual Meeting of the Association for Computational Linguistics: Human Language Technologies, Proceedings of the Conference, 19–24 June, 2011, Portland, Oregon, USA, pp. 1375–1384 (2011). https://www.aclweb.org/anthology/P11-1138/

18. Wu, J., Zhang, R., Mao, Y., Guo, H., Soflaei, M., Huai, J.: Dynamic graph convolutional networks for entity linking. In: Huang, Y., King, I., Liu, T., van Steen, M. (eds.) WWW 2020: The Web Conference 2020, Taipei, Taiwan, April 20–24, 2020, pp. 1149–1159. ACM / IW3C2 (2020). https://doi.org/10.1145/3366423.3380192

19. Yaghoobzadeh, Y., Schütze, H.: Corpus-level fine-grained entity typing using contextual information. CoRR abs/1606.07901 (2016). http://arxiv.org/abs/1606.07901

20. Yamada, I., Shindo, H., Takeda, H., Takefuji, Y.: Joint learning of the embedding of words and entities for named entity disambiguation. In: Proceedings of the 20th SIGNLL Conference on Computational Natural Language Learning, CoNLL 2016, Berlin, Germany, August 11–12, 2016, pp. 250–259 (2016). https://doi.org/10.18653/v1/k16-1025

21. Yang, X., et al.: Learning dynamic context augmentation for global entity linking. In: Proceedings of the 2019 Conference on Empirical Methods in Natural Language Processing and the 9th International Joint Conference on Natural Language Processing, EMNLP-IJCNLP 2019, Hong Kong, China, November 3–7, 2019, pp. 271–281 (2019). https://doi.org/10.18653/v1/D19-1026

Spatial Temporal Data Analysis

Accurate Indoor Positioning Prediction Using the LSTM and Grey Model

Xuqi Fang[1], Fengyuan Lu[1], Xuxin Chen[1], and Xinli Huang[1,2](✉)

[1] School of Computer Science and Technology, East China Normal University,
Shanghai 200062, China
xlhuang@cs.ecnu.edu.cn
[2] Shanghai Key Laboratory of Multidimensional Information Processing,
Shanghai 200062, China

Abstract. The indoor positioning prediction technologies are developed to locate and predict actual positions of the objective indoors, and can be applied to smart elderly-caring application scenarios, helping to discover and reveal irregular life routines or abnormal behavior patterns of the elderly living at home alone. In this paper, we focus on accurate indoor positioning prediction and introduce an improved prediction model for IoT sensing data based on the LSTM and Grey model. In order to enhance the prediction ability of nonlinear samples in IoT sensing data and improve the prediction accuracy of the model, we propose to incorporate into and utilize the advantages of the LSTM model in dealing with nonlinear time series data of different spans, and the ability of the Grey model in dealing with incomplete information and in eliminating residual errors generated by LSTM. To demonstrate the effectiveness and performance gains of the model, we setup experiments based on the indoor trajectory dataset. Experimental results show that the model proposed in this paper outperforms its competitors, producing an arresting increase of the positioning prediction accuracy, with the RSME for the next day and the next week being 63.39% and 54.86%, respectively, much lower than that of the conventional models.

Keywords: Internet of Things · Indoor positioning prediction · Long short-term memory network (LSTM) · Grey model (GM) · Smart elderly-caring

1 Introduction

At present, the aging problem has become an important social problem facing and concerned by the whole world. With the rapid increase of the elderly population, how to provide sufficient resources and high-quality care services for the elderly has become a big burden for the government and society. The World Health Organization advocates concepts such as ageing in place [1] and active ageing [2], to encourage the elderly to spend their retirement in their growing environment, such as family, rather than in nursing homes.

© Springer Nature Switzerland AG 2020
Z. Huang et al. (Eds.): WISE 2020, LNCS 12342, pp. 357–368, 2020.
https://doi.org/10.1007/978-3-030-62005-9_26

However, if the elderly choose to live alone at home, they are often unable to take good care of themselves, especially for the severely disabled elderly. By deploying a set of intelligent monitoring and care system in the home of the elderly living alone, the health status of the elderly can be monitored and analyzed. When an emergency or abnormal health event occurs, it is particularly important to give an alarm or early warning timely and accurately.

In order to provide intelligent monitoring and care services for the elderly, researchers in many countries have done a lot of exploring. Acousto-optic sensing devices connected to the medical center were installed in the home to solve the problem of the elderly falling in Singapore [3]. In the European Union, Mobi-Health technology has been implemented in member countries, to collect the physiological parameters of the elderly by wireless sensors and actuators [4]. Japan launched the Sukoyaka Family 21 plan to measure the blood pressure and electrocardiogram of the elderly by wireless sensor devices [5]. Although the above schemes have achieved the comprehensive physical signs monitoring, they either ignore the position information or just record the position as auxiliary information. Further, the position information can be utilized to take care of the elderly, such as indoor positioning prediction, which contributes to identifying daily activities, discovering living habits and diagnosing the chronic diseases.

In this paper, we focus on the aforementioned problems, investigate the issues of accurate indoor positioning prediction, and manage to introduce new ideas, methods and solutions to locate and predict actual positions of the objective indoors, and to ultimately apply them to smart elderly-caring applications, helping to discover and reveal irregular life routines or abnormal behavior patterns of the elderly living at home alone, with the aim of anticipating and preventing the occurrence of emergency or health risks. The work and main contributions of this paper are summarized as follows:

- We propose an improved indoor positioning prediction model for IoT sensing data based on the LSTM and Grey model (LSTM-GM), and apply it to smart elderly-caring application scenarios. In order to enhance the prediction ability of nonlinear samples in IoT sensing data and improve the prediction accuracy of the model, we propose to fully incorporate into and utilize the advantages of LSTM model in dealing with nonlinear time series data of different spans, and the ability of GM model in dealing with incomplete information and eliminating residual errors generated by LSTM.
- We setup experiments to demonstrate the effectiveness and performance of the LSTM-GM model by the real-world indoor trajectory dataset. The results show that our model outperforms the conventional LSTM model, producing an arresting increase of the positioning prediction accuracy, with the RSME for the next day and the next week being 63.39% and 54.86% lower than that of the conventional LSTM model, respectively.

The remainder of the paper is organized as follows. We review related work in Sect. 2, and then present in detail the problem formulation in Sect. 3, the model design in Sect. 4, respectively. We describe experimental setup and report the results and analysis in Sect. 5, followed by the conclusion and future work of this paper in the last section.

2   Related Work

In recent years, many researchers have proposed effective prediction models based on various positioning technologies. These models can be broadly divided into two categories: (i) Individual-based prediction models; (ii) Population-based prediction models.

Individual-based prediction models consider that the activities of each individual are independent from each other, and the user's next position only depends on himself or herself. Therefore, such models try to mine the periodic behaviors of individual users by using their own migration data records. For instance, Wang et al. [6] proposed a hybrid Markov model for positioning prediction that integrates a long short-term memory model (LSTM). Li et al. [7] discussed the impact of the granularity and duration of stay of target positions, as well as different behavioral features on the prediction accuracy. Yang et al. [8] proposed a novel approach DestPD for destination prediction, which first predicted the most probable future position and then reported the destination based on that. Kong et al. [9] proposed a Spatial-Temporal Long-Short Term Memory (ST-LSTM) model which combines spatial-temporal influence into LSTM and employ a hierarchical extension of the proposed ST-LSTM (HST-LSTM) in an encoder-decoder manner which models the contextual historic visit information. Although these models perform well in the prediction of mobility with dense data records, they have high requirements on data quality, requiring long-term movement trajectory of users, which is difficult to obtain in practical applications. The sparsity of data often leads to the occurrence of cold startup problems, or even the failure of prediction.

To overcome the difficulty of individual movement prediction caused by the sparsity of moving records, some scholars proposed population-based prediction models, which consider that the behavior pattern of users has certain commonness, and the behavior of users can be predicted by the correlation pattern among users. The basic idea is to use the history of other users to predict their next position. For example, Jie Feng et al. [10] propose DeepMove to capture the multiple factors that govern the transition regularities of human movements. Mou et al. [11] proposed a new R-FP-growth (tuple-relation frequent pattern growth) algorithm for mining association rules. Zhao et al. [12] proposed a new Spatio-Temporal Gated Network (STGN) by enhancing LSTM network, where Spatio-temporal gates are introduced to capture the Spatio-temporal relationships between successive check-ins. Liu et al. [13] extended RNN and proposed a new method called spatial time recurrent neural network, which can model the local temporal and spatial context in each layer with time-specific transition matrices for different time intervals and distance-specific transition matrices for different geographic distances. Ying et al. [14] proposed a novel mining-based positioning prediction approach called Geographic-Temporal-Semantic-based Location Prediction (GTS-LP). While using other users' historical movement data can help improve the accuracy of the predictions, it requires a large number of user trajectories. What's more, they are faced with significant privacy issues as users are reluctant to share their position data.

However, these models focus on positioning prediction in outdoor environment or relatively open indoor environment, with few studies on smaller environments such as family quarters. Besides, most of these models are used in business, with little applied to elder care. Therefore, in this paper, we propose the LSTM-GM model which combines the advantage of LSTM and Grey Model methods to predict the position of elderly people at home.

3   Problem Formulation

The purpose of this paper is to locate and predict the position of the elderly, helping to detect the abnormal behaviors or irregular life routines of the elderly who live at home alone. In this section, we formally describe the related definitions and formulate the objective positioning prediction tasks in this paper:

Definition 1 (Position): Position p is defined as a connected area, representing the position of the elderly at a certain moment. Each position is identified by a unique digital ID.

Definition 2 (Sensor Records): A sensor record r is described as a triple $\{ID_i, V_i, X_i\}$, where ID_i is the number of the sensor, $V_i \in \{0,1\}$ indicates whether the sensor is in the fired status or not, and X_i is the sensor position. The position triggered indicates the position of the old man at that moment.

Definition 3 (Trajectory): The trajectory $traj = \{p_{t_i}\}_{i=1}^n$ represents the position of elderly at time t_i.

Definition 4 (Positioning Prediction Problem): When the set of trajectories at the past n moments $traj_{past} = \{p_1, p_2, \cdots, p_n\}$ of the elderly is given, the positioning prediction task is to predict the set of trajectories $traj_{future} = \{p_{n+1}, p_{n+2}, \cdots, p_m\}$ in the next m moments. Thus the problem that we focus on is to find a function f to improve the accuracy of the predictions $traj_{future}$ as much as possible. The positioning prediction problem is as follows:

$$traj_{future} = f(traj_{past}). \tag{1}$$

4   Model Design

In this section, we propose to transform the indoor positioning data collected by various sensors into an effective time series [15], and to apply the LSTM model to predict the elderly indoor positions in smart elderly-caring application scenarios. We also propose to correct the residuals of the LSTM model by grey model to improve the accuracy of the prediction model.

4.1   Long Short-Term Memory Network

Long short-term memory network (LSTM) is a special type of recurrent neural networks (RNN) [16]. Its core idea is the setting of input gate, output gate and forgotten gate. Figure 1 respectively shows the structure of the LSTM model [17]. In the figure, σ is the Sigmoid function. *Tanh* is the activation function that processes the data on the state and output. LSTM realizes the function of forgetting and memory through a structure composed of a sigmoid function and a dot product operation. The output value of the sigmoid function is in the interval [0, 1], with 0 indicating that it is not allowed to pass at all and 1 indicating that it is allowed to pass at all.

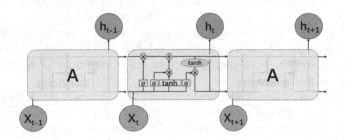

Fig. 1. LSTM Model

4.2   Grey Model

The grey model is to establish a grey differential prediction model with a small amount of incomplete information and make a long-term description of the ambiguity of the development of things. The grey model is usually represented as $GM(n, m)$, where n denotes the order of the differential function and m is the number of parameters. In this paper, we adapt $GM(1, 1)$, for position p is the only variable and the law of prediction is not very complicated.

$GM(1, 1)$ prediction model is used to generate a set of new data sequences with obvious trend by means of accumulation of data sequences, so as to strengthen the influence of known factors and weaken the influence of unknown factors. The relevant parametric equations are constructed and the values of the parameters are determined by means of mathematical solution, so as to realize the prediction of the model. When the set of original sample data $x^{(0)} = \{x_1^{(0)}, x_2^{(0)}, \cdots, x_n^{(0)}\}$ is given, the same kind of data are accumulated in original sequence to form a new sequence $x^{(1)} = \{x_1^{(1)}, x_2^{(1)}, \cdots, x_n^{(1)}\}$. For each variable $x_i^{(1)}$ in the sequence, it is represented as follow:

$$x_i^{(1)} = \sum_{k=1}^{i} x_k^{(0)}. \tag{2}$$

According to the grey theory, the differential equation of $x^{(1)}$ with respect to t is established as follow:

$$\frac{dx^{(1)}(t)}{dt} + ax^{(1)}(t) = u , \tag{3}$$

where a is called development coefficient, and u is called grey action.

4.3   LSTM-GM

Although the LSTM model could learn the regularity and periodicity of the time series and the results perform well, the residuals are still present in the predicted results. The residual discussed in this paper are mathematically defined as the difference between the measured value and the predicted value. And the residual analysis is the analysis of the reliability, periodicity or other interference of the data by the residual. When the residual is large, it will affect the accuracy of the prediction, so it is necessary to correct the residual so as to improve the accuracy of the prediction results.

Indoor positioning prediction based on LSTM-GM model is mainly composed of the conventional LSTM model and grey model, correcting the residual of the LSTM model by setting up grey model. The grey forecast model is a prediction method that builds a mathematical model and makes predictions through a small amount of incomplete information. Using grey mathematics to process uncertainties and quantify them, the internal laws can be discovered in disorderly phenomena. In the LSTM-GM model, the data in the training set is trained by LSTM to obtain a training model to predict the position. The residual corresponding to each predicted value is calculated, and each residual is added to the residual sequence. Then the grey model analyzes the residual sequence at the time t to obtain the residual at the time $t + 1$, in turn, the residual of the predicted value is corrected. According to the problem defined in Sect. 2, the problem can be further formalized as follow:

$$traj_{future} = LSTM(traj_{past}) . \tag{4}$$

And the correction formula for the predicted value is:

$$x_t^{'} = x_t - e_t , \tag{5}$$

where x_t is the LSTM prediction at t period, e_t is the correction of the grey prediction model and $x_t^{'}$ is the final prediction.

The flow chart of the LSTM-GM model is shown in the Fig. 2. And the specific steps of the LSTM-GM model are as follows, and the more details are shown in Algorithm 1:

- **Data set acquisition:** Clean and pre-process the data set, and filter out invalid null data.
- **LSTM-GM model training:** Firstly, the data is trained and preliminary predicted by LSTM model, and the parameters of the LSTM are optimized for the overall effect of the new model. After getting the results of the preliminary

predictions, they are compared with the measured values in the dataset and the residuals generated in the preliminary prediction are also calculated. Then the residual is corrected by the grey model to get the final predicted value.

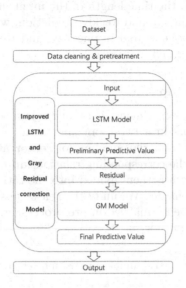

Fig. 2. The flow chart of the LSTM-GM model

Algorithm 1. LSTM-GM Algorithm
___

Input:

  D_0: raw data set, l: size of data set, D_0': LSTM format data set, d_i: the $i - th$ data in the data set, d_i': the $i - th$ data of the LSTM format data set, E: residual sequence, e_i: the $i - th$ residual, d_{ti}: predicted value at time t, d_{ti}':corrected the predicted value at time t;

Output:

  D: output sequence

1: **for** $i = 1; i < n; i + +$ **do**
2:    $d_i' = \text{reshape}(d_i)$;
3:    $d_{ti} = \text{lstm}(d_i')$;
4:    $e_i = \text{compare}(d_i, d_i')$;
5:    $E \leftarrow e_i$;
6:    $e_{(i+1)} = \text{greymodel}(E, d_{ti})$;
7:    $d_{t(i+1)}' = d_{t(i+1)} - e_{(i+1)}$;
8:    $D \leftarrow d_{t(i+1)}'$;
9: **end for**
10: **return** D
___

5 Experimentation and Evaluation

In this section, we setup experiments to compare the accuracy and fitting of the LSTM model prediction results with the conventional LSTM model prediction results. And according to the time length of the input data set, the experiment is divided into daily prediction and weekly prediction, where the input data set of the daily prediction is generated in one day, and the input data set of the weekly prediction is generated in one week.

5.1 Experimental Setup

The data set used in the experiments is an open data set from [18], which was collected from the sensors in a two-room apartment. The sensors are located in 14 different positions, and the numbers of sensor at different position are different. Table 1 shows the corresponding relationship between the sensors and the numbers. And the data is then divided into the ratio of 2:1 and 67% of the data is used to train the model. We unify the sensor position into a time series, and the time series will be normalized in pre-processing to eliminate the effects of different orders of magnitude.

Table 1. Number of position sensors.

Number	Sensor	Number	Sensor	Number	Sensor
1	Microwave oven	9	Sideboard	18	Pan
5	Toilet door	12	Front door	20	Washing machine
6	Bathroom door	13	Dishwasher	23	Grocery cupboard
7	Cupboard	14	Toilet flush	24	Hall bedroom door
8	Refrigerator	17	Freezer		

We use the root-mean-square error (RMSE) to represent the accuracy of the prediction results of the two models, that is, the square root of the ratio of the differences between the predicted value p and the measured value m and the number of observations n. The calculation formula is as follows:

$$RMSE = \sqrt{\frac{1}{n}\sum_{i=1}^{n}(p_i - m_i)^2} \tag{6}$$

In the conventional LSTM model and the LSTM-GM model, the parameters of the LSTM part will be set to the same. In this experiment, the number of hidden neurons in each layer of num_units is set as 128. Table 2 lists the RMSEs for different hidden meta numbers. In the experiments, we compare the RMSE of the predicted value of the LSTM model when the number of hidden elements is 4, 16, 32, 64, 128, 256. It can be seen that the root mean square error is minimal when the number of hidden neurons is 128.

Table 2. RMSE of different hidden neurons

Number of hidden neurons	RSME	Number of hidden neurons	RSME
4	1.31	64	0.89
16	1.00	128	0.84
32	1.09	256	ERROR

5.2 Positioning Prediction by Conventional LSTM

Figure 3 shows the results of training and predicting for one day's data, and Fig. 4 shows the results of training and predicting for one week's data. We can observe that the prediction effect of the LSTM model on the daily prediction is better than the weekly prediction. The reason should be that the variability of activities in a week is greater than that in a day. In addition, there is a large gap between the predicted curve and the actual curve in the figure. We can find that although the LSTM model has good time series trend learning ability, it is still not enough to cope with the positioning prediction problem.

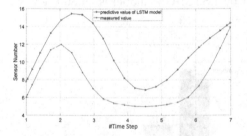

Fig. 3. Daily prediction results of the LSTM model

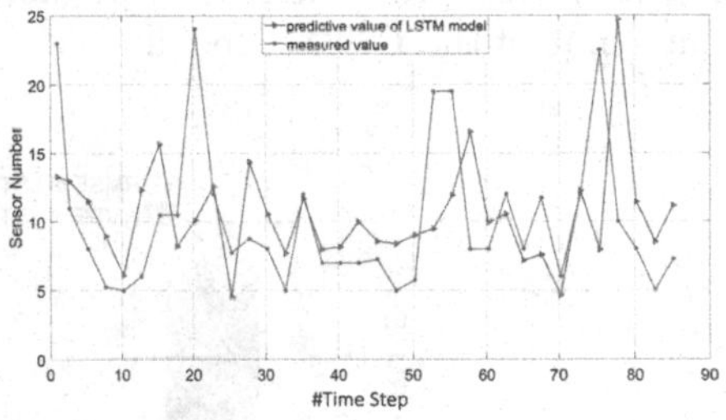

Fig. 4. Weekly prediction results of the LSTM model

5.3 Positioning Prediction by LSTM-GM

The experiments are also divided into two parts. Figure 5 shows the results of training and predicting for one day's data. Figure 6 shows the results of training and predicting for one week's data. In this experiment, like the LSTM model, the daily prediction effect of the LSTM-GM model is better than the weekly prediction. It can be seen from the figure that its predicted curve for the position of the elderly is more fitted to the measured curve, which means that the LSTM-GM model works better than the LSTM model in the experiment. This will be described in detail in the comparison of LSTM and LSTM-GM.

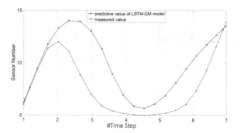 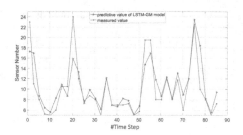

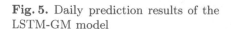

Fig. 5. Daily prediction results of the LSTM-GM model

Fig. 6. Weekly prediction results of the LSTM-GM model

5.4 Comparisons of LSTM and LSTM-GM

According to above experiments, it is obvious that the prediction effect of the LSTM-GM model proposed in this paper is better than the conventional LSTM model under the same parameters. It can be seen from Fig. 7 that, in the same dimension, the RMSE value of the daily prediction value predicted by LSTM-GM model is 63.39% lower than that of the LSTM model, and the RMSE value of the weekly prediction value predicted by LSTM-GM model is 54.86% lower than that of the LSTM model.

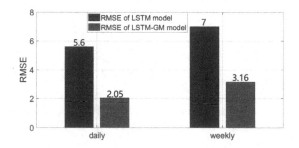

Fig. 7. RMSE of the LSTM and LSTM-GM model

It can be seen from Fig. 8 and Fig. 9 that the predicted value of the LSTM-GM model is closer to the measured value by the residual correction than the predicted value of the conventional LSTM model, and the predicted trend is not weakened. The reason is that although the LSTM model has mastered the regularity and periodicity of the time series, there are still residuals in the prediction. When the residual is large, it will affect the accuracy of prediction. In our model, the residual corresponding to each predicted value is calculated by introducing the GM model, and adds every residual to the residual sequence to correct the residual value of the predicted value, which makes the prediction results perform well in predicting trends and greatly improve the accuracy of the prediction.

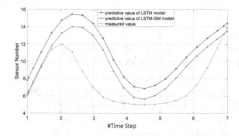

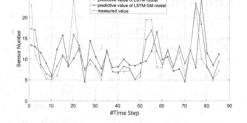

Fig. 8. Comparisons of the LSTM and the LSTM-GM Model for Daily Prediction

Fig. 9. Comparisons of the LSTM and the LSTM-GM Model for Weekly Prediction

6 Conclusions and Future Work

In this paper, we propose to utilize the LSTM model to predict the elderly indoor positions in smart elderly-caring application scenarios, helping to discover and reveal irregular life routines or abnormal behavior patterns of the elderly living at home alone, with the aim to anticipate and prevent the occurrence of emergency or health risks. Considering that indoor positioning data grows exponentially over time, it is transformed into an effective time series for further analysis. However, the conventional LSTM model has achieved the desired effect in the prediction trend, but the existence of the residual greatly affects the accuracy of the prediction. Therefore, the LSTM-GM model is proposed to correct the residual which improves accuracy of the predicted value. To demonstrate the effectiveness and performance gains of the model, we setup experiments based on the indoor trajectory dataset collected by real deployed sensors. And the experiment results show that the RMSE value of the daily predicted value of the LSTM-GM model is 63.39% lower than that of the LSTM model, and the RMSE value of the weekly predicted value is 54.86% lower than that of the LSTM model. With the model, one can have a more accurate and comprehensive understanding of the regularity and periodicity of the positioning time series for the elderly living alone.

In the future, we plan to carry out our research work in the following two directions: (i) further refine our prediction model, with the aim of enhancing the prediction ability and improving the prediction accuracy of the model, and (ii) consider applying the model to the real-world application scenarios with a large number of sensors deployed in a much more complex indoor environment, to verify the actual performance of the model.

Acknowledgements. This work was supported in part by the National Key Research and Development Plan of China under Grant 2019YFB2012803, in part by the Key Project of Shanghai Science and Technology Innovation Action Plan under Grant 19DZ1100400 and Grant 18511103302, in part by the Key Program of Shanghai Artificial Intelligence Innovation Development Plan under Grant 2018-RGZN-02060, and in part by the Key Project of the "Intelligence plus" Advanced Research Fund of East China Normal University.

References

1. Andrews, G., Phillips, D.: Ageing and Place. Routledge, Abingdon (2004)
2. Kalache, A., Gatti, A.: Active ageing: a policy framework. Aging Male **5**(1), 1–37 (2003)
3. Nyan, M.N., Francis, E.H.T., Matthew, Z.E.: MAH: application of motion analysis system in pre-impact fall detection. Jo. Biomech. **41**(10), 2297–2304 (2008)
4. Konstantas, D., Herzog, R.: Continuous monitoring of vital constants for mobile users: the MobiHealth approach. In: International Conference of the IEEE Engineering in Medicine and Biology Society IEEE (2017)
5. Elichegaray, C., Bouallala, S., Maitre, A., et al.: Development and current status of atmospheric pollution. Revue des Maladies Respiratoires **26**(2), 191–206 (2009)
6. Wang, P., Wang, H., Zhang, H., Lu, F., Wu, S.: A hybrid markov and LSTM model for indoor location prediction. In: IEEE Access (2019)
7. Li, H.: A Systematic Analysis of Fine-Grained Human Mobility Prediction with On-Device Contextual Data (2019)
8. Yang, Z., et al.: An efficient destination prediction approach based on future trajectory prediction and transition matrix optimization. IEEE Trans. Knowl. Data Eng. **1**, 1 (2018)
9. Dejiang, K., Fei, W.: HST-LSTM: a hierarchical spatial-temporal long-short term memory network for location prediction. In: Proceedings of the Twenty-Seventh International Joint Conference on Artificial Intelligence (2018)
10. Feng, J., et al.: DeepMove: predicting human mobility with attentional recurrent networks. In: 2018 World Wide Web Conference (2018)
11. Mou, N., Wang, H., Zhang, H., et al.: Association rule mining method based on the similarity metric of tuple-relation in indoor environment. In: IEEE Access (2020)
12. Zhao, P., Zhu, H., Liu, Y., et al.: Where to go next: a spatio-temporal gated network for next POI recommendation. In: Thirty-Third AAAI Conference on Artificial Intelligence (2019)
13. Liu, Q., Wu, S., Wang, L., Tan, T.: Predicting the next location: a recurrent model with spatial and temporal contexts. In: Proceedings of the Thirtieth AAAI Conference on Artificial Intelligence, pp. 194–200 (2016)
14. Ying, J.C., Lee, W.C., Tseng, V.S.: Mining geographic-temporal-semantic patterns in trajectories for location prediction. ACM Trans. Intell. Syst. Technol. 1–33 (2013)
15. Antunes, C., Arlindo, L.: Temporal data mining-an overview. In: KDD Workshop on Temporal Data Mining, pp. 1–13 (2001)
16. Schuster, M., Paliwal, K.: Bidirectional recurrent neural networks. Proc. IEEE Trans. Signal Process. **45**(11), 2673–2681 (1997)
17. Yan, S: Understanding LSTM networks (2015). Accessed 11 Aug 2015
18. Kasteren, T., Noulas, A., Englebienne, G., Ben, J.: Accurate activity recognition in a home setting. In: Proceedings of the ACM International Conference on Ubiquitous Computing, pp. 1–9 (2008)

Bus Travel-Time Prediction Based on Deep Spatio-Temporal Model

Kaixin Zhang[1], Yongxuan Lai[1(✉)], Liying Jiang[1], and Fan Yang[2]

[1] Shenzhen Research Institute/School of Informatics,
Xiamen University, Xiamen, China
laozhng@icloud.com, laiyx@xmu.edu.cn, jiangliying@stu.xmu.edu.cn
[2] Department of Automation,
Xiamen University, Xiamen, China
yang@xmu.edu.cn

Abstract. Bus travel time estimation in urban city is of great importance, which reduces passengers' waiting time and improves the quality of service of bus transportation. However, the travel time estimation is affected by various factors, including spatio-temporal dependencies (e.g. traffic conditions and road networks) and external factors (e.g. weather). Moreover, the bus dwelling and transit time are predominantly affected by different factors and hence have different patterns, with a fact that there are not so much study on how to divide the dwelling and transit areas and to build independent models for them. In this paper, we propose an end-to-end deep learning framework for Bus Travel Time Estimation (called DeepBTTE) where the target path is of arbitrary length. Two independent spatio-temporal components that use 1D-CNN and LSTM are adopted to estimate the dwelling time and transit time separately, which are then combined for the final estimation. We conduct experiments to evaluate our model using a real-world dataset. The experimental results show that our approach significantly outperforms other existing methods.

Keywords: Bus travel time estimation · Spatio-temporal model · Deep learning

1 Introduction

With the development of urban transportation and data sensing technologies, a large amount of data is created every day within transportation, so there is a trend of using data for the Intelligent Transportation Systems (ITS) [13,14] where travel time estimation plays a critical role [5]. Through travel time

This research is in part supported by the Natural Science Foundation of China (61672441, 61872154, 61862051), the Shenzhen Basic Research Program (JCYJ20170818141325209, JCYJ20190809161603551), Natural Science Foundation of Fujian (2018J01097), Special Fund for Basic Scientific Research Operation Fees of Central Universities (20720200031).

ⓒ Springer Nature Switzerland AG 2020
Z. Huang et al. (Eds.): WISE 2020, LNCS 12342, pp. 369–383, 2020.
https://doi.org/10.1007/978-3-030-62005-9_27

estimation, it becomes possible, to reduce passenger's waiting time and bring about various ITS applications. Among different transportation modes, the bus system has the largest coverage and the lowest overall cost. Hence, bus travel and arrival time estimation are of crucial importance for the efficiency and the quality of public transportation service.

The traditional methods for predicting travel time or arrival time are based on historical data. There are research that treated the traffic trajectory as time series data, and used the Autoregressive Integrated Moving Average (ARIMA) model to predict the travel time of vehicles on the highway [4]. And Wu et al. [22] used Support Vector Machine (SVM) to predict the travel time in next time step by the data from past time steps. These methods enable relatively accurate predictions of arrival time for taxis or private cars, but are not very effective for the buses. The bus runs on a fixed route and needs to stop at various bus stations. Therefore, the dwelling time of a station has a great influence on the total travel time. Moreover, the bus dwelling and transit time are affected by different factors and hence have different patterns. Using existing methods to directly predict the travel time similar to those of taxis or private cars may lead to inaccurate results. Actually, the estimation of dwelling time and transit time of buses could be viewed as two different tasks which need further optimization and integration, yet few studies have focused on how to divide the dwelling and transit areas and how to build independent models for them [18].

To address the limitations of previous work, in this paper we propose an end-to-end framework for Bus Travel Time Estimation based on 1D-CNN and LSTM, called DeepBTTE. The primary contributions are summarized as follows:

- We use a spatio-temporal component to learn the spatio-temporal dependencies from raw GPS sequences. The component consists of two parts: 1) a convolutional layer that transforms the GPS sequences to a series of feature maps and captures the spatial dependencies from consecutive GPS points directly; 2) a recurrent layer that learns the temporal dependencies of the obtained features and the embedded external features.
- We adopt an attribute component to embed the external factors, including temperature, weather condition, day of the week, distance of the path, vehicle conditions, the habits of drivers, etc. The latent representations are fed into the model to enhance the representation capability of these external factors.
- We propose a model for the bus travel time estimation which handles arbitrary length of bus trajectories. Given any two stations p_a and p_b in bus route, the model outputs the estimation of the cost of time from p_a to p_b. Two separated components are used to estimate the transit time and the dwelling time respectively.
- We conduct extensive experiments on real-world datasets to verify the performance of the proposed model. The mean absolute percentage error on real-world dataset is 6.82%, which significantly outperforms existing methods.

The remainder of this paper is organized as follows. Section 2 describes the related works. Section 3 introduces our problem and the data preprocessing process. Section 4 shows three components of our model. Section 5 shows the experimental setup, and Sect. 6 compares the estimate performance between our model and other methods on a real-world dataset. Section 7 presents the conclusion and future work.

2 Related Work

There are a large number of studies on the travel time estimation. Various models, such as models based on historical data, statistical models, Kalman Filtering models and models based on Deep Learning have been proposed [1]. Wall et al. [20] presented a model to predict the travel time based on historical average. Chen et al. [7] used a dynamic model for bus travel time predicting. The model consists of two elements, an artificial neural network model and a Kalman Filter based model. Balasubramanian et al. [3] proposed a bus travel time prediction model which took the periodicity of data into consideration and it also used the real-time bus information to improve the prediction accuracy.

There are some researches on bus travel time prediction based on deep learning. The followings describe the travel time prediction methods for taxis or private cars used deep learning. Existing solutions could be divided into two categories, route-based methods and OD-based methods [19]. The former predicts the travel time with path route given and the latter predicts the travel time only with the origin and destination given.

Route-based methods [2,17] can be categorized as segment-based methods and sub-path-based methods. Segment-based methods treat a path as a sequence of independent road segments and then predict the travel time of each segment individually. Segment-based methods ignore the correlation between segments and it will accumulate the prediction error of each segment, which leads to inaccurate predictions. Different from segment-based methods, sub-path-based methods consider the correlation between different segments and treat a path as a set of sub-paths. For example, Zhang et al. [24] proposed a method called DEEPTRAVEL to predict the travel time based on neural network with auxiliary supervision, which used dual interval loss mechanism to optimize loss function both forward and backward. Duan et al. [8] and Liu et al. [16] predicted the travel time based on LSTM, and the results showed that LSTM has a great ability on capturing the temporal dependencies of temporal sequences. However, these two methods didn't consider the influence of weather and other external factors, so they are limited in improving the accuracy of travel time prediction.

Jindal et al. [11] proposed a model called ST-NN(Spatio-Temporal Neural Network) based on deep neural network, which can predict the travel time when only the origin and destination are given. Then the paper developed a carpool system based on reinforcement learning (RL). In ECML/PKDD Discovery Challenge 2015 competition, Lam et al. [15] estimated the travel time of taxis in real-time based on trip matching and ensemble learning without knowing the whole path. Wang et al. [21] presented a method called DeepTTE using Attribute Component, Spatio-Temporal Component and Multi-task Learning Component to capture the external features and spatio-temporal correlations individually, and finally combining them to predict travel time. In this paper, we propose an end-to-end deep learning framework to learn the pattern of bus travel time. In this way, bus travel time can be effectively estimated.

3   Preliminaries and Data Preprocessing

In this section, we briefly introduce the bus travel time prediction problem and the data preprocessing process.

3.1   Bus Travel Time Prediction Problem

The travel time of a bus are composed of the dwelling time at stations and transit time between stations. Therefore, we defined some concepts as follows.

Historical Trajectory. The trajectory of one bus from station i to station j is defined as $T = \{p_i, ..., p_j\}$, $p_i = \{p_i.lat, p_i.lng, p_i.in, p_i.out\}$, containing the latitude, longitude, inbound time and outbound time of station i. Moreover, for each trajectory T, we add *weather, temperature, carID, driverID, timeslotID* and *weekday* as external features.

Dwelling Time. The dwelling time of station i is defined as

$$p_i.dt = p_i.out - p_i.in \tag{1}$$

Transit Time. The transit time of station $i - 1$ to station i is defined as

$$p_i.tt = p_i.in - p_{i-1}.out \tag{2}$$

Arrival Time. The arrival time of station i is defined as

$$p_i.at = p_i.dt + p_i.tt \tag{3}$$

Previous Dwelling Time. The previous dwelling time of station i is the dwelling time of previous time slot (length of time slot is 10 min) at station i, which is written as

$$p_i^t.pdt = p_i^{t-1}.dt \tag{4}$$

The notations and their meanings are shown in Table 1. We first construct deep models on the estimation of the transit time and dwelling time of the given path and the corresponding external factors, based on the spatio-temporal features extracted from trajectories. Then given a departure time and any two stations p_a and p_b in bus route, the model outputs the estimation of the cost of time from p_a to p_b. We assume that the travel path from p_a to p_b is the sub-path of bus route with arbitrary path length.

Table 1. Notations and their meanings

Notation	Meaning	Notation	Meaning
T	A trajectory of a bus	p_i	Station i
$p_i.lat$	The latitude of station i	$p_i.lng$	The longitude of station i
$p_i.in$	Bus arrival time at station i	$p_i.out$	Bus departure time at station i
$carID$	Identity of the vehicle	$driverID$	Identity of the driver
$timeslotID$	The time slot of day	$weekday$	The day of week
$weather$	Weather information	$p_i.tt$	Transit time of station
$p_i.dt$	Dwelling time of station i	$p_i.at$	Arrival time of station i
$p_i.pdt$	Previous dwelling time of station i		

3.2 Data Preprocessing Process

Trajectory Data Processing. The raw bus-to-departure data includes all inbound and outbound information. Before performing feature extraction, we need to drop duplicate and invalid data. Then, group the data by date, direction(uplink or downlink) and $carID$. For each trajectory, the dwelling time and transit time of each station are calculated. We fill in the dwelling time and transit time according to two situations. In weekdays, we populate them with weekdays' average value. And on weekends, we fill them with weekends' average value. If more than 50% station data of a trajectory is lost, then drop this trajectory.

External Feature Processing. External features include weather condition, $driverID$, $carID$, $timeslotID$ and $weekday$. We fill the weather data with the latest hour data if there is any missing data. And we choose temperature and weather description as two features from weather information which have great impact on bus travel time. We also associate each trajectory with a driver according to the schedule data.

4 Model Description

As shown in Fig. 1, this model consists of three components. The attribute component is used to process the external factors (e.g. weather) and basic information (e.g. start time). The output of the attribute component will be part of the input of other components. The spatio-temporal component is the main component that learns the spatial correlations and temporal dependencies from the GPS sequences. Finally, the fusion estimation component forecasts the bus travel time of a given path based on the previous two components.

4.1 Attribute Component

The travel time of a bus is affected by many factors. For instance, in the peak hour there are many vehicles on the road and numerous passengers waiting at the station, so buses have to spend more time to finish a route. But in non-peak hour, the travel time is shorter. So the travel time is time-varying. Furthermore,

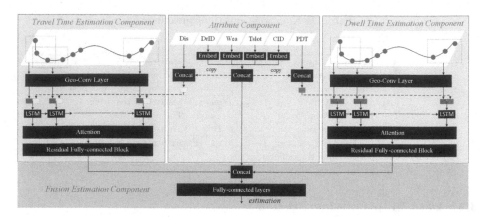

Fig. 1. Overall Framework. Dis denotes distance, DrID denotes driverID, Wea denotes weather, Tslot denotes time slot, CID denotes carID, PDT denotes previous dwelling time, Concat denotes concatenation operation.

the travel time is also affected by many external factors, like weather, driver habit, bus running information and day of week. We use a simple and efficient component to incorporate such factors into our model, which we call attribute component.

As shown in Fig. 1, we use weather, driverID, carID, weekday to represent external factors respectively, the timeslot represents the start time. These attributes are categorical values that can not be fed into the neural network directly. So we need to use the embedding method [9] to transform each categorical attribute into the low-dimensional real vector. More specifically, the embedding method maps each categorical value $x \in [V]$ to a real space $R^{E \times 1}$, where V represents the vocabulary size of categories value and E represents the dimension of embedding space. Usually, $E << V$. The embedding method effectively reduces the input dimensions and thus reduces the computational complexity.

Besides the embedded attributes, other important attributes (e.g. travel distance and dwelling time of the previous time slot) are incorporated. We use $\Delta d_{p_a} \to d_{p_b}$ to denote the total distance traveling from station p_a to p_b along the path, i.e., $\Delta d_{p_a} \to d_{p_b} = \sum_{i=a}^{b-1} Dis\,(p_i, p_{i+1})$ where Dis is the geographical distance between two bus stations. Similarly, we use $\Delta t_{p_a} \to t_{p_b}$ to denote the total dwelling time of previous time slot between p_a and p_b, i.e., $\Delta t_{p_a} \to t_{p_b} = \sum_{i=a}^{b-1} Time\,(p_i, p_{i+1})$ where $Time$ is the total dwelling time of previous time slot between two bus stations. Then, we concatenate the obtained embedded vector together with $\Delta d_{p_1} \to d_{|T|}$ and $\Delta t_{p_1} \to t_{|T|}$, respectively. We denote the concatenation as the output of attribute component.

4.2   Spatio-Temporal Component

The spatio-temporal component has two sub-components. Dwelling time estimation component is used to predict the total dwelling time of each station, and the transit time estimation component is used to predict the total transit time between stations. These two components have the same structure, they are composed of two parts. The first part is a geo-convolutional neural network which transforms raw station GPS sequences to a series of feature maps. The second part is the recurrent neural network which learns the temporal correlations of obtained feature maps from the previous part.

Geo-Conv Layer. As we mentioned before, a historical trajectory T is a sequence of GPS points $\{p_1, ..., p_{|T|}\}$ where each p_i represents one bus station which contains the corresponding longitude, latitude, arrival and departure time. Extracting spatial dependencies from GPS sequences is critical to dwelling time and transit time estimation. The convolutional neural network has been widely used for capturing spatial relationships. A typical convolutional network consists of multiple convolutional filters and the filters learn the spatial relationships in the input by applying the convolution operation. Since the spatial dependency cannot be obtained directly from GPS coordinates, Zhang et al. [25] found a way to transform the GPS coordinates to capture the spatial feature. They first partitioned the area where all GPS records are located into $I \times J$ grids and then map each GPS coordinate into a grid cell, and finally, they used 2D-CNN to extract the spatial correlation from the number of the grid cell. However, in our case, directly mapping the GPS coordinates into grid cells is unsuitable. Firstly, the GPS coordinates of each bus station are fixed and limited, and they don't need to be mapped. Secondly, if two stations are mapped to the same cell we can't distinguish them. Thus, we use a Geo-Conv layer which can capture the spatial dependency in geo-location sequences while retains the location information.

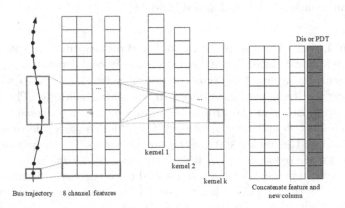

Fig. 2. Illustration of Structure of the Geo-Conv Layer. Dis denotes distance, PDT denotes previous dwelling time.

The architecture of Geo-Conv layer is shown in Fig. 2. We first use a non-linear mapping to transform each station GPS point p_i in the sequence into vector. $loc_i \in R^8$.

$$loc_i = tanh\left(W_{loc} \cdot [p_i.lat \circ p_i.lng]\right) \tag{5}$$

where $\circ$ denotes the concatenate operation and W_{loc} is a learnable weight matrix. Thus, the output sequence $loc \in R^{8 \times |T|}$ indicates the non-linear mapping for each GPS point. loc_i can be seen as an 8-channel input. Each channel represents the geographical features of the original GPS sequence. We use a 1D-CNN [21], with parameter matrix $W_{conv} \in R^{k \times 8}$, to extract spatial features where k is the kernel size of a filter. It applies the convolutional operation on the sequence loc, along with a one-dimensional sliding window. The i-th dimension of its output is denoted as:

$$loc_i^{conv} = \sigma\left(W_{conv} * loc_{i:i+k-1} + b\right) \tag{6}$$

where $*$ denotes the convolutional operation, b is the bias term, $loc_{i:i+k-1}$ is the subsequence of loc from i to $i + k - 1$ and the σ is an activation function.

The sub-sequence from station p_i to p_{i+k-1} is defined as the i-th local path. The loc_i^{conv} represents the spatial feature of the i-th local path. We get the feature map of local paths with shape $R^{c \times (|T|-k+1)}$, where c is the number of filters.

However, in our task, the transit time and dwelling time are highly correlated with the total distance of the path and the total dwelling time of previous time slot, respectively. We can't extract the geometric distance and previous dwelling time from latitude/longitude through 1D-CNN. Therefore, we further append a column to the previously obtained feature map in Geo-Conv layer. As shown in Fig. 2, in transit time estimation component, the i-th element of new appended column is the distance of the i-th local path, i.e., $\sum_{j=i+1}^{j+k-1} Dis\left(p_{j-1}, p_j\right)$. Similarly, in dwelling time estimation component, the i-th element of new column is the previous dwelling time of the i-th local path, i.e, $\sum_{j=i+1}^{j+k-1} Time\left(p_{j-1}, p_j\right)$. Finally, we obtain the feature map of shape $R^{(c+1) \times (|T|-k+1)}$ by Geo-Conv layer. These feature maps are denoted as loc^d and loc^t.

Recurrent Layer. The feature maps loc^d and loc^t capture the spatial dependencies from all local paths. To further capture temporal dependencies, we introduce the recurrent layers. The recurrent neural network (RNN) is widely used for capturing the temporal dependency in sequential learning. The RNN can 'memorize' the history in the processed sequence. When processing sequences in each time step, it updates its memory (i.e., hidden state) according to the current input and the hidden state of the previous step. The output of the RNN is the hidden state sequence of the input sequence at all time steps.

In both sub-components of spatio-temporal component, the feature map (i.e., loc^d and loc^t) by the Geo-Conv layer is set as the input of the recurrent neural networks. As mentioned before, the length of feature map is $|T|-k+1$. To further improve the estimating ability of recurrent layers, we incorporate the attributes' information (i.e., $attr^d$ and $attr^t$) obtained from the attribute component. A simple updating rule of the RNN can be expressed as:

$$h_i = \sigma \left(W_x \cdot loc_i^f + W_h \cdot h_{i-1} + W_a \cdot attr \right) \tag{7}$$

where h_i represents the hidden state after the processing of the i-th local path and σ_{rnn} is the activation function. W_x, W_h and W_a are learnable parameters' matrix of spatial features. However, the RNN usually fails when processing the long sequence due to the vanishing gradient and exploding gradient problem [10]. To overcome this issue, we use Long Short-term Memory [10] layer to replace it. Compared with RNN, the LSTM uses the input gate and forget gate to control the in/out information flow. Such gates enable LSTM to retain important information and filter out the unimportant information which effectively mitigates the gradient vanishing/exploding problem.

Now, we obtain the spatio-temporal sequence $\{h_1, h_2, ..., h_{|T|-k+1}\}$ from raw data by utilizing the Geo-Conv layer and the recurrent layer.

Estimation. Since the length of each bus trajectory is variable, the length of spatio-temporal feature sequence $\{h_i\}$ is also variable. To estimate the transit time/dwelling time of the entire path directly, we need an attention model that can calculate sets of hidden states of different sizes. The attention mechanism is essentially the weighted sum of sequence $\{h_i\}$, where the weights are parameters learned by model. Thus, we have that:

$$h_{att} = \sum_{i=1}^{|T|-k+1} \alpha_i \cdot h_i \tag{8}$$

where α_i is the weight for the i-th local path, and the summation of all weights equals to 1. In our case, the weight of each local path is related to the spatial information of the local paths, as well as external features such as start time slot, the weekday and the weather condition. The attention mechanism can be expressed as:

$$z_i = \langle \sigma_{att}(attr), h_i \rangle$$
$$\alpha_i = \frac{e^{z_i}}{\sum_j e^{z_j}} \tag{9}$$

where $attr$ represents the external factors captured by attribute component, $\{h_i\}$ represents the spatial-temporal feature of local path, $\langle \cdot \rangle$ represents the inner operator and the σ_{attr} is a non-linear mapping which maps $attr$ to a vector with the same length as h_i. Substituting Eq. (9) into Eq. (8), we obtain the attention based vector h_{attr}.

Finally, we pass h_{attr} to several Residual fully-connected layers with equal size. In our model, we use σ_{f_i} to denote the i-th layer. For the first layer, the output of this layer is $\sigma_{f_1}(x)$. For the rest of the residual fully-connected layers, we use x to denote the output of i-th layer. Then, the output of the $(i+1)$-th express as $x \oplus \sigma_{f_{i+1}}(x)$ where the $\oplus$ is the element-wise add operation. At the last layer, we use a single neuron to obtain the estimation.

4.3 Fusion Estimation Component

In fusion estimation component, we combine the previous components and estimate the total arrival time of the input path:

$$total = t_{\hat{transit}} \circ t_{\hat{dwell}} \circ attr \qquad (10)$$

where $total$ represents the input of fusion estimation component, $t_{\hat{transit}}$ represents the estimation of transit time, $t_{\hat{dwell}}$ represents the estimation of dwelling time, and $attr$ represents the embedded vector of external features. We pass $total$ to several fully-connected layers. At the last layer, we use a single neuron to obtain the estimation.

5 Experimental Setup

5.1 Data Description

We evaluated our model on a real-world dataset. This experiment is based on the bus-to-departure data and the corresponding bus schedule data of Xiamen 22 bus from September 1, 2018, to February 28, 2019. The bus-to-departure data contains 301,130 records. After data preprocessing, as mentioned in Data Preprocessing Process, we obtain 9523 trajectories information. The route map of 22 bus is shown in Fig.3, where the total route length is about 14 km. Due to the departure interval of 22 bus is 10 min, we set the length of time slot as 10 min. Each trajectory information contains the longitude, latitude, dwelling time, transit time of each station. Finally, we intercept arbitrary-length paths from complete trajectory information. The shortest trajectory contains 10 stations and the longest trajectory contains 25 stations (Number of bus station on Route 22).

The trajectories in Xiamen dataset are associated with the corresponding weekday, timeslot, driverID, and carID.

5.2 Parameter Setting

The parameters we used in our experiment are described as follows:

- In attribute component, we embed driverID to R^3, carID to R^3, weekID to R^3, timeslotID to R^4, temperature to R^3 and corresponding weather condition to R^3.
- We set the same parameters of Geo-conv for both sub-components of spatio-temporal component. The number of filters $c = 32$ and we use the ELU as the activation function. The ELU is defined as $ELU(x) = e^x - 1$ for $x \le 0$ and $ELU(X) = x$ for $x > 0$. For kernel size k, which represents the length of local path, we set the size of k to 3 to evaluate our model.
- We set the same parameters of the recurrent layer for both sub-components. We use $tanh$ as the activation function and set the size of the hidden vector as 128.

Fig. 3. Route of No. 22 Bus in Xiamen City.

- In fusion estimation component, the ReLU function is as the activation function of the fully-connected layers. We set the number of the fully-connected layers as 2 and the size of each layer as $[64, 32]$.

In the Xiamen dataset, we divided the trajectories generated in February 2019 into two parts, one for evaluation and the other for test. Except for the trajectories generated in February 2019, the rest of trajectories are used as the training set. We adopt an Adam optimization algorithm to train the parameters. The learning rate of Adam is le^{-4}, the batch size during training is 64 and the epochs of training are 50.

Our model is implemented with PyTorch 1.1.0, a widely used Deep Learning Python library. We train/test our model on the server with one NVIDIA TITAN X GPU.

6 Results and Analyses

6.1 Performance Comparison

Because there is little research on bus arrival time estimation, to demonstrate the strength of our model, we use Root Mean Square Error (RMSE) [6], Mean Absolute Error (MAE) [6], and Mean Absolute Percentage Error (MAPE) [12] as metrics to compare our model with other baseline methods, including:

- **AVG:** AVG simply calculates the average transit time and dwelling time of each station on the training set, and estimates the arrival time of given trajectory based on the average transit time/dwelling time calculated on training set.

- **GBDT:** Gradient Boosting Decision Tree (GBDT) [23] is a powerful ensemble method. In our case, the input of GBDT is the same as the input of our model, including all the inputs of attribute component and the raw GPS sequences. We use a GBDT to estimate the arrival time. Because of the lengths of GPS sequences are variable, GBDT can't handle the sequences directly. Thus, we transform each GPS sequence to fixed length of 25.
- **MLPBTTE:** Multi Layer perception, a 5-layer Perception is used to estimate the time of arrival. The activation function of Mlp is ReLU, the input of Mlp is the concatenation of GPS sequences and embedded external features. The size of hidden layers in Mlp is fixed as 128.
- **Single Spatio-Temporal Model:** In the case of not estimating the dwelling time and transit time separately, a single spatio-temporal component is used to estimate the bus arrival time. We concatenate the output of spatio-temporal component and attribute component and the concatenation is passed to a residual fully-connected layer to obtain the estimation.
- **RnnBTTE:** RnnBTTE is a simplified model of our model. We replace the LSTM in the recurrent layer with RNN and use mean pooling to process the output of the recurrent layer into a 128-dimensional feature vector. We concatenate the feature vector and the output of attribute component. Finally, the concatenation is passed to the fusion estimation component to obtain the estimation.

Table 2. Performance comparison

	$RMSE$	MAE	$MAPE$
AVG	602.3859	423.2324	25.7433
GBDT	506.1589	389.7922	21.6339
MlpBTTE	281.6376	200.4457	10.3773
RnnBTTE	195.2742	144.9747	7.5667
DeepBTTE	**172.5082**	**127.8042**	**6.8258**
Single Spatio-Temporal Model	203.4458	147.6104	7.7000

As shown in Table 2, simply using the average transit time and dwelling time of each station leads to a very inaccurate result. The ensemble method GBDT is much better than AVG. MlpBTTE shows a better performance than AVG and GBDT. However, it does not consider the spatio-temporal dependencies in data. The RnnBTTE uses the RNN in recurrent layers to capture the spatio-temporal dependencies in data. It achieves 7.57% on Xiamen dataset which is much better than above mentioned methods. Our DeepBTTE model further significantly outperforms RnnBTTE and other methods. The mean absolute percentage error of Xiamen dataset is only 6.82%.

6.2   Effect of Separate Estimation

We use single spatio-temporal model to estimate bus arrival time. Compared with DeepBTTE, single spatio-temporal model makes an overall estimation of arrival time. As shown in Table 2, Our DeepBTTE model outperforms single spatio-temporal model. More specifically, the bus dwelling time and transit time are affected by different factors and hence have different patterns. In our model, we use two independent components to estimate dwelling and transit time separately which can enhance the estimation performance.

6.3   Effect of External Factors

To evaluate the effectiveness of different external factors, We devise a series of controlled experiments on Xiamen dataset. For each experiment, we eliminate one attribute. As shown in Table 3, we find that weekdays affect the estimation significantly. This conforms to reality that the traffic condition of 22 bus line is completely different on weekdays and weekends. The average arrival time on weekends is longer than that of weekdays. Eliminating the weather causes an error growth of 0.47%. This also confirms to reality that the average arrival time will be longer under bad weather condition. Eliminating the driver and vehicle information have little effect on the results.

Table 3. Effect of different attributes

	$RMSE$	MAE	$MAPE$
Eliminate carID	175.0516	131.7047	6.9885
Eliminate driverID	173.0027	128.1601	6.8483
Eliminate weekday	194.3426	145.1981	7.5775
Eliminate weather	181.0990	136.6376	7.3041
Without eliminate	**172.5082**	**127.8042**	**6.8258**

6.4   Training and Predicting Time

Table 4 shows the training and predicting time of each method (i.e., AVG, GBDT, MlpBTTE, RnnBTTE and DeepBTTE), the GPU we used to train our model is a TITAN X. Despite the training time of RnnBTTE and DeepBTTE are longer, it is acceptable in offline training. In practical applications, we can obtain a lot of offline resources for pre-training. During the test phase, estimating about 1000 paths takes DeepBTTE 0.0015 s, RnnBTTE 0.0018 s, MlpBTTE 0.0005 s on GPU.

Table 4. Comparative training time and testing time

	Training time (s)	Predicting time (s)
GBDT	3505	0.0132
MlpBTTE	161	0.0005
RnnBTTE	306	0.0018
DeepBTTE	355	0.0015
Single Spatio-Temporal Model	231	0.0016

7   Conclusion and Future Work

In this paper, we study the problem of estimating bus travel time for any given path in one route. We propose an end-to-end framework based on 1D-CNN and LSTM network. This model can effectively capture the spatio-temporal dependencies from raw GPS sequences. Our model also considers various external factors such as weather condition and time which may affect the arrival time. We evaluate our model on a real-world dataset. The results show that our method achieves a high estimation accuracy and outperforms the other methods significantly. For future work, we will enhance the performance of our model and add additional data as model inputs, such as GPS points of intersections and traffic lights and the number of passengers getting on and off at each station.

References

1. Altinkaya, M., Zontul, M.: Urban bus arrival time prediction: a review of computational models. Int. J. Recent Technol. Eng. (IJRTE) **2**(4), 164–169 (2013)
2. Asif, M.T., et al.: Spatiotemporal patterns in large-scale traffic speed prediction. IEEE Trans. Intell. Transp. Syst. **15**(2), 794–804 (2013)
3. Balasubramanian, P., Rao, K.R.: An adaptive long-term bus arrival time prediction model with cyclic variations. J. Public Transp. **18**(1), 6 (2015)
4. Billings, D., Yang, J.S.: Application of the ARIMA models to urban roadway travel time prediction - a case study. In: IEEE International Conference on Systems, Man and Cybernetics, SMC 2006 (2006)
5. Birr, K., Jamroz, K., Kustra, W.: Travel time of public transport vehicles estimation. Transp. Res. Proc. **3**, 359–365 (2014)
6. Chai, T., Draxler, R.R.: Root mean square error (RMSE) or mean absolute error (MAE)? - arguments against avoiding RMSE in the literature. Geosci. Model Dev. **7**(3), 1247–1250 (2014)
7. Chen, M., Liu, X., Xia, J., Chien, S.I.: A dynamic bus-arrival time prediction model based on APC data. Comput.-Aided Civ. Infrastruct. Eng. **19**(5), 364–376 (2004)
8. Duan, Y., Lv, Y., Wang, F.Y.: Travel time prediction with LSTM neural network. In: 2016 IEEE 19th International Conference on Intelligent Transportation Systems (ITSC), pp. 1053–1058. IEEE (2016)
9. Gal, Y., Ghahramani, Z.: A theoretically grounded application of dropout in recurrent neural networks. In: Advances in Neural Information Processing Systems, pp. 1019–1027 (2016)

10. Hochreiter, S., Schmidhuber, J.: Long short-term memory. Neural Comput. **9**(8), 1735–1780 (1997)
11. Jindal, I., Qin, Z.T., Chen, X., Nokleby, M., Ye, J.: Optimizing taxi carpool policies via reinforcement learning and spatio-temporal mining. In: 2018 IEEE International Conference on Big Data (Big Data), pp. 1417–1426. IEEE (2018)
12. Khair, U., Fahmi, H., Hakim, S.A., Rahim, R.: Forecasting error calculation with mean absolute deviation and mean absolute percentage error. J. Phys.: Conf. Ser. **930**, 012002 (2017)
13. Lai, Y., Lv, Z., Li, K.C., Liao, M.: Urban traffic coulomb's law: a new approach for taxi route recommendation. IEEE Trans. Intell. Transp. Syst. **20**, 3024–3037 (2018)
14. Lai, Y., Yang, F., Zhang, L., Lin, Z.: Distributed public vehicle system based on fog nodes and vehicular sensing. IEEE Access **6**, 22011–22024 (2018)
15. Lam, H.T., Diaz-Aviles, E., Pascale, A., Gkoufas, Y., Chen, B.: (Blue) taxi destination and trip time prediction from partial trajectories. arXiv preprint arXiv:1509.05257 (2015)
16. Liu, Y., Wang, Y., Yang, X., Zhang, L.: Short-term travel time prediction by deep learning: a comparison of different LSTM-DNN models. In: 2017 IEEE 20th International Conference on Intelligent Transportation Systems (ITSC), pp. 1–8. IEEE (2017)
17. Lv, Y., Duan, Y., Kang, W., Li, Z., Wang, F.Y.: Traffic flow prediction with big data: a deep learning approach. IEEE Trans. Intell. Transp. Syst. **16**(2), 865–873 (2014)
18. Ma, J., Chan, J., Ristanoski, G., Rajasegarar, S., Leckie, C.: Bus travel time prediction with real-time traffic information. Transp. Res. Part C: Emerg. Technol. **105**, 536–549 (2019)
19. Veres, M., Moussa, M.: Deep learning for intelligent transportation systems: a survey of emerging trends. IEEE Trans. Intell. Transp. Syst. **21**, 3152–3168 (2019)
20. Wall, Z., Dailey, D.: An algorithm for predicting the arrival time of mass transit vehicles using automatic vehicle location data. In: 78th Annual Meeting of the Transportation Research Board, pp. 1–11. Citeseer (1999)
21. Wang, D., Zhang, J., Cao, W., Li, J., Zheng, Y.: When will you arrive? Estimating travel time based on deep neural networks. In: Thirty-Second AAAI Conference on Artificial Intelligence (2018)
22. Wu, C.H., Ho, J.M., Lee, D.T.: Travel time prediction with support vector regression. Intell. Transp. Syst. **5**, 276–281 (2003)
23. Lai, Y., Yang, X., Cao, Q., Cao, H., Wang, T., Yang, F.: A bus running length prediction method based on gradient boosting. Big Data Res. **5**, 58–78 (2019)
24. Zhang, H., Wu, H., Sun, W., Zheng, B.: DeepTravel: a neural network based travel time estimation model with auxiliary supervision. arXiv preprint arXiv:1802.02147 (2018)
25. Zhang, J., Zheng, Y., Qi, D., Li, R., Yi, X.: DNN-based prediction model for spatio-temporal data. In: Proceedings of the 24th ACM SIGSPATIAL International Conference on Advances in Geographic Information Systems, p. 92. ACM (2016)

TraSP: A General Framework for Online Trajectory Similarity Processing

Zhicheng Pan[1], Pingfu Chao[2], Junhua Fang[1(✉)], Wei Chen[1], Zhixu Li[1], and An Liu[1]

[1] Department of Computer Science and Technology,
Soochow University, Suzhou, China
zcpan28@stu.suda.edu.cn, {jhfang,robertchen,zhixuli,anliu}@suda.edu.cn
[2] The University of Queensland, Brisbane, Australia

Abstract. Trajectory similarity is one of the most fundamental operations in spatial-temporal data analysis. Although many recent works focus on improving the efficiency on single machine, their solutions are not directly applicable to DSPEs (*Distributed Stream Processing Engine*) in an online manner. On one hand, the similarity processing on DSPEs is always susceptible to data skew and completeness issue. On the other hand, their methods only support a single trajectory similarity measure which could not serve for adaptive adjustment strategies in different scenario. In this paper, we propose a new general framework for online *Trajectory Similarity Processing*, named TraSP. Specifically, our proposal includes a matrix-based data dispatcher to provide balance and completeness guarantee for stream join, an atomic table generator to accommodate different similarity criteria and a lightweight filter to shed irrelevant workloads. Empirical studies on real world trajectory data sets validate the usefulness of our proposals and the comparison experiment shows the high performance of our framework.

Keywords: Trajectory similarity · Distributed stream join · Load balance · Online processing

1 Introduction

The proliferation of trajectory data in various application domains has inspired tremendous research efforts to analyze large scale trajectory data [13]. A fundamental ingredient of these analysis tasks and applications are similarity processing for effectively determining how similar two trajectories are. In the meantime, more and more applications nowadays rely on online analysis, which require the calculations to be done once the trajectories are generated, such as the peer monitoring during the epidemic and ride-sharing recommendation. Therefore, it is necessary to handle the large-scale trajectory similarity in an online scenario.

As a viable platform for such task, the DSPE is designed for massive stream data processing with low latency. However, the traditional solutions for trajectory similarity cannot be directly extended to DSPE systems due to two main

© Springer Nature Switzerland AG 2020
Z. Huang et al. (Eds.): WISE 2020, LNCS 12342, pp. 384–397, 2020.
https://doi.org/10.1007/978-3-030-62005-9_28

reasons. First, the existing solutions mainly focus on how to improve the performance on single-machine, which are unable to guarantee the computation correctness in a parallel shared-nothing environment. This is because the trajectory similarity is a stateful operation where distributed environment affects the order and integrity of trajectory data. Second, the workload skew and variance in distributed environment hinders the existing centralized solutions to achieve competitive performance. When massive stream data flood into a distributed system for processing and analysis, even slight distribution change on the incoming data stream may significantly affect the system performance.

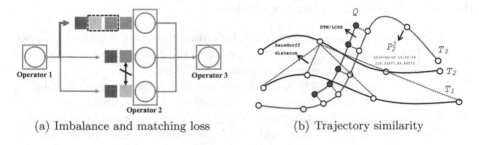

(a) Imbalance and matching loss   (b) Trajectory similarity

Fig. 1. Challenges and trajectory concepts (Color figure online)

Motivation. Although several distributed or parallel solutions [4,6,11,16,17] are proposed for large-scale trajectory similarity, they cannot be directly adapted to the online processing where an accurate and low-latency similarity result needs to be obtained. Besides, they are specific to single similarity measure and often has its disadvantages in special scenarios.

Challenges. Based on our observations, we divide the challenges of solving this problem into two categories. *natural:* As shown in Fig. 1(a), there should be two complete matching of blue data and yellow data, but a match is missing due to the distribution strategy. In addition, the load of the first task node is twice that of other task nodes. This is the inherent data imbalance and matching loss in DPSEs. *purposeful:* To illustrate the *purposeful* challenge, we give a simple example in Fig. 1(b). There are four trajectories, and the similarity measure of trajectory trajectories Q and T_3 is suitable for DTW/LCSS. The trajectory similarity measure between T_1 and T_2 is more suitable for *Hausdorff distance*. Note that this does not mean using different measures on the same data set. Such measure differentiation problems are very common.

Our Goal. Our proposal should have the following characteristics: (1) *Robust:* our framework should be robust to data skew of varying degrees; (2) *Accurate:*

our solution should guarantee the computation correctness in a parallel shared-nothing environment; (3) *Efficient:* our framework should achieve competitive performance regardless of throughput, latency and data shedding; (4) *Generic:* our framework should be generic to accommodate different similarity criteria. Besides, the processing should be implemented in an online scenario.

Contribution. We propose TraSP, a general framework for online trajectory similarity processing, which uses a matrix-based partition scheme. The proposal achieves competitive performance, by allowing trajectory similarity processing in a general way and facilitating more accurate and adaptive stream join. Overall, the major contributions of this work include:

(1) We propose a new general framework for online trajectory similarity processing. The proposed framework thus provides the balance and completeness guarantee in the processing of trajectory similarity.
(2) We devise a matrix-based partition scheme and propose a generic solution to accommodate different existing trajectory similarity criteria.
(3) We design a lightweight filtering mechanism. The strategy supports the distributed *Hausdorff distance* perfectly, which discards the partial matches that have no contributions to the final result.
(4) We implement TraSP on top of *Apache Flink* and conduct an empirical evaluation by using a real-world trajectory data set to measure its performance.

The remainder of this paper is organized as follows. In Sect. 2, we define relevant concepts for trajectory and stream join problems, then we formalize the goal of TraSP. Section 3 describes the details of our TraSP and instantiates it with three similarity criteria. Then, Sect. 4 discusses the experimental study with real-world data sets. The related work is introduced from three aspects in Sect. 5. Finally, Sect. 6 concludes the paper with remarks on future work (Table 1).

Table 1. Summary of notations

Notation	Description
P_ε^{id}	The εth point of T_{id}
$T_{id} = \langle t_1, t_2, \cdots, t_m \rangle$	Trajectory T_{id}
$Q_{id} = \langle q_1, q_2, \cdots, q_n \rangle$	Trajectory Q_{id}
$\mathcal{T} = \langle T_1, T_2, \cdots, T_N \rangle$	Data set consisting of N trajectories
m_{ij}	Node/cell of ith row jth column
$\mathcal{H}$	Trajectory similarity measure criterion
$\mathcal{A}$	Trajectory-to-trajectory table
$\mathcal{D}$	Similarity value

2   Preliminaries

2.1   Trajectory Similarity

Definition 1 *(Sample Point). A sample point is obtained from GPS devices and each point has four attributes, which are expressed as a 4-dimensional vector: $P_\varepsilon^{id} = (id,\ longitude,\ latitude,\ timestamp)$.*

Definition 2 *(Trajectory). A trajectory is a sequence of m sample points, denoted as: $T_{id} = \langle t_1, t_2, \cdots, t_m \rangle$ and the trajectory set consisting of N trajectories is represented as $\mathcal{T} = \langle T_1, T_2, \cdots, T_N \rangle$.*

Definition 3 *(Trajectory Similarity Query). Given a set of trajectories $\mathcal{T}$, a similarity criterion $\mathcal{H}$, and a distance threshold ϑ, the trajectory similarity query for the query trajectory Q is defined as $\Gamma(Q, \mathcal{T}, \mathcal{H}, \vartheta)$. The query result is a set $\mathcal{G} \in \mathcal{T}$, s.t., $\forall T_i \in \mathcal{G}$, $\mathcal{H}(T_i, Q) \leqslant \vartheta$.*

We use Fig. 1(b) to understand these definitions. $p_2^3 = (3,\ 116.32977,\ 39.89573,\ 2020\text{-}02\text{-}02\ 13{:}32{:}16)$. T_1 is a sequence of 4 points. It also describes a query $\Gamma(Q, \mathcal{T}, DTW, \vartheta)$ where $\mathcal{T} = \langle T_1, T_2, T_3 \rangle$.

Based on these definitions, we introduce an example to illustrate the diversity of trajectory similarity criterion. As shown in Fig. 2, there are two example trajectories and the corresponding point-to-point distance table.

	P_1^3	P_2^3	P_3^3	P_4^3	P_5^3	P_6^3
P_6^1	5.66	4.12	2.24	1	1.41	1
P_5^1	5	4	2	0	1	1.41
P_4^1	4.24	3	1	1	2	2.24
P_3^1	2.24	1.41	1.41	3.16	4.12	4.47
P_2^1	1	3.16	3.16	4.24	5	5.66
P_1^1	0	3	3.61	5	5.83	6.40

Notation	Trajectory
T_1	(1, 1), (1, 2), (3, 2), (4, 4), (4, 5), (5, 5)
T_3	(1, 1), (4, 1), (4, 3), (4, 5), (4, 6), (5, 6)

Fig. 2. Example trajectory and point-to-point distance table

Table 2 describes several trajectory similarity criteria used in this paper and their calculated values between T_1 and T_3 in this example. We give the example calculation process of *Hausdorff distance* and DTW in Fig. 3. For *Hausdorff distance*, we traverse each point from T_1 and keep the shortest distance from each point to T_3 (framed in blue). Then, the maximum value from them is $D_H(T_1, T_3) = 1.41$. For DTW, we use a matrix (w) to store distance values, where $w_{i,j}$ represents $dist(t_i, q_j)$. DTW$(T_1, T_3) = w_{1,1} + w_{2,1} + w_{3,2} + w_{4,3} + w_{5,4} + w_{5,5} + w_{6,6} = 5.41$. In a word, Fig. 3(a) is a *maxmin* algorithm and Fig. 3(b) is a dynamic programming algorithm. Based on these, we found that different trajectory similarity criteria on the same trajectory data set are completely different in computational complexity and specific results.

Table 2. Different similarity measures yield different values on the same data

Criterion	Description	Example
Hausdorff	$D_H(T,Q) = \max\left\{ \max\limits_{t_i\in T}\min\limits_{q_j\in Q} d(t_i,q_j),\ \max\limits_{q_j\in Q}\min\limits_{t_i\in T} d(q_j,t_i)\right\}$	$D_H(T_1,T_3) = 1.41$
DTW	$DTW(T,Q) = \min\left\{ \frac{1}{K}\left[\sum_{k=1}^{K} w_k\right]^{\frac{1}{2}}\right\}$	$DTW(T_1,T_3) = 5.41$
LCSS	$LCSS_{\gamma,\delta}(T,Q) = \max\left\{L_{common}\right\}_{\gamma,\delta}$	$LCSS_{2,5}(T_1,T_3) = 3$

	P_1^3	P_2^3	P_3^3	P_4^3	P_5^3	P_6^3
P_6^1	5.66	4.12	2.24	1	1.41	1
P_5^1	5	4	2	0	1	1.41
P_4^1	4.24	3	1	1	2	2.24
P_3^1	2.24	1.41	1.41	3.16	4.12	4.47
P_2^1	1	3.16	3.16	4.24	5	5.66
P_1^1	0	3	3.61	5	5.83	6.40

(a) Hausdorff distance

	P_1^3	P_2^3	P_3^3	P_4^3	P_5^3	P_6^3
P_6^1	18.14	13.54	7.65	4.41	4.83	5.41
P_5^1	12.48	9.41	5.41	3.41	4.41	5.83
P_4^1	7.48	5.41	3.41	4.41	6.41	8.65
P_3^1	3.24	2.41	3.83	6.99	11.11	15.59
P_2^1	1	3.16	6.16	10.40	15.40	21.06
P_1^1	0	3	6.61	11.61	17.44	23.84

(b) DTW

Fig. 3. Example trajectory similarity measures (Color figure online)

2.2 Stream Join

The problem of stream join is to identify all pairs of tuples from stream R and stream S meeting the user specified join predicate, denoted by $R \bowtie S$. In this paper, we use a matrix model to process web-scale trajectory stream join.

The matrix model splits $R \bowtie S$ into several smaller parallel join processing units which are the cells in the matrix decided by rows and columns. In Fig. 4(a), e.g., stream R (S respectively) has two substreams r_0 and r_1 (s_0 and s_1 respectively). The join computation between R and S is decomposed into four tasks, each of which takes a pair of substreams from R and S and calculates the join result over these substreams.

Partition Scheme. As shown in Fig. 4(b), the specific partition scheme is employed to randomly split the incoming data stream into a number of non-overlap substreams. By randomly dividing stream R into $|R|$ nonoverlap substreams (i.e., $R = r_0 \cup \ldots \cup r_{|R|-1}$) and stream S into $|S|$ non-overlap substreams (i.e., $S = s_0 \cup \ldots \cup r_{|S|-1}$), the complete join result is the union of join results on all pairs of substreams(i.e., $R \bowtie S = \bigcup_{ij} r_i \bowtie s_j$).

2.3 Framework Overview

We make a simple overview of TraSP. Specifically, We divide TraSP's workflow into three phases, which is shown in Fig. 5.

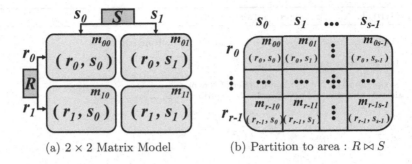

(a) 2 × 2 Matrix Model   (b) Partition to area : $R \bowtie S$

Fig. 4. An introduction to matrix-based partition scheme

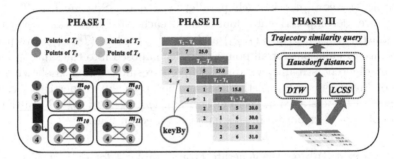

Fig. 5. Overview workflow of TraSP

3   TraSP

3.1   Matrix-Based Partitioning

In the first phase, we implement the partition scheme which is discussed in Sect. 2.2. But in the implementation, key-based map operations cannot complete such strategy naturally. So we use a list to store the set of task nodes that the current substream should reach, which is generated by a random factor.

Figure 5 illustrates a specific example of the partition scheme, and we can see that it generates a complete cartesian product match with the support of such strategy. The whole process is described as Algorithm 1, and finally stream T is randomly routed through the *keyBy* operator in line 6.

3.2   Atomic Table Generation

After the mapping operation of the trajectory data, we process the distance-join within the shared-nothing task node. We aim to generate a state of the trajectory pair consisting of the trajectory pair and its corresponding distance, i.e., *trajectory-to-trajectory table* (Atomic Table). The process can be decomposed into the following two steps.

Algorithm 1. Matrix-based partitioning.

Input:
    Trajectory data stream $\mathcal{T}$ (generated by data set), default matrix size λ
Output:
    Updated trajectory data stream $\mathcal{T}_u$;
1: **for** P in $\mathcal{T}$ **do**
2:     $\mathcal{L} = random(\lambda)$; /* A random number of row or column */
3:     Calculate the assigned list of nodes L according to $\mathcal{L}$;
4:     $P \leftarrow < P, L >$;
5: **end for**
6: $\mathcal{T}_u \leftarrow \mathcal{T}.keyBy(L.next())$;
7: **return** $\mathcal{T}_u$;

Step1: Complete Table. The input stream $\mathcal{T}_{row}$ and $\mathcal{T}_{col}$ generate a complete table of five-tuples in each parallel task node. Specifically, $(T_1, T_2) \bowtie (T_3, T_4)$ are decomposed into four tasks, each of which takes a pair of substreams($p_1^1, p_2^1, p_3^2, p_4^2$ in (T_1, T_2) and $p_5^3, p_6^3, p_7^4, p_8^4$ in (T_3, T_4)). We process the distance calculations for all points that decomposed into the same task node and from different streams, yielding complete results, which is shown in the middle of Fig. 6. For example, the node in the upper left corner is the calculation result of the distance of $(p_1^1, p_5^3), (p_1^1, p_6^3), (p_3^2, p_5^3), (p_3^2, p_6^3)$. Note that the example of data distribution in this paper is based on Fig. 5.

Step2: Atomic Table. We generate atomic table $\mathcal{A}$ for each trajectory pair by aggregating the complete table with the trajectory number as key. e.g., the $T_1 - T_3$ table is based on the key (T_1, T_3). Since such table respectively matches the trajectories, triples consisting of pairwise point-matching and their distance become the elements of $\mathcal{A}$. Based on this, we can shed redundant calculation in subsequent operator for trajectory similarity as it reduce the number of candidates for trajectory similarity calculation. Besides, the atomic table can provide a generic solution to accommodate any existing similarity criteria based on the alignment between trajectories. This is because table-based solution can be easily adapted to similarity criteria.

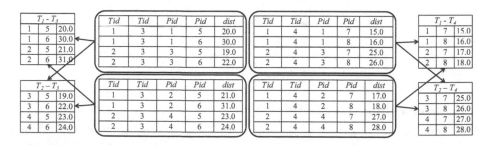

Fig. 6. The result of the join is generated as a table and matched against the corresponding trajectories, e.g., the table $T_1 - T_3$ has all the information needed to calculate the similarity between T_1 and T_3.

3.3 Generic Solution

Trajectory Alignment. Trajectory similarity calculation in a distributed environment usually encounters such difficulties that the alignment between trajectories must be restrict to the ordering of the trajectory segments [16]. To make the matter worse, calculating similarity measures between two trajectories is often expensive to restore the ordering, which may potentially cause a heavy straggler problem. In TraSP, we assign each sample point an arrival number on entering the framework, e.g., in the five-tuple table generated in the previous section, Pid reflects the ordering.

Generic Similarity Processing. As shown in Algorithm 2, we generate a two-dimensional matrix $\mathcal{M}$ from $\mathcal{A}$ to prepare for subsequent dynamic programming process (which is adopted by LCSS or DTW) and final similarity calculation.

Algorithm 2. Alignment-based similarity calculation on $\mathcal{A}$

Input:
  Table $\mathcal{A}$, parameters of $LCSS$ γ, δ;
Output:
  The similarity value of the two trajectories $\mathcal{D}$
1: Define a matrix $\mathcal{M}$ for subsequent DP
2: $\mathcal{M} = generation(\mathcal{A})$;
3: Select the similarity criterion $\mathcal{H}$, DTW goes to line 4 and $LCSS$ goes to line 5;
4: $\mathcal{D} \leftarrow DTW(\mathcal{M})$;
5: $\mathcal{D} \leftarrow LCSS(\mathcal{M}, \gamma, \delta)$;
6: **return** $\mathcal{D}$;

3.4 Filtering Mechanism

Based on the *maxmin* characteristics of *Hausdorff distance*, we propose a concise filtering mechanism to greatly improve the overall performance of our framework. Such filtering mechanism is described in Algorithm 3. In each parallel node, we only transmit the minimal distance between P_T and P_Q to the next task.

In Fig. 7, we use a specific example to introduce this filtering technique. In the transformation from complete table to atomic table, we discard the transmission of eight tuples. This is because the Hausdorff distance only focuses on the minimal distance from the point of interest to the other trajectory. So we circumvent half the calculation in this example. However, in practical applications, if P is matched to $\mathcal{Q}_N$, we will discard the transmission of $N - 1$ tuples, which will result in considerable performance optimization.

Algorithm 3. Hausdorff distance with a filtering mechanism

Input:
  Trajectory data stream $\mathcal{T}$ and $\mathcal{Q}$, default matrix size λ;
Output:
  The similarity value of the two trajectories $\mathcal{D}$
1: Matrix-based partition $(\mathcal{T}, \lambda)$;
2: Matrix-based partition $(\mathcal{Q}, \lambda)$;
3: **for** $P_{\mathcal{T}}$ in $\mathcal{T}$ **do**
4:    **for** $P_{\mathcal{Q}}$ in $\mathcal{Q}$ **do**
5:      calDistance($P_{\mathcal{T}}, P_{\mathcal{Q}}$);
6:      Transmit the minimum distance and store as triples $\langle id,\ id,\ d(P_{\mathcal{T}}, P_{\mathcal{Q}}) \rangle$
7:    **end for**
8: **end for**
9: $\mathcal{D}_1 \leftarrow \max_{t_i \in \mathcal{T}} \min_{q_j \in \mathcal{Q}} d(t_i, q_j)$
10: $\mathcal{D}_2 \leftarrow \max_{q_j \in \mathcal{Q}} \min_{t_i \in \mathcal{T}} d(q_j, t_i)$
11: $\mathcal{D} \leftarrow \max\{\mathcal{D}_1, \mathcal{D}_2\}$;
12: **return** $\mathcal{D}$;

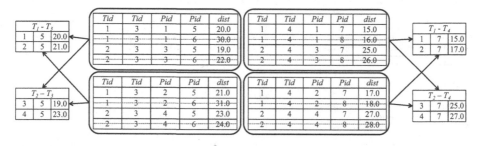

Fig. 7. Filtering mechanism to discard tuples that are not minimum values

4   Experiments

Environment. We implement the approaches and conduct all the experiments on top of *Apache Flink* [1], and all algorithms are implemented in *Java*. To the best of our knowledge, none of the prior studies associate the join-matrix model with *Flink*. The *Flink* system is deployed on a cluster which runs CentOS 7.4 operating system and is equipped with 128 processors (Intel(R) Xeon(R) CPU E7-8860 v3 @ 2.20 GHz). Overall, our cluster provides 120 computing nodes and a 512-core environment for the experiment.

Data Set. We test our model using a real-world GPS trajectory data set of Beijing Taxi, which contains over 13000 trajectories. Each trajectory record includes the taxi ID, the timestamp of the event with its GPS location at that time. Besides, we clean and complete the data set before feeding them to the streaming system, so that we can better simulate the real-time processing environment by *Apache Kafka*.

Performance Metric. Given the need for experiments on the performance of our framework, we select the following characteristics for experiments:

(1) Delay: the length of time from the inflow of T, Q to the generation of D.
(2) Accuracy: the correct rate of the output in our framework.
(3) Throughput: the average number of tuples processed per second.
(4) Volume: the total number of tuples in a task operator.

Performance Experiment. TraSP is our original 3-phase framework for *Hausdorff distance* without a filtering technique; TraSP-F is the framework with filtering mechanism discussed in Sect. 3.4. Based on theses, comparison between TraSP and TraSP-F can show the degree of performance optimization.

4.1 Efficiency

Latency. The size of matrix varies and the time window is adjusted from 100 to 500 (s). We analyze the presented experimental results from two perspectives shown in Fig. 8. Firstly, (a)–(c) presents that the latency effect of TraSP and TraSP-F is very different. This is because the TraSP-F keeps only one tuple of the pairwise trajectory-matching passed into the downstream during the filtering process. Secondly, according to the overall volatility of these figures, the overall delay is decreasing when the matrix size increases.

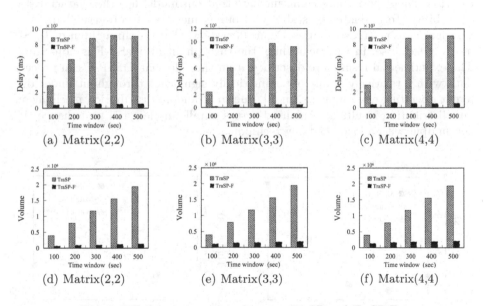

Fig. 8. Efficiency comparison between TraSP and TraSP-F

Volume. As can be seen from Fig. 8(d)–(f), when the matrix size of the TraSP-F increases, there is a slight increase in the amount of data, which is due to the shared-nothing optimization operation of each node, while the results presented by the TraSP increase obviously when the time window increases. The data volume can sideways reflect the current node workloads, where TraSP and TraSP-F are all balanced, this is because they are naturally load-balancing frameworks, but the load shedding technique in TraSP-F dramatically reduces the high load caused by the large amount of data.

Throughput. Figure 9(a), (b) draw the throughput of our model, where (a) is the TraSP algorithm and (b) is the TraSP-F. On the whole, the throughput of the TraSP-F is higher than that of the TraSP, e.g., the minimal throughput of TrasSP-F is the same as the maximum throughput of TraSP, because the data records of the abrupt decrease in the optimized operator has a positive effect on the throughput. At the same time, the effect on the TraSP-F is relatively obvious, and this is because, compared to the TraSP that computes redundant data multiple times, TraSP-F shed the partial matches which make no contribution to the final result.

4.2 Robustness

Stability. Our framework is robust to data skew, as shown in Fig. 9, the average throughput performance can indicate that our model has the characteristic of stability. To evaluate the stability of our framework, we observe changes in throughput as matrix size varies. Figure 9(a) shows that the throughput of TraSP has a steady rise. In the meanwhile, the throughput of TraSP-F is non-rising. The experimental results are described from the side that TraSP-F can process data with a relatively stable but numerically significant throughput under different operating environments. In all, due to a random distribution mechanism, our distribution results are balanced across parallel nodes, and we enhance the throughput with a load-shedding technique.

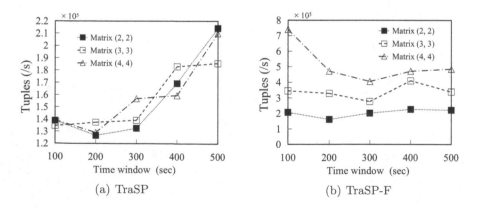

(a) TraSP   (b) TraSP-F

Fig. 9. Throughput evaluation

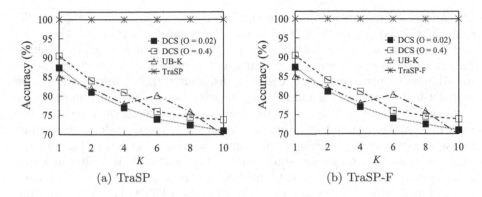

(a) TraSP   (b) TraSP-F

Fig. 10. Accuracy evaluation

Accuracy. The corresponding experiment is based on the current epidemic of *Top-k* [2] application, to study the effect of the change of k for accurate rate. We compare accurate rates with state-of-the-art distributed frameworks. As can be seen in the Fig. 10(a) (b), based on the matrix-based partition scheme, the complete information is transmitted in each operator without omission, which means that we consider each pairwise point-matching and end up with results that are consistent with correct case.

5   Related Work

Trajectory Similarity. In the past decades, there have been many efforts put into the studying of different distance measures. [3] introduces a novel distance function, *Edit Distance on Real sequence* (EDR), to measure the similarity between two trajectories. EDR is based on edit distance on strings, and removes the noise effect by quantizing the distance between a pair of elements to two values. [7] addresses the issue of trajectory similarity by defining a similarity metric, proposing an efficient approximation method to reduce its calculation cost. Ranu et al. [10] utilize a universe set of vantage points as a fast estimation to their self-proposed non-metric EDwP distance function.

Stream Join Algorithms. For stream join processing, much effort has been put into designing non-blocking algorithms. Wilschut presents the symmetric hash join SHJ [15]. It is a special type of hash join and assumes that the entire relations can be kept in main memory. For each incoming tuple, SHJ progressively creates two hash tables, one for each relation, and probes them mutually to identify joining tuples. XJoin [14] and DPHJ [8] both extend SHJ by allowing parts of the hash tables to be spilled out to the disk for later processing, greatly enhancing the applicability of the algorithm. Dittrich [5] designed a non-blocking algorithm called PMJ that inherits the advantages of sorting-based algorithms and also supports inequality predicates. The organic combination of XJoin and PMJ is conductive to the realization of HMJ [9], presented by Mokbel.

Distributed Framework. This naturally brings challenges in designing efficient distributed similarity processing algorithms and frameworks for the large-scale trajectory data. [11] introduces a temporal-first matching and two-phase framework to process the trajectory similarity join efficiently on very large trajectory datasets. Furthermore, [16] designs a distributed framework that leverages segment-based partitioning and indexing to answer similarity search queries over large trajectory data. Besides, [4] proposes a new framework DCS, which is the real-time parallel search framework that finds the k most similarity trajectory for each trajectory, by utilizing locally upper bounds on LCSS. [17] proposes a road-network-aware trajectory similarity function and designs a filtering-refine framework to solve trajectory similarity search and join problem. [12] proposes a distributed in-memory trajectory analytics system DITA, DITA provides the trajectory similarity search and join operations with a user-friendly SQL and DataFrame API. [6] designs a new computing model ART for real-time trajectory similarity and a series of data partition functions designated for ART, which ensure the high throughput and low processing latency under workload variance.

6   Conclusion and Future Work

In this paper, we present a general framework for online trajectory similarity processing, named TraSP. In such framework where experiments prove it effective to handle the problem of data skew and low accuracy as opposed to the traditional distributed similarity calculation. Additionally, we design a generic solution to accommodate different existing similarity criteria. To improve the performance of TraSP, we propose a lightweight filtering technique provided for *Hausdorff distance*. Empirical studies show that our proposal is robust, accurate, efficient and generic. In the future, we will explore the possibility on more flexible and scalable mapping and migration strategies with sampled matrix in pursuit of a smaller migration cost.

Acknowledgements. This work is partially supported by NSFC (No.61802273), the Postdoctoral Science Foundation of China under Grant (No. 2017M621813), the Postdoctoral Science Foundation of Jiangsu Province of China under Grant (No. 2018K029C), and the Natural Science Foundation for Colleges and Universities in Jiangsu Province of China under Grant (No. 18KJB520044).

References

1. Apache Flink Project. http://flink.apache.org/
2. Cao, P., Wang, Z.: Efficient top-k query calculation in distributed networks. In: PODC, pp. 206–215 (2004)
3. Chen, L., Özsu, M.T., Oria, V.: Robust and fast similarity search for moving object trajectories. In: SIGMOD, pp. 491–502 (2005)
4. Ding, J., Fang, J., Zhang, Z., Zhao, P., Xu, J., Zhao, L.: HPCC/SmartCity/DSS, pp. 1398–1405. IEEE (2019)

5. Dittrich, J., Seeger, B., Taylor, D.S., Widmayer, P.: Progressive merge join: a generic and non-blocking sort-based join algorithm. In: VLDB, pp. 299–310 (2002)
6. Fang, J., Zhao, P., Liu, A., Li, Z., Zhao, L.: Scalable and adaptive joins for trajectory data in distributed stream system. J. Comput. Sci. Technol. **34**(4), 747–761 (2019). https://doi.org/10.1007/s11390-019-1940-x
7. Frentzos, E., Gratsias, K., Theodoridis, Y.: Index-based most similar trajectory search. In: ICDE, pp. 816–825 (2007)
8. Ives, Z.G., Florescu, D., Friedman, M., Levy, A.Y., Weld, D.S.: An adaptive query execution system for data integration. In: SIGMOD, pp. 299–310 (1999)
9. Mokbel, M.F., Lu, M., Aref, W.G.: Hash-merge join: a non-blocking join algorithm for producing fast and early join results. In: ICDE, pp. 251–262 (2004)
10. Ranu, S., Deepak, P., Telang, A.D., Deshpande, P., Raghavan, S.: Indexing and matching trajectories under inconsistent sampling rates. In: ICDE, pp. 999–1010 (2015)
11. Shang, S., Chen, L., Wei, Z., Jensen, C.S., Zheng, K., Kalnis, P.: Parallel trajectory similarity joins in spatial networks. VLDB J. **27**(3), 395–420 (2018). https://doi.org/10.1007/s00778-018-0502-0
12. Shang, Z., Li, G., Bao, Z.: DITA: distributed in-memory trajectory analytics. In: SIGMOD, pp. 725–740 (2018)
13. Su, H., Liu, S., Zheng, B., Zhou, X., Zheng, K.: A survey of trajectory distance measures and performance evaluation. VLDB J. **29**(1), 3–32 (2019). https://doi.org/10.1007/s00778-019-00574-9
14. Urhan, T., Franklin, M.J.: Dynamic pipeline scheduling for improving interactive query performance. In: VLDB, pp. 501–510 (2001)
15. Wilschut, A.N., Apers, P.M.G.: Dataflow query execution in a parallel main-memory environment. Distrib. Parallel Databases **1**(1), 103–128 (1993). https://doi.org/10.1007/BF01277522
16. Xie, D., Li, F., Phillips, J.M.: Distributed trajectory similarity search. Proc. VLDB Endow. **10**(11), 1478–1489 (2017)
17. Yuan, H., Li, G.: Distributed in-memory trajectory similarity search and join on road network. In: ICDE, pp. 1262–1273 (2019)

Indicators for Measuring Tourist Mobility

Sonia Djebali$^{(\boxtimes)}$, Nicolas Loas$^{(\boxtimes)}$, and Nicolas Travers$^{(\boxtimes)}$

Leonard de Vinci, Research Center, 92 916 Paris La Défense, France
{sonia.djebali,nicolas.travers}@devinci.fr, nicolas.loas@edu.devinci.fr

Abstract. Digital traces left by active users on social networks have become a popular means of analyzing tourist behavior. The large amount of data generated by tourists provides a key indicator for understanding their behavior according to various criteria. Analyses of tourists' movement have a crucial role in tourism marketing to build decision-making tools for tourist offices. Those actors are faced with the need to discern tourists' circulation both quantitatively and qualitatively. In this paper, we propose a measure to capture tourist mobility on various areas which relies on a flow network of data from TripAdvisor into a Neo4j graph database. Thanks to this representation, we produce aggregated graphs at various scales and apply deep tourists' analysis. One centrality aspect of graphs is used to propose a key indicator of tourists mobility.

Keywords: Location and trajectory analytics · Mobile data analytics · Mobile location-based social networks · Graph databases · Neo4j

1 Introduction

In present days, tourism is considered as one of the widest and fastest growing industries [5]. Tourism is a displacement phenomenon that fully participates in the global traffic of people's experiences, norms and indeed, tourism represent a major vector of globalization, mobility and traffic. The tourist's mobility brings into play relations that need to be analyzed and understood.

Studying tourism through the "circulation flow" object means taking into account the diversity of contemporary mobility with Web technologies. In fact, e-Tourism becomes a means to identify tourism circulation flow, via digital traces. Tourism takes advantage of the social network like TripAdvisor, Booking, Facebook, Instagram, Flickr, etc. It needs to take into account both space and time. With millions of comments and photos on locations, it becomes a real challenge for tourism actors to analyze enormous volumes of data to understand how tourist circulation evolves [15]. Analyzing tourist travel behavior and knowledge of travel motivation plays a key part in tourism marketing to create a broader vision and assist tourists in decision-making [17].

By modeling the tourist flow as a graph of areas' interconnections, it becomes possible to analyse and measure the quality or capacity of the network by applying graph theory concepts like centrality, modularity, ranking, etc. However, those methods make absolute assumptions about the manner that a graph

© Springer Nature Switzerland AG 2020
Z. Huang et al. (Eds.): WISE 2020, LNCS 12342, pp. 398–413, 2020.
https://doi.org/10.1007/978-3-030-62005-9_29

behaves and not on a precise flow in the network like counting shortest paths [8], or multiple paths like information of infections [2].

By applying a measure to a given set of circulation flow characteristics to another different flow will consequently generate a loss of ability to fully interpret results and get poor and inappropriate answers. In this context, it becomes a real challenge to identify a correct key indicator that turns out to be appropriate to a given graph. As a matter of fact, it becomes important to produce measures based on the network structure while it witnesses a continuous evolution.

This paper proposes an approach to extract and interpret tourist mobility on geographic areas. Based on a graph-oriented database, we model tourists' reviews from social networks as a circulation graph which can be scaled at various levels of granularity over a geographic area. We propose the Circulation Factor which captures locally and globally how populations behave over a given area. Our contributions can be resumed as the following:

- A circulation graph data model which can be aggregated on time and space,
- The `Circulation Factor` to capture tourists flow on the circulation graph,
- The implementation and the analysis of tourists mobility at various scales.

This paper is organized as follows. We first detail in Sect. 2 the related work on flow modeling with graphs. Then, we formalize our graph data model and explain how to aggregate it in Sect. 3. Section 4 presents the *Circulation Factor* to highlight mobility in the graph. To finish with, Sect. 5 details its integration in `Neo4j` and analyze the factor on the TripAdvisor dataset.

2   State of the Art

Many graph theory algorithms and concepts are used in network analysis to measure the importance of nodes, to understand interactions in the network, to show information circulation or to deduce communities of nodes that share some characteristics. Each measure follows a specific definition and rules to target important nodes in order to have a network understanding.

In literature, most used concepts for social network analysis which lead to tourists' indicators are based on the identification of nodes importance, clustering nodes into communities or extracting patterns as trajectories in the graph. Identifying nodes importance in graphs is highly dependent on its definition. In our context, nodes importance can be used in order to comprehend how tourists circulate all over a territory represented as a graph.

A first family focuses on clustering algorithms to identify collections of nodes which share some characteristics and produce communities. HCS [11] focuses on the maximization of connectivity within clusters HCS and is used to detect communities of individuals. The *Louvain* algorithm [1] is a hierarchical algorithm that maximizes cluster modularity by merging nodes into high level in a hierarchical tree. *Label Propagation* [18] binds a unique label to each node which tries to spread their own label to neighbor nodes. *Chameleon* [14] overcomes the limitations of existing agglomerative hierarchical clustering algorithms by adapting to the cluster characteristics. Even if clustering methods could identify groups of behaviors, it does not target the circulation issue or groups too many nodes between each other which do not help to identify nodes independently.

A second family of algorithms focuses on spanning trees [10]. They are used defining the cheapest subset of edges that keeps the graph in one connected component or finding frequent patterns in a graph like with Mining Maximal Frequent sub-graphs from Graph Databases [12]. However, in our context spanning trees will only give the main path and not a global sight on a territory.

Last but not least, centrality algorithms aim at providing relevant analytical information about nodes in the graph and then nodes importance into a graph. *Closeness centrality* [6] scores each node based on their closeness to all other nodes in the network. *Betweenness centrality* [6] measures how much a node lies on the shortest path between other nodes in a network which helps to find the node that influences the data flow. *Degree centrality* [6] assigns an importance score based simply on the number of links held by each node, which shows very connected, popular or informational nodes. *EigenCentrality* [6] measures nodes influence based on the number of links but consider nodes connectivity as well as the neighborhood. However, those measures focus on single nodes to identify most representative ones but do not integrate this notion of circulation. One variant of the EigenCentrality is the well-known *PageRank* [16] mostly used by the Google search engine to rank web pages to have more accurate search results. Even if the PageRank score is a good indicator of circulation, it can only be compared with other nodes of a given graph and not between different graphs. In fact, PageRank scores are highly dependent on the composition of the graph and two different graphs even with the same set of nodes can produce very different scores and are hardly comparable which is the main objective of our circulation measure. However, PageRank will be integrated further in our measure since it enhances the circulation.

All methods presented above can be applied to any graph. However, those methods make absolute hypotheses about the manner of the graph behave and not on specific flows in the network. Applying a measure to a given set of circulation flow characteristics to another different flow will consequently generate a loss of ability to fully interpret results and get poor and inadequate answers.

On a broader analysis aspect of tourists flows extraction, some researches have been proposed like [4] for flow visualization, pattern mining [21], extraction of Point-of-Interests [19], or Kernel density estimations [20]. However, those solutions are either static focusing on hot spots hardly comparable with other points, or hardly flexible to compare various densities of paths.

To offset this problem, the main object is the formalism of graphs in a graph-oriented database model that takes into account tourists' circulation flow. Smart circulation graphs that will be able to scale out various levels of granularity can be produced. Then, we propose a measure that helps to identify the mobility of each node for a given population in order to develop a valuable indicator in terms of mobility and centrality.

3   A Tourism Circulation Graph

Modeling tourism data requires to take into account locations information, users' properties and their interactions. We propose the circulation graph data model in order to deal with interactions on locations. Graphs rely on links between users and locations through their reviews. A circulation graph is thus modeled with all the properties associated to the users.

3.1   Graph Data Models

Data Types. Our database is composed of geolocalized locations, reviews and users. A location is composed of a type (hotel, restaurant, attraction), localization (lat, long) and a rating ($r \in \mathbb{R} \land r \in [1.0, 5.0]$). To characterize localization, each location has been aligned with administrative areas (GADM)[1].

Each location l is linked to an area a if its geo-localization is contained into the area's shape (*SpatialPolygon* function SP), such that the $SP(l.lat, l.long) = a$. This administrative area is composed of a country, a region, a department, a district and a city. Thus, each location l is identified by: $l \in \mathcal{L}(type, r, a)$.

A review represents a note ($n \in \mathbb{N} \land n \in [1, 5]$) given by a user u on a location l at time t (t is in the discrete time domain $\mathcal{T}$). Each review is then defined by an event r_t such that: $r_t = (l, u, n)$.

Graph Data Model. To understand tourists' behavior and mobility in the study of a given destination (*e.g.*, department, region, country), we need to target tourists. For this, we focus only on users u who visit a destination at least once. Then we get all the reviews r_t they made, even elsewhere, in order to gather their circulation all over the world.

The initial graph data model T is a natural bipartite graph which links users to locations, as illustrated in Fig. 1 with dotted edges.

Even if this huge graph contains relevant information in order to produce analyzes, it will not ease the way to manipulate it or extract circulation of tourists. Therefore, we need to provide a new graph data model based on T that will allow the analyzes.

Circulation Graph Data Model. Since our analyzes require several levels of studies (*i.e.*, international to local), we need providing a generic graph data model. To achieve this, based on graph T a new graph data model is built by focusing on circulation between locations.

The circulation graph model $C(V', E')$ relies on the fact that tourists can review several locations during their trip. Consequently, the sequence of reviews from user u can generate new edges between locations. However, we consider that a trip is composed of reviews r_{t_1} and r_{t_2} written at most at 7 days apart [9]

---

[1] GADM: https://gadm.org/index.html - 386,735 administrative areas (country, region, department, district, city and town).

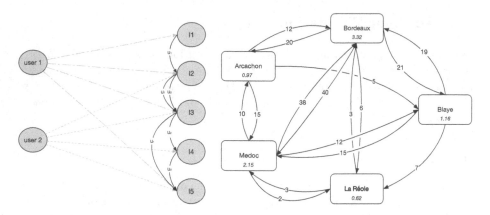

Fig. 1. Circulation graph   **Fig. 2.** Filtered circulation graph

$(t_2 - t_1 < 7d)$. Two consecutive events from user u that occur within 7 days generate an edge e between nodes l_1 and l_2. Figure 1 illustrates the transformation of graph T by connecting directly locations on users' trip (plain edges).

3.2   Aggregated Graphs

Tourism actors need to focus their studies in a geodesic point of view, on both time and space. According to that, we need to provide fine grain studies on the circulation graph C by aggregating vertices and edges, while filtering on properties (*i.e.,* users, locations, time). For this, we produce new graphs where nodes are aggregated according to a property P on areas (*e.g.,* district, city) and produced edges give the number of edges in E' between aggregated nodes [7].

We can notice that time is discretized on both years and months. It will enable the focus on long time periods for studies, at a minimum at month scale and show the evolution over time of tourists behavior.

Moreover, we can aggregate nodes and edges from the circulation graph C to obtain an aggregated graph on other shared properties (between V' nodes). This aggregated circulation graph will be denoted by AC. Then, the study will focus on circulation between groups of locations (*e.g.,* districts, cities) on a given zone. Figure 2 illustrates this aggregation between cities of the Gironde's department.

In the following, the AC aggregate circulation graph is denoted as $ac_{n,e}^{p,f}$ where indices give nodes and edges aggregating properties and exponent as the filtering predicate on edges. Here an example of graph AC on aggregated on city nodes, years and nationality edges focused on Americans in 2018:

$$ac_{cities,year\&nat}^{USA,2018} = AC(cities, year\&nat, \sigma_{nat="USA" \land year=2018})$$

4   The Circulation Factor

The goal of our study is to provide a novel way to characterize the flow of touristic circulation with a valuable indicator. To achieve this, we propose the Circulation Factor CF which relies on both the circulation graph AC and the combination of PageRank computations [22].

As we saw in Sect. 2, the PageRank score of a node represents the current best solution in our context to represent the fact that tourists tend to go through an area during their journey. However, even if this score is a good indicator of circulation, it can only be compared with other nodes of a given graph and not between different graphs. In fact, PageRank scores are highly dependent on the composition of the graph and two different graphs even with the same set of nodes can produce very different scores and are hardly comparable.

To give an example, we wish to study the score's evolution of the city of Bordeaux for American tourists over years. Thus, we can compute PageRank scores on the extracted graph from $ac_{cities,year}^{2018,USA}$. Thus, "Bordeaux" PageRank score $PR_{Bordeaux}(ac_{cities,year}^{USA,2018})$ has a meaning according to other nodes (cities) in the graph like $PR_{Blaye}(ac_{cities,year}^{USA,2018})$. But the comparison is useless with $PR_{Bordeaux}(ac_{cities,year}^{USA,2017})$. We propose in this article a new measure which helps to compare various flows of circulation in AC.

4.1   The Transient Circulation Factor

To cope with this issue, we propose the Transient Circulation Factor which gives for a node, a value that represents how much a population circulates in an area compared to the whole population.

Definition 1. *The Transient Circulation Factor $TCF_{n,e}^{p,f}$ is a factor applied on an aggregate graph $AC = ac_{n,e}^{-,f}$. The factor $TCF_{n,e}^{p,f}(AC,\nu)$ is the comparison between the PageRank PR of a node $\nu \in AC$ where edges are only filtered out by the context f, with its PageRank in AC where edges are filtered out by both the population p and the context f:*

$$TCF_{n,e}^{p,f}(AC,\nu) = \frac{PR_\nu(ac_{n,e}^{p,f})}{PR_\nu(ac_{n,e}^{-,f})}$$

The Transient Circulation Factor is an impact factor of a population circulation flow over a directed graph. It represents how much a population circulates compared to other populations. The following example illustrates the *Transient Circulation Factor* of the city of "Bordeaux" for American in 2018.

$$TCF_{cities,year}^{USA,2018}(AC,Bordeaux) = \frac{PR_{Bordeaux}(ac_{cities,year}^{USA,2018})}{PR_{Bordeaux}(ac_{cities,year}^{-,2018})}$$

Remind that the Weighted PageRank [22] is based on the computation of navigation probabilities with weighted for both in and out links.

In our approach, TCF compares the same set of nodes and edges but updates weights according to a given population. Therefore, the comparison can be summarized to the variation of ratios of weights from a given population with all populations.

Thus, the computation of focused in/out weights on a given population can be higher or lower than 1. The TCF of the whole population is equal to 1. Consequently, a TCF value over 1 says that a given population tends to circulate through this node more than others. At the opposite a value below 1 says that the node is less central in the circulation of this population.

Thanks to this **Transient Circulation Factor**, we can compare populations on a given area, but also for all the areas on the whole graph. Thus, we can give the circulation profile of a population to say if they are more mobile than others or remain central in a narrow area.

Moreover, this factor can also be used to study the evolution of a population over years. In fact, the evolution of proportions from incoming and outgoing arcs' weight computed by PageRanks in the TCF can be compared between two years for a given population. The following statement means that Americans focus more on Bordeaux in 2018 than 2017.

$$TCF_{cities,year}^{USA,2017}(AC, Bordeaux) < TCF_{cities,year}^{USA,2018}(AC, Bordeaux)$$

We can also compare two populations in AC to identify if the first one tends to circulate more through a given node than the second population.

$$TCF_{cities,year}^{French,2018}(AC, Bordeaux) < TCF_{cities,year}^{USA,2018}(AC, Bordeaux)$$

4.2 The Global Circulation Factor

Since the TCF captures the intrinsic flow value of tourists circulation on a given area, we need to produce an indicator of a global sight on the behavior of a given population. The **Global Circulation Factor** computes all TCFs of a given population on the whole graph to show how much this population circulates on the territory compared to other people.

Definition 2. *The* **Global Circulation Factor** *$GCF_{n,e}^{p,f}$ is a factor which computes the mean value of TCF values for all nodes $\nu \in AC$ with aggregated nodes on property n and edges e, and filtered out by the context f and the population p:*

$$GCF_{n,e}^{p,f}(AC) = \frac{\sum_{\nu \in AC} TCF_{n,e}^{p,f}(AC, \nu)}{|AC|}$$

The mean of all TCF values integrates all local behaviors to provide a broad sight of all populations p in a given zone (*i.e.,* graph AC). The capability to manipulate the level of aggregation on nodes n and edges e helps to capture different kinds of behavior: local circulation (city) to global (district or department), evolution (years) to seasons (months), etc. To finish with, filter f on edges allows us to focus on a specific aggregated contexts (*e.g.,* years, months).

Section 5 will validate our approach with various settings. It will focus especially on the capabilities to enlighten behaviors with *TCF* and *GCF* at various scales and aggregations. Even if the number of possibilities is really wide, we have produced most significant observations which enhance our contributions.

5   Experiments

5.1   The Neo4Tourism Implementation

Neo4Tourism [3] is a framework which helps to manipulate graphs by aggregating and filtering them by taking into account geographic data. Graphs are stored in a *Neo4j* server dedicated to the tourist circulation characterization. It transforms bipartite graphs in circulation graphs *C* and its aggregations *AC*. Aggregated graphs are also materialized as new graphs for optimization purposes. Since data are stored incrementally (time dependency), materialized graphs do not have to be updated, only new edges are added to graphs.

In our circulation characterization, we will focus especially on aggregated graphs *AC*. The *Cypher* query language used in Neo4j helps to manipulate graphs and produces TCFs and GCFs values.

Aggregated Graphs Materialization. TCF and GCF require several granularities of studies (*e.g.*, regions, districts, cities, etc.), it is then necessary to compute queries on various aggregate graphs. The first *AC* graph focuses on aggregated nodes with the smallest area: towns. Every location belonging to a town is merged in a single node. All edges which share the same properties (*i.e.*, nodes, year, month, nationality, age) produce an edge with a new property NB that represents the number of merged edges. This first graph at town scale is built by a Java program and stored in Neo4j. It reads all series of reviews from each user to generate the circulation between locations and consequently towns.

Other aggregated graphs are built from the first one, they merge nodes that share a same property (department, district, city), so do the edges. To achieve this in Cypher, the MERGE clause is used to produce derived graphs:

```
MATCH (t1:Town) -[t:trip]-> (t2:Town)
MERGE (c1:City{name:t1.city})  MERGE (c2:City{name:t2.city})
MERGE (c1)-[ct:trip{year:t.year,month:t.month,nat:t.nat,age:t.age}]->(c2)
ON CREATE SET ct.NB=t.NB        ON MATCH SET ct.NB=ct.NB+t.NB
```

This query relies on the first graph composed of *Town* nodes where edges are typed as *trip*. Since each node is labeled with GADM administrative zones, we can merge them according to various areas, here city names. Then, edges are merged when they share same properties and NB values are summed.

All graphs are generated at all scales: city, district, departments, regions and countries. Each time, nodes from a given scale contains areas information to be filtered out and to focus on a specific zone. Thus, we can extract subgraphs on a given zone like a department or a region.

Table 1. Tripadvisor dataset

Table	# instances
Locations	4.8×10^4
Users	1.31×10^6
Reviews	3.58×10^6

Table 2. Different AC graphs characteristics

AC Graphs	Aggregation	# nodes	# edges
Nouvelle-Aquitaine	Cities	482	382,266
	Districts	41	170,403
Hauts-de-France	Cities	297	153,345
	Districts	27	75,429
Gironde	Cities	55	94,032
	Districts	6	26,704

Advanced Manipulations. Now we have circulation graphs at all scales; we can compute *PageRanks* on them by applying prepared statement queries. The following query integrates a callable function from the *graph data science* package[2] to produce PageRank scores for each district within the "Gironde" department.

```
CALL gds.graph.create.cypher("CypherProjection",
 "MATCH (c:City{department:'Gironde'}) RETURN id(c) as id",
 "MATCH (c1:City)-[t:trip{year:2018,nationality:'USA'}]->(c2:City)
  RETURN id(c1) as from, id(c2) as to, sum(toFloat(t.NB)) as weight")
CALL gds.pageRank.stream("CypherProjection",
 {dampingFactor:0.85,iterations:50,weightProperty:true}) YIELD node, score
RETURN node.city, sum(score) as score;
```

We can see that the *AC* graph uses a Cypher projection to get *Gironde*'s graph (Bordeaux's department in France) where only edges in 2018 done by Americans are kept and merged (sums of NBs). Then, the PageRank is computed on this sub-graph "CypherProjection" with PageRank scores for each node/city of the sub-graph. Figure 2 gives an example of this result where PageRank scores are above each node.

We must notice that we keep edges that link a node to itself. In fact, this edge represents the reality that tourists circulate within an area.

To provide $TCF_{n,e}^{p,f}(AC, \nu)$, it requires to compute two PageRanks. The first one is given by the query above, and the second one just removes the filter on nationality. TCF values are then associated to each node in AC. Then, GCF values are computed with a simple mean query on considered nodes.

Graphs Characteristics. To support our approach, we need constituting a dataset which represents the best notion of circulation over a territory by visiting various locations. Several e-tourism websites were considered. *Booking* focuses only on accommodation and cannot be used at small scales (between cities). *Flickr* is really interesting for precise locations. However, the public dataset is

---

[2] Neo4j 4, GDS: https://neo4j.com/docs/graph-data-science/1.2/.

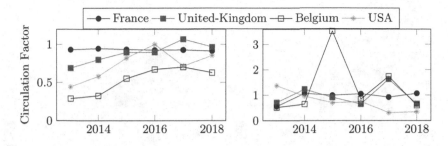

Fig. 3. TCFs of 1) Bordeaux and 2) Blaye (city-scale)

too small to be representative for diverse populations. Finally, we chose *TripAdvisor* which gives precise information on locations, populations and constitutes a sufficient amount of data to begin to be representative.

Table 1 gives the initial dataset gathered from *TripAdvisor* focused on two French regions: *Nouvelle-Aquitaine* and *Hauts-de-France*. This dataset contains 3.58×10^6 reviews on 4.8×10^4 locations.

The setting of our experiments tries to enhance both graph data manipulations at various scales and the capability of TCF and GCF to witness the circulation of tourists. To achieve this we have extracted six AC graphs (Table 2) with three zones aggregated on both cities and districts: *Nouvelle-Aquitaine* (region), *Hauts-de-France* (region) and *Gironde* (department). This will help to understand local and global behaviors. Notice that the number of edges here is the sum of edges' weight within the graph. This loss between the number of reviews and the number of edges' weight corresponds to the fact that we focused only tourists circulation (seven-day trip as mentioned previously).

5.2   Transient Circulation Factor's Evaluation

TCF at City-Scale. Figure 3 shows the TCF evolution of *Bordeaux* and *Blaye* cities for different nationalities. We can see that the ratio of PageRanks for *Bordeaux* is almost equal to one for French while witnessing a small decrease of interest. At the same time, British and American populations grow significantly to reach 1 in 2016. This effect is correlated to the opening of the new high-speed train line between Paris and Bordeaux making this city more central in tourist trips. Belgians have a lower TCF but tend to grow in the past years.

The city of *Blaye* known for its castle and wines witnesses an interesting aspect of the TCF: event detection. The French population is on average more represented in this area which confirms the fact that they do prefer the countryside. But interestingly Belgian people in 2015 has a factor reaching almost 4. This anomaly is explainable by an event organized by a tour operator that occurred in 2015 with a consequent group of belgians. Consequently, this score grows up significantly at once. This fact is also observed locally for various populations.

Maps that are shown in Fig. 4 correspond to TCF values of *Gironde* cities in 2014 for respectively French and British. The colors scale helps to highlight

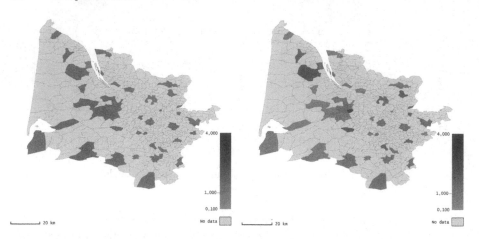

Fig. 4. 1) French and 2) British TCFs for Gironde's cities in 2014

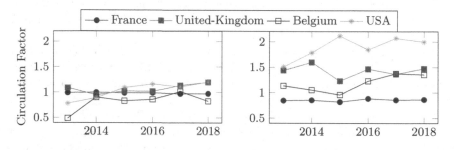

Fig. 5. TCFs of 1) Bordeaux and 2) Libourne (district-scale)

areas where those populations are more mobile. Of course we identify touristic zones easily like *Bordeaux* (hypercenter), *Arcachon* (South-west), *Lespar-Médoc* (North), *Blaye* (South), *Langon* (South-East) or *Libourne* (North-East).

For French, we can see that most cities have a score around 1 which means that they homogeneously circulate all over cities. At the opposite, British are less uniform while they are focusing mainly on the area of *Lespar-Médoc* and *Libourne*. *Arcachon* and *Bordeaux*'s suburbs are less central in their journey.

It is interesting to bring out mobility differences of local behaviors on the territory. The map identifies clearly where each population concentrates their journey. More importantly those distributions are comparable to each other.

TCF at District-Scale. We now aggregate the graph at district scale with six significant areas in Gironde. It produces a smaller graph in which mobility is also concentrated between those nodes.

As we can see, scores are more homogeneous in Fig. 5 for the Bordeaux district. It is due to the fact that nodes and edges are aggregated which leads to less variability on mobility. We can confirm the fact that French people are less rep-

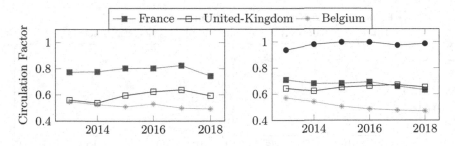

Fig. 6. GCFs of 1) *Nouvelle-Aquitaine* and 2) *Hauts-de-France* (city-scale)

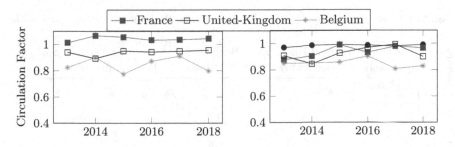

Fig. 7. GCFs of 1) *Nouvelle-Aquitaine* and 2) *Hauts-de-France* (district-scale)

resentative of the circulation around *Bordeaux*. Belgians witness a better score than at city scale, this means that the mobility is higher with more exchanges between districts (back and forth to *Bordeaux*'s area).

However, the attractiveness growth of the *Libourne* district is really significant with higher scores for all populations except for the French one. The whole area concentrates many castles, wine tasting and tour operator activities and thus attractive for foreign tourism.

It is interesting to see that the TCF brings different conclusions at each scale. The city-scale helps to extract events for a given population and more local mobility. At the opposite, the district-scale gives tendencies for populations but also the typology of places that are considered. We observed similar results on other departments and the graph on "Hauts-de-France".

5.3   Global Circulation Factor's Evaluation

While TCF shows hotspots within a circulation graph, GCF focuses on the whole graph in order to characterize a population all over the territory.

To enhance this circulation indicator, we now apply the computation on wider graphs and try to compare both scaling of areas and two different zones. To achieve this we apply the GCF on region graphs on *Hauts-de-France* and *Nouvelle-Aquitaine*. Having two different geographic zones will be useful to study differences and common points between them. And for each graph we have two scales: city scale and district scale.

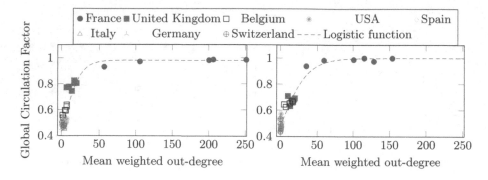

Fig. 8. Logistic function of 1) *Nouvelle-Aquitaine* & 2) *Hauts-de-France* (city-scale)

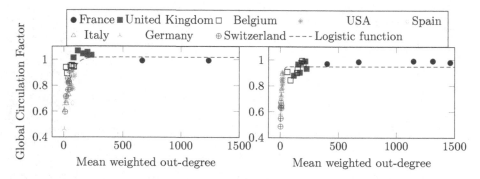

Fig. 9. Logistic function of 1) *Nouvelle-Aquitaine* & 2) *Hauts-de-France* (district-scale)

Figure 6 gives the GCF evolution on regions at city scale. This scale is interesting since tourists usually do not circulate in such a wide zone (region) with small stops (cities). Consequently, this evolution shows the global interest of a population to visit a region. From *Nouvelle-Aquitaine* (left) and *Hauts-de-France* (right) we can especially see that British tourists are more interested in the first one (old British country). At the opposite Belgians tend to visit more places in *Hauts-de-France* which is next to their own country, even though the British circulate more than the latter in this region.

At district scale in Fig. 7, since cities are aggregated into bigger areas, the circulation effect is higher and less fluctuating. Differences between British and Belgians are less visible. However, Americans witness a significant growth from the previous analysis. It is due to the fact that Americans target specific zones within each district, especially on Second World War memorials (*i.e.,* memory tourism [13]).

5.4   The GCF Property

As we saw, GCF allows highlighting populations behavior at various scales. One can argue that some other centrality measures can bring out similar results. Remind that the most similar measure is based on the PageRank. However

Table 3. Logistic parameters and errors

Parameters	NA Cities	HF Cities	NA District	HF District
L	9.791×10^{-1}	9.984×10^{-1}	1.017×10^{0}	9.488×10^{-1}
k	9.215×10^{-2}	5.611×10^{-2}	2.250×10^{-2}	1.062×10^{-1}
x_0	2.667×10^{0}	1.598×10^{0}	-1.653×10^{1}	-1.937×10^{0}
MSE	1.620×10^{-1}	7.229×10^{-2}	2.871×10^{-1}	1.231×10^{-1}
MAE	4.540×10^{-2}	3.165×10^{-2}	5.660×10^{-2}	4.366×10^{-2}
MAPE	8.204 %	5.688 %	7.136 %	5.753 %

as stated previously, it cannot compare various scales, zones, populations or evolution.

Another similar solution is the degree centrality [6] by computing the average out-degree centrality of all nodes for different populations as well as its evolution.

This centrality gives an approximate solution (for space reason we only present this one). We tried to find a correlation between the weighted out-degree centrality and the GCF. Figures 8 (city scale) and 9 (district scale) show a logistic regression of mean weighted out-degrees x: $f(x) = \frac{L}{1+e^{-k(x-x_0)}}$ where L denotes the upper bound, k the growth rate and x_0 its midpoint.

Since the PageRank is a logarithmic measure, it is natural to follow this law. But this scaling effect is all the more important since it gives an exploitable indicator of circulation. In fact, we can see that French out-degrees can be very high, fluctuating (axes are cut for *Nouvelle-Aquitaine*) and hardly comparable, likewise low out-degrees make all small populations packed all together. The GCF helps to differentiate them in a logarithmic scale.

Table 3 gives parameters which belong to each distribution (Nouvelle-Aquitaine/Hauts-de-France, cities-districts). L values are bounded to 1.02 which corresponds to British and French circulations and is as expected higher for district scale. k is the tendency of the curve which is the opposite of the growth, then values tend to grow faster at city scale. x_0 gives the midpoint, the lower the value is, the lower is the minimum GCF. To validate our correlation between the weighted out-degree centrality and the GCF, we used three forecasting accuracy techniques MAE, MSE and MAPE[3]. MAPE is the most precise measure to compare the accuracy between different items since it measures the relative performance. In our case MAPE values are lower than 8.2% which is an excellent accurate.

6   Conclusion

We have formalized in this article a methodology to produce and manipulate circulation graphs from a digital trace of users on Social Networks. This approach

---

[3] MAE: Mean Absolute Error, MSE: Mean Squared Error, MAPE: Mean Absolute Percentage Error.

helps to produce various aggregated graphs by zooming on geographic areas and filtering on population characteristics. We also proposed the *Circulation Factor* which enables the mobility comparison from a population to another either on space and time. Our approach has been integrated in the *Neo4j* database which easily produces various aggregated graphs and applies graph theory algorithms on them. Our experiments showed that the TCF can highlight events and local mobility while the GCF enhances global tendencies and population behavior.

For future works, we wish to propose a prediction model that takes into account for each population its tendency and predicts the next circulation factor on a zone. On the other hand, it should be interesting to focus on detecting global propagation patterns for a given population like spanning trees by taking into account coverage. To finish with, we wish to focus on community extraction on the graph to compare how much linked cities of a cluster can be correlated to an administrative zone and thus representative of its impact in the area.

References

1. Blondel, V.D., Guillaume, J.L., Lambiotte, R., Lefebvre, E.: Fast unfolding of communities in large networks. J. Stat. Mech.: Theory Exp. **2008**(10), P10008 (2008)
2. Borgatti, S.P.: Centrality and network flow. Soc. Netw. **27**(1), 55–71 (2005)
3. Chareyron, G., Quelhas, U., Travers, N.: Tourism analysis on graphs with Neo4Tourism. In: U, L.H., Yang, J., Cai, Y., Karlapalem, K., Liu, A., Huang, X. (eds.) WISE 2020. CCIS, vol. 1155, pp. 37–44. Springer, Singapore (2020). https://doi.org/10.1007/978-981-15-3281-8_4
4. Chua, A., Servillo, L., Marcheggiani, E., Moere, A.V.: Mapping Cilento: using geotagged social media data to characterize tourist flows in southern Italy. Tour. Manag. **57**, 295–310 (2016)
5. Cooper, C., Hall, C.M.: Contemporary Tourism. Routledge, Abingdon (2007)
6. Das, K., Samanta, S., Pal, M.: Study on centrality measures in social networks: a survey. Soc. Netw. Anal. Min. **8**(1), 13 (2018). https://doi.org/10.1007/s13278-018-0493-2
7. Endriss, U., Grandi, U.: Graph aggregation. Artif. Intell. **245**, 86–114 (2017)
8. Freeman, L.C., Borgatti, S.P., White, D.R.: Centrality in valued graphs: a measure of betweenness based on network flow. Soc. Netw. **13**(2), 141–154 (1991)
9. Gössling, S., Scott, D., Hall, C.M.: Global trends in length of stay: implications for destination management and climate change. J. Sustain. Tour. **26**(12), 2087–2101 (2018)
10. Graham, R.L., Hell, P.: On the history of the minimum spanning tree problem. Ann. Hist. Comput. **7**(1), 43–57 (1985)
11. Hartuv, E., Shamir, R.: A clustering algorithm based on graph connectivity. Inf. Process. Lett. **76**(4–6), 175–181 (2000)
12. Huan, J., Wang, W., Prins, J., Yang, J.: Spin: mining maximal frequent subgraphs from graph databases. In: ACM SIGKDD 2004, pp. 581–586 (2004)
13. Jacquot, S., Chareyron, G., Cousin, S.: Le tourisme de mémoire au prisme du "Big Data". Cartographier les circulations touristiques pour observer les pratiques mémorielles. Mondes du Tourisme, June 2018

14. Karypis, G., Han, E.H., Kumar, V.: Chameleon: hierarchical clustering using dynamic modeling. Computer **32**(8), 68–75 (1999)
15. Keng, S.S., Su, C.H., Yu, G.L., Fang, F.C.: AK tourism: a property graph ontology-based tourism recommender system. In: KMICe, pp. 83–88. UUM (2018)
16. Langville, A.N., Meyer, C.D.: Deeper inside pagerank. Internet Math. **1**(3), 335–380 (2004)
17. March, R., Woodside, A.G.: Tourism Behaviour: Travellers' Decisions and Actions. Cabi, Wallingford (2005)
18. Raghavan, U.N., Albert, R., Kumara, S.: Near linear time algorithm to detect community structures in large-scale networks. Phys. Rev. E **76**(3), 1–11 (2007)
19. Spyrou, E., Korakakis, M., Charalampidis, V., Psallas, A., Mylonas, P.: A geo-clustering approach for the detection of areas-of-interest and their underlying semantics. Algorithms **10**(1), 35 (2017)
20. Sun, Y., Fan, H., Helbich, M., Zipf, A.: Analyzing human activities through volunteered geographic information: using Flickr to analyze spatial and temporal pattern of tourist accommodation. In: Krisp, J. (ed.) Progress in Location-Based Services. LNGC, pp. 57–69. Springer, Heidelberg (2013). https://doi.org/10.1007/978-3-642-34203-5_4
21. Vu, H.Q., Li, G., Law, R., Ye, B.H.: Exploring the travel behaviors of inbound tourists to Hong Kong using geotagged photos. Tour. Manag. **46**, 222–232 (2015)
22. Xing, W., Ghorbani, A.: Weighted pagerank algorithm. In: CNSR 2004, pp. 305–314, May 2004

Modeling Local and Global Flow Aggregation for Traffic Flow Forecasting

Yuan Qu, Yanmin Zhu$^{(\boxtimes)}$, Tianzi Zang, Yanan Xu, and Jiadi Yu

Shanghai Jiao Tong University, Shanghai 200240, China
{quyuan,yzhu,zangtianzi,xuyanan2015,jiadiyu}@sjtu.edu.cn

Abstract. Traffic flow forecasting is significant to traffic management and public safety. However, it is a challenging problem, because of complex spatial and temporal dependencies. Many existing approaches adopt Graph Convolution Networks (GCN) to model spatial dependencies and recurrent neural networks (RNN) to model temporal dependencies, simultaneously. However, the existing approaches mainly use adjacency matrix or distance matrix to represent the correlations between adjacent road segments, which fail to capture dynamic spatial dependencies. Besides, these approaches ignore the lag influence caused by propagation times of traffic flows and cannot model the global aggregation effect of traffic flows. In response to the limitations of the existing approaches, we model local aggregation and global aggregation of traffic flows. We propose a novel model, called the Local and Global Spatial Temporal Network (LGSTN), to forecast the traffic flows on a road segment basis (instead of regions). We first construct time-dependent flow transfer graphs to capture dynamic spatial correlations among the local traffic flows of the adjacent road segments. Next, we adopt spatial-based GCNs to model local traffic flow aggregation. Then, we propose a Lag-gated LSTM to model global traffic flow aggregation by considering free-flow reachable time matrix. Experiments on two real-world datasets have demonstrated our proposed LGSTN considerably outperforms state-of-the-art traffic forecast methods.

Keywords: Traffic flow forecasting · Local and global flow aggregation · Graph convolution · Transfer flow graphs · Lag-gated LSTM

1 Introduction

Urban transport is the lifeblood of a city [4], which plays an essential role in every resident's daily life. With high flexibility and fast speed, automobiles have gradually become the first choice for people to travel in cities [26]. Meanwhile, traffic congestion has become one of the factors restricting the healthy operation of modern economy and society. In this paper, we focus on the task of urban traffic flow forecasting, one of the typical tasks in Intelligent Transportation Systems (ITS) [6]. To be specific, we forecast the number of vehicles passing

© Springer Nature Switzerland AG 2020
Z. Huang et al. (Eds.): WISE 2020, LNCS 12342, pp. 414–429, 2020.
https://doi.org/10.1007/978-3-030-62005-9_30

through each road segment for the next time interval based on the historical traffic data. This task is widely used in transportation services, such as vehicle flow control [1], road trip planning [8] and prospective traffic navigation [21].

Many traditional methods [7,10,15,23] have been widely used in the field of traffic flow forecasting. In recent years, the development of deep learning provides new direction for traffic flow forecasting. Convolutional neural networks (CNN) [9] is effective to capture spatial dependencies between pixels in image processing. Recurrent neural network (RNN) and its variants, long short-term memory (LSTM) [14] and Gated Recurrent Unit (GRU) [2], has achieved great success in sequence learning. Based on that, Shi et al. [18] proposed ConvLSTM and Yao et al. [24] proposed STDN to model spatial and temporal dependencies simultaneously by integrating CNN and LSTM. These methods with traditional CNN can only be applied to the traffic data based on 2-D grid regions, but not applicable to complex road network data with topological graph structure.

Recently, substantial research has focused on graph convolution which extends traditional convolution to data of graph structure. Graph convolution can be divided into two categories: spectral-based GCNs and spatial-based GCNs. Spectral-based GCNs define graph convolution with the perspective of graph signal processing [22]. Representatively, Yu et al. [25] propose STGCN to learn spatio-temporal features of traffic data. However, due to the symmetry of Laplacian matrix, spectral-based GCNs are only applicable to undirected graph, which cannot reflect the real road network. Spatial-based GCNs propagate node information along edges according to adjacency matrix. Li et al. [11] regard graph convolutions as a diffusion process and combines diffusion convolution with RNN to incorporate both spatial and temporal dependencies in the traffic flow. Song et al. [12] simultaneously extract localized spatio-temporal correlation information based on multiple spatial-temporal synchronous graph convolutional layers.

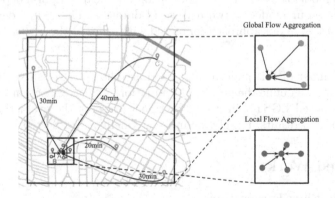

Fig. 1. Local and global flow aggregation.

However, two limitations exist in above methods. First, they ignore that the local spatial dependencies between adjacent road segments are highly dynamic. For example, during the morning of weekday, traffic flow transfers more along the direction from residential areas to work areas. But at night, this transfer pattern is opposite. Second, these methods do not consider that lag influence from distant roads caused by the propagation time of traffic flow. The travel time of vehicles from one road segment to another is ignored in previous research.

We notice that the current traffic flow of one road segment is the local aggregation result from adjacent upstream road segments in the short term, and in the meanwhile, it is the global aggregation from other distant roads after a period of travel time. As Fig. 1 shown, the flow of Road A comes from the adjacent roads represented by Road B in a short period. And due to the delay of traffic flow propagation, the distant roads represented by Road C will have lag influences on Road A after 20 to 40 minutes delay.

Inspired by our insights, we proposed a novel deep learning model Local and Global Spatio-Temporal Network(LGSTN) for urban traffic flow forecasting by modeling the local and global flow aggregation. We construct time-dependent transfer flow graphs from the historical trajectories to represent the local flow aggregation from adjacent upstream road segments at each time interval. Due to the aggregation pattern of traffic flow is similar to the propagation of information in spatial-based GCN. A modified spatial-based GCN is applied to realize the local aggregation of information and capture the dynamic evolution process of transfer flow graphs. At the same time, we propose an Lag-Gated LSTM to model the global flow aggregation and near temporal dependency by explicitly filtering the global aggregation information at each time step.

Our contributions can be summarized as follows:

- We propose to construct time-dependent transfer flow graphs and feed them into modified spatial-based GCNs to model the dynamic local flow aggregation.
- We take propagation time of traffic flow into account and propose a Lag-gated LSTM to model the global flow aggregation and near temporal dependencies.
- We evaluate our model on two real-world datasets. The results demonstrate the advantages of our approach compared with state-of-the-art traffic forecast methods.

The remainder of this paper is organized as follows. After surveying related work in Sect. 2, we define the problem formulation in Sect. 3. Section 4 will introduce our proposed LGSTN in detail. Section 5 presents the evaluation dataset and experimental results. Finally, we make a conclusion in Sect. 6.

2   Related Work

2.1   Traffic Flow Forecasting

Traffic flow forecasting is an important problem that has been intensively studied in the last few decades. Traditional methods mostly treated traffic forecasting as a simple time series forecasting task of every target. For example, Kalman filtering [23], ARIMA and its variations [10,16,17] have been widely applied in the problem of traffic flow forecasting. However, these methods are usually realized on the stationary assumption on sequential data, while it is hard to hold in complex traffic conditions that vary over time. Some machine learning methods such as SVM [7] and KNN [15] can model complex traffic data, but they need more costs for feature extraction.

Since deep learning has achieved great success in many domains, such as image processing and sequence learning, more researchers apply deep learning to traffic flow forecasting. Ma et al. [14] successfully applied LSTM to the traffic speed forecasting to capture long-term temporal dependencies. Ma et al. [13] utilized CNN on images of traffic speed for the speed forecasting problem. Zhang et al. [27] proposed to use residual CNN on the images of traffic flow. However, while these methods model only one of temporal dependencies and spatial dependencies, all of them overlook both aspects simultaneously.

To overcome the limitation as discussed previously, Shi et al. [18] proposed convolutional LSTM to capture spatial and temporal dependencies simultaneously. To further improve the performance of traffic forecasting, Yao et al. [24] proposed STDN to model spatial and temporal dependencies simultaneously by integrating CNN and LSTM. However, these methods with traditional CNN can only be applied to the grid-based traffic data, but cannot solve complex road network data with graph structure.

2.2 Graph Convolution Network

In recent years, the graph convolution successfully generalizes traditional CNN from grid-structured data to graph-structured data. Two mainstreams of graph convolution are spectral-based GCN and spatial-based GCN. The spectral-based GCN is developed from the perspective of spectral domain with Laplacian matrix to represent the structure of a graph. Yu et al. [25] used spectral-based GCN to learn spatio-temporal features of traffic data applicable to undirected graph. Guo et al. [5] added attention mechanisms to capture the dynamic spatial and temporal correlation of traffic flow. However, spectral-based GCNs are only applicable to undirected graph due to the symmetry of Laplacian matrix.

Spatial-based GCNs perform convolution filters on a graph's nodes and their neighbors with adjacency matrix. Li et al. [11] have combined diffusion convolution and RNN to incorporate both spatial and temporal dependency in the traffic flow. Song et al. [19] proposed STSGCN to simultaneously extract localized spatio-temporal correlation information based on the adjacency matrix of localized spatio-temporal graph. But these methods do not consider the dynamic spatial dependencies of traffic data.

3 Problem Formulation

In this section, we give some related definitions and formulate the problem of traffic flow forecasting.

Definition 1. (Road Network): A road network of a city can be represented as a directed graph, $\mathcal{G} = (V, E, A)$, where V is a set of road segments with $|V| = N$, E is a set of edges representing the adjacency among the road segments. For each $r \in V$, we use $r.s$ and $r.e$ to present the start and end vertexes of r. $A \in \mathbb{R}^{N \times N}$ is the adjacency matrix, e.g., $A[i, j] = 1$ if r_i and r_j have $r_i.e = r_j.s$, and $A[i, j] = 0$ if not.

Definition 2. (Trajectory): In original data, a point-based trajectory of one vehicle is recorded as a series of points with latitude, longitude and timestamp. According to our needs, we map trajectories to road network. For all points in a trajectory, we only keep the first point on each road segment to represent the time when a vehicle is transferred to a new road segment. We define a road-based trajectory as $T = \{tr^1, tr^2, ..., tr^M\}$. For each $tr \in T$, $tr = (t, r)$ denotes that the vehicle transfer from an upstream road segment to r at time interval t.

Definition 3. (Traffic Flow): For each road segment $r \in V$, the traffic flow X_t^i is defined as the amount of vehicles passed road segment r_i at time interval t. Given a set of historical trajectory records $\mathcal{T} = \{T_1, T_2, ..., T_H\}$, the traffic flow of road segment r_i at time interval t can be calculated by:

$$X_t^i = |\{T \in \mathcal{T} | \exists tr \in T \wedge tr.t = t \wedge tr.r = i\}|$$

Definition 4. (Traffic Flow Forecasting): Given the historical traffic flow of each road segment $\{X_i \in \mathbb{R}^N | i = 1, 2, ..., t-1\}$ before the time interval t, predict X_t.

4   Proposed Model

In this section, we introduce our proposed LGSTN model. We first give overall structure of the model, and then introduce the specific implementation of each component.

4.1   Overview of Model

The architecture of our proposed LGSTN is shown in Fig. 2. With the traffic flow of near time intervals as input, first, we process the historical trajectories data and construct transfer flow graphs for each time interval. Then we use a stack of modified spatial-based GCNs to model dynamic local flow aggregation. After that, we apply an LSTM to model the near temporal dependencies and add a gating-mechanism to control the global flow aggregation in each time step. At the meanwhile, with the traffic flow of the same time interval of previous days as input, we learn periodic representation for each road segment through an LSTM. Finally, we fuse the periodic and global representations of all road segments and obtain forecasting results.

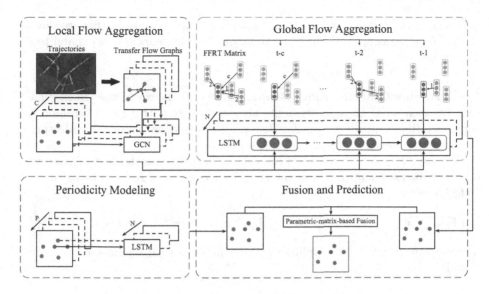

Fig. 2. The overview of LGSTN model.

4.2 Modeling Local Flow Aggregation

We notice that the traffic flow of a road segment is aggregated by the flow of adjacent upstream road segments in a short period as Fig. 1 shown. Convolution Neural Network can effectively capture local spatial dependencies, but it can only be applied for the 2D-grid data. In literatures, spectral-based and spatial-based graph convolution are two main approaches to generalize the traditional CNN to data of graph structures. Spectral-based graph convolution [22] is to transfer the data into spectral domain by graph Fourier transforms. However, this approach can only be applied in undirected graphs and road network is a directed graph. Inspired by diffusion convolution [11], we apply a modified spatial-based graph convolution for directed road network to achieve the information propagation from surrounding road segments to the central road segment and model the process of local flow aggregation.

Spatial-Based Graph Convolution. Our modified spatial-based graph convolution operation over the input of l-layer $X^{l-1} \in \mathbb{R}^{N \times F}$ and an adjacency matrix A is defined as:

$$X^l = GConv(X^{l-1}, A) = f(W^l \circ (D_I^{-1} A^T) X^{l-1} + b^l),$$

where $GConv()$ denotes the spatial-based graph convolution operation. $W^l \in \mathbb{R}^{N \times F}$ and $b^l \in \mathbb{R}^N$ are the learnable parameters of l-th graph convolution layer, and $\circ$ is Hadamard product. $D_I = diag(A^T 1)$ is the in-degree diagonal matrix, and $1 \in \mathbb{R}^N$ denotes an all one vector.

Compared with diffusion convolution, our modified spatial-based graph convolution only considers in-degree matrix (the traffic flow of a road segment depends on the inflow from upstream road segments, not the outflow to down-stream road segments). At the same time, our training parameters expand to N dimensions.

Constructing Transfer Flow Graphs. In reality, however, the transfer pattern between adjacent road segments always changes with time. Therefore, we construct time-dependent transfer flow graphs based on historical trajectories for each time interval.

We use $TF_t \in \mathbb{R}^{N \times N}$ to record the volume of traffic transfer between any two adjacent road segments at t time interval. Each element of TF_t can be calculated as following:

$$TF_t[i,j] = |\{T \in \mathcal{T} \wedge tr^{m-1}, tr^m \in T | tr^m.t = t \wedge tr^{m-1}.r = i \wedge tr^m.r = j\}|,$$

where $TF_t[i,j]$ denotes the volume transfering from road segment r_i to r_j at t time interval.

Generating Local Representation. For each input X_t, we replace the fixed adjacency matrix with the TF_t matrix and feed them into a spatial-based graph convolution layer. Then we stack L graph convolution layers to aggregate local information from L-hop neighbors. To avoid the vanishing and exploding gradient problems, we adopt residual learning [27] and linear shortcut connections after each graph convolution layer and choose $ReLU(x) = max(x, 0)$ as activation function. The formulation is defined as:

$$X_t^l = ReLU(GConv(X_t^{l-1}, TF_t) + X_t^{l-1}).$$

Finally, the output of L-th layer $X_t^L \in \mathbb{R}^{N \times F}$ is the local representation of each road segment at time interval t.

4.3   Modeling Global Flow Aggregation

From the perspective of whole city, the traffic flow of a road segment is also the aggregation of flow from distant roads after different propagation times, because it takes time for vehicles to travel from one road segment to another [20]. Thus, we design a Lag-Gated LSTM (LG-LSTM) and take C time intervals as input to explicitly add lag influence between road segments to every time step of LSTM. This gating mechanism helps to control the global aggregation at each input time interval.

Constructing Free-Flow Reachable Time Matrix. To measure the lag influence between any two road segments, we construct a free-flow reachable time matrix to represent the time for vehicles to travel between any two road segments. Free-flow speed is a term in ITS to describe the average speed that a vehicle would travel if there were no congestion or other adverse conditions (such as bad weather) [3]. We can estimate the average free-flow speed $FFS[i, j]$ between r_i and r_j through the historical trajectories and obtain free-flow speed matrix $FFS \in \mathbb{R}^{N \times N}$. Then we construct a distance matrix $Dist \in \mathbb{R}^{N \times N}$ calculated by the length of each road segment in road network, and each element $Dist[i, j]$ denotes the real roadway distance from r_i to r_j. Then the free-flow reachable time from r_i to r_j can be calculated as:

$$FFRT[i, j] = \frac{Dist[i, j]}{FFS[i, j]}.$$

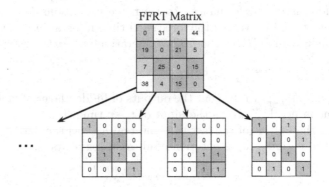

Fig. 3. Generating look-up matrices from FFRT matrix.

Lag-Gating Mechanism. Lag-gating Mechanism control global aggregation in each time step by looking up the FFRT matrix. For ease of calculation, we generate a 01 matrix G for each time step. Take Fig. 3 as an example, the size of time interval is set to $\Delta = 10\,min$. $FFRT[1, 0] = 19$ indicates that the traffic flow on Road 0 is influenced by Road 1 with a lag of two time intervals. Therefore, when we forecast the traffic flow of Road 0 at t time interval, the information of Road 1 at $t - 2$ time interval is needed. The look-up matrix of $t - c$ time step can be formulated as:

$$G_{t-c}[i, j] = \begin{cases} 1, & FFRT[i, j] \in ((c - 1)\Delta t, c\Delta t] \\ 0, & otherwise. \end{cases}$$

Finally, we add an identity matrix to each G to represent the influence of the road on itself.

Generating Global Representation. Based on the length of input sequences, we get C look-up matrices $\{G_{t-C}, ..., G_{t-2}, G_{t-1}\}$. The input of LSTM after lag-gating mechanism can be calculated as $\{G_{t-C}X_{t-C}^L, ..., G_{t-2}X_{t-2}^L, G_{t-1}X_{t-1}^L\}$. The LG-LSTM operation at $t - c$ time step is denoted as:

$$h_{i,t-c}^{glo} = LSTM(h_{i,t-c-1}^{glo}, G_{t-c}X_{t-c}^L),$$

where $h_{i,t-c}^{glo}$ and $h_{i,t-c-1}^{glo}$ represents the output of hidden layer of r_i at $t-c$ and $t-c-1$ time interval, $G_{t-c}X_{t-c}^L$ is the input of $t-c$ time step.

We regard the output of the last time interval $h_{i,t}^{glo}$ as the global representation of r_i. The global representation of all road segments at t time interval $S_{glo} = \{h_{1,t}^{glo}, h_{2,t}^{glo}, ..., h_{N,t}^{glo}\}$.

4.4   Modeling Periodic Dependencies

One of the common methods for forecasting near future traffic flow is to use the latest time intervals. Besides, we can also use the periodic patterns to improve traffic flow forecasting performance. To learn daily periodic dependencies, we select the same time intervals of the forecasted time interval in previous P days and feed them into an $LSTM$. The $LSTM$ operation at $t-p$ time step is denoted as:

$$h_{i,t-p}^{per} = LSTM\left(h_{i,t-p-1}^{per}, X_{i,t-p}\right),$$

where $h_{i,t-p}^{per}$ and $h_{i,t-p-1}^{per}$ represent the outputs of hidden layer of r_i at $t-p$ and $t-p-1$ time interval, $X_{i,t-p}$ is the input of $t-p$ time step.

We regard the output of the last time interval $h_{i,t}^{per}$ as the periodic representation of i. The periodic representation of all road segments is:

$$S_{per} = (h_{1,t}^{per}, h_{2,t}^{per}, ..., h_{N,t}^{per})^T.$$

4.5   Fusion and Forecasting

As mentioned previously, traffic flows are affected by near and periodic temporal dependencies. The degree of impact on traffic flows may be different. We fuse the global representation S_{glo} and periodic representation S_{per} by a parameter-matrix-based fusion matrix:

$$\hat{X}_t = \tanh\left(W_{glo} \circ S_{glo} + W_{per} \circ S_{per}\right),$$

where W_{glo} and W_{per} are learnable parameters, $\circ$ is element-wise multiplication.

Our model can be trained via backpropagation to forecast X_t by minimizing mean squared error between the forecasted value and the true value:

$$L(\theta) = \left\| X_t - \hat{X}_t \right\|_2^2,$$

where θ are all learnable parameters in our model.

5   Experiments

We conduct experiments on two real-world datasets in Chengdu and Xi'an to evaluate the effectiveness of our model respectively. We first describe the dataset used in our experiments followed by the baselines. Then we show the experimental settings including hyperparameters and evaluation metrics. At last, we present the experimental results in detail.

5.1   Datasets Description

We conduct experiments on two real-world large scale trajectory datasets collected by Didi Chuxing[1] in cities: **Chengdu** and **Xi'an**. Both of them are collected within the period from Nov 1st, 2016 to Nov 30th, 2016. In the preprocessing stage, we map the trajectory data to road network which is provided by OpenStreetMap[2]. Using Definition 3, we obtain traffic flows of each road segment. After removing road segments with sparse data, Chengdu has 528 and Xi'an has 221 road segments. Besides, we select time interval $\Delta t = 5\,\text{min}/10\,\text{min}/15\,\text{min}$ in two datasets to evaluate the universality of our model. We choose data from the last 7 days as the testing set, and all data before that as training set.

5.2   Experimental Settings

Baselines. We compared our model with the following methods.

- **HA** (Historical Average): Historical average predicts the demand using average values of previous demands at the location given in the same relative time interval (i.e., the same time of the day).
- **SVR** (Support Vector Regression) [10]: A well-known analysis method that uses linear support vector machine for the regression task.
- **LSTM** (Long Short-term Memory) [14]: As an improvement of RNN, FC-LSTM has a good performance in dealing with long temporal dependencies and reduce gradient vanishing and exploding.
- **DCNN** (Deep Convolution Neural Network) [13]: DCNN combines convolution, pooling and fully-connected for traffic prediction.
- **STDN** (Spatial-Temporal Dynamic Network) [24]: STDN handles spatial and temporal information via local CNN and LSTM. Flow gating mechanism and periodically shifted attention mechanism are proposed to learn spatial dynamic similarity and temporal dynamic periodic similarity respectively.
- **DCRNN** (Diffusion Convolution Recurrent Neural Network) [11]: Diffusion Convolutional regards graph convolutions as a diffusion process. DCRNN is a method to combine diffusion convolution and RNN for capturing spatio-temporal dependencies.
- **STSGCN** (Spatial-Temporal Synchronous Graph Convolutional Networks) [19]: STSGCN simultaneously extracts localized spatio-temporal correlation information based on the adjacency matrix of localized spatio-temporal graph.

Hyperparameter. For spatial-based GCN, we set $L = 3$ (number of graph convolution layers), $F = 16$ (number of filters in each graph convolution layer). We set the sequence length $C = 4$ to model flow aggregation and $P = 3$ for periodicity modeling. The dimension of hidden output of LSTM is 32. We train our

---

[1] https://gaia.didichuxing.com.
[2] https://www.openstreetmap.org.

model with the following hyperparameters: batch size (8), learning rate (0.0002) with adam optimizer. We also set early-stopping in all the experiments. The hyperparameters of baselines are same as the setting in their paper.

Evaluation Metrics. In our experiment, we use Root Mean Square Error (RMSE) and Mean Absolute Errors (MAE) as evaluation metrics. They are defined as the following equations:

$$RMSE = \sqrt{\frac{1}{N} \sum_{i=1}^{N} (X_{i,t} - \hat{X}_{i,t})^2}, \qquad MAE = \frac{1}{N} \sum_{i=1}^{N} \left| X_{i,t} - \hat{X}_{i,t} \right|,$$

where $X_{i,t}$ and $\hat{X}_{i,t}$ represent the real value and forecasting value of r_i at t time interval, N is the total number of road segments in test set.

5.3   Experimental Results

Table 1. Performance comparison of different methods on Chengdu dataset.

Chengdu	$\Delta t = 5\,min$		$\Delta t = 10\,min$		$\Delta t = 15\,min$	
	RMSE	MAE	RMSE	MAE	RMSE	MAE
HA	13.86	6.43	25.14	12.01	32.86	15.67
SVR	12.26	5.20	20.66	9.81	26.43	13.19
LSTM	11.84 ± 0.30	5.42 ± 0.09	21.13 ± 0.60	10.21 ± 0.23	30.44 ± 2.25	14.84 ± 0.87
DCNN	9.92 ± 0.56	5.15 ± 0.33	17.30 ± 0.73	9.49 ± 0.24	26.12 ± 1.14	14.07 ± 0.59
STDN	7.56 ± 0.39	4.45 ± 0.25	14.86 ± 0.71	8.85 ± 0.55	22.34 ± 0.94	13.76 ± 0.77
DCRNN	6.58 ± 0.22	3.85 ± 0.09	13.87 ± 0.84	8.28 ± 0.27	20.44 ± 0.62	11.78 ± 0.44
STSGCN	6.35 ± 0.12	3.84 ± 0.06	12.92 ± 0.52	7.84 ± 0.28	20.11 ± 0.63	**11.01 ± 0.64**
STN	7.45 ± 0.45	4.40 ± 0.29	14.25 ± 0.37	8.08 ± 0.35	21.37 ± 1.02	12.77 ± 0.76
LSTN	6.21 ± 0.24	3.60 ± 0.14	13.85 ± 1.10	7.99 ± 0.68	20.91 ± 0.64	12.04 ± 0.37
GSTN	6.25 ± 0.28	3.59 ± 0.13	13.51 ± 0.76	7.82 ± 0.34	19.71 ± 0.80	11.36 ± 0.58
LGSTN	**5.18 ± 0.14**	**3.35 ± 0.15**	**12.36 ± 1.12**	**7.70 ± 0.36**	**19.19 ± 0.30**	11.60 ± 0.18

Performance of Different Methods. We compare our models with the 7 baseline methods on Chengdu and Xi'an datasets with 5 min/10 min/15 min time intervals, as shown in Table 1 and Table 2. We report the performance on the testing set over multiple runs. LGSTN achieves the lower RMSE (5.18/12.36) and the lower MAE (3.35/7.70) compared with 7 baseline methods on Chengdu dataset with 5 min/10 min time intervals. As for 15 min time interval, the RMSE of our method is best, but STSGCN has the lower MAE. Nevertheless, our method is more stable than STSGCN. In the meanwhile, LGSTN also has a good performance on Xi'an dataset. For example, it reduces RMSE by 21.2% and MAE by 12.9% compared with the best baseline method on Chengdu dataset with 5 min interval, RMSE by 10.8% and MAE by 3.2% on Xi'an dataset with 10 min interval.

Table 2. Performance comparison of different methods on Xi'an dataset.

Xi'an	$\Delta t = 5\,min$		$\Delta t = 10\,min$		$\Delta t = 15\,min$	
	RMSE	MAE	RMSE	MAE	RMSE	MAE
HA	6.33	3.46	15.10	7.49	27.36	12.69
SVR	5.47	3.09	11.51	6.19	18.28	9.53
LSTM	5.52 ± 0.06	3.14 ± 0.03	11.67 ± 0.09	6.32 ± 0.03	18.91 ± 0.61	9.92 ± 0.22
DCNN	5.63 ± 0.23	3.35 ± 0.09	11.97 ± 0.33	7.02 ± 0.31	18.72 ± 0.42	10.82 ± 0.33
STDN	3.91 ± 0.24	2.50 ± 0.12	9.30 ± 0.61	5.72 ± 0.47	14.86 ± 0.79	$\mathbf{8.94 \pm 0.53}$
DCRNN	5.63 ± 0.07	3.40 ± 0.03	11.42 ± 0.20	6.51 ± 0.13	16.12 ± 0.49	8.96 ± 0.18
STSGCN	4.56 ± 0.09	2.58 ± 0.05	9.50 ± 0.34	5.11 ± 0.17	16.59 ± 0.65	9.53 ± 0.28
STN	4.43 ± 0.06	2.63 ± 0.10	11.25 ± 0.30	6.97 ± 0.39	16.54 ± 0.81	9.85 ± 0.58
LSTN	4.22 ± 0.18	2.49 ± 0.16	9.88 ± 0.62	5.93 ± 0.47	15.81 ± 0.72	9.24 ± 0.45
GSTN	3.88 ± 0.08	2.40 ± 0.04	9.57 ± 0.66	5.90 ± 0.47	15.24 ± 0.57	9.50 ± 0.53
LGSTN	$\mathbf{3.87 \pm 0.16}$	$\mathbf{2.38 \pm 0.06}$	$\mathbf{8.29 \pm 0.12}$	$\mathbf{4.95 \pm 0.07}$	$\mathbf{14.27 \pm 1.37}$	9.02 ± 1.03

Besides, we observe some phenomena in both datasets. Specifically, we can see that SVR achieves an acceptable performance because it has relatively many parameters. LSTM and DCNN can not consider spatial and temporal dependencies simultaneously. Although they have a similar or better effect as SVR, the performance of these methods is much worse than our LGSTN. STDN integrates CNN with LSTM and outperforms above neural-network-based methods. Nevertheless, its shortcoming is also obvious that it ignores the topology of road network. DCRNN combines diffusion convolution and RNN to capture short-term temporal dependencies. STSGCN uses same module to capture the localized spatial-temporal correlations. They achieve better performances in both datasets, and MAE of STSGCN and STDN is even lower than that of LGSTN in Chengdu and Xi'an dataset respectively with 15 min interval. However, in other cases, our proposed LGSTN can achieve better performances.

Effectiveness of Local and Global Aggregation. To verify the effectiveness of local and global aggregation, we evaluate some variants with different following components:

- **STN**: A basis model integrating spatial-based GCN with LSTM to capture spatial dependencies and near temporal dependencies. At the same time, we model the periodic dependencies and fuse the outputs.
- **LSTN**: Based on STN, our modified spatial-based GCN is adopted and we use time-dependent transfer flow graphs to model local dynamic flow aggregation.
- **GSTN**: Based on STN, we build FFRT matrix and utilize LG-LSTM with Lag-gating mechanism to model global flow aggregation.

Table 1 and Table 2 also show the results of different variants on both datasets. Through observation, we can find that adding any components causes a significant error reduction which illustrates the effectiveness of each component. Take Chengdu dataset with 5 min interval as an example, after adopting our modified spatial-based GCN and replacing fixed adjacency matrix by transfer flow matrices we construct from trajectories, RMSE reduces by 16.6% and MAE reduces by 18.1%. By constructing FFRT matrix and adopting LG-LSTM based on STN, RMSE reduces by 16.1% and MAE reduces by 18.4%. Once we realize local and global aggregation simultaneously, our proposed LGSTN has a greater improvement with 30.4% on RMSE and 23.8% on MAE.

Performance on Different Days and Road Segments. We analyze the detailed results on both datasets with 10 min interval. Figure 4(a) and Fig. 4(c) show the average MAE of different methods on each road segment. Due to the large scale of road network, we select 20 road segments with the largest traffic flow in Chengdu and Xi'an datasets respectively to display. We think that these road segments can better reflect the local and global aggregation patterns. It is observed that LGSTN model outperforms other baseline methods, illustrating that the proposed model achieves great success in spatial domain. Figure 4(b) and Fig. 4(d) show the average MAE of different methods during one week in test set. In the figure, our proposed LGSTN achieves best results consistently in all days except Thursday on Chengdu dataset, demonstrating that LGSTN also perform excellently in temporal domain.

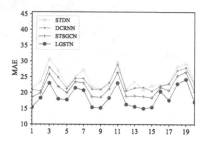 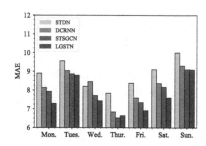

(a) MAE on 20 road segments on Chengdu dataset.   (b) MAE during one week on Chengdu dataset.

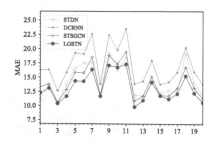 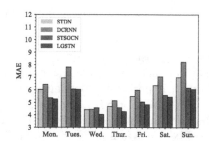

(c) MAE on 20 road segments on Xi'an dataset.   (d) MAE during one week on Xi'an dataset.

Fig. 4. Detailed results of Chengdu and Xi'an datasets.

6   Conclusion

In this paper, a novel local and global spatio-temporal network(LGSTN) is proposed and successfully applied to traffic flow forecasting. Our approach captures the spatial dependencies of urban road segments from the perspective of traffic flow aggregation. First, we construct time-dependent transfer flow graphs and apply a modified spatial-based GCN to model local flow aggregation. Then we propose a Lag-gated LSTM by considering propagation time through a free-flow reachable time matrix to model global flow aggregation. We also learn periodic patterns using LSTM to improve forecasting accuracy. The experimental results on two real-world datasets demonstrate the effectiveness of LGSTN model for traffic flow forecasting problem.

For future work, we plan to improve our research by the following aspects: (1) take some external factors into account such as weather and holidays; (2) extend our proposed model for multiple-step forecasting.

Acknowledgements. This research is supported in part by the 2030 National Key AI Program of China 2018AAA0100503 (2018AAA0100500), National Science Foundation of China (No. 61772341, No. 61472254, No. 61772338 and No. 61672240), Shanghai Municipal Science and Technology Commission (No. 18511103002, No. 19510760500, and No. 19511101500), the Innovation and Entrepreneurship Foundation for oversea high-level talents of Shenzhen (No. KQJSCX20180329191021388), the Program for Changjiang Young Scholars in University of China, the Program for China Top Young Talents, the Program for Shanghai Top Young Talents, Shanghai Engineering Research Center of Digital Education Equipment, and SJTU Global Strategic Partnership Fund (2019 SJTU-HKUST).

References

1. Chen, C., Li, K., Teo, S.G., Zou, X., Wang, K., Wang, J., Zeng, Z.: Gated residual recurrent graph neural networks for traffic prediction. Proc. AAAI Conf. Artif. Intell. **33**, 485–492 (2019)
2. Cho, K., et al.: Learning phrase representations using RNN encoder-decoder for statistical machine translation. Computer Science (2014)
3. Cui, Z., Henrickson, K., Ke, R., Wang, Y.: High-order graph convolutional recurrent neural network: a deep learning framework for network-scale traffic learning and forecasting. CoRR abs/1802.07007 (2018)
4. Engström, R.: The roads' role in the freight transport system. Transp. Res. Procedia **14**, 1443–1452 (2016)
5. Guo, S., Lin, Y., Feng, N., Song, C., Wan, H.: Attention based spatial-temporal graph convolutional networks for traffic flow forecasting. Proc. AAAI Conf. Artif. Intell. **33**, 922–929 (2019)
6. Jabbarpour, M.R., Zarrabi, H., Khokhar, R.H., Shamshirband, S., Choo, K.-K.R.: Applications of computational intelligence in vehicle traffic congestion problem: a survey. Soft Comput. **22**(7), 2299–2320 (2017). https://doi.org/10.1007/s00500-017-2492-z
7. Jeong, Y.S., Byon, Y.J., Castro-Neto, M.M., Easa, S.M.: Supervised weighting-online learning algorithm for short-term traffic flow prediction. IEEE Trans. Intell. Transp. Syst. **14**(4), 1700–1707 (2013)

8. Khetarpaul, S., Gupta, S.K., Subramaniam, L.V.: Analyzing travel patterns for scheduling in a dynamic environment. In: Cuzzocrea, A., Kittl, C., Simos, D.E., Weippl, E., Xu, L. (eds.) CD-ARES 2013. LNCS, vol. 8127, pp. 304–318. Springer, Heidelberg (2013). https://doi.org/10.1007/978-3-642-40511-2_21
9. Krizhevsky, A., Sutskever, I., Hinton, G.E.: Imagenet classification with deep convolutional neural networks. In: International Conference on Neural Information Processing Systems (2012)
10. Li, X., et al.: Prediction of urban human mobility using large-scale taxi traces and its applications. Front. Comput. Sci. China $6(1)$, 111–121 (2012)
11. Li, Y., Yu, R., Shahabi, C., Liu, Y.: Diffusion convolutional recurrent neural network: data-driven traffic forecasting. arXiv preprint arXiv:1707.01926 (2017)
12. Lv, Z., Xu, J., Zheng, K., Yin, H., Zhao, P., Zhou, X.: LC-RNN: A deep learning model for traffic speed prediction. In: IJCAI, pp. 3470–3476 (2018)
13. Ma, X., Dai, Z., He, Z., Ma, J., Wang, Y., Wang, Y.: Learning traffic as images: a deep convolutional neural network for large-scale transportation network speed prediction. Sensors $17(4)$, 818 (2017)
14. Ma, X., Tao, Z., Wang, Y., Yu, H., Wang, Y.: Long short-term memory neural network for traffic speed prediction using remote microwave sensor data. Transport. Res. Part C 54, 187–197 (2015)
15. May, M., Hecker, D., Korner, C., Scheider, S., Schulz, D.: A vector-geometry based spatial KNN-algorithm for traffic frequency predictions. In: IEEE International Conference on Data Mining Workshops (2008)
16. Moreira-Matias, L., Gama, J., Ferreira, M., Mendes-Moreira, J., Damas, L.: Predicting taxi-passenger demand using streaming data. IEEE Trans. Intell. Transport. Syst. $14(3)$, 1393–1402 (2013)
17. Pan, B., Demiryurek, U., Shahabi, C.: Utilizing real-world transportation data for accurate traffic prediction. In: 12th IEEE International Conference on Data Mining, ICDM 2012, Brussels, Belgium, 10–13 December 2012, pp. 595–604 (2012)
18. Shi, X., Chen, Z., Hao, W., Yeung, D.Y., Wong, W., Woo, W.: Convolutional LSTM network: a machine learning approach for precipitation nowcasting. In: International Conference on Neural Information Processing Systems (2015)
19. Song, C., Lin, Y., Guo, S., Wan, H.: Spatial-temporal synchronous graph convolutional networks: a new framework for spatial-temporal network data forecasting. In: AAAI Conference on Artificial Intelligence (2020)
20. Tao, Y., Sun, P., Boukerche, A.: A novel travel-delay aware short-term vehicular traffic flow prediction scheme for vanet. In: 2019 IEEE Wireless Communications and Networking Conference (WCNC), pp. 1–6. IEEE (2019)
21. Wu, Z., Li, J., Yu, J., Zhu, Y., Xue, G., Li, M.: L3: Sensing driving conditions for vehicle lane-level localization on highways. In: IEEE INFOCOM 2016-The 35th Annual IEEE International Conference on Computer Communications, pp. 1–9. IEEE (2016)
22. Wu, Z., Pan, S., Chen, F., Long, G., Zhang, C., Yu, P.S.: A comprehensive survey on graph neural networks. arXiv preprint arXiv:1901.00596 (2019)
23. Xie, Y., Zhang, Y., Ye, Z.: Short?term traffic volume forecasting using kalman filter with discrete wavelet decomposition. Comput. Aided Civil Infrastructure Eng. $22(5)$, 326–334 (2010)
24. Yao, H., Tang, X., Wei, H., Zheng, G., Li, Z.: Revisiting spatial-temporal similarity: a deep learning framework for traffic prediction. In: AAAI Conference on Artificial Intelligence (2019)

25. Yu, B., Yin, H., Zhu, Z.: Spatio-temporal graph convolutional networks: a deep learning framework for traffic forecasting. arXiv preprint arXiv:1709.04875 (2017)
26. Yu, J., et al.: Sensing human-screen interaction for energy-efficient frame rate adaptation on smartphones. IEEE Trans. Mob. Comput. **14**(8), 1698–1711 (2014)
27. Zhang, J., Zheng, Y., Qi, D.: Deep spatio-temporal residual networks for citywide crowd flows prediction. In: Thirty-First AAAI Conference on Artificial Intelligence (2017)

NSA-Net: A NetFlow Sequence Attention Network for Virtual Private Network Traffic Detection

Peipei Fu[1,2], Chang Liu[1,2], Qingya Yang[1,2], Zhenzhen Li[1,2], Gaopeng Gou[1,2], Gang Xiong[1,2], and Zhen Li[1,2(✉)]

[1] Institute of Information Engineering, Chinese Academy of Sciences, Beijing, China
lizhen@iie.ac.cn
[2] School of Cyber Security, University of Chinese Academy of Sciences, Beijing, China

Abstract. With the increasing attention on communication security, Virtual private network(VPN) technology is widely used to meet different security requirements. VPN traffic detection and classification have become an increasingly important and practical task in network security management. Although a lot of efforts have been made for VPN detection, existing methods mostly extract or learn features from the raw traffic manually. Manual-designed features are often complicated, costly, and time-consuming. And, handling the raw traffic throughout the communication process may lead to the compromise of user privacy. In this paper, we apply bidirectional LSTM network with attention mechanism to the VPN traffic detection problem and propose a model named NetFlow Sequence Attention Network (NSA-Net). The NSA-Net model learns representative features from the NetFlow sequences rather than the raw traffic to ensure the user privacy. Moreover, we adopt the attention mechanism, which can automatically focus on the information that has a decisive effect on detection. We verify our NSA-Net model on the NetFlow data generated from the public ISCXVPN2016 traffic dataset. And the experiment results indicate that our model can detect VPN from non-VPN traffic accurately, and achieve about 98.7% TPR. Furthermore, we analyze the performance of our model in the presence of sampling and our model still achieves over 90% TPR and Accuracy at low sampling rates.

Keywords: NetFlow data · Vpn detection · Attention mechanism · Deep learning · Network security

1 Introduction

A virtual private network(VPN) [1] is a virtual network, which establishes secure and encrypted connections to help ensure that sensitive data is safely transmitted. At present, with the increasing emphasis on communication security, VPN technology is widely used in network communications to meet different security requirements. However, with the widespread application of VPN technology,

© Springer Nature Switzerland AG 2020
Z. Huang et al. (Eds.): WISE 2020, LNCS 12342, pp. 430–444, 2020.
https://doi.org/10.1007/978-3-030-62005-9_31

they also bring some challenges to network security and management. On the one hand, VPN is easy to be exploited by attackers or hackers to hide their malicious behaviors, making them difficult to detect [7]. On the other hand, VPN adopts tunneling protocols and encryption techniques, which make it difficult to detect VPN traffic from other encrypted non-VPN traffic [8]. In addtion, encrypted VPN traffic detection is also a huge challenge to traditional port-based and rule-based methods. Therefore, how to effectively identify VPN traffic has already become an increasingly important and practical task in network management and cyberspace security.

At present, VPN traffic detection attracts the widespread attention of academia [3–8]. Currently, the typical methods are based on machine learning [3–5] or deep learning [6–8]. The machine learning methods generally need to select effective features to detect VPN traffic. However, the features are often extracted artificially, which depends on professional experience heavily. And, these methods have a great dependence on these features. Once the features have changed, the model will fail.. In order to reduce the cost of manually constructing features, deep learning is gradually applied, which could learning features automatically. Although current deep learning models have achieved excellent results, they all incline to use the raw encrypted traffic or information from the raw encrypted traffic as input to learn features. It will inevitably lead to huge input and time-consuming problems of the model. At the same time, capturing and using raw traffic can also cause user privacy problems to some extent. To solve these problems, an alternative approach that can be taken into account is to use the NetFow data [9], which is proposed by Cisco and only contains session-level statistical information. Compared with state-of-the-art methods, little research has been found for VPN detection using NetFlow data.

In this paper, we propose a model named NetFlow Sequence Attention Network (NSA-Net) for VPN traffic classification. The NSA-Net learns representative features from the NetFlow sequences rather than the raw traffic to ensure the user privacy. As bidirectional LSTM is able to exploit sequential information from both directions and attention mechanism is able to focus on the important information, the bidirectional LSTM (bi-LSTM) network with attention mechanism [10] is adopted to capture the most important representative information in a NetFlow sequence. Specifically, the NSA-Net model includes four layers: input layer to input the NetFlow sequence, bi-LSTM layer, and attention layer to generate the features, output layer to output the predicted labels. The features for detection are learned automatically from the raw NetFlow sequences by the bi-LSTM layer and are boosted by the attention layer.

Our contributions can be briefly summarized as follows:

- We propose a NSA-Net model for VPN traffic detection. And, to our best knowledge, in the field of VPN detection, it is the first attempt that an attention mechanism is used to obtain features, and the NetFlow sequence is used as input to protect user privacy.

- The optimal NetFlow inputs that can discriminate VPN from non-VPN are demonstrated. Meanwhile, we improve the performances of our model by adjusting the class_weight parameter to solve the problem of data imbalance.
- Our NSA-Net produces excellent results on the NetFlow data generated from the public VPN-nonVPN traffic dataset (ISCXVPN2016) [4], outperforming other deep learning models and the state-of-the-art method. Even at low sampling rates, our model still works very well.

The rest of the paper is organized as follows. Section 2 summarizes the related work. The methodology is elaborated in Sect. 3, and the experiments are presented in Sect. 4. Finally, we conclude this paper in Sect. 5.

2   Related Work

In this section, we present some current related work in the area of VPN detection, NetFlow application, as well as the previous works on attention mechanism.

2.1   VPN Detection

Over the years, a great deal of researches have focused on how to detect VPN traffic on the Internet. Pattern recognition and machine learning are proposed to solve VPN detection problem. Abideen M.Z. et al. proposed a lightweight approach to detect VPN activity by extracting features from plain information [3]. Gerard D. et al. studied the effectiveness of flow-based time-related features with C4.5 and KNN to detect VPN traffic [4]. However, the feature selection of these methods requires rich experiences and lots of human effort.

Nowadays, since deep learning can learn features automatically, researchers pay more attention to the methods based on deep learning [6–8]. However, these methods mostly tend to learn features from: the raw traffic [6], or statistical information extracted from raw traffic [7], or image converted from raw traffic [8]. Learning from raw traffic could improve the results to a certain degree. But it also will lead to huge input and time-consuming problems. Especially, it may invade users' privacy, as the raw traffic data often contains sensitive information about network users.

2.2   NetFlow Application

NetFlow represents a high level summary of network conversations, which is an alternative to the raw traffic. With the development of NetFlow technology, the protocol is extremely attractive to academic research and has been widely used in many research fields [11–14], especially in the field of anomaly detection, traffic classification, network measurement and network monitoring. In recent years, NetFlow has also been gradually applied to the field of deep learning. Liu et al. [15] presented a method to detect network attacks and intrusions with CNN by constructing images from NetFlow data. Yang et al. [16] proposed RNN deep learning method to analyze NetFlow information for attack detection.

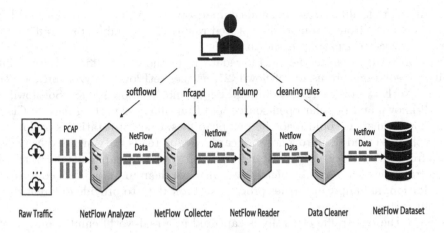

Fig. 1. The data preprocessing process of our work.

Although research on NetFlow has already very abundant, most of them are still based on machine learning, and very little is based on deep learning. Therefore, the application of deep learning on NetFlow is still in its early stage and needs to be explored more deeply.

2.3 Attention Mechanism

In recent years, the study of deep learning has become more and more in-depth, and there have been many breakthroughs in various fields. Neural networks based on attention mechanisms have become a recent hot topic in neural network research. Attention-based neural networks have recently been successful in a wide range of tasks [17–20], especially in image classification and natural language processing.

To the best of our knowledge, we have not found a deep learning algorithm based on the attention mechanism applied to NetFlow data for VPN detection. Therefore, in this paper, we explore the RNN with an Attention mechanism to learn NetFlow sequence features for VPN detection.

3 Methodology

In this work, we develop a NetFlow Sequence Attention Network, called NSA-Net, which based on NetFlow data for VPN detection tasks. In this section, the data preprocessing and the architecture of the proposed NSA-Net will be presented in the following.

3.1 Data Preprocessing

Before training the NSA-Net model, we need to prepare the NetFlow dataset to feed into the model. As it is very hard to label the real-world NetFlow data accurately, to facilitate research and comparison, we consider using the NetFlow data generated from the manual-captured raw traffic or public raw traffic. Our data preprocessing process is shown in Fig. 1.

In order to generate the NetFlow data from the raw traffic, the following software solutions are used: softflowd [21] as the NetFlow analyzer, nfcapd [22] as the NetFlow collector, and nfdump [22] as the NetFlow reader. Softflowd is a well-known and popular open-source tool that allows exporting data in Cisco NetFlow format. It could read the raw pcap files, calculate statistics of flow on its expiry and send them to a network card. Nfcapd reads the generated NetFlow data from the network card and stores them into files. Nfdump reads the NetFlow data from the files stored by nfcapd and displays them in a user-defined format. In addition, nfdump can further process NetFlow data to provide different data output formats.

Since the raw traffic generally is captured in a real-world emulation, it may contain some irrelevant packets that we don't care about. And the NetFlow data is transferred from the raw traffic, there are inevitably some outliers and noise in the data. Therefore, before using these data, data cleansing process is required. First, some irrelevant service data should be discarded, such as DNS service(port 53), which is used for hostname resolution and not relevant to either application identification or traffic characterization. Therefore, the NetFlow sequence based on port 53 is omitted from the dataset. Second, sequences with one-way flows and few packets are deleted. With no data interaction between the two sides of the communication, and the number of one-way packets is so small that such data sequence is difficult to characterize the traffic.

Finally, after processing the raw traffic, we collect NetFlow data sequences, including flow packets, flow bytes, flow duration, protocol, TCP flag and so on. To evaluate the NSA-Net model comprehensively and to explore which NetFlow data sequences work better, we generate four kinds of NetFlow data sequences at different sampling rates, which includes the unidirectional original NetFlow data sequence (U_ONData), unidirectional extended NetFlow data sequence (U_ENData), bidirectional original NetFlow data sequence (B_ONData) and bidirectional extended NetFlow data sequence(B_ENData). Bidirectional Net-Flow data sequences contains more information than unidirectional data. In addition, to verify the sensitivity and validity of our model, we take three sampling rates: 1, 1/10 and 1/100. Table 1 shows these NetFlow data sequences in detail. The *bps, pps, bpp* indicate the "bytes per second", "packets per second" and "bytes per packet" respectively.

3.2   The NetFlow Sequence Attention Network

In this section, we present our NetFlow Sequence Attention Network (NSA-Net) in detail. The NSA-net considers both feature learning and classification together. As shown in Fig. 2, NSA-Net in this paper contains four layers. In the following of this section, we will describe each layer in detail.

Table 1. The NetFlow dataset description.

Data type	Information description
U_ONData	{duration, protocol, sport, dport, TCP-flag, pktnum, bytnum}
U_ENData	U_ONData + {bps, pps, bpp}
B_ONData	{duration, protocol, sport, dport, TCP-flag, uplink-pktnum, uplink-bytnum, downlink-pktnum, downlink-bytnum, flow-num}
B_ENData	B_ONData + {uplink{bps, pps, bpp}, downlink{bps, pps, bpp} }

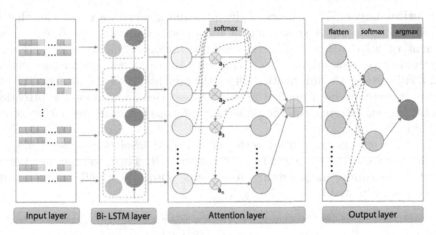

Fig. 2. The framework of our work.

Input Layer. In order to make the data fit the model better, we need to standardize our data to avoid outliers and extremes. In addition, classification labels encoded with one-hot coding. As we evaluate different kinds of NetFlow sequences on our proposed model, each input needs to be reshaped into a form of tensor to meet the input format of bidirectional LSTM. Then, the converted NetFlow sequences will be fed to the bi-LSTM layer.

Bi-LSTM Layer. This layer takes the converted vectors of a NetFlow sequence as input and generates high level features. To further improve the representation of the model, we use the bidirectional LSTM(bi-LSTM) in our model, which is an artificial recurrent neural network (RNN) architecture.

Bi-LSTM is an extension of traditional LSTM, which can connect two hidden layers of opposite directions to the same output. Therefore, it is able to exploit sequential information from both directions and provide bidirectional context to the network and result in fuller learning on the problem.

Given an input NetFlow sequence $S = [x_1, x_2, \cdots, x_n]$, the bi-LSTM contains a forward LSTM network which reads S from x_1 to x_n and a backward LSTM network which reads S from x_n to x_1. At time step $t(t \in [1, n])$, the $\overrightarrow{h}_t$ [Eq.(1)] and $\overleftarrow{h}_t$[Eq.(2)] are the forward and backward hidden states respectively. Then, the states h_t [Eq.(3)] is obtained from concatenating the forward hidden state and the backward hidden state at the time step t.

$$\overrightarrow{h}_t = \overrightarrow{LSTM}(\overrightarrow{h}_{t-1}, x_t) \tag{1}$$

$$\overleftarrow{h}_t = \overleftarrow{LSTM}(\overleftarrow{h}_{t-1}, x_t) \tag{2}$$

$$h_t = [\overrightarrow{h}_t, \overleftarrow{h}_t] \tag{3}$$

Attention Layer. The essence of the attention mechanism is the idea that it selectively filters out a small amount of important information from a large amount of information and focuses on the important information, ignoring mostly unimportant information. Attention can assign weights to information and finally weight the information. Since the appearence of the attention mechanism, it has now been applied in many areas. Here, we adopt the attention mechanism for filtering out the important information from the output of bi-LSTM layer.

Let H represent the output vector of bi-LSTM layer $[h_1, h_2, h_3, ..., h_n]$, we can get the vector e by a linear transformation of H. Then, softmax function is adopted to compute the weightings α of vector e(Eq.(4)). Finally, the attention vector c is calculated by Eq.(5).

$$\alpha = softmax(e), (\alpha_t = \frac{exp(e_t)}{\sum_{k=1}^{n} exp(e_k)}, t \in [1, n]) \tag{4}$$

$$c = \sum_{t=1}^{n} \alpha_t h_t \tag{5}$$

Output Layer. In the output layer, after flattening the output vector of the attentional layer, we use another softmax classifier to obtain the distribution of different categories (similar to Eq.(4)). And we take the category with the maximum probability as the prediction label y.

4   Experiments

In this part, we give a comprehensive introduction to our experiments. We mainly present our dataset, evaluation criteria, model setting, experiments and sensitivity analysis.

4.1   Dataset

In this paper, the public VPN-nonVPN traffic dataset (ISCXVPN2016) [4], which was published by the University of New Brunswick, is selected to evaluate our proposed model. There are two data formats in this traffic dataset, flow features and raw traffic (i.e. pcap format). We use the raw traffic dataset as our test dataset, which contains about 150 pcap files. The raw traffic is about 25 GB in the pcap format, which includes regular encrypted traffic (such as Skype and Facebook) and the corresponding traffic with VPN (such as VPN-Skype and VPN-Facebook). Data preprocessing, which is described in detail in Sect. 3.1, is conducted based on ISCXVPN2016 raw traffic. U_ONData, U_ENData, B_ONData and B_ENData represent the unsampled NetFlow data sequence. B_ENData_10S and B_ENData_100S represent the bidirectional extended NetFlow data sequence at 1/10 and 1/100 sampling rates. Finally, we use the datasets (Table 2) to test and verify our NSA-Net model.

Table 2. The statistical information of NetFlow datatset.

Data type	VPN flows	Non-VPN flows
U_ONData and U_ENData	5530	49625
B_ONData and B_ENData	2795	42695
B_ENData_10S	1453	8370
B_ENData_100S	510	863

4.2   Evaluation Critera

In this paper, we use True Positive Rate (TPR), False Positive Rate (FPR), and Accuracy(Acc) as our assessment indicators. Acc means the ratio of all correctly predicted samples and total samples, which is used to evaluate the overall performance of the model. Due to the existence of unbalanced data, the accuracy can not completely evaluate whether a model is good or not. Therefore, TPR and FPR indicators are selected to evaluate the model. TPR means the ratio of predicting positive sequences as positive and all positive sequences. And FPR means the ratio of predicting negative sequences as positive and all negative sequences.

4.3   NSA-Net Setting

To implement our proposed model, we choose Keras Library [24], with Tensorflow [23] as its backend, which runs on Redhat 7.2 64bit OS. We take the NetFlow sequences as the input of the NSA-Net. 1/5 of data were randomly selected as test data, and the rest is training data. The mini-batch size is 60 and the cost function is cross entropy. Moreover, we set the dimension of hidden states of each LSTM as 128, and take dropout with 0.4 ratio to avoid over-fitting, and the Adam optimizer [25] with learning rate 0.005 is used. Totally, the learning procedure adopts early-stopping.

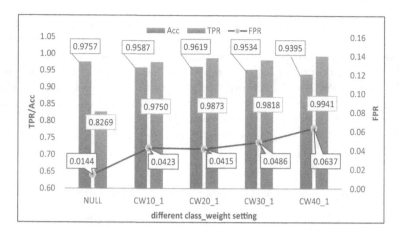

Fig. 3. Results of NSA-Net with different class_weight.

Class_weight Setting. As shown in Table 2, our non-VPN sequences number is much larger than that of the VPN sequence, which indicates samples of our NetFlow dataset are out-of-balance. On the problem of binary classification of unbalanced data, the accuracy indicator cannot fully reflect the performance of a model. In the paper, our goal is to get a higher TPR and a lowever FPR, while guaranteeing a certain accuracy. Keras provides a **class_weight** parameter for model fitting. In the process of model training, class_weight parameter can make the loss function pay more attention to the data with insufficient samples, by providing a weight or bias for each output class.

In order to solve the inbalance problem, different values of class_weight (i.e., 10:1, 20:1, 30:1, 40:1) are set, which means VPN samples are provided a bigger weight. As shown in Fig. 3, the higher the weight of the positive sample(VPN sample), the better the TPR obtained by the model. The parameter class_weight can effectively promote the TPR of our model. When we do not set the class_weight parameter, the TPR is only 82%. However, the TPR achieves more than 98% when class_weight is in {20:1, 30:1, 40:1}, and increases about 16%. Although it leads to increased FPR and reduced accuracy, the impact is small. Finally, considering the overall performance, we choose {20:1} as class_weight parameter by considering the performance of all metrics.

Optimal Inputs Selection. We have four kinds of NetFlow data, which contain U_ONData, U_ENData, B_ONData, and B_ENData. The comparison results are shown in Fig. 4. From the comparative results, we can obtain the following conclusions:

1. Bidirectional NetFlow sequence outperform the unidirectional sequence. Although TPR increases by only 1%, Acc increases by almost 15%, and FPR decreases by about 16%. The main reason is that the bidirectional NetFlow

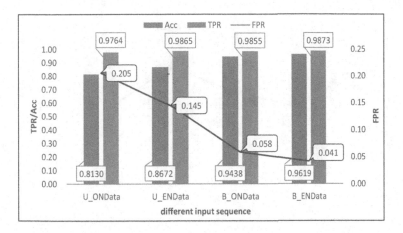

Fig. 4. Results of NSA-Net with different input sequences.

sequence aggregates two-way flow information, therefore it has better feature representation than the unidirectional sequence.

2. Extended sequence, which adds *bps, pps, bpp* information, have enhanced the performance of the model. For the bidirectional data, the Acc of extended sequence improves almost 2%, and the FPR decreases by 1%. As a result, the richer the information, the more expressive the model.

Therefore, we choose the bidirectional extended sequence (B_ENData) as our default input data.

4.4 Comparison Experiments

Different Deep Learning Models Comparison. Here, a number of typical deep learning models, including 1-layer bidirectional LSTM(1biLSTM), 2-layer bidirectional LSTM with Attention mechanism(2biLSTM+Att), 1D-CNN in [6] are applied to our NetFlow sequence to detect VPN.

Figure 5 respectively describes the experimental results of TPR, FPR, and Acc of these classifiers, which indicates that our proposed NSA-Net model outperforms three alternatives methods in VPN detection. The Accuracy and TPR of NSA-Net are higher than 1biLSTM by 1%, which indicates that the attention mechanism is able to enhance the performance of the model. Compared to our model, 2biLSTM+Att has no significant improvement, but it has one more bi-LSTM layer, which needs more time to train the model. In addition, 1D-CNN models in [6], which presents the best performance in VPN detection, is applied to our dataset. The results reveal that the ability of the 1D-CNN model to learn features from NetFlow data is not as good as our NSA-Net model. Therefore, the best overall performance is our proposed NSA-Net model.

Existing Methods Comparison. The state-of-the-art method [4] is compared with our proposed NSA-Net model comprehensively. As we focus on detecting

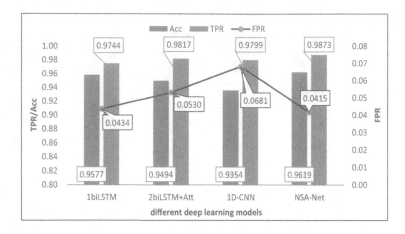

Fig. 5. Comparison results of different deep learning models.

VPN from non-VPN traffic, we compare the two-category identification results of the method. We make a comparison from the following aspects: (1) C45-TRF: the existing work in [4], (2) C45-BEND: the existing model used in [4] applied to our NetFlow data(B_ENData), (3) NSA-TRF: our NAS-Net model applied to the time-related features in [4], and (4) NSA-BEND: our work. Table 3 shows the comparison results between our proposed method and the state-of-the-art method in [4]. As their paper only showed the precision and recall(TPR) results, so we compare performance using the recall(Rec), precision(Pre), and F1-score(F1), which presents an overall evaluation of precision and recall.

As shown in Table 3, NetFlow sequence shows better than time-related features at these two methods, which indicates that flow sequence features have great potential in traffic detection. Obviously, the NSA-BEND performs the best, which indicates that our NSA-Net model performs very well in VPN detection based on our NetFlow data. Although our method needs a long time than existing C4.5 method, our overall performance is better than the other methods.

Table 3. Comparison results with existing method.

Method	Model	Data	Rec	Pre	F1
C45-TRF	C4.5 [4]	Time-related features [4]	0.9200	0.89	0.9048
C45-BEND	C4.5 [4]	B_ENData	0.8539	0.9818	0.9133
NSA-TRF	NSA-Net	Time-related features [4]	0.7388	0.7569	0.7477
NSA-BEND	NSA-Net	B_ENData	**0.9873**	**0.9735**	**0.9804**

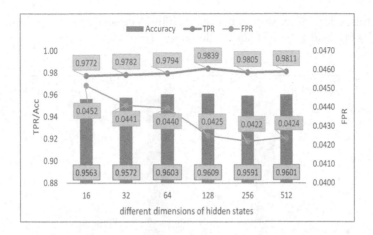

Fig. 6. The results of different dimensions of hidden states.

4.5   Sensitivity Analysis

The Dimension of Hidden States: Hidden states of bi-LSTM layer in the NSA_Net model is used to extract the latent usful information of the NetFlow data sequences. The larger units number, the stronger the ability to learn potential information, and the longer training time. The dimension of hidden states is vital to the performance of our model.

Therefore, we select different dimensions of bi-LSTM hidden states (i.e., 16, 32, 64, 128, 256, 512) to train our model. From the results shown in the Fig. 6, the performance of the model is improving as the dimensions of hidden states increases. Even the dimension of hidden state is small, our model still performs well. When the dimensions of hidden states exceeds 128, the training time becomes longer and the enhancement of detection effects becomes less and less obvious. Therefore, considering the balance between performance and training time, we choose 128 as the dimension of our paper.

The Sampling Rates: In this section, we analyze the influence of different sampling rates on the detection results. Due to the explosive growth of network traffic, packet sampling is widely used as a means of network measurement and network management to reduce resource consumption. The lower the sampling rate, the less useful information could be extracted and used.

Therefore, we focus on analyzing the result based on the bidirectional extended NetFlow sequence at different sampling rates(p = 1, 1/10, 1/100), which is presented in Fig. 7. Obviously, as the sampling rate decreases, the TPR and Acc decrease while the FPR rises. The main reason is that the sampled data is only a part of raw traffic data, and some information of traffic characteristics will be lost. The datasets(Table 2) reveals that as the sampling rate decreases, the amount of data decreases dramatically. The decrease in available information will inevitably lead to a decrease in detection rate.

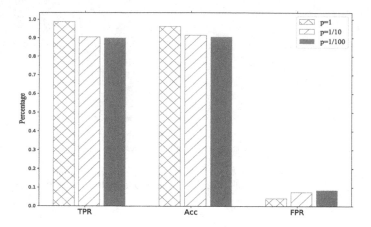

Fig. 7. The overall performance at different sampling rates.

Despite the decrease in available information, our model still achieves over 90% accuracy and TPR at low sampling rates. It reveals that our NSA-Net model still has great potential in sparse data environment.

From the above comprehensive comparison experiments, we can get that our proposed NSA-Net outperforms other deep learning models and the state-of-the-art method. While considering the user privacy and lightweight input, our model still achieves excellent performance. Therefore, the NSA-Net model is a very promising VPN detection method based on the NetFlow sequence, even at low sampling rate situation. In our future work, we will apply our NSA-Net model on a real-world Netlfow dataset. I believe that our work will still work very well.

5   Conclusion

In this paper, considering the feature automatic learning and user privacy protection, we propose a model named NAS-Net to detect VPN traffic. The NSA-Net takes the bidirectional LSTM model with attention mechanism to learn features from the NetFlow sequences rather than the raw traffic. As far as we know, this is the first time to combine attention mechanism and NetFlow data for VPN traffic detection. The model combines the advantages of a recurrent neural network and attention mechanism to learn representative information from Net-Flow data automatically and saves human effort to design features. In addition, we demonstrate that the bidirectional NetFlow data could improve the performance of detection. Thorough and comparative experiments on the NetFlow data generated from the public ISCXVPN2016 traffic dataset are conducted to validate the effectiveness of the NSA-Net model. The experiment results show that our NSA-Net can achieve an excellent performance(0.987 TPR, 0.041 FPR, and 0.962 Acc). Even at low sampling rate, our NSA-Net model still has great potential.

Acknowledgments. This work is supported by The National Key Research and Development Program of China (No. 2020YFE0200500 and No.2016QY05X1000) and The Key research and Development Program for Guangdong Province under grant No. 2019B010137003 and The National Key Research and Development Program of China (No. 2018YFB1800200). Zhen Li is the corresponding author.

References

1. Harmening, J.T.: Virtual private networks. In: Vacca, J.R. (ed.) Computer and Information Security Handbook, pp. 843–856. Morgan Kaufmann, Burlington (2017)
2. Lotfollahi, M., Siavoshani, M.J., et al.: Deep packet: a novel approach for encrypted traffic classification using deep learning. Soft Comput. **24**(3), 1999–2012 (2020). https://doi.org/10.1007/s00500-019-04030-2
3. Zain ul Abideen, M., Saleem, S., Ejaz, M.: VPN traffic detection in SSL-protected channel. Secur. Commun. Netw. **2019**(5), 1–17 (2019)
4. Draper-Gil, G., Lashkari, A.H., Mamun, M.S.I., et al.: Characterization of encrypted and vpn traffic using time-related. In: ICISSP, pp. 407–414 (2016)
5. Bagui, S., Fang, X., et al.: Comparison of machine-learning algorithms for classification of VPN network traffic flow using time-related features. J. Cyber Secur. Technol. **1**(2), 108–126 (2017)
6. Wang,W., Zhu, M., Wang, J., et al.: End-to-end encrypted traffic classification with one-dimensional convolution neural networks. In: 2017 IEEE International Conference on Intelligence and Security Informatics (ISI), pp. 43–48. IEEE (2017)
7. Miller, S., Curran, K., Lunney, T.: Multilayer perceptron neural network for detection of encrypted VPN network traffic. In: 2018 International Conference On Cyber Situational Awareness, Data Analytics And Assessment, pp. 1–8. IEEE (2018)
8. Guo, L., Wu, Q., Liu, S., et al.: Deep learning-based real-time VPN encrypted traffic identification methods. J. Real-Time Image Proc. **17**(1), 103–114 (2020). https://doi.org/10.1007/s11554-019-00930-6
9. Claise, B.: Cisco systems neflow services export version 9 (2004)
10. Zhou, P., Shi, W., Tian, J., et al.: Attention-based bidirectional long short-term memory networks for relation classification. In: Proceedings of the 54th Annual Meeting of the Association for Computational Linguistics, pp. 207–212 (2016)
11. Hofstede, R., Hendriks, L., Sperotto, A., et al.: SSH compromise detection using NetFlow/IPFIX. ACM SIGCOMM Comput. Commun. Rev. **44**(5), 20–26 (2014)
12. Schatzmann, D., Mühlbauer, W., Spyropoulos, T., et al.: Digging into HTTPS: flow-based classification of webmail traffic. In: Proceedings of the 10th ACM SIGCOMM Conference on Internet Measurement, pp. 322-327 (2010)
13. Manzoor, J., Drago, I., Sadre, R.: How HTTP/2 is changing Web traffic and how to detect it. In: 2017 Network Traffic Measurement and Analysis Conference (TMA), pp. 1–9. IEEE (2017)
14. Lv, B., Yu, X., Xu, G., et al.: Network traffic monitoring system based on big data technology. In: Proceedings of the International Conference on Big Data and Computing 2018, pp. 27–32 (2018)

15. Liu, X., Tang, Z., Yang, B.: Predicting network attacks with CNN by constructing images from NetFlow Data. In: BigDataSecurity, pp. 61–66. IEEE (2019)
16. Yang, C.T., Liu, J.C., Kristiani, E., et al.: NetFlow monitoring and cyberattack detection using deep learning with Ceph. IEEE Access 8, 7842–7850 (2020)
17. Mnih, V., Heess, N., Graves A.: Recurrent models of visual attention. In: Advances in Neural Information Processing Systems, pp. 2204–2212 (2014)
18. Bahdanau, D., Cho, K., Bengio, Y.: Neural machine translation by jointly learning to align and translate. Comput. Sci. arXiv preprint arXiv:1409.0473 (2014)
19. Chorowski, J., Bahdanau, D., Serdyuk, D., et al.: Attention-based models for speech recognition. Comput. Sci. 10(4), 429–439 (2015)
20. Luong, M.T., Pham, H., Manning, C.D.: Effective approaches to attention-based neural machine translation. Comput. Sci. arXiv preprint arXiv:1508.04025 (2015)
21. Softflowd. http://www.mindrot.org/projects/softflowd/
22. Nfdump. http://nfdump.sourceforge.net/
23. Abadi, M., Agarwal, A., et al.: Tensor-flow: large-scale machine learning on heterogeneous distributed systems, arXiv preprint arXiv:1603.04467 (2016)
24. Chollet, F., et al.: Keras (2017). https://github.com/fchollet/keras
25. Kingma, D.P., Ba, J.: Adam: a method for stochastic optimization. CoRR, vol. abs/1412.6980 (2014)

Spatial and Temporal Pricing Approach for Tasks in Spatial Crowdsourcing

Jing Qian[1], Shushu Liu[2], and An Liu[1(✉)]

[1] School of Computer Science and Technology, Soochow University, Suzhou, China
jqian197@stu.suda.edu.cn, anliu@suda.edu.cn
[2] Department of Communications and Networking, Aalto University, Espoo, Finland
liu.shushu@aalto.fi

Abstract. Pricing is an important issue in spatial crowdsourcing (SC). Current pricing mechanisms are usually built on online learning algorithms, so they fail to capture the dynamics of users' price preference timely. In this paper, we focus on the pricing for task requesters with the goal of maximizing the total revenue gained by the SC platform. By considering the relationship between the price and the task, space, and time, a spatial and temporal pricing framework based task-transaction history is proposed. We model the price of a task as a three-dimensional tensor (task-space-time) and complete the missing entries with the assistant of historical data and other three context matrices. We conduct extensive experiments on a real taxi-hailing dataset. The experimental results show the effectiveness of the proposed pricing framework.

Keywords: Spatial crowdsourcing · Pricing · Task assignment

1 Introduction

Spatial crowdsourcing (SC) employs workers to accomplish tasks at specific locations and has many important applications including information collection (e.g., OSM [4]), micro-tasks (e.g., Gigwalk [2]) and ride sharing (e.g., Uber [5] and DiDi [1]). One of the key challenges in SC is to determine the price of tasks reasonably, which is illustrated by the following taxi-hailing example. Upon receiving a request of taxi-hailing, the taxi-hailing platform decides a price for it. The price should be as high as possible to maximize the profit of the platform, but on the other hand should be also affordable by requesters who will cancel their requests otherwise. When it comes to hiring workers, overpricing leads to high labor cost, but underpricing may prevent workers from accepting tasks. Therefore, a suitable price is very important to the SC platform.

Nevertheless, existing pricing mechanisms [10,11,18,19] are not quite ready for dynamic and online SC scenarios. On one hand, most of them work interactively with the environment and thus can provide a satisfying result only after a long term training. These slow starters are not suitable for some real-time SC applications. On the other hand, they fail to capture the dynamics of users

© Springer Nature Switzerland AG 2020
Z. Huang et al. (Eds.): WISE 2020, LNCS 12342, pp. 445–457, 2020.
https://doi.org/10.1007/978-3-030-62005-9_32

preference of price. In a taxi-hailing system, for example, requesters will accept higher price in rush hours than in off-peak hours, or they are willing to pay higher prices when visiting new places with many friends than going home alone.

To overcome the above weakness, we propose a spatial and temporal pricing framework. As a first step, we focus on pricing only for requesters, that is, to determine how much requesters should pay when their tasks are completed. In the model, we try to learn the price preference of requesters from historical transaction data directly. To accurately capture the dynamics of price preference, we consider historical transactions from three aspects: the category feature, the temporal feature and the spatial feature. More specifically, we model the price of a task by a three-dimensional tensor. Considering the tensor is too sparse to be learned effectively, we further generate a historical tensor and three context matrices that are much denser. With the help of these data, we complete missing entries of the original tensor through tensor decomposition. To empirically validate the performance of our spatial and temporal pricing framework, we conduct extensive experiments on a real-world taxi-hailing dataset.

In the rest of the paper, Sect. 2 reviews related work and Sect. 3 formally presents the problem definition. Section 4 shows the details of our model and Sect. 5 reports the evaluation result. Finally Sect. 6, concludes this paper.

2   Related Work

Existing pricing mechanisms for crowdsourcing can be divided into two groups: one is pricing for requesters, the other is for workers. For requesters, [19] explored and boosted requesters price acceptance estimation in grid areas using upper confidence bound (UCB [6]). Based on the price acceptance estimation, they dynamically decided the price for each grid by considering worker supply and dependency between grids to maximize total revenue. For workers, [12] converted the pricing problem into an equivalent multi-arm bandit (MAB) problem without setting a price range, then developed an algorithm to decide the price for each coming worker by considering the budget. [18] discretized the range of price into K candidates and used a heuristic policy that selects the price with the highest upper confidence bound of the expected revenue under the budget constraint. [11] considered both the recruiting cost and task completion quality by formulating it as a nonlinear integer programming problem, and proposed three algorithms by using their discoveries on Poisson binomial distributions. [10] proposed an online pricing mechanism to stimulate multi-minded workers with unknown cost distributions in mobile crowdsourcing. They solved the problem by modeling it as a dummy semi-bandits with multiple knapsacks and random cost problem.

The above work concentrated on online learning algorithms that explore price acceptance. There are many other pricing mechanisms proposed in the literature of crowdsourcing. These mechanisms focused on a different scenario. For example, procurement auction determines price based on workers' truthful bids [7,17,21]. Due to a large number of workers, this architecture will significantly increase the communication cost, but our pricing mechanism does not have this problem. Another direction is to determine payment according to the quality of work [8,16,20]. [16] and [20] considered the offline scenarios that all workers will stay from the beginning until the end while [8] considered the online scenarios that workers may arrive and leave anytime. They evaluated the workers reliability or work quality and then select workers and decide monetary compensation. However, the quality of the task and worker selection problems are beyond the scope of our problem. [13] and [9] solved the pricing problem with a completely different approach. They considered the dynamics of worker arrival, modeled and solved the pricing problem by using game theory and decision theory. Research in this direction considers pricing problems over a long period of time, while we aim to handle the pricing problems in a short time slot in which workers and requesters are static.

3   Problem Definition

In this section, we first define some terms used throughout the paper and then define our target problem.

Definition 1 (Spatial Task). A spatial task (or task for short) s is defined as a tuple $\langle t, \phi, ori, des, c, v \rangle$. More specifically, it is published at time t, and will be expired at time $t + \phi$. To complete s, a worker must travel from an origin location ori to a destination location des. Task s has a category c, which is a label used to represent the characteristics of the task (e.g., taxi-hailing tasks for economic and luxury cars belong to different categories). Besides, s is submitted by a requester who has a private valuation v representing the maximum price she is willing to pay.

It should be noted that the above valuation v is private, that is, the value is only known to the requester.

Definition 2 (Worker). A worker w is defined as a tuple $\langle t, \phi, l, r \rangle$ where the range $[t, t + \phi)$ indicates the time that w is available, l is the place that w is currently in, and r is the working radius, that is, w is willing to perform a task s if and only if the distance (e.g., Euclidean distance) between them is not larger than r.

In our model, a worker can handle only one task at a time and will definitely accept the task assigned to her.

Definition 3 (Pricing). Given a set S of tasks, pricing is a process during which the SC platform gives for each task $s \in S$ a quote p.

After pricing, every task has an offer p. Recall that every task requester has a private valuation v. If p is larger than v, the task requester will cancel her task and notify the SC platform of her decision.

Definition 4 (Task Assignment Instance Set). Given a worker set $W = \{w_1, w_2, ...\}$ and a task set $S = \{s_1, s_2, ...\}$, a task assignment instance set A

is composed of tuples in the form of $\langle w, s, p \rangle$, meaning the requester of task s should pay p to the SC platform which arranges a valid worker w for s. Note a worker w is said to be valid for s if the distance between them is not larger than the worker's working radius $w.r$.

Definition 5 (Revenue). The revenue of a task assignment instance set A is the sum of the amount of money that the SC patform receives by completing all tasks in A.

Problem Statement: Given a set of workers W_i and a set of spatial tasks S_i that are available at time t_i, our problem is to determine the quote for the task in S_i and then find a task assignment instance set A that maximizes the revenue.

4   Spatial and Temporal Pricing Framework

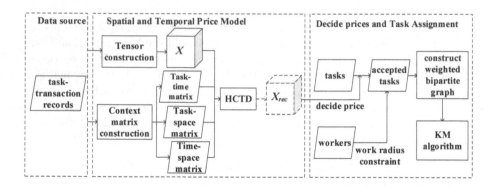

Fig. 1. Framework of our model

To solve the two problems mentioned in the introduction, our approach recovers price information for all different tasks through historical data by using a three-dimensional tensor. We comprehensively consider the dynamics of price from three aspects: the temporal feature, the category feature and the spatial feature. In the widely used NYC dataset, the three-dimensional tensor has only 0.061% non-zero entries. To address the data sparsity problem, we propose a History-based Context-aware Tensor Decomposition (HCTD) method. Based on the recovered price information, we decide quote for each requester and then assign workers to requesters that accept quotes to maximize the platform's revenue. Our framework (see Fig. 1) consists of two components: 1) Spatial and Temporal Price Model (STPM) using History-based Context-aware Tensor Decomposition (HCTD); and 2) Decide Prices for tasks (requesters) and Task Assignment (DPTA) component.

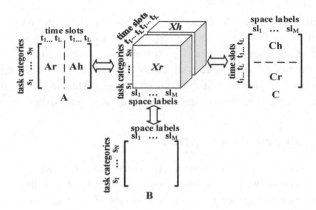

Fig. 2. Tasks' Spatial and Temporal Price Model

4.1 Tasks' Spatial and Temporal Price Tensor Construction

We build a task-space-time tensor, $\mathcal{X}_r \in \mathbb{R}^{N \times M \times L}$, based on task-transaction history in the most recent L time slots (e.g., L time slots in recent two days). Task-transaction history has recorded historical transaction information denoted as $H = \{h_1, h_2, ..., h_n\}$, each element is a transaction $h_i = \langle s_i, w_i, up_i \rangle$ representing that task s_i has been completed by a worker w_i at unit price up_i. Note that the unit price up_i is obtained by dividing the transaction price of task s_i by the distance from the $s_i.ori$ $s_i.des$. $\mathcal{X}_r$ is composed of three dimensions, i.e., task categories, space labels and time slots. The task category represents the category feature. The space label represents the spatial features, determined by the space types of task's origin and destination. The space types are generated from the functionality of the area, such as park and school. A time slot is a specific time interval of a day, for example, 14:00–14:10. Tensor $\mathcal{X}_r$ depicts the spatial and temporal unit price of tasks with different categories, as illustrated in Fig. 2. Given task category c, space label sl, time slot $\mathbb{T}$ and task-transaction history H, the value in tensor, denoted by $p(c, sl, \mathbb{T})$, is the average unit price of tasks in task-transaction history that category is c, space label is sl and begin time is in $\mathbb{T}$. The formula is as follows.

$$p(c, sl, \mathbb{T}) = \frac{\sum_{h_i \in H} \eta(h_i) up_i}{\sum_{h_i \in H} \eta(h_i)}$$

$$\eta(h_i) = \begin{cases} 1, s_i.c = c, s_i.sl = sl, s_i.t \in \mathbb{T} \\ 0, otherwise \end{cases}$$

(1)

To make $\mathcal{X}_r$ accurately capture unit price preferences, it is necessary to use the most recent data. However, only a few entries in $\mathcal{X}_r$ have transaction records in a short period, so there are many missing entries in the tensor. The unit price of all tasks can be easily obtained once these missing entries are completed. However, it is not accurate enough to perform decomposition only on its non-zero entries because $\mathcal{X}_r$ is very sparse. Thus we introduce another tensor $\mathcal{X}_h$, based on records during a longer period (e.g., two weeks). The tensor $\mathcal{X}_h$ is also aggregated from time slot 1 to L, so the structure is the same as $\mathcal{X}_r$. The tensor $\mathcal{X}_h$ is much denser as it represents the requesters' long-term expected unit price. By decomposing $\mathcal{X}_r$ together with $\mathcal{X}_h$, the error can be reduced. We divide it into two tensors rather than using all historical data. The main reason is that the change in price is not only a cyclical change by the day. It is possible that price in a longer time and in a more recent time are very different.

4.2 Context Matrix Construction

To make the decomposition of tensor $\mathcal{X}_r$ more effective, we further construct three denser matrices, which are task-time matrix A, task-space matrix B and time-space matrix C. They capture the similarity of unit prices in different task categories, as well as spatial and temporal correlation of unit price changes, so not only do they better solve the data sparsity problem due to higher density but also the additional information improve the accuracy of the recovered information.

Task-time Matrix. As shown in Fig. 2, matrix A consists of A_r and A_h, capturing the average unit price of different tasks over time. Each row denotes a task category and each column denotes a time slot. The matrix A_r and A_h respectively represent the recent and historical average unit price over the same span of time of a day. An entry $A_r(k, j)$ of $A_r \in \mathbb{R}^{N \times L}$, represents the average unit price of task category k in time slot j in recent time (e.g., two days) in task transaction history. The similarity of two different columns indicates the correlation of different time slots, for example, the price may be similar in time 8:30–9:30 and 17:00–18:00. The similarity of two different rows, likewise, indicates the similarity patterns of task categories.

Task-space Matrix. Matrix $B \in \mathbb{R}^{N \times M}$ captures the average unit price of different tasks with different space labels as shown in Fig. 2. Each entry $B(k, j)$ represents the average unit price of the k-th task in the j-th space label in the task transaction history. The similarity of two different columns indicates the similarity of two different space labels, for example, travel from a school to a well-known restaurant and a shopping mall.

Time-space Matrix. The time-space matrix $C \in \mathbb{R}^{2L \times M}$ consists of two parts: recent time-space matrix C_r and historical matrix C_h as shown in Fig. 2, capturing the average unit price of all tasks in a certain time slot and space label in the task transaction history, which represents unit price determined by space and time. For example, in rush hours the expected unit price of traveling from residential areas to office zones is higher than suburban areas. The matrix C_h has the same structure as C_r and the matrix C_h can be seen as the historical unit price that decided by time and space while C_r may show the short term change like facilities renovation.

4.3 Tensor Decomposition and Completion

We model tasks' spatial and temporal unit price with tensor $\mathcal{X}_r$ and estimate the missing entries. A straightforward way is using the tucker decomposition model [14]. Tucker decomposition decomposes tensor into a core tensor multiplied by three matrices. However, only decomposing tensor $\mathcal{X}_r$ is not accurate enough because tensor $\mathcal{X}_r$ is over sparse. Thus we combine $\mathcal{X}_r$ and $\mathcal{X}_h$ together as $\mathcal{X}$, and then decompose $\mathcal{X} \in \mathbb{R}^{N \times M \times 2L}$ with the aid of context matrices A, B and C, which is illustrated in Fig. 2. We leverage tucker decomposition to decompose $\mathcal{X}$ as a core tensor O and three matrices W, S and T.

$$\mathcal{X} \approx O \times_W W \times_S S \times_T T \tag{2}$$

where $O \in \mathbb{R}^{d_W \times d_S \times d_T}$ is a core tensor that shows the level of interaction between three components; $W \in \mathbb{R}^{N \times d_W}, S \in \mathbb{R}^{M \times d_S}$ and $T \in \mathbb{R}^{2L \times d_T}$ are the low-rank latent factor matrices; d_W, d_S and d_T are dimensions.

The three context matrices are factorized as follows: $A \in \mathbb{R}^{N \times 2L}$ can be factorized as $A = WFT^T$ in which $F \in \mathbb{R}^{d_W \times d_T}$; $B \in \mathbb{R}^{N \times M}$ can be factorized as $B = WGS^T$, in which $G \in \mathbb{R}^{d_W \times d_S}$ and $C \in \mathbb{R}^{2L \times M}$ can be factorized as $C = THS^T$, in which $H \in \mathbb{R}^{d_T \times d_S}$. Based on the tensor $\mathcal{X}$ and three context matrices A ,B and C, we decompose $\mathcal{X}$ and the loss function is Eq. 3.

$$\mathcal{L}(O,W,S,T,F,G,H) = \frac{1}{2}\|\mathcal{X} - O \times_W W \times_S S \times_T T\|^2$$
$$+ \frac{\lambda_1}{2}\|A - WFT^T\|^2 + \frac{\lambda_2}{2}\|B - WGS^T\|^2 + \frac{\lambda_3}{2}\|C - THS^T\|^2 \tag{3}$$
$$+ \frac{\lambda_4}{2}(\|O\|^2 + \|W\|^2 + \|S\|^2 + \|T\|^2 + \|F\|^2 + \|G\|^2 + \|H\|^2)$$

where $\|\cdot\|$ denotes the Frobenius norm, $\frac{\lambda_4}{2}(\|O\|^2 + \|W\|^2 + \|S\|^2 + \|T\|^2 + \|F\|^2 + \|G\|^2 + \|H\|^2)$ is a regularization of penalties to avoid over-fitting. And parameters λ_1, λ_2, λ_3 and λ_4 are used to control the contribution of different parts. We apply gradient descent algorithm to minimize the loss function. The missing entries are filled in the recovered tensor $\mathcal{X}_{rec}$, and $\mathcal{X}_{rec} = O \times_W W \times_S S \times_T T$.

4.4 Decide Prices and Task Assignment

For a given time slot, the platform decides to quote for each requester based on the unit prices in $\mathcal{X}_{rec}$. Then the platform receives reject-or-accept responses and

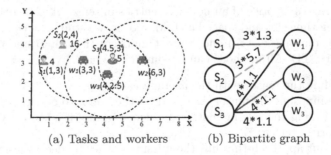

(a) Tasks and workers   (b) Bipartite graph

Fig. 3. An example of pricing process (Color figure online)

assigns workers to tasks that accept quotes. To consider the spatial distribution of workers and maximize the total revenue, we construct a bipartite graph between tasks and workers according to the working radius. The weights of the edges on the bipartite graph are set to the revenue of corresponding tasks. Then we use the typical Kuhn-Munkras (KM) algorithm [15] to do the allocation. As shown in Fig. 3, according to the distribution of workers and tasks in Figure (a), the bipartite graph can be constructed as Figure (b). The private valuation v of each requester is denoted as a red number in (a). In Figure (b), according to the task's category, space label, publish time and the $\mathcal{X}_r$ part in $\mathcal{X}_{rec}$, the unit price for s_1, s_2 and s_3 are 3, 3 and 4, respectively. Meanwhile, the travel distances are 1.3, 5.7, 1.1, respectively. So the quotes for s_1, s_2, and s_3 are 3.9, 17.1, and 4.4, respectively. Except for s_2, all requesters accept the quotes. After the KM algorithm makes the assignment, the allocation result is shown by the red line and the maximum total revenue is 8.3.

5   Experiment

5.1   Experimental Setup

We use a real-world dataset from NYC [3] to validate the proposed pricing model. We select the data from the 1st to the 30th day of January 2018 as the historical data, and the data on the 31st day to simulate online task requests. Based on the data on the 29th and 30th day, we build tensor $\mathcal{X}_r$. Since there is no worker information, we generate different numbers of workers. We focus on two time spans of a day, one is 0:00–3:00 as dataset1, and the other is 7:00–10:00 as dataset2. We generate task categories from the number of passengers (e.g., 1,2) and payment methods (e.g., cash, credit card). We separate the whole space into 263 areas according to space functionality. The parameters (e.g., λ_1, λ_2 and λ_3) of loss function are set to 0.007. The length of time period is set to 60 seconds in the task assignment phase. The parameters used in our experiments are summarized in Table 1 where the bold is the default setting. In the experiments of task assignment, we report the average total revenue.

Because there is no similar method to fill the tasks' price based on historical data, we proposed the following methods to evaluate our approach. A baseline algorithm: Average Value Filling (AVF) algorithm, which fills missing entries with the average unit price of all non-zero entries in tensor $\mathcal{X}$ that belong to the corresponding time slot and space label. Moreover, we study the contribution of the historical tensor (i.e., $\mathcal{X}_h$) and context matrices (i.e., A, B and C) by TD, TD+H and HCTD. All methods can be summarized as:

- **AVF:** Filling the tensor with the average value.
- **TD:** Decomposing the tensor $\mathcal{X}_r$ based on the non-zero entries of itself.
- **TD+H:** Filling the missing entries by decomposing the tensor $\mathcal{X}_r$ together with historical tensor $\mathcal{X}_h$.
- **HCTD:** Our approach that fills the missing entries by decomposing the tensor with historical tensor and three context matrices.

To evaluate the accuracy of filling missing entries, we adopt the widely-used measures, Root Mean Square Error (RMSE) and Mean Absolute Error (MAE). We randomly remove 20% of non-zero entries from tensor $\mathcal{X}_r$, which are used as testing set and the remaining 80% are used as training set.

Table 1. Parameter settings

Parameter	Settings
Valid time of tasks $s.\phi$(min)	5
Valid time of workers $w.\phi$(min)	5
Workers' radius $w.r$(km)	2
Time span of historical data h(week)	4,3,2,1
Span of a time instance ts(min)	10,30,60,120,180
Ratio of number of workers to tasks	**0.8**,1.2,1.6,2,2.4,2.8,3,3.5,4,5,6,7,8

Table 2. Performance of different methods for STPM

Methods	MAE	RMSE
AVF	4.476	39.304
TD	4.243	21.522
TD+H	3.472	20.731
HCTD	2.962	15.903

5.2 Experimental Results

Performance of Spatial and Temporal Pricing Model. Table 2 shows the results with default parameters on dataset1, in which AVF achieves the worst performance while HCTD performs best. The improvement between TD and AVF shows that tensor decomposition makes it more accurate to fill missing entries. The result of TD+H is better than TD, which illustrates the role of historical tensor. The result of HCTD indicates that our approach provides higher accuracy because of considering historical data and contextual information.

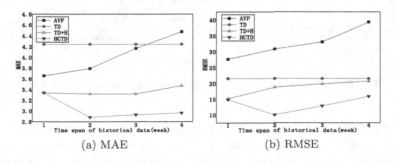

(a) MAE   (b) RMSE

Fig. 4. Performance of STPM: change the value of h

As shown in Fig. 4, our approach HCTD performs better than all the algorithms, which confirms the contributions of historical tensor and context matrices. Moreover, except TD the accuracy of all algorithms increases gradually as the time span decreases. However, it shows reversely in the first week. This can be explained by the following reasons. First, the accuracy of TD is not affected by the historical data since it decomposes the tensor without historical information. Second, since the time span grows, the values in the historical tensor $\mathcal{X}_h$ are more and more different from the recent tensor $\mathcal{X}_r$. Thus resulting in the accuracy of estimation decreases. Last but not least, note that accuracy of estimation is not only affected by the similarity between tensor $\mathcal{X}_r$ and $\mathcal{X}_h$, but also the density of tensor $\mathcal{X}_h$. So when the time span reduces into one week, both RMSE and MAE in HCTD no longer decrease.

Performance of Deciding Price and Task Assignment. In this section, we apply STPM for price decision and task allocation. The performance is evaluated by the average total revenue in multiple trials. We compare HCTD, TD+H, TD, AVF and MAPS through this part. The MAPS was proposed in [19] which estimates the price acceptance ratio with an online learning method. The key point in their approach is to consider the supply and demand of workers and tasks. Beyond that, there is no longer a method that focuses on pricing for requesters. Moreover, we show the result of the optimal pricing algorithm which means the price is set as requesters' private valuations v and we refer to it as OPT. To simulate different supply and demand, we generate different numbers of workers and denoted it by the ratio of the number of workers to tasks. We first show the performance of all algorithms by using default parameters on dataset1 and dataset2 and then show the effect of time span of historical data on dataset1 and the effect of the span of a time instance on dataset2.

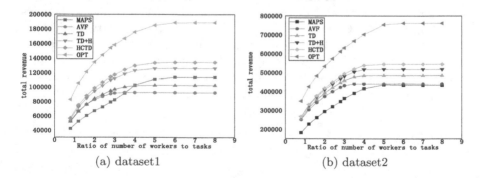

(a) dataset1   (b) dataset2

Fig. 5. Performance of DPTA:total revenue in default parameters

As Fig. 5 shows, our method HCTD outperforms all other models except OPT, since OPT is the upper bound. With the number of workers increases, the total revenue increases and eventually no longer changes. This is because as the number of workers increases, more tasks that accept price can be completed, and eventually, there are no tasks that accept price but no worker is available. That is, when workers are sufficient, total revenue is determined by the price model.

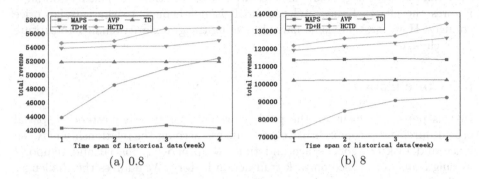

(a) 0.8   (b) 8

Fig. 6. Performance of DPTA: Effect of h

Effect of h. We evaluate the effect of the time span of historical data h by comparing the total revenue and the ratio of the number of workers to tasks is set to 0.8 and 8. The ratio is 0.8 and 8, respectively, indicating insufficient and sufficient workers. As shown in Fig. 6, our approach HCTD outperforms others. However, unlike estimation accuracy, which decreases with time span of historical data increases, the total revenue of all algorithms increases except TD for TD has nothing to do with historical data. This is because tensor $\mathcal{X}_r$ is too sparse, although the estimation accuracy of non-zero entries is improved, there are still a large number of zero entries with unknown accuracy. When the time span of historical data grows, the historical tensor becomes denser and denser, and the possibility of similar data appears greatly increases. When similar data exists in historical data, the estimation of zero entries will be more accurate.

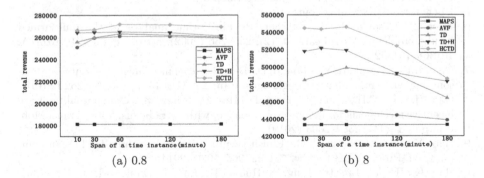

(a) 0.8   (b) 8

Fig. 7. Performance of DPTA: Effect of ts

Effect of ts. We evaluate the effect of the span of a time instance on dataset2 and the ratio of the number of workers to tasks is set to 0.8 and 8. As shown in Fig. 7, our approach HCTD performs best. We can see that with the span of a time instance grows, total revenue increases first and then decreases. The reason may be that as time granularity becomes larger, the change of unit price over time is more neglected. As a result, setting the time granularity too large ignores the effect of time and too small is not enough to characterize price changes. However, there is no harm in setting the granularity smaller to capture price changes.

6   Conclusion

In this paper, we focus on the pricing problem for task requesters in spatial crowdsourcing. By comprehensively taking into account the relationship between price and task category, space and time, we propose a spatial and temporal pricing framework based on task-transaction history. We address the challenges arising from data sparsity by proposing a History-based Context-aware Tensor Decomposition (HCTD) method. In our next work, we will focus on the problem of two side pricing in spatial crowdsourcing, that is, pricing task requesters and workers at the same time.

Acknowledgments. This paper is partially supported by Natural Science Foundation of China (Grant No. 61572336), Natural Science Research Project of Jiangsu Higher Education Institution (No. 18KJA520010), and a Project Funded by the Priority Academic Program Development of Jiangsu Higher Education Institutions.

References

1. Didi chuxing. http://didichuxing.com/
2. Gigwalk. https://www.gigwalk.com
3. Nyc data. https://www1.nyc.gov/site/tlc/about/tlc-trip-record-data.page
4. Openstreetmap. https://www.openstreetmap.org
5. Uber. https://www.uber.com/
6. Auer, P., Cesabianchi, N., Fischer, P.: Finite-time analysis of the multiarmed bandit problem. Mach. Learn. **47**(2), 235–256 (2002). https://doi.org/10.1023/A:1013689704352
7. Chandra, P., Narahari, Y., Mandal, D., Dey, P.: Novel mechanisms for online crowdsourcing with unreliable, strategic agents. In: AAAI 2015, pp. 1256–1262 (2015)
8. Gao, H., Liu, C.H., Tang, J., Yang, D., Hui, P., Wang, W.: Online quality-aware incentive mechanism for mobile crowd sensing with extra bonus. IEEE Trans. Mob. Comput. **18**(11), 2589–2603 (2019)
9. Gao, Y., Parameswaran, A.G.: Finish them!: pricing algorithms for human computation. Proc. VLDB Endow. **7**(14), 1965–1976 (2014)
10. Han, K., He, Y., Tan, H., Tang, S., Huang, H., Luo, J.: Online pricing for mobile crowdsourcing with multi-minded users. In: Mobihoc 2017, pp. 1–10 (2017)
11. Han, K., Huang, H., Luo, J.: Quality-aware pricing for mobile crowdsensing. IEEE/ACM Trans. Netw. **26**(4), 1728–1741 (2018)
12. Hu, Z., Zhang, J.: Optimal posted-price mechanism in microtask crowdsourcing. In: IJCAI 2017, pp. 228–234 (2017)

13. Jin, H., Guo, H., Su, L., Nahrstedt, K., Wang, X.: Dynamic task pricing in multi-requester mobile crowd sensing with markov correlated equilibrium. In: INFOCOM 2019. pp. 1063–1071. Paris, France, April 2019
14. Kolda, T.G., Bader, B.W.: Tensor decompositions and applications. SIAM Rev. **51**(3), 455–500 (2009)
15. Munkres, J.: Algorithms for the assignment and transportation problems. J. Soc. Ind. Appl. Math. **5**(1), 32–38 (1957)
16. Peng, D., Wu, F., Chen, G.: Pay as how well you do: A quality based incentive mechanism for crowdsensing. In: Mobihoc 2015, pp. 177–186 (2015)
17. Singer, Y., Mittal, M.: Pricing mechanisms for crowdsourcing markets. In: WWW 2013, pp. 1157–1166 (2013)
18. Singla, A., Krause, A.: Truthful incentives in crowdsourcing tasks using regret minimization mechanisms. In: WWW 2013, pp. 1167–1178 (2013)
19. Tong, Y., Wang, L., Zhou, Z., Chen, L., Du, B., Ye, J.: Dynamic pricing in spatial crowdsourcing: a matching-based approach. In: SIGMOD 2018, pp. 773–788 (2018)
20. Zhang, Y., Jiang, C., Song, L., Pan, M., Dawy, Z., Han, Z.: Incentive mechanism for mobile crowdsourcing using an optimized tournament model. Electron. Lett. **35**(4), 880–892 (2017)
21. Zhao, D., Li, X., Ma, H.: Budget-feasible online incentive mechanisms for crowdsourcing tasks truthfully. IEEE/ACM Trans. Netw. **24**(2), 647–661 (2016)

Service Computing and Cloud Computing

Attention-Based High-Order Feature Interactions to Enhance the Recommender System for Web-Based Knowledge-Sharing Service

Jiayin Lin[1]([⊠]), Geng Sun[1]([⊠]), Jun Shen[1]([⊠]), Tingru Cui[2]([⊠]), David Pritchard[3]([⊠]), Dongming Xu[4]([⊠]), Li Li[5]([⊠]), Wei Wei[6]([⊠]), Ghassan Beydoun[7]([⊠]), and Shiping Chen[8]([⊠])

[1] School of Computing and Information Technology, University of Wollongong, Wollongong, Australia
jl461@uowmail.edu.au, {gsun,jshen}@uow.edu.au
[2] School of Computing and Information Systems, The University of Melbourne, Melbourne, Australia
tingru.cui@unimelb.edu.au
[3] Research Lab of Electronics, Massachusetts Institute of Technology, Cambridge, MA, USA
dpritch@mit.ed
[4] UQ Business School, The University of Queensland, Brisbane, Australia
d.xu@business.uq.edu.au
[5] Faculty of Computer and Information Science, Southwest University, Chongqing, China
lily@swu.edu.cn
[6] School of Computer Science and Engineering, Xi'an University of Technology, Xi'an, China
weiwei@xaut.edu.cn
[7] School of Information, System and Modelling, University of Technology Sydney, Sydney, Australia
ghassan.beydoun@uts.edu.au
[8] Data 61 CSIRO, Sydney, NSW, Australia
shiping.chen@csiro.au

Abstract. Providing personalized online learning services has become a hot research topic. Online knowledge-sharing services represents a popular approach to enable learners to use fragmented spare time. User asks and answers questions in the platform, and the platform also recommends relevant questions to users based on their learning interested and context. However, in the big data era, information overload is a challenge, as both online learners and learning resources are embedded in data rich environment. Offering such web services requires an intelligent recommender system to automatically filter out irrelevant information, mine underling user preference, and distil latent information. Such a recommender system needs to be able to mine complex latent information, distinguish differences between users efficiently. In this study, we refine a recommender system of a prior work for web-based knowledge sharing. The system utilizes attention-based mechanisms and involves high-order feature interactions. Our experimental results show that the system outperforms known benchmarks and has great potential to be used for the web-based learning service.

© Springer Nature Switzerland AG 2020
Z. Huang et al. (Eds.): WISE 2020, LNCS 12342, pp. 461–473, 2020.
https://doi.org/10.1007/978-3-030-62005-9_33

Keywords: Recommender system · Neural network · Web-based learning · Machine learning · Information retrieval

1   Introduction

Web-based learning services aim to effectively utilize mobile devices to conduct real-time personalized learning activities [1]. Online knowledge sharing is one representative and popular informal learning style [2]. Online users post and answer questions from various discipline in a knowledge-sharing platform (like Quora[1] and Stackoverflow[2]), and the platform engages users with new questions based on their profile and historical activities. However, the plethora of user interests and backgrounds could easily result massive volumes of options and can induce disengagement, i.e. producing questions that have the opposite effect. Hence, a web-based knowledge-sharing platform has to rely on a sophisticated recommender system to filter out irrelevant information to truly create a personalized learning service.

An effective recommender system needs to handle and merge different types and formats of information from both the users' profiles and historical activities and the resources profiles. Higher-order feature interaction (combination) is also crucial for good performance [3]. Generating high-order feature interaction manually requires strong domain background, and is very time consuming and labour-intensive, which make it impractical for the large-scale online system, in the context of big data. Furthermore, different features have various importance levels for a personalized recommendation task [4]. How to precisely distinguish the importance differences of different features for a specific user is also vital for a personalized web-based learning service. Conventional recommendation strategies such as simple collaborative filtering and content-based filtering [5] are no longer adequate to handle massive, complex, dynamic data due to their drawbacks in scalability and modelling higher order features.

In this paper, we combine high-order feature interactions and the attention mechanisms to refine a recommender system proposed in one previous work [6]. This enables automatic exploration of high-order feature interactions, differentiating the important degrees for different features, and mining latent features from the original input.

The rest of this paper will be organized as follows. The prior work of recommendation strategies used for online learning services will be firstly reviewed in Sect. 2. Different state-of-the-art mechanisms proposed in recent research of deep learning will also be introduced in Sect. 2. In Sect. 3, the architecture of the proposed model and the relevant technical details of each component will be presented and explained. The experimental results will be presented and analysed in Sect. 4. The summary of this research and the future plan will finally be discussed in Sect. 5.

---

[1] https://www.quora.com/.
[2] https://stackoverflow.com/.

2   Related Work

2.1   Recommendation in Online Learning

Recommender systems have been studied for many years based in various application areas. However, due to the pedagogical considerations [7, 8], a recommendation strategy used in other domains cannot be transformed to fulfil the delivery of online learning services. This is particularly true for the delivery of micro learning services. Hence, this knowledge sharing task still lacks a sophisticated solution to enable recommending personalized online resources to target users.

A blended model was proposed in prior study [9], which combined a learning management system, a set of web 2.0 tools and the e-learning recommender system to enhance personalized online learning. However, this study did not provide technical innovations of a recommender system for online learning service. In an early survey [8], several recommendation approaches for e-learning service are listed and analysed. However, they were all too preliminary to be applied to web-based informal learning services. Another study [10] proposed a hybrid recommendation algorithm which combined the collaborative filtering and sequential pattern mining together for a peer-to-peer learning environment. Learning path recommendation is investigated in [11] and [12]. However, the proposed models were constructed mainly based on the demographic information, which does not provide much scope for exploring individual preferences.

2.2   Factorization Machine and Deep Network

In many recommendation scenarios, features involved impact each other (called feature interaction). For example, the feature pair (learning interests, knowledge level) determines the difficulty of a specific course for a learner. A classic work Factorization machine (FM) [13] use the inner product to model the feature interaction. This alleviates the data sparsity problem by using *embeddings* to represent the user and the item. However, due to computational and space complexity, only up to second-order feature interaction can be applied to many real-world applications.

Deep learning has been used in many application areas [14–16] and has demonstrated its outstanding ability to model complex problems. Hence, many researchers are investigating combining deep learning technique with conventional recommendation strategies. One representative model proposed by Google is 'Wide&Deep'. This combines the benefits of memorization and generalization by using a linear model and a deep network [17]. Both low-order and high-order feature interactions are also investigated in [18] through FM and DNN components. As the DNN models high-order interactions in an implicit manner, the learned results can be arbitrary; in another work, an extreme deep factorization machine (xDeepFM) is proposed for generating and modelling feature interactions of a bounded degree [19]. Similarly, model [19] also contains two components, a compressed interaction network (CIN) and a DNN. The CIN and DNN learn the explicit and implicit feature interaction, respectively.

2.3 Mining Latent Information

With the successes of 'Wide&Deep' and DeepFM [18], multi-component network structure are increasingly popular. Such structure shows outstanding performance in applying different techniques to mining latent information from different perspectives simultaneously. Among them, feature interaction and weighting strategies are mostly benefitting from such multi-component structure network.

Feature Interaction. This is a fundamental problem and plays a significant role in a recommendation task. There are also many prior studies [3, 20] mainly focusing on feature interaction strategies. The study [20] proposed a novel cross-network, which explicitly applied feature interaction in each layer; and the cross-network consisted of all the cross-terms of degree up to the highest. In another study [3], key-value attention mechanism was used for determining which feature combinations were meaningful.

Attention and Residual Mechanism. Attention mechanism has been widely used in many areas, such as computer vision and natural language processing. This allows the network to pay different degrees of attention to different parts. Attentional factorization machine was proposed in [21], which could distinguish the importance differences of various feature combinations. Instead of simple attention network, multi-head attention mechanism [22] was also used another recommendation strategy [3]. That showed the ability to explore meaningful feature combinations through different non-linear transformations. Squeeze-and-Excitation network (SENET) [23] was used in the study of Feature Importance and Bilinear feature Interaction network (FiBiNET) [4]. SENET was used to make the model pay more attention to the important features and decrease the weight of uninformative features through inner product and Hadamard product.

Demonstrated in [16], the idea of residual shows outstanding performance in stabilizing the optimization process of a deep network. Moreover, the residual function can also improve the model performance by providing sufficient information from previous layers of the network. Hence, for the recommendation task, as the network becomes deeper, many researchers start involving the residual connection (unit) in some components of the network. One prior study [24] used residual units to implicitly perform specific regularizations leading to better stability. Similarly, in the crossing component of the 'Deep&Cross' Network [20], the residual unit was used in each crossing layer to add the current input information back. In [3], standard residual connections were added in both interaction and output layers to achieve a hierarchical operation manner.

3   Solution

In this section, we firstly propose three hypotheses which might be significant to web-based knowledge sharing service. The design of the model and technical details of each component is then presented and discussed.

3.1 Hypotheses

In this study, the proposed model is designed based on the following hypotheses:

1. High-order feature interaction is vital to further improve the performance of a recommender system which used for a web-based big data application. Low-order feature interaction cannot sufficiently mine and model the underlying complex feature interactions for informal learning service.
2. The features involved in a learning platform have different important degrees. Precisely differentiate the feature importance is vital to a personalized learning service, and it can further improve the recommendation results.
3. The proposed model also holds moderate efficiency. For a web-based learning service in the big data era, efficiency is also an important indicator.

For example, the proposed model manages to recommend the question to the user that might be interested in. The give question is about machine learning, the difficult level of this question is entry-level. And we have two users with the following features:

User_1: (*Interested topic: computer science, Occupation: student, Gender: male, Age:23, Location: Australia*)
User_2: (*Interested topic: computer science, Occupation: research fellow, Gender: male, Age 32, Location: China*)

For the example, the proposed model should effectively and automatically generate meaningful feature combinations such as *(Interested topic, Occupation)* and *(Gender, Age)*, distinguish the importance difference between different features such as for a give question the feature *(Interested topic, Occupation)* might be more important than the feature *(Gender, Age)* and decide that the *User_1* might be more interested in this question.

3.2 Model Architecture

Based on the above hypotheses and one previous proposed initial idea [6], the general architecture of our model is shown in Fig. 1. Our proposed model contains three significant networks: a cross-network for exploring feature interactions, a deep network for mining latent information, and an attention network for distinguishing the importance of different features. The input of the model is high dimensional vectors which contain both user and item relevant information. The output of the model are decimals range from 0–1 indicate how much a user will be interested in a given question.

3.3 Embedding

For a recommendation task, the input contains many highly sparse categorical data, such as the genre describing the discipline of the educational resource and it may be multi-valued (for example, subject of 'machine learning' could belong to both disciplines of mathematics and computer science). In our proposed model, we apply an embedding

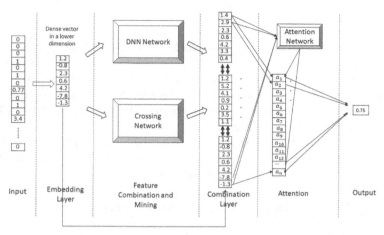

Fig. 1. The overall network structure of the proposed cross attention boosted recommender system

layer to reduce the dimensionality and sparsity of the raw data. The raw input contains history of interaction between user and item and side information of the user and item. The embedding operation could not only reduce the computational workload, but also boost the model performance. This process can be formulated as follows:

$$X_{embed,i} = W_{embed,i} X_i \tag{1}$$

Where $X_{embed,i}$ is the embedding result of the i-th categorical feature, $W_{embed,i}$ is the embedding matrix that maps the i-th original categorical feature into the low dimensional space, and X_i is the i-th feature.

3.4 Cross Network

The cross-network used in this study is based on the method proposed in [20]. The cross network is used to automatically generate the high-order feature interaction. Such network consists of several layers, and each layer is an operation of feature interaction. For each interaction layer, the operation of feature crossing can be simply formulated as follows:

$$X_{l+1} = X_0 X_l^T W_l + b_l + X_l \tag{2}$$

Where X_l is the output of the l-th crossing layer, W_i and b_i are weights and bias parameters of each crossing-layer. As demonstrated in [20], such special structure of network can increase the interaction degree as the network goes deep, with the highest $n + 1$ polynomial degree of the n-th layer. Moreover, regarding the efficiency of the network [20], the time and space complexity are both linear in input dimension.

3.5 Deep Network

For the simplicity and generalization, a conventional fully connected neural network is used in the proposed model as the deep component. The deep network implicitly captures the latent information and feature combinations. Each layer of the DNN network can be formulated as follow:

$$h_{l+1} = f(W_l h_l + b_l) \tag{3}$$

Where h_l denotes the output of the l-th layer of the deep component, $f(\cdot)$ is the activate function, where ReLU is used in this study. W_l and b_l are parameters of the l-th layer of the deep network.

3.6 Residual Connection and Attention Network

Before providing transformed information to the attention network, the original input information is continuously added to the output of the deep network and the cross network by using residual. This aims to maintain the original input information, which might suffer information loss after going through several layers of neural network.

An attention network is applied right after the combination layer to interpret the important difference of various features. The attention mechanism can be formulated as follows:

$$a_i' = ReLU(WX_i + b) \tag{4}$$

$$a_i = \frac{\exp\left(a_i'\right)}{\sum_i \exp\left(a_i'\right)} \tag{5}$$

where, both W and b are model parameters. The attention score is calculated through Softmax function. The calculated attention scores are projected back to the output of the combination layer. The process can be formulated as follows:

$$X_s = a_i WX \tag{6}$$

Where, a_i is the calculated attention score, W is the weight adopted in the network, and the X is the output of the former combination layer, and X_s is the final value after attention mechanism is applied. The demonstration of the attention operation is shown in Fig. 2. Where f_1 to f_n are the latent features passed into the attention network, the outputs of the network are attention scores for latent features with Softmax function. Lastly, each score is assigned to the feature by Hadamard product.

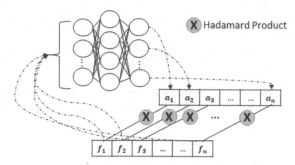

Fig. 2. Visualization of the attention operation

4   Experiments and Analysis

In this section, we compare our proposed model with several state-of-the-art recommendation strategies.

4.1   Evaluation Metrics and Baselines

Evaluation Metrics. In the experiment, we used the Area Under Curve (AUC) as the main criteria to evaluate the performance of each model. The proposed is a binary classifier which predicts whether a user will be interested in a given question, and the AUC can measure the capability of a model in distinguishing two labels. The calculation of the AUC used in this study is calculated as follow:

$$\text{AUC} = \frac{\sum_{i \in positiveClass} rank_i - \frac{M(1+M)}{2}}{M \times N} \tag{7}$$

Where M and N are the number of positive and negative samples respectively, $rank_i$ is the location of the i-th sample. We also used mean square error (MSE) and binary cross entropy to reflect the errors that made by each model. The binary cross entropy used in this study is formulate as:

$$\text{H} = -\frac{1}{N} \sum_{i=1}^{N} y_i \log(p(y_i)) + (1 - y_i) \log(1 - p(y_i)) \tag{8}$$

Baselines. AutoInt [3], FM, DeepFM, Deep&Cross Network (DCN), Attention Factorization Machine (AFM) are used as baselines in the experiment. The overall architecture comparison of AutoInt, DeepFM, AFM, and DCN is shown in Fig. 3. Each of these models contains an embedding layer, a feature interaction layer, and uses Softmax function to make a prediction. The main difference among these models is the choice of techniques for feature interaction, where multi-head self-attention is used in AutoInt, the combination of the FM and the DNN is used in DeepFM, the combination of FM and simple forward attention network is used in AFM, and combination of the cross-network and the FM is used in DCN.

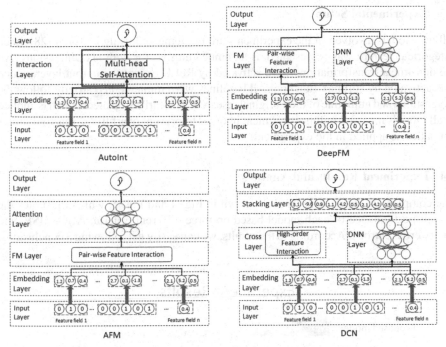

Fig. 3. Overall model structure of AutoInt, DeepFM, AFM, and DCN

4.2 Dataset

The dataset used in the experiment contains 10 million (question, user) pair, which was collected from Zhihu[3]. The user here stands for an online learner who is or is not a participant of a certain question. This dataset also contains other side-information about the users and the questions, such as the answers to the question, categorical information (such as gender) about the user, and the user's learning interests. Moreover, the label of the dataset is unbalanced (the ratio of negative label to positive label is around 4), which reflects the real and typical online application scenario, where for most users, only a small amount of questions they would like to answer due to various reasons, such as 'pedagogical lurking'. A negative sample stands for recommending a question to a user, but the user does not participate any learning activities; while a positive sample stands for recommending a question to a user and the user participate a certain of learning activity which could be answering the question or commenting the answers given by other users. As discussed in many pedagogical studies [25–27], for the online learning service, it is very difficult to enable learners interacting with each other like offline learning, even if the online learners have great and similar interests in the current learning session.

---

[3] https://www.zhihu.com/.

4.3 Experimental Setup

All the models involved in the experiments are implemented using PyTorch [28]. Each categorical feature was represented as an embedding vector with six dimensions. All the non-linear transformations were activated by ReLU except the output layer, which was activated by softmax function. All baselines were implemented strictly follow the suggestions and guideline of their original research. The early-stop mechanism is applied to all models involved in this experiment in order to prevent overfitting. Ten-fold-cross-validation is applied in the experiment.

4.4 Experiment Results and Analysis

The comparative experiment result is shown in Fig. 4 and Fig. 5, which illustrates the overall performance of each model based on three different criteria, AUC, MSE, and binary cross entropy. According to the results, we can easily get following 3 conclusions:

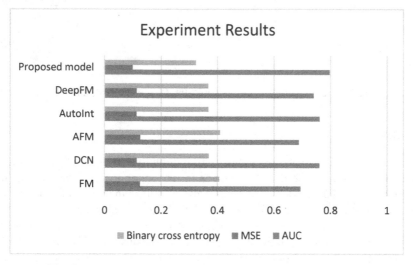

Fig. 4. The experiment result of different recommendation models

The Importance of the High-Order Feature Interaction. According to the results in Fig. 4, we can clearly see that the AUC scores of FM and AFM are the lowest ones, and the MSE and binary cross entropy values of these two models are the highest. These two models are the only two of which involves up to second-order feature interactions. Other baseline methods and our proposed model all involved high-order feature interactions, even though the ways of feature interaction are different. Hence, we argue that the high-order feature interaction (complex feature combination) do reflect how online learners make their decisions and involving the high-order feature interaction is useful and necessary to the large-scale web-based learning recommendation task. This finding proves the first hypothesis that made in Sect. 3.

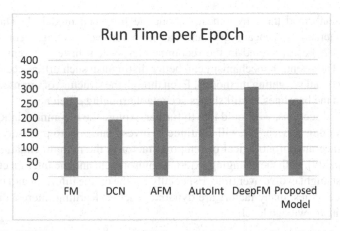

Fig. 5. Efficiency comparison of different models in terms of rum time (s/epoch)

The Significance of Attention Mechanism. Another conclusion we can get from Fig. 4 is the models (AFM and out proposed model) involved the attention mechanism have higher AUC scores than the models which do not. One possible explanation is AFM and our proposed model refine the results of high-order feature interaction via the attention mechanism. Such result approves the second hypothesis that we made in Sect. 3. The main difference between AutoInt and the proposed model is that they use different techniques to explore the feature interactions. According to the experiment result, we can see that our proposed model outperforms AutoInt when handling the recommendation problem in the online knowledge sharing scenario.

The Efficiency of the Proposed Model. To investigate the third hypothesis that we made in Sect. 3, we also evaluate the computation efficiency of our proposed model and various state-of-the-art recommendation models. The result is also shown in Fig. 5. The proposed model is in the third place, outperforming the FM, AutoInt and DeepFM, and very close to the second one (AFM). However, the AFM does not involve high-order feature interactions. The most efficient model is DCN, which only takes around 195 s under our experiment setting, while our proposed model takes 261 s on average for each training epoch. As the network architecture of our proposed model is extended from DCN and much more complex than other baselines, considering the improvements in recommendation performance and other models have been verified their efficiency on real-world applications, the slight increase of the training time is reasonable and acceptable. The possible reason for this is many operations involved in the proposed model run simultaneously (such as the crossing network and deep network). Hence, we trust under proper configurations our model can reach efficiency requirements.

5   Summary and the Future Work

In this study, we refine an existing recommender system [6] using attention mechanism together with high-order feature interaction methods to boost the performance of a web-based knowledge sharing service. By comparing with the-state-of-the-art recommender system, we confirmed three hypotheses about the proposed model: 1. The involved high-order interaction is meaningful and can help further boosting recommendation performance. 2. Features used in the recommender system have different degrees of importance. The attention mechanism can better distinguish such difference comparing to the conventional weighting method. 3. Even though the structure of our model is more complex than the baselines, it still shows acceptable running efficiency.

For future directions, we would like to further explore the recommendation strategy for online learning service. We will continue to investigate how to precisely represent, model and integrate chronological or temporal factor in the recommendation task. As highlighted in [29], with the changing external environment and the internal cognition, a user's interest might evolve over time. Especially for the online informal and non-formal educational activities, many factors are dynamic, such as learning interest drifting and the changes in knowledge level.

Acknowledgments. This research has been carried out with the support of the Australian Research Council Discovery Project, DP180101051, and Natural Science Foundation of China, no. 61877051.

References

1. Lin, J., et al.: From ideal to reality: segmentation, annotation, and recommendation, the vital trajectory of intelligent micro learning. World Wide Web **23**(3), 1747–1767 (2019). https://doi.org/10.1007/s11280-019-00730-9
2. Eshach, H.: Bridging in-school and out-of-school learning: formal, non-formal, and informal education. J. Sci. Educ. Technol. **16**(2), 171–190 (2007). http://dx.doi.org/10.1007/s10956-006-9027-1
3. Song, W., et al.: Autoint: automatic feature interaction learning via self-attentive neural networks. In: Proceedings of the 28th ACM International Conference on Information and Knowledge Management, pp. 1161–1170. ACM (2019)
4. Huang, T., Zhang, Z., Zhang, J.: FiBiNET: combining feature importance and bilinear feature interaction for click-through rate prediction. In: Proceedings of the 13th ACM Conference on Recommender Systems, pp. 169–177 (2019)
5. Pazzani, M.J.: A framework for collaborative, content-based and demographic filtering. Artif. Intell. Rev. **13**(5–6), 393-408 (1999)
6. Lin, J., et al.: Deep-cross-attention recommendation model for knowledge sharing micro learning service. In: Bittencourt, I.I., Cukurova, M., Muldner, K., Luckin, R., Millán, E. (eds.) AIED 2020. LNCS (LNAI), vol. 12164, pp. 168–173. Springer, Cham (2020). https://doi.org/10.1007/978-3-030-52240-7_31
7. Wu, D., Lu, J., Zhang, G.: A fuzzy tree matching-based personalized e-learning recommender system. IEEE Trans. Fuzzy Syst. **23**(6), 2412–2426 (2015). https://doi.org/10.1109/TFUZZ.2015.2426201
8. Sikka, R., Dhankhar, A., Rana, C.: A survey paper on e-learning recommender system. Int. J. Comput. Appl. **47**(9), 27–30 (2012). https://doi.org/10.5120/7218-0024
9. Hoic-Bozic, N., Dlab, M.H., Mornar, V.: Recommender system and web 2.0 tools to enhance a blended learning model. IEEE Trans. Educ. **59**(1), 39–44 (2015)
10. Chen, W., Niu, Z., Zhao, X., Li, Y.: A hybrid recommendation algorithm adapted in e-learning environments. World Wide Web **17**(2), 271–284 (2012). https://doi.org/10.1007/s11280-012-0187-z

11. Rusak, Z.: Exploitation of micro-learning for generating personalized learning paths. In: DS 87–9 Proceedings of the 21st International Conference on Engineering Design (ICED 17), 21-25.08, 2017, vol 9, pp. 129-138. Design Education, Vancouver, Canada (2017)
12. Zhao, Q., Zhang, Y., Chen, J.: An improved ant colony optimization algorithm for recommendation of micro-learning path. In: 2016 IEEE International Conference on Computer and Information Technology (CIT), pp. 190–196. IEEE (2016)
13. Rendle, S.: Factorization machines. In: 2010 IEEE International Conference on Data Mining, pp. 995–1000. IEEE (2010)
14. Fischer, T., Krauss, C.: Deep learning with long short-term memory networks for financial market predictions. Eur. J. Oper. Res. **270**(2), 654–669 (2018)
15. Liu, J., Chang, W.-C., Wu, Y., Yang, Y.: Deep learning for extreme multi-label text classification. In: Proceedings of the 40th International ACM SIGIR Conference on Research and Development in Information Retrieval, pp. 115–124 (2017)
16. He, K., Zhang, X., Ren, S., Sun, J.: Deep residual learning for image recognition. In: Proceedings of the IEEE Conference on Computer Vision and Pattern Recognition, pp. 770–778 (2016)
17. Cheng, H.-T., et al.: Wide and deep learning for recommender systems. In: Proceedings of the 1st Workshop on Deep Learning for Recommender Systems, pp. 7–10. ACM (2016)
18. Guo, H., Tang, R., Ye, Y., Li, Z., He, X.: DeepFM: a factorization-machine based neural network for CTR prediction. In: Proceedings of the Twenty-Sixth International Joint Conference on Artificial Intelligence, pp. 1725–1731 (2017)
19. Lian, J., Zhou, X., Zhang, F., Chen, Z., Xie, X., Sun, G.: xDeepFM: combining explicit and implicit feature interactions for recommender systems. In: Proceedings of the 24th ACM SIGKDD International Conference on Knowledge Discovery and Data Mining, pp. 1754–1763. ACM (2018)
20. Wang, R., Fu, B., Fu, G., Wang, M.: Deep and cross network for ad click predictions. In: Proceedings of the ADKDD'17, p. 12. ACM (2017)
21. Xiao, J., Ye, H., He, X., Zhang, H., Wu, F., Chua, T.-S.: Attentional factorization machines: learning the weight of feature interactions via attention networks. In: Proceedings of the Twenty-Sixth International Joint Conference on Artificial Intelligence, pp. 3119–3125 (2017)
22. Vaswani, A., et al.: Attention is all you need. In: Advances in Neural Information Processing Systems, pp. 5998–6008 (2017)
23. Hu, J., Shen, L., Sun, G.: Squeeze-and-excitation networks. In: Proceedings of the IEEE Conference on Computer Vision and Pattern Recognition, pp. 7132–7141 (2018)
24. Shan, Y., Hoens, T.R., Jiao, J., Wang, H., Yu, D., Mao, J.: Deep crossing: web-scale modeling without manually crafted combinatorial features. In: Proceedings of the 22nd ACM SIGKDD International Conference on Knowledge Discovery and Data Mining, pp. 255–262. ACM (2016)
25. Dobozy, E.: University lecturer views on pedagogic lurking. In: 2017 IEEE 17th International Conference on Advanced Learning Technologies (ICALT), pp. 1–2. IEEE (2017)
26. Beaudoin, M.F.: Learning or lurking?: tracking the "invisible" online student. Internet High. Educ. **5**(2), 147–155 (2002)
27. Dennen, V.P.: Pedagogical lurking: student engagement in non-posting discussion behavior. Comput. Hum. Behav. **24**(4), 1624–1633 (2008)
28. Paszke, A., et al.: PyTorch: an imperative style, high-performance deep learning library. Adv. Neural Inf. Process. Syst., pp. 8024–8035 (2019)
29. Zhou, G., et al.: Deep interest evolution network for click-through rate prediction. In: Proceedings of the AAAI Conference on Artificial Intelligence, pp. 5941–5948 (2019)

Designing Context-Based Services for Resilient Cyber Physical Production Systems

Ada Bagozi, Devis Bianchini$^{(\boxtimes)}$, and Valeria De Antonellis

Department of Information Engineering, University of Brescia, Via Branze 38,
25123 Brescia, Italy
{a.bagozi,devis.bianchini,valeria.deantonellis}@unibs.it

Abstract. Dynamicity and complexity of Cyber Physical Production Systems (CPPS) increase their vulnerability and challenge their ability to react to faults. Ensuring resilience of CPPS is even more difficult when on-field workers operating on the machines are engaged in the production line, according to the Human-in-the-Loop paradigm. In this scenario, we present a service-oriented approach to model resilient CPPS. Specifically, we design runtime recovery actions as a set of services. Services are context-based, i.e., they are associated with the steps of the production process as well as with the CPPS involved in the recovery actions. In the context model, different factors that could influence CPPS resilience, namely the functional degradation of one or more parts of the CPPS and changes in the environment in which the production process is executed, are monitored through proper parameters. The approach provides runtime selection of services among the ones associated with the process steps and CPPS involved in the recovery procedure. The values of monitored parameters are used as service inputs, while service outputs are displayed to the operators supervising the CPPS on which recovery actions must be performed, enabling fast and effective resilience also in a Human-In-the-Loop scenario. The approach is validated in a food industry case study.

Keywords: Resilient cyber physical production system ·
Context-aware resilience · Service-oriented architecture ·
Human-In-the-Loop

1 Introduction

Cyber Physical Systems (CPS) are hybrid networked cyber and engineered physical elements that record data (e.g., using sensors), analyse them using connected services, influence physical processes and interact with human actors using multichannel interfaces. Examples of CPS interacting with humans include smart cities (where citizens are surrounded by thousands of IoT-based devices) and industrial production environments (where workers supervise the operations

© Springer Nature Switzerland AG 2020
Z. Huang et al. (Eds.): WISE 2020, LNCS 12342, pp. 474–488, 2020.
https://doi.org/10.1007/978-3-030-62005-9_34

of industrial machines or work centers, according to the Human-In-the-Loop paradigm [11]). The latter application of CPS is also referred to as Cyber Physical Production Systems (CPPS), where single CPS corresponding to work centers are connected to form the production line. Given the pervasiveness of CPS in humans' everyday life, reinforcing their resilience is a key challenge. The Oxford English dictionary defines resilience as the ability of a system to recover quickly from a disruption that can undermine the normal operation and consequently the quality of the system. Resilience in CPPS is even more challenging since it must be addressed both on single work centers and at the shop floor on the production line and surrounding environment. Moreover, in modern digital factory work centers are fully connected, therefore changes in one of them may require recovery actions on the others. Finally, ensuring resilience of CPPS is made more difficult when on-field operators are engaged in the production line, to perform manual tasks according to the Human-in-the-Loop paradigm and to supervise the physical system.

To address the above mentioned complexity of the domain, we propose a service-oriented approach for modelling resilient CPPS, in line with recent approaches for the design of resilient CPS [5]. Services form a middleware at the cyber-side to implement recovery strategies. They can be deployed at different architectural layers (e.g., on the edge, fog or cloud layers) to detect failures or disturbances either on single work centers or on the whole production line. Their language independent and modular nature speed up integration of different services, as well as substitution between services implementing comparable recovery actions or addition of new smart components that bring their own services at edge layer and must be easily plugged in the service-oriented middleware. In this paper, we propose a step forward by designing services that display recovery actions to human workers who supervise those part(s) of the production line where such actions must be promptly confirmed and executed, thus ensuring efficiency also in a Human-In-the-Loop scenario. To this aim, the approach includes: *(a)* a *context model*, designed to capture the hierarchical organisation of the CPPS in modern digital factories and the association between each component and the services that have been designed to implement runtime recovery actions over the steps of the production process; *(b)* a set of *system monitoring and recovery phases*, that provide runtime support for the human supervisors, easing the identification of critical conditions and the selection of proper recovery services. Failures and disruptions on one or more components of the CPPS and on the whole production process environment are monitored through proper parameters, whose values are used as service inputs, while service outputs are displayed nearby the work centers on which recovery actions must be performed. An anomaly detection component [2] has been included to promptly detect anomalies through the identification of variations in Big Data streams collected from monitored parameters.

The paper is organised as follows: in Sect. 2 related work are discussed; in Sect. 3 we introduce the motivating scenario in a food industry case study, in which the approach is being validated; Sect. 4 describes the context model; in

Sect. 5 the system monitoring and recovery phases are presented; in Sect. 6 and 7 we describe implementation and experimental validation; finally, Sect. 8 closes the paper.

2   Related Work

Monitoring of industrial plants for anomaly detection purposes is aimed at ensuring timely recovery actions. In [3] we integrated an anomaly detection approach with a CMMS (Computerised Maintenance Management System) to provide time-limited maintenance interventions. However, a recent ever growing interest has been devoted to resilience challenges, often related with the notion of self-adaptation, as witnessed by an increasing number of surveys on this topic [8, 10, 13]. This will be the focus of this paper. Time-limited solutions can be considered if the identification of recovery actions fails. CPS resilience approaches are mainly focused on security issues at runtime on the communication layer or to implement resource balancing strategies between the edge and cloud computing layers. In [9] a model-free, quantitative, and general-purpose evaluation methodology is given, to evaluate the systems resilience in case of cyber attacks, by extracting indexes from historical data. This kind of approaches is out of the scope of our paper. However, resilience/self-adaptation specifically designed for CPPS has been addressed in few works [4, 14], where ad-hoc solutions are provided, focusing on single work centers or on the production line, without considering the effects of resilience across connected components. In [6] a decision support system is proposed to automatically select the best recovery action based on KPIs (e.g., overall equipment efficiency) measured on the whole production process.

Our approach contributes to the state of the art by introducing a context model, apt to relate recovery services with work centers organised in the fully connected hierarchy of smart components (from connected devices up to the whole production line at shop floor level), according to the IEC62264/IEC61512 standards of the RAMI 4.0 reference architectural model [7]. Work centers are in turn associated with supervising operators. This model enables, on the one hand, to take into account the propagation of recovery effects throughout the hierarchy of connected work centers and, on the other hand, to personalise the visualisation of recovery actions nearby the involved work centers, supporting the operativity of workers who supervise the production line.

The adoption of context-awareness to implement resilient CPS has been investigated in the Cyber Physical Systems (CAR) Project [1]. In the project, resilience patterns have been identified via the empirical analysis of practical CPS systems and implemented as the combination of recovery services. Our approach enables a better flexibility, given by the adoption of service-oriented technologies, and a continuous evolution of the service ecosystem is realised through the design of new services in case of unsuccessful recovery service identification. Authors in [5] share with us the same premises regarding the adoption of service-oriented technologies, modelling processes as composition of services that can be invoked to ensure resilience. With respect to them, we add here the context model and we propose a set of context-driven phases to support the human operator in the identification of critical conditions and in the confirmation of recovery services.

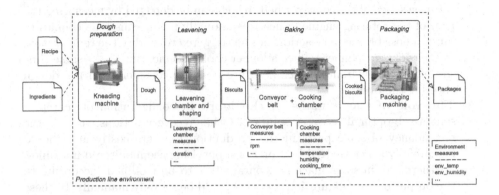

Fig. 1. Production process for the food industry case study.

3 Motivating Example

In this section we introduce a motivating case study in the food industry to validate our approach. The production line of the case study is summarised in Fig. 1. Specifically, we consider a company that bakes biscuits, starting from the recipe and the ingredients to the finished product (ready-to-sell biscuits). In the production process, the dough is prepared by a kneading machine and let rise in a leavening chamber. Once the biscuits are ready to be baked, they are placed in the oven. Indeed, the oven is composed of a conveyor belt, mounting a rotating engine, and the cooking chamber. By regulating the velocity of the belt through the rpm of the rotating engine, it is possible to setup the cooking time of the biscuits. Moreover, also the temperature and the humidity of the cooking chamber can be regulated. Finally, other measures can be gathered at the shop floor level, such as the temperature and the humidity of the whole production line environment. The considered case study presents the characteristics of any fully connected digital factory, with a specific focus on the following requirements.

R1) **A multi-level hierarchy of smart components.** According to RAMI 4.0 reference architectural model [7] (IEC62264/IEC61512 standards), modern digital factories present a hierarchical organisation, from connected devices up to the fully connected work centers and factory. Measures can be gathered from components in this hierarchy and from the operating environment at shop-floor level; this allows to detect anomalous working conditions, that may propagate over the whole hierarchy, and making the component smarter. In Fig. 1, only a subset of measures is showed and used as a running example in this paper.

R2) **Recovery actions across different components.** Identification of anomalous working conditions on a smart component in the hierarchy may trigger recovery actions either on the component itself or on a different one. This is enabled by the strong connection between components over the whole production line. Let's consider, for example, an anomaly on the rotating speed of the conveyor belt, which can be detected by measuring the rpm of the rotating engine. Since this anomaly might cause the cookies to burn, a possible recovery action can be triggered to modify the temperature of the cooking chamber to face a longer cooking time. But at the same time the effects of environment temperature must be considered on the dough entering the oven.

R3) **The role of the humans in the loop.** Re-configuration often represent a sub-optimal recovery solution. Other recovery actions, such as the redundancy of some parts in the production line, are costly and should be carefully acknowledged by workers supervising the monitored machines. Furthermore, in some machines, parameters to be modified (according to the output of re-configuration services) could be set only manually. In these cases, recovery actions must be modelled as a support for operators. In particular, the proper suggestion must be visualised nearby the involved work centers only, in order to let them manage the complexity of the system and speed up recovery actions also in those cases where the human intervention is required. This enforces the importance of the Human-In-the-Loop paradigm [11] also to realise resilient CPPS.

Smart components are equipped with an edge computing layer to connect them to the rest of the digital factory and to communicate with the components themselves. Either the collection of data from the components or the invocation of recovery actions on them can be implemented according to As-a-Service technological infrastructure. Services may implement functions that relate the controlling variables as service inputs (e.g., the rpm of the rotating engine) with controlled ones (e.g., the temperature of the cooking chamber). Services may also be used to send messages to the edge device installed nearby the monitored component, to support on-field operators, e.g., to suggest manual operations on the machines such as configurations or substitutions of corrupted parts.

4   Context Model

In Fig. 2 we report the context model to support resilience in CPPS. In the proposed model, the *Context* is described by the *Product* that is being produced (e.g., a certain type of biscuits), the *Process* to produce a certain product (e.g., biscuits baking process) and the *Environment Parameters* that may influence the production (e.g., the environment temperature and humidity). A *Process* is associated with *Services* and is performed by one or more *components*, that cooperate to successfully complete the production. For example, the biscuits baking process involves the kneading machine to prepare the dough, the leavening chamber to prepare biscuits, the oven to bake the biscuits, and so on.

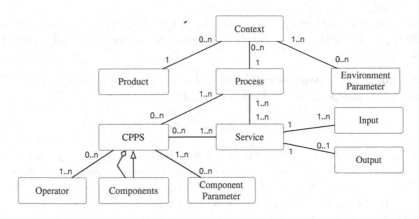

Fig. 2. Resilient CPPS context model.

Components can be organized hierarchically (see requirement R1). For example, the oven is composed of the conveyor belt and the cooking chamber. *Services* represent recovery actions to be executed on a component or the whole CPPS to ensure resilience (see requirement R2). A component is supervised by at least one *Operator* and can be monitored and controlled through a set of *Component Parameters* (e.g., the oven temperature).

Both *Environment Parameters* and *Component Parameters* are used to monitor the behaviour of a *CPPS* in a given *Context*. Indeed, parameters are used to observe physical phenomena of the monitored system and are formally defined as follows.

Definition 1 (Parameter). *A parameter represents a monitored variable that can be measured. A parameter P_i is described as a tuple*

$$P_i = \langle n_{P_i}, u_{P_i}, \tau_{P_i}, down^e_{P_i}, up^e_{P_i}, down^w_{P_i}, up^w_{P_i} \rangle \tag{1}$$

where: (i) n_{P_i} is the parameter name; (ii) u_{P_i} represents the unit of measure; (iii) τ_{P_i} is the parameter type, either environment or component parameter; (iv) $down^e_{P_i}$, $up^e_{P_i}$, $down^w_{P_i}$, $up^w_{P_i}$ are the parameter bounds, that delimit admissible values for P_i measurements. We define a measurement for the parameter P_i as a scalar value $X_i(t)$, expressed in terms of the unit of measure u_{P_i}, taken at the time t. Let's denote with $\mathbf{P} = \{P_1, P_2 \ldots P_n\}$ the overall set of parameters.

Parameters bounds are used during anomaly detection to establish if a critical condition has occurred on the monitored CPPS. In particular, we distinguish between two kinds of critical conditions: *(a) warning* conditions, that may lead to breakdown or damage of the monitored system in the near future; *(b) error* conditions, in which the system can not operate. Since the measurement $X_i(t)$ usually belongs to a range of values, we consider four types of parameters bounds: lower error/warning bound, upper error/warning bound.

During parameter monitoring, if an anomaly is detected, recovery actions are required. As mentioned above, actions on a certain *CPPS* can be performed by invoking *Services*. Indeed, in our model, for a *CPPS* it is possible to define one or more, possibly alternative, *Services*. A recovery service is formally defined as follows.

Definition 2 (Recovery service). *A recovery service S_j associated to a CPPS (or one of its components) is described as a tuple*

$$S_j = \langle n_{S_j}, IN_{S_j}, out_{S_j}, type_{S_j}, CPPS_{S_j} \rangle \tag{2}$$

where: (i) n_{S_j} is the service name; (ii) IN_{S_j} is the set of input parameters of S_j; (iii) out_{S_j} is an optional service output; (iv) $type_{S_j}$ is the service type; (v) $CPPS_{S_j}$ is the component or the whole CPPS associated to the service. Service I/O can be either Component or Environment Parameters. Let's denote with $\mathbf{S} = \{S_1, S_2 \ldots S_n\}$ the overall set of recovery services.

Flexibility of service-oriented architectures enables to include and dynamically add different types of services. For instance, a recovery service may implement the function that relates one or more input parameters with the output one. We refer to this type of service as "re-configuration". For example, the following service.

```
setOvenTemperature(ConveyorBelt.rpm) → CookingChamber.temperature
```

is a re-configuration service to set the cooking chamber temperature when the conveyor belt rpm changes, to avoid cookies overheating. When a re-configuration is not an applicable solution (e.g., if the service returns a cooking chamber temperature out of an acceptable range of values), other recovery actions must be applied, such as to replace or repair the conveyor belt. An example of "component substitution" service would be the following:

```
replaceConveyorBeltRotatingEngine(ConveyorBelt.rpm) → void
```

that has no output parameter to modify. This service is associated to the conveyor belt. The function implemented within the re-configuration services, as well as other service information (e.g., execution cost, time) that can be used to choose among different kinds of services, are based on the knowledge of the domain and are set on the service when it is added. The examples of recovery service types considered here is not exhaustive and may be extended [12]. Recovery services can be exposed in different ways, for example as web services, invoked from a local library, etc.

5   Context-Driven Monitoring and Recovery Phases for Resilience Enforcement

In this section, we present each phase of the proposed context-driven approach to support the on-field operators in the identification of critical conditions and in the runtime selection of services that implement proper recovery actions.

Figure 3 reports the five phases of the proposed approach. As a pre-condition, the designer, who has the domain knowledge about the production plant, is in

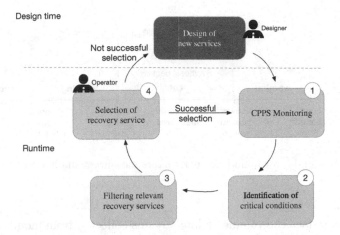

Fig. 3. Five context-driven monitoring and recovery phases to design resilient Cyber Physical Production Systems.

charge of preparing the context model described in the previous section, as well as an initial set of recovery services. At runtime, the CPPS is monitored by relying on IDEAaS [2], an innovative system for data stream exploration and anomaly detection on Cyber Physical Production Systems (Phase 1). An overview of the anomaly detection procedure within the scope of the present approach will be given in Sect. 5.1. Critical working conditions are detected by inspecting data streams collected from monitored CPPS. Anomalies are propagated over the hierarchy of connected CPPS (Phase 2). When critical conditions are detected, the relevant recovery services are identified (Phase 3). Service filtering takes into account the involved component, on which anomalies have been detected, and parameters whose values exceed warning or error bounds. Selected recovery services are suggested to the on-field operators working on the involved component (Phase 4). If no recovery services have been found, a feedback is stored for planning the design of additional recovery services in the future. The filtering and selection phases of relevant recovery services will be detailed in Sect. 5.2.

5.1   Identification of Critical Conditions

The goal of the anomaly detection is to identify critical conditions and send alerts concerning the *system status*. Three different values for the status are considered: *(a)* ok, when the system works normally as established by domain experts according to their expertise; *(b)* warning, when the system works in anomalous conditions that may lead to breakdown or damage (proactive anomaly detection); *(c)* error, when the system works in unacceptable conditions or does not operate (reactive anomaly detection). Therefore, the warning status is used to perform an early detection of a potential deviation towards an error status. The identification of the system status is based on the comparison between parameters measurements and warning/error bounds, according to Definition (1).

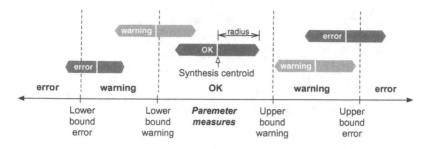

Fig. 4. Warning and error bounds for parameters measurements in order to detect the system status.

However, to face the volume of data streams collected from monitored CPPS and to avoid misleading anomaly detection due to noise and false outliers, that may affect single measures, anomaly detection is performed over a summarised representation of collected measures, incrementally built, called *syntheses*. Each synthesis corresponds to an aggregation of measures that are close each others in the parameter space. In other words, a synthesis contains measures collected when the observed system is operating in the same working conditions. A synthesis is described by a centroid, that is the representative measure in the synthesis, and a radius, corresponding to the standard deviation of measures in the synthesis. Syntheses generation is performed by executing an incremental clustering algorithm, that is run every Δt seconds on the stream of parameters measurements. Every Δt seconds, after updating the clustering results, syntheses are compared against bounds as shown in Fig. 4, in order to classify the status of a CPPS, starting from the status of its monitored parameters, among the following options:

- ok, if the status for all its parameters is ok;
- warning, if the status of at least one parameter is warning;
- error, if the status of at least one parameter is error.

Once the status of a single work center has been established, it is propagated over the hierarchy of connected components (see requirement R1) as follows:

- ok, if the status for all its components is ok;
- warning, if the status of at least one component is warning;
- error, if the status of at least one component is error.

When the status of a component, after applying the propagation rules, changes towards warning or error, an alert is raised by the system. Alerts generation is referred to as *data relevance evaluation*. In literature, data relevance is defined as the distance from an expected status. In our approach, the unexpected status is the raising of warning or error events. This makes syntheses as relevant and helps focus anomaly detection on those syntheses only.

5.2   Filtering and Selection of Relevant Recovery Services

Once an anomalous event (corresponding to a critical condition) is detected on one of the CPPS components, the event is used to identify relevant services that implement recovery actions on the involved component or connected ones (see requirement R2). Relevant recovery services are identified by inspecting their inputs. In particular, a recovery service is candidate to be identified as relevant if one of its input parameters have been classified in the **error** (reactive resilience) or **warning** status (proactive resilience). On the other hand, before including the service among relevant ones, the candidate is checked to verify the following conditions: *(a)* if the service type is "re-configuration", the value of its output parameters must not exceed any parameter bound; *(b)* if the service type is "component substitution", the component associated to the service should have an alternative machinery or component ready to be used in substitution of the one affected by the anomaly. For all the other options (e.g., repair), ad-hoc procedure must be engaged. Formally, a relevant recovery service is defined as follows.

Definition 3 (Relevant recovery service).   *A recovery service $S_k \in \mathbf{S}$ is defined as relevant if:*

- *the status of at least one input parameter in IN_{S_k} has been set to* **warning** *or* **error***;*
- *if $type_{S_k}$ =* **re-configuration***, then the value of the output parameter out_{S_k} resulting from service execution must fall inside the admissible parameter bounds;*
- *if $type_{S_k}$ =* **component_substitution***, then there must exist a component ready to be substituted to the $CPPS_{S_k}$ associated to S_k.*

The **setOvenTemperature** service described above is a candidate to be identified as relevant if an anomaly has been detected on the values of rotating engine rpm in the conveyor belt. Since the service type in this case is "re-configuration", before proposing the service to the operator, the value of the service output must be compared against parameter bounds of the cooking chamber temperature.

Once relevant recovery services have been identified, they are suggested to the operator assigned to the $CPPS_{S_k}$ in order to be confirmed and executed on the component or the whole CPPS (for re-configuration services) or to proceed with a substitution of the affected part (for component substitution services). In particular, the second type of services again highlights the role of human operators as the final actuators of recovery actions on the production line (see requirement R3). The service-oriented architecture does not exclude that in the future some kinds of services (such as the re-configuration ones) that require the human intervention can be made fully automatic. This aspect is at the moment out of the scope of the approach. We remark here that another feature of the approach is that the information about the recovery actions to undertake, as results of recovery services execution, are proposed only to the operators supervising the involved CPPS component and visualised on the edge computing device. This means that *"the right information is made available on the right place only"*, avoiding useless data propagation if not necessary and information flooding towards operators that may hamper their working efficiency.

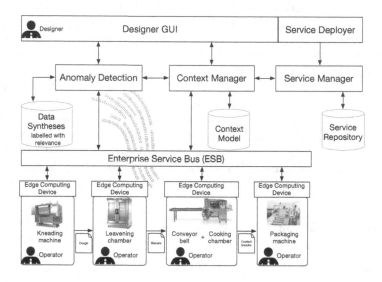

Fig. 5. Approach architecture.

6   Approach Architecture

The approach described in this paper has been integrated with the anomaly detection module and the resulting architecture is presented in Fig. 5. During anomaly detection, the `Context Manager` is invoked in order to contextualise the incoming data. To this purpose, the `Context Manager` will provide the following information: *(i)* an identifier for the context; *(ii)* a set of parameters, either *Environment or Component parameters*, to be analysed; *(iii)* the observed CPPS; *(iv)* the product that is being produced; *(v)* the running process. Such information is extracted from the *Context Model* database.

Collected measurements of the parameters in the context are properly summarised as syntheses by applying IDEAaS data summarisation techniques. Furthermore, syntheses are processed in order to detect anomalies, label the syntheses with the relevance information and save them in the *Data Syntheses* database. Labelled summarised data are visualised: *(a)* on the `Designer GUI` to let the designer monitor the overall evolution of the monitored system; *(b)* on the `Edge Computing Device` of the involved component, to let the on-field operator to better understand the behaviour of the component. Moreover, when anomalous conditions are detected, the `Context Manager` is notified with the identifier of the context and the list of critical parameters on which the anomaly occurred, together with their measurements. The `Context Manager` will search for relevant recovery services, associated to the component in the context. Once relevant recovery services have been identified, the Context Manager launches the execution of the services by interacting with the Service Manager, which is responsible for services registration in the *Service Repository* and for their execution. The result of the services execution is sent to the operator assigned to the component, in order to let the operators choose the most suitable service.

On the other hand, when the identification of relevant recovery services fails, the Context Manager reports the unsuccessful service selection to the designer, thought the Designer GUI, including all data of the context and available services. The designer will take into account such report for future services design. The new service, when available, will be registered in the *Service repository* through an interaction between the Service Deployer and the Service Manager. The Context Manager is notified as well, in order to update the *Context model*.

7   Preliminary Validation of the Approach

In this section we present a proof-of-concept validation of the approach to demonstrate its applicability. In particular, we report: *(i)* processing time required to promptly detect anomalies and activate recovery actions services (being this aspect a potential bottleneck for the whole approach); *(ii)* a proposal of dashboards to be used in real case studies in order to show the effectiveness of the context model for structuring and filtering the right information displayed on the right location within the digital factory.

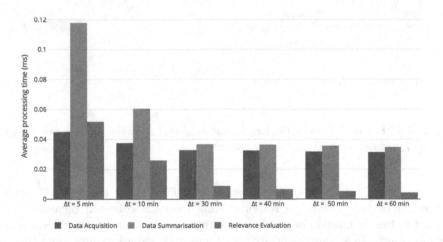

Fig. 6. Average response times per measure for anomaly detection that activates recovery activation services.

Figure 6 reports the average time required for data acquisition, summarisation and relevance evaluation. We run experiments on a MacBook Pro Retina,

Fig. 7. Dashboards for context-driven monitoring and recovery phases: (a) designer view; (b) operator view. (Color figure online)

with an Intel Core i7–6700HQ processor, at 2.60 GHz, 4 cores, RAM 16GB. Figure 6 reports the average response time for each collected measure, with respect to the Δt interval in which data summarisation and relevance evaluation are performed. In a counter-intuitive way, lower Δt values require more time to process data. This is due to the nature of the incremental data summarisation algorithm. In fact, for each algorithm iteration (i.e., every Δt seconds), a certain amount of time is required for some recurring operations such as opening/closing connection to database and retrieval of the set of syntheses previously computed. Therefore, enlarging Δt values means better distribution of this overhead over more measures. On the other hand, higher Δt values decrease the promptness in identifying anomalous events, as the frequency with which the data relevance is evaluated is lower. According to these conditions, we quantified the capability to detect anomalies on the collected measures using the Pearson Correlation Coefficient (PCC) $\in [-1, +1]$, that estimates the correlation between the real variations and the detected ones. In the experiment, the best PCC value is higher than 0.85 in the case $\Delta t = 10\,\mathrm{min}$, that represents a strong correlation. Further evaluation is being performed.

To demonstrate the effectiveness of the context model for structuring and filtering information in case of faults, we designed proof-of-concepts dashboards, one for the designer (Monitored System Dashboard, MSD) and one for the operator who supervises the CPPS or one of its components. Figure 7 reports the MSD *(a)* and the Oven Dashboard *(b)*. Let's consider now the following scenario. Rpm values that exceed the error bounds have been detected on the rotating engine of the conveyor belt. To recover from the anomaly, three services have been selected as relevant ones, namely `setOvenTemperature`, `setOvenCookingTime` and `replaceConveyorBeltRotatingEngine`. However, `setOvenCookingTime` service is not applicable, because this service controls the cooking time by changing the rpm of the belt rotating engine, which is already in error state. On the MSD it is visualised the hierarchy of the CPPS and, for each component, the parameters and the available services, according to our model. The parameters measurements that exceed their error bounds are highlighted in red, the ones that exceed their warning bounds in yellow and the others in green. On the list of services, the blue stars identify services that have been proposed to recover from the anomaly and the red cross on the `setOvenCookingTime` means that the service is not applicable. On the other hand, on the Oven Dashboard shown in Fig. 7(b), only warning or error events that require recovery actions on the oven are displayed. Not applicable services are not shown to avoid hampering the efficiency of the operator (see requirement R3). The operator can also refuse the suggestions by pushing on "Request new service" button and, in that case, a notification "Service not found" will be properly displayed on the MSD for designing future countermeasures.

8 Concluding Remarks

In this paper, we proposed a context-driven approach to model resilient Cyber Physical Production Systems. Specifically, we proposed: *(a)* a *context model*, designed to capture the hierarchical organisation of the CPPS in modern digital factories and the association between the CPPS and the services that have been designed to implement runtime recovery actions over the steps of the production process; *(b)* a set of *system monitoring and recovery phases*, that provide runtime support for the operators supervising the system, easing the identification of critical conditions and the selection of proper recovery services. A preliminary validation of approach performances has been performed in a food industry case study, but other scenarios will be investigated to provide a generalization of the approach. Nevertheless, evaluation of performances under functional degradation requires additional time and is being performed with a new set of experiments by fixing component parameters (e.g., the oven temperature) and varying environmental parameters (e.g., the environment temperature and humidity). Future efforts will be also devoted to the improvement of service selection criteria, for example defining a cost model and using simulation-based modules to predict the effects of recovery actions on the production process, like the one described in [6]. Finally, an extensive validation of the dashboard with human operators will be performed.

References

1. Context-Active Resilience in Cyber Physical Systems (CAR) European Project. http://www.msca-car.eu (2018)
2. Bagozi, A., Bianchini, D., De Antonellis, V., Garda, M., Marini, A.: A Relevance-based approach for Big Data Exploration. Future Gener. Comput. Syst. **101**, 51–69 (2019)
3. Bagozi, A., Bianchini, D., De Antonellis, V., Marini, A., Ragazzi, D.: Big data summarisation and relevance evaluation for anomaly detection in cyber physical systems. In: Proceedings of 25th International Conference on Cooperative Information Systems (CoopIS 2017), pp. 429–447 (2017)
4. Barenji, R.V., Barenji, A.V., Hashemipour, M.: A multi-agent RFID-enabled distributed control system for a flexible manufacturing shop. Int. J. Adv. Manuf. Technol. 1773–1791 (2014). https://doi.org/10.1007/s00170-013-5597-2
5. Bicocchi, N., Cabri, G., Mandreoli, F., Mecella, M.: Dynamic digital factories for agile supply chains: an architectural approach. J. Ind. Inf. Integrat. **15**, 111–121 (2019)
6. Galaske, N., Anderl, R.: Disruption management for resilient processes in cyber-physical production systems. Proc. CIRP Design Conf. **50**, 442–447 (2016)
7. Hankel, M., Rexroth, B.: The reference architectural model industrie 4.0 (RAMI 4.0). In: ZVEI (2015)
8. Moura, J., Hutchison, D.: Game theory for multi-access edge computing: survey, use cases, and future trends. IEEE Commun. Surv. Tutorials **21**(1), 260–288 (2019)
9. Murino, G., Armando, A., Tacchella, A.: Resilience of cyber-physical systems: an experimental appraisal of quantitative measures. In: Proceedings of 11th International Conference on Cyber Conflict (CyCon), vol. 900, pp. 1–19 (2019)
10. Musil, A., Musil, J., Weyns, D., Bures, T., Muccini, H., Sharaf, M.: Multi-disciplinary engineering for cyber-physical production systems. Patterns for Self-Adaptation in Cyber-Physical Systems, pp. 331–368 (2017)
11. Nunes, D., Silva, J.S., Boavida, F.: A Practical Introduction to Human-in-the-Loop Cyber-Physical Systems. Wiley IEEE Press (2018)
12. Pumpuni-Lenss, G., Blackburn, T., Garstenauer, A.: Resilience in complex systems: an agent-based approach. Syst. Eng. **20**(2), 158–172 (2017)
13. Ratasich, D., Khalid, F., Geissler, F., Grosu, R., Shafique, M., Bartocci, E.: A roadmap toward the resilient internet of things for cyber-physical systems. IEEE Access **7**, 13260–13283 (2019)
14. Vogel-Hauser, B., Diedrich, C., Pantförder, D., Göohner, P.: Coupling heterogeneous production systems by a multi-agent based cyber-physical production system. In: Proceedings of 12th IEEE International Conference on Industrial Informatics (INDN), pp. 713–719 (2014)

A Novel Repair-Based Multi-objective Algorithm for QoS-Constrained Distributed Data-Intensive Web Service Composition

Soheila Sadeghiram$^{(\boxtimes)}$, Hui Ma$^{(\boxtimes)}$, and Gang Chen$^{(\boxtimes)}$

Victoria University of Wellington, Wellington, New Zealand
{Soheila.Sadeghiram,Hui.Ma,Aaron.Chen}@ecs.vuw.ac.nz

Abstract. *Distributed Data-intensive Web service composition (DWSC)* implements new applications by using existing Web services distributed over the Internet. DWSC has multiple conflicting objectives, e.g. minimising response time and cost, and often needs to satisfy user-defined QoS constraints. While various approaches are proposed to handle this NP-hard problem, they either ignore the impact of the distributed nature of services on QoS, or treat DWSC as a single-objective optimisation problem assuming users can provide quantified preferences for each QoS, or do not consider QoS constraints while searching for composite solutions. To solve QoS-constrained multi-objective distributed DWSC problems, we propose a knowledge-based repair method for NSGA-II algorithm to effectively search for Pareto-optimal service compositions that satisfy all QoS constraints. This repair method facilitates the construction of constraint-obeying composite services. Experimental results verify that our algorithm outperforms existing NSGA-II based approaches for this problem on standard benchmark datasets.

Keywords: Distributed composition · Data-intensive Web services · Multi-objective optimisation · User constraints · Repair · NSGA-II

1 Introduction

Web services are software modules that provide reusable business functionalities. They can be discovered, located, invoked and loosely coupled across the Web to facilitate the integration of newly established applications. However, existing services often cannot satisfy the personalised and diversified needs of users. *Web Service Composition (WSC)* creates new composite services via integrating existing services to accomplish new functionalities. For a given service request, there are many possible composite services. Non-functional properties, i.e. *Quality of Service (QoS)*, are significant criteria for judging the quality of composite services. Meanwhile, companies such as Google and Amazon provide access to some of their internal business resources in the form of *Data-intensive Web services*, e.g. Amazon Web Services (AWS). A large volume of data often needs to

© Springer Nature Switzerland AG 2020
Z. Huang et al. (Eds.): WISE 2020, LNCS 12342, pp. 489–502, 2020.
https://doi.org/10.1007/978-3-030-62005-9_35

be transmitted among services to be executed in sequence within a composite service. This gives rise to *data-intensive Web service composition (DWSC)*, for which communication time and cost must be considered because they greatly affect the QoS of any composite services. More specifically, WSC falls under the category of NP-hard problems. Although some algorithms might find optimal solutions for specific problem instances, algorithms to efficiently find optimal solutions to all problem instances do not exist.

Existing work on WSC can be grouped into fully- and semi-automated (*service selection*) categories. The former designs an appropriate workflow, i.e. a directed acyclic graph (DAG) and selects services from a large service repository to fulfil each functional unit of the DAG. The latter assumes that a workflow of abstract services has already been provided and selects concrete services to satisfy the relevant abstract service [4,6]. However, the workflow design and service selection depend on each other and therefore should be jointly managed to construct suitable workflow and effective solutions. In this paper, the fully-automated strategy will be considered.

DWSC is a naturally multi-objective optimisation problem and QoS requirements are usually specified by users as constraints [2,4]. For example, a deadline and a budget can be set on the overall QoS by a service requester [2,4]. Evolutionary Computation (EC) algorithms are particularly effective at finding near-optimal solutions with high efficiency, which makes them popular choices for WSC [16,17,21]. Among all EC algorithms, *Non-dominated Sorting Genetic Algorithm-II (NSGA-II)* [3] has leading performance on many multi-objective WSC problems [14,20]. However, the direct use of these approaches may not guarantee the satisfaction of any QoS constraints. Evolutionary search for Pareto optimal solutions that satisfy given QoS constraints is challenging since the goal to search for valid solutions may pose a negative impact on the optimisation of QoS [2,12] Existing approaches for WSC have several limitations. Some approaches have mostly ignored the importance of service distributions and inter-service communications [20,21], while they are vital in determining the overall QoS of composite services. Further, QoS constraints have only been considered by semi-automated WSC [4,6], rather than by fully-automated approaches. Most of those approaches address the single-objective WSC. The only existing approach on multi-objective constrained WSC [6] follows constraint handling techniques proposed for other problems, e.g. penalising the fitness value of invalid solutions. However, repairing invalid solutions can help the search maintain more solutions in the feasible region. Solutions that violate any QoS constraints may be easily repaired. For example, random repair operators have been proposed, paving the way for discovering high-quality solutions in future generations [19], but they are not effective since most of the random changes cannot fix the solutions and also might affect the average quality of solutions in the population. Therefore, to effectively address constrained distributed multi-objective DWSC problems, effective repair operators must be designed.

The above-mentioned issues motivate us to develop an effective multi-objective method supported by a repair method that enjoys a set of heuristics obtained from domain knowledge. We propose a model to represent domain knowledge, including the distribution of services. The newly developed repair method can be integrated with any effective multi-objective approach for the constrained fully-automated DWSC. Therefore, we combine the repair method with NSGA-II algorithm enhanced by a set of selection rules enabling the algorithm to compare valid and invalid solutions. An extensive empirical analysis of the proposed knowledge-based NSGA-II and comparison with an existing repair-based method as well as a cutting-edge constrained handling method proposed for the composition problem [6, 23] verify the effectiveness of our methods.

2   Literature Review

In this section, we provide a literature review on constrained WSC problems.

While constrained WSC is studied in the literature [2,4,6,23], they only address the semi-automated WSC. In some cases, a constrained single-objective problem has been formulated as an unconstrained multi-objective problem [2,4], i.e. the weighted sum of QoS attributes is regarded as one objective and constraints are formulated as the second objective. The formulation in those methods does not need any consideration of constraint handling during the search.

For addressing a single-objective constrained WSC problem [23], a repair technique has been combined with a memetic approach (i.e. combination of the EC algorithm with a local search technique), where a crossover operator is designed to repair invalid solutions. This method addresses conflict constraints (conflict constraints are restrictions in the mutual implementation of some Web services) for a semi-automated WSC in a centralised environment. However, repair techniques can be designed more effectively by considering the domain-knowledge of WSC problems.

Multi-objective QoS-Constrained WSC [6] has been addressed by employing a penalty-based technique, where the fitness of an infeasible solution is degraded by a penalty, depending on the amount of the constraint violation they display. However, setting parameters of the penalty function might be difficult and therefore, making these methods highly sensitive to the choice of penalty parameters. The ideal penalty factor to be adopted in a penalty function cannot be known *a priori* for an arbitrary problem. Furthermore, simply penalising invalid solutions restricts the flexibility of the search process for generating good offspring solutions in the valid region. The information carried by invalid solutions can enable speedy movements into the valid region, especially when this region is only a small part of the entire search space [24]. Finally, optimal solutions of most engineering problems lie on the constraint boundary [22], which means that an invalid solution near the constraint boundary can be easily converted to a high-quality valid solution with a repair technique.

In this paper, we develop an effective repair technique to handle QoS constraints for fully-automated WSC in a distributed environment. Our proposed technique will not only allow invalid solutions to participate during the search to improve the convergence rate but also repair invalid solutions to create compositions that obey user-defined QoS constraints.

3 Preliminary

This section presents basic terms and the representation method that are necessary for the problem formulation and fully-automated composition, respectively.

A *Web service* is defined as a tuple $S_i = \langle I(S_i), O(S_i), QoS(S_i), h(S_i)\rangle$, where $I(S_i)$ is a set of inputs required by S_i and $O(S_i)$ is a corresponding set of outputs produced by S_i. $QoS(S_i) = \langle Q^1(S_i), Q^2(S_i), ..Q^n(S_i)\rangle$ is an associated tuple of quality attributes describing the non-functional properties of S_i. $h(S_i)$ is the geographical location of Web service S_i. A *service repository* $\mathcal{R}$ is a collection of n Web services ($i \in \{1 \ldots n\}$). Two QoS attributes will be considered: execution time and cost, as they are the most frequently and commonly used QoS in the literature [16, 20, 21].

Definition 1. *Given a tuple of QoS attributes $\langle Q^1(Si), Q^2(S_i), ..Q^n(S_i)\rangle$, global QoS constraints are denoted as $\mathcal{G} = \{k_1, k_2, ..k_n\}$, where each k_j is an upper bound for a negative (i.e. an attribute that should be minimised) or a lower bound for a positive QoS attribute $Q^j(S_i)$. Each global QoS constraint in $\mathcal{G}$ is used to restrict its corresponding QoS attribute of a composition solution and needs to be satisfied.*

Definition 1 has been obtained from the definition of user constraints on end-to-end QoS of the composite solution in the literature [2, 4]. For example, a user requests a flight within a limited cost budget, e.g. k_{cost}.

A *service request* (or a *composition task*) is defined as a triple $T = (I(T), O(T), \mathcal{G}(T))$ where $I(T)$ indicates inputs provided to the composition, $O(T)$ represents outputs required from the composition. For task T, we should find a composition ($\mathcal{CS}$) that takes $I(T)$, produces $O(T)$ with optimised QoS and satisfies global QoS constraint $\mathcal{G}(T)$. A *composite service*, $\mathcal{CS}$, can be represented as a directed acyclic graph (DAG), specifying the order of component services in which they should be run, and the necessary communication between them as well. A DAG has a set of P component services jointly accomplishing the given task and must be *functionally correct (feasible)*, i.e. it should be executable at run-time using the inputs, $I(T)$, to successfully produce the desired outputs ($O(T)$). Two special services are used to represent the overall inputs and outputs of $\mathcal{CS}$: *Start*, where $I(Start) = \varnothing$ and $O(Start) = I(T)$, and *End* where $I(End) = O(T)$ and $O(End)= \varnothing$. $O(Start) = I(T))$, and *End* ($I(End) = O(T)$, $O(End) = \varnothing$).

In a composite Web service, point-to-point communication happens between S_i and S_j if a direct edge in the DAG connects them. A *communication link* is a link joining two services and is defined as $l = \langle T_L(l), C_L(l)\rangle$. The communication link has a communication time (T_L) and a communication cost (C_L) attributes.

We use a permutation representation for searching and a DAG representation for evaluation. A *decoding* process is applied to each permutation to convert it to a DAG and must ensure the functional correctness [21] to transform sequences into a corresponding executable service composition, i.e., a feasible workflow [16]. The following example clarifies the decoding process:

Figure 1 shows a task with $\{a, b\}$ as inputs and $\{g\}$ as the output, as well as inputs and outputs of each service. The decoding process checks the sequence from left to right until a service is found which has the output matching the input of service End (It starts with the composition's output, $End(g, \varnothing)$). Since service S_4 outputs $\{g\}$, which is required by End, we connect it to End. Afterwards, the decoding process continues to scan the sequence for services that can satisfy $I(S_4) = \{d, e\}$. Since $O(S_1) = \{e\}$, and $O(S_5) = \{d\}$, the union of the two services S_1 and S_5 satisfies the input of service S_4, both S_1 and S_5 are separately connected to service S_4 as the predecessor services of S_4. Afterward, the decoding process looks for service(s) whose outputs satisfy the inputs of service S_1 and S_5. Since service S_2 outputs $\{c, b\}$, which satisfy the inputs required by both services S_1 and S_5, service S_2 is connected to both S_1 and S_4 as their predecessor. Finally, the output of service S_6 is connected to the input of service S_2, where S_6 can be satisfied by the composition's inputs and $Start$ is reached.

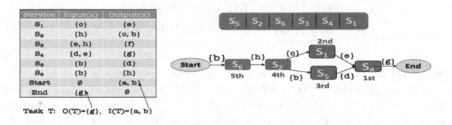

Fig. 1. Decoding of a permutation

The decoding process automatically generates a composition based on a given sequence. It takes inputs from the task and generates outputs requested by the task, with all the inputs of all the nodes on the graph fulfilled by predecessors. If a functionally correct workflow does not exist for the corresponding sequence, the decoding process will not be able to reach $Start$ and build the workflow. The sequence will, therefore, be identified as "infeasible" and be discarded.

4   Problem Formulation

The overall cost of a composite service $\mathcal{CS}$ is the cost of all component services aggregated with communication costs, as shown in Eq. (1):

$$C_{total}(\mathcal{CS}) = \sum_{i \in P} C_S(S_i) + \sum_{k \in \mathcal{L}} C_L(l_k) \tag{1}$$

where P and $\mathcal{L}$ are numbers of component services and communication links in $\mathcal{CS}$, respectively. Suppose that a composition $\mathcal{CS}$ has m paths in its DAG from $Start$ to End, and $\mathcal{T}_i$ is the execution time of path i. In fact, $\mathcal{T}_i$ includes time for executing component services and communication time of links on path i. $\mathcal{T}_i$ can be obtained through Eq. 2:

$$T_i = \sum_{q \in Q} T_S(S_q) + \sum_{k \in \mathcal{L}} T_L(l_k) \tag{2}$$

where Q and $\mathcal{L}$ are numbers of component services and communication links on that path, respectively. Hence, T_{total}, the overall response time of CS, is calculated as the time of the most time-consuming path in the composition as it is shown in Eq. (3):

$$T_{total}(CS) = \max_{i \in \{1...m\}} \{T_i\} \tag{3}$$

Therefore, we can define the distributed DWSC with *objective functions*, $f_1(CS) = \hat{T}_{total}(CS)$ and $f_2(CS) = \hat{C}_{total}(CS)$, where $\hat{T}_{total}$ and $\hat{C}_{total}$ are normalised values of T_{total} and C_{total}, respectively. We will consider two constraints with each constraint corresponding to one QoS attribute. The QoS-constrained multi-objective distributed DWSC problem can be written as follows:

$$\min_{CS} f(CS) = \{f_1(CS), f_2(CS)\},$$
$$s.t. \quad f_1(CS) \le k_1 \ \ and \ \ f_2(CS) \le k_2. \tag{4}$$

where Eq. 4 capture constraints on cost and time, respectively, and k_1 and k_2 are constants provided by the user ($\mathcal{G} = \{k_1, k_2\}$). The objective space in which $f_1(CS) \le k_1$ and $f_2(CS) \le k_2$ is called the *valid region*. This definition and the constraints model for the WSC problems has been employed in the literature for WSC problem [2,4].

In order to measure the quality of solutions regarding constraint satisfaction, we use *constraint violation degree* introduced in Definition 2.

Definition 2. *Constraint violation degree is a measure to indicates to which extent a composite service violates constraints and is obtained by Eq. (5):*

$$CV = max((f_1 - k_1), 0) + max((f_2 - k_2), 0) \tag{5}$$

that is, if $CV = 0$, no constraint is violated.

Definition 3. *A service composition solution is invalid if $CV \ne 0$ for that service composition.*

5   Proposed Repair-Based Multi-objective Method

This section presents the proposed NSGA-II with knowledge-based repair approach to QoS-constrained multi-objective DWSC, including the main algorithm, selection rules, operators, and our proposed repair technique.

Algorithm 1 shows an overview of the proposed method. A random initial population of solutions in the form of service sequences is created (line 1). These solutions are then converted into DAGs to be evaluated (line 2). The following process continues for the maximum number of generations: parent solutions are

Algorithm 1. NSGA-II with knowledge-based repair to the DWSC

INPUT: Task (T), Service Repository ($\mathcal{R}$), population size (N)
OUTPUT: Solution set
1: Randomly initialise population P from $\mathcal{R}$ by creating N solutions;
2: Decode the individuals into DAGs and evaluate them;
3: **while** the maximum number of generations is not reached **do**
4:    Select solutions using a tournament selection based on the *Selection rules* (see Section 5.1);
5:    Apply crossover and mutation on selected solutions to generate offspring;
6:    Decode and evaluate offspring solutions;
7:    **Identify invalid solutions (see Definition 3);**
8:    **Apply selection heuristics to identify the service to be repaired, and then repair invalid solutions (Algorithm 2);**
9:    Perform non-domination sorting using *Selection rules*;
10:    Select N individuals for the next-generation.
11: **end while**
12: Return non-dominated set of solutions in the final generation;

selected for the crossover and mutation operators (line 4) to produce offspring solutions which are then evaluated. Invalid solutions are subsequently identified (line 7). Among all invalid solutions, a portion of solutions will be chosen through a heuristic method (see Subsect. 5.2) for repair (line 9). All solutions are sorted according to the front and crowding distance, and the best solutions are transferred to the next generation (line 11). Each key component of this algorithm will be explained in detail in the following subsections.

5.1 Selection Rules, Crossover and Mutation

A tournament selection, based on the pairwise comparison of solutions is proposed in [10] for single-objective optimisation. That selection uses *selection rules* to rank solutions in the population where both valid and invalid solutions are included, as follows: 1) any valid solution is preferred to any invalid solution; 2) between two valid solutions, the one with better objective function value is preferred; and 3) between two invalid solutions, the one with the smallest violation of constraints, $\mathcal{CV}$ (Definition 5), is preferred. If used lonely, these rules may discard many invalid solutions which might carry useful information in their structure. However, optimal solutions to the unconstrained problems often lie on constraint boundaries [13]. Hence simply discarding close-to-boundary solutions will make it difficult to discover optimal solutions and reduce the solution diversity in the population. To tackle the above-mentioned shortcoming, we design a repair method that enhances these rules, in Sect. 5.2. Based on the selection rules, two parents for crossover and one parent for mutation are selected. The crossover operation is designed to prevent service duplication in any offspring solutions. The mutation operator is applied to one sequence by randomly exchanging two services of that sequence (Detailed discussions on the mutation and crossover have been provided in [21]).

5.2 Knowledge-Based Repair Method

In this Subsection, a technique is proposed to repair invalid solutions and create enhanced neighbouring solutions without destroying valuable genetic information. Further, a set of novel heuristics are defined to locate services in the work-flow for repair. It is proven that solutions with moderate violation of constraints enjoy the nice chance of producing good valid solutions in their neighbourhood [10,11]. As shown in the example in Fig. 2, only a small portion of non-dominated solutions lies in the valid region. However, two invalid solutions are located in a closer distance to the valid region on the Pareto-front. If repaired properly, e.g., in the direction of minimising one of the objective functions in this example, good valid non-dominated solutions will be produced through exploiting these invalid solutions, which will help the algorithm to discover high-quality solutions. These solutions will be used to evolve high-quality solutions in future populations.

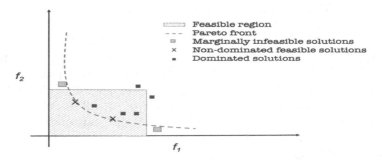

Fig. 2. Feasible region

We are interested in using structural information of solutions, such as the communication links used to connect two services, or the QoS of component services to repair and improve invalid solutions with minor violation of constraints. The proposed repair method aims to reduce the constraint violation degree ($\mathcal{CV}$) of any invalid solution (see Definition 2).

Our new repair method tries to address both time and cost constraints. This way, neighbour solutions can be generated to replace the current invalid one if their $\mathcal{CV}$ is smaller than that of the current invalid solution. This is specifically achieved by swapping a Web service S with all other services in the same layer as S (details on layers can be found in [21]). This is an effective neighbourhood generation strategy for distributed DWSC in [16]. In this paper, we define a set of seven novel criteria for choosing the Web service for replacement. Hence, for each invalid composite service, up to seven component services can be selected. This set is called the *selection heuristics set*. This selection heuristic set jointly covers the majority of chances for repair is as follows: 1) the longest execution time among all services in the workflow; 2) the highest cost among all services in the workflow; 3) the highest sum of execution time and cost; 4) the longest link in the workflow as its output; 5) the longest link in the workflow as its input; 6) the most expensive link in the workflow as its output; 7) the most time-consuming link in the workflow as its output.

Algorithm 2 presents the pseudo-code of the proposed repair. A simple example of creating neighbours in the repair method is illustrated in Fig. 3. Suppose that service S_1 is one of the selected services by the *selection heuristic set* and

---

Algorithm 2. Knowledge-based repair method.

1: **while** Not all invalid solutions are investigated for the repair **do**
2:     Select an invalid composite solution CS;
3:     Select a set of component services in CS using the *selection heuristic set*;
4:     **for all** selected services in CS **do**
5:         Swap the selected service with other services in the same layer;
6:     **end for**
7:     Decode and evaluate the new permutations;
8:     Calculate CV for each newly generated solution;
9:     Select a solution from the neighbours with the smallest CV to replace the original solution.
10: **end while**

---

there exist two other services, S_3 and S_4, in the same layer of S_1 [21]. Therefore, two solutions will be generated by swapping S_1 with S_3 and S_4, respectively. Neighbour solutions will be decoded and evaluated, and the one with the smallest CV will be chosen to replace the original solution if the corresponding CV is smaller than or equal to that of the original solution. Simultaneous enhancing both objectives in Eq. (4) during the local optimisation is achieved by eliminating an expensive communication link appeared in the composite solution, and/or a costly and time-consuming component (single Web service).

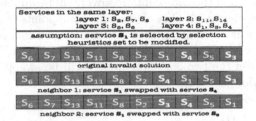

Fig. 3. Example of the repair operator

6   Experiments

In this section, we present benchmark datasets, experimental settings, test problems and results.

The experiments have been carried out using WSC-2008 [1] and WSC-2009 [7] benchmark datasets, which contain eight and five service repositories of varying sizes, respectively. One associated composition task per repository of services are given in advance [1,7]. These datasets are chosen because of being the largest benchmarks that have been broadly exploited in the WSC literature, e.g. [5,9,15–18]. QoS information for any service and distance between each pair of Web services are obtained from [15].

We compare algorithms according to their execution time, inverted generational distance (IGD) and hypervolume (HV) metrics, which provide a comprehensive measure of the performance of each method.

6.1   Baseline Algorithms

To verify the effectiveness of our novel repair technique combined with NSGA-II, we choose two constraint handling techniques, *Penalty-baseline* [6] and *Crossover repair* [23], and combine them with NSGA-II to compare with our proposed method. They have been applied in a very similar problem (Web service selection). 1) *Penalty-baseline*: a penalty mechanism employed in [6] for multi-objective WSC and has outperformed two other penalty functions. We will use it in our problem as in Eq. (6):

$$
f_1'(\mathcal{CS}) = (f_1(\mathcal{CS}) - k_1) + f_1(\mathcal{CS}) \\
and \quad f_2'(\mathcal{CS}) = (f_2(\mathcal{CS}) - k_2) + f_2(\mathcal{CS})
\tag{6}
$$

where $f_1 - k_1$ and $f_2 - k_2$ are the amount of constraint violations for objective functions f_1 and f_2, respectively. f_1' and f_2' are new fitness values for the solution (Note that all values have been normalised between 0 and 1). 2) *Crossover repair (CrRepair)*: generally, a repair can be performed through crossover operators. A crossover repair operator has been employed in [23] for the constrained Web service selection problem to reduce the chance for a composition solution to violate any constraints. Following [23], two invalid solutions sequences will be selected as parents $p1$ and $p2$, and the two-point crossover will be applied to generate two offspring sequences $c1$ and $c2$. One of the crossover points is determined using $p1$, based on the location of the longest communication link in $p1$. The second crossover point is determined using $p2$ in the same way as $p1$. The sub-sequence of $p1$ and $p2$ between the two crossover points will be swapped while ensuring no service duplication (see [21]).

6.2   Parameter Settings

For all methods, a population of size 500 was evolved for 51 generations, as established in [20] for the multi-objective WSC problem. Tournament selection strategy, with size 2, selects candidate solutions for the mutation and crossover operators. The tournament selection is based on the selection rules (Sect. 5.1). The probability of crossover and mutation operators are 0.95 and 0.05, respectively. Repair probability in *Crossover-Repair* method is 0.95. The above parameter settings were based on popular settings discussed in the literature [8,16,20]. We determined constraints on each objective. First, we carried out the experiments using the unconstrained version of our problem to obtain the constraints on f_1 and f_2. The constraint value, i.e. k_1 and k_2 were then set 0.75 of the corresponding QoS's worst value. For example, if the worst response time for the task is 0.5 when running without constraints, we set the time constraint equal to 0.375. This means the response time should not be higher than 0.375.

6.3   Evaluation Results

Table 1 shows the IGD and HV observed for *PenaltyBaseline* [6], *CrRepair* [23] and our proposed method. A pairwise Wilcoxon signed statistical test with a significance level of 5% was conducted on the average of these values over the 30 runs for the three algorithms at 0.05 significance level ($p\text{-}value = 0.05$).

Table 1. Mean and standard deviations over IGD and HV for 30 runs.

Task	IGD			HV		
	Penalty [6]	CrRepair [23]	Proposed method	Penalty [6]	CrRepair [23]	Proposed method
WSC08-1	0.21 ± 0.04	0.2 ± 0.027	$\mathbf{0.05 \pm 0.001}$	0.2992 ± 0.01	0.3195 ± 0.01	$\mathbf{0.3321 \pm 0.01}$
WSC08-2	0.2 ± 0.04	0.19 ± 0.03	$\mathbf{0.12 \pm 0.02}$	0.3215 ± 0.01	0.3254 ± 0.002	$\mathbf{0.3563 \pm 0.001}$
WSC08-3	0.09 ± 0.001	0.1 ± 0.001	$\mathbf{0.01 \pm 0.002}$	0.3315 ± 0.001	0.3281 ± 0.003	0.3616 ± 0.001
WSC08-4	0.05 ± 0.002	0.07 ± 0.002	$\mathbf{0.01 \pm 0.001}$	0.3199 ± 0.004	0.3165 ± 0.02	$\mathbf{0.3492 \pm 0.007}$
WSC08-5	0.19 ± 0.01	0.23 ± 0.1	$\mathbf{0.036 \pm 0.01}$	0.3033 ± 0.001	0.3006 ± 0.004	$\mathbf{0.3439 \pm 0.003}$
WSC08-6	0.0548 ± 0.003	0.0664 ± 0.002	$\mathbf{0.046 \pm 0.001}$	0.3241 ± 0.004	0.3122 ± 0.01	$\mathbf{0.3438 \pm 0.02}$
WSC08-7	0.102 ± 0.044	0.11 ± 0.052	0.092 ± 0.01	0.3212 ± 0.006	0.3221 ± 0.004	0.3328 ± 0.003
WSC08-8	0.063 ± 0.002	0.061 ± 0.003	0.06 ± 0.002	0.3113 ± 0.002	0.3113 ± 0.003	0.3142 ± 0.012
WSC09-1	0.03 ± 0.01	0.02 ± 0.02	$\mathbf{0.01 \pm 0.01}$	0.3525 ± 0.01	0.3539 ± 0.002	$\mathbf{0.3542 \pm 0.001}$
WSC09-2	0.091 ± 0.005	0.098 ± 0.004	$\mathbf{0.03 \pm 0.001}$	0.3297 ± 0.009	0.3284 ± 0.001	$\mathbf{0.337 \pm 0.04}$
WSC09-3	0.001 ± 0.0002	0.01 ± 0.0001	0.001 ± 0.0001	0.3191 ± 0.01	0.3166 ± 0.04	$\mathbf{10.3272 \pm 0.05}$
WSC09-4	0.08 ± 0.001	0.13 ± 0.001	$\mathbf{0.03 \pm 0.002}$	0.3311 ± 0.03	0.2716 ± 0.06	$\mathbf{0.3402 \pm 0.02}$
WSC09-5	0.11 ± 0.002	0.24 ± 0.003	$\mathbf{0.07 \pm 0.005}$	0.3283 ± 0.012	0.2529 ± 0.012	$\mathbf{0.3311 \pm 0.04}$

Table 2. Mean execution time [seconds] over 30 runs.

Task	Penalty [6]	CrRepair [23]	Proposed method
WSC08-1	$\mathbf{3.4 \pm 0.09}$	4.14 ± 0.07	5.19 ± 0.17
WSC08-2	$\mathbf{3.1 \pm 0.08}$	3.92 ± 0.023	5.023 ± 0.52
WSC08-3	$\mathbf{7.2 \pm 1.3}$	11.38 ± 0.9	18.13 ± 1.79
WSC08-4	$\mathbf{4.93 \pm 0.09}$	5.15 ± 0.21	6.94 ± 0.13
WSC08-5	$\mathbf{12.43 \pm 0.021}$	15.41 ± 1.11	22.06 ± 1.13
WSC08-6	$\mathbf{14.33 \pm 2.35}$	29.55 ± 2.23	51 ± 3.42
WSC08-7	$\mathbf{59.03 \pm 4.79}$	101.53 ± 6.03	128.8 ± 11.33
WSC08-8	$\mathbf{78.8 \pm 5.39}$	134.51 ± 16.64	201 ± 20.87
WSC09-1	$\mathbf{1.84 \pm 0.12}$	2.1 ± 0.3	3.93 ± 0.17
WSC09-2	$\mathbf{32.88 \pm 4.07}$	38.33 ± 6.24	67.12 ± 5.33
WSC09-3	$\mathbf{57.36 \pm 8.27}$	124.33 ± 8.06	163.15 ± 27.87
WSC09-4	$\mathbf{108.5 \pm 10.33}$	312.5 ± 42.91	442.36 ± 55.39
WSC09-5	$\mathbf{112 \pm 13.54}$	199 ± 38.18	279 ± 41.21

It is evident from Table 1 that the proposed method achieves significantly better values of IGD and HV for all of the tasks, and *CrRepair* performs better than *PenaltyBaseline* merely for 30% of tasks. Unlike *CrRepair* which performs a relatively random repair, the proposed knowledge-based repair method explicitly considers the distribution of services and effectively remove the longest communication links from those solutions that violate any QoS constraints. It also noticeably improves the performance of NSGA-II by properly balancing the trade-off between exploration and exploitation during the evolutionary search. The average execution time and the corresponding standard deviation in seconds over 30 independent runs for each approach are shown in Table 2. Table 2 indicates that the proposed method has the longest execution time since it performs more evaluations while applying the repair operator on all invalid solutions. *PenaltyBaseline* executes in a shorter time because, unlike repair techniques, it does not require extra evaluations or corresponding decoding.

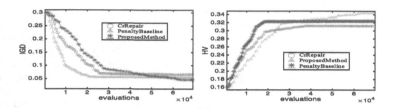

Fig. 4. Mean IGD (left) and HV (right) on task WSC8-6.

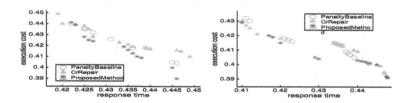

Fig. 5. Pareto solutions for WSC8-6 (left) and WSC9-2 (right) (note that only valid solutions and region are shown in this figure).

6.4   Further Analysis

We further evaluate the convergence behaviours concerning both IGD and HV over 30 runs using task WSC08-6 as an example. Figure 4 demonstrates the evolution of the mean values of the IGD and HV over the number of evaluations for the proposed method, *CrRepair* and *PenaltyBaseline*, where, the proposed method converges slower than the two other methods but continues to improve the solutions. We present a plot of the Pareto front valid solutions of WSC08-6 and WSC09-2 obtained by the three methods over 30 independent runs in Fig. 5. The best Pareto optimal solutions are identified based on the combined results of all 30 runs of each method. It is evident from Fig. 5 that the Pareto front generated by the *PenaltyBaseline* is more widely distributed for both tasks.

The solutions generated by our proposed approach dominate the solutions from the other two approaches in the central region. This demonstrates that our proposed approach is effective in generating all quality solutions that satisfied all QoS constraints. To be specific, our proposed method produces valid solutions with better quality than that of other methods. For example, our proposed method has found several extreme solutions, in particular, for task WSC08-6.

7 Conclusions

We studied the problem of constrained multi-objective DWSC in a distributed environment. We proposed multi-objective approach based on NSGA-II which handles QoS constraints effectively. We designed a novel knowledge-based repair method which takes advantage of a set of heuristics for the distributed DWSC, obtained using domain knowledge. Results verified our proposed method can achieve significantly better performance in all of the benchmarks compared to the existing relevant DWSC methods while searching valid composition solutions.

References

1. Bansal, A., Blake, M.B., Kona, S., Bleul, S., Weise, T., Jaeger, M.C.: WSC-08: continuing the Web services challenge. In: 2008 10th IEEE Conference on E-Commerce Technology and the Fifth IEEE Conference on Enterprise Computing, E-Commerce and E-Services, pp. 351–354. IEEE (2008)
2. Chen, F., Dou, R., Li, M., Wu, H.: A flexible QoS-aware Web service composition method by multi-objective optimization in cloud manufacturing. Comput. Ind. Eng. **99**, 423–431 (2016)
3. Deb, K., Pratap, A., Agarwal, S., Meyarivan, T.: A fast and elitist multiobjective genetic algorithm: NSGA-II. IEEE Trans. Evol. Comput. **6**(2), 182–197 (2002)
4. Deng, S., Huang, L., Wu, H., Wu, Z.: Constraints-driven service composition in mobile cloud computing. In: 2016 IEEE International Conference on Web Services (ICWS), pp. 228–235. IEEE (2016)
5. Gabrel, V., Manouvrier, M., Moreau, K., Murat, C.: QoS-aware automatic syntactic service composition problem: complexity and resolution. Future Gener. Comput. Syst. **80**, 311–321 (2018)
6. Han, X., Yuan, Y., Chen, C., Wang, K.: QoS-aware multiobjective optimization algorithm for Web services selection with deadline and budget constraints. Adv. Mech. Eng. **6**, 361298 (2014)
7. Kona, S., Bansal, A., Blake, M.B., Bleul, S., Weise, T.: WSC-2009: a quality of service-oriented Web services challenge. In: IEEE Conference on Commerce and Enterprise Computing, 2009, CEC 2009, pp. 487–490. IEEE (2009)
8. Koza, J.R.: Genetic Programming: On the Programming of Computers by Means of Natural Selection, vol. 1. MIT Press, Cambridge (1992)
9. Li, J., Yan, Y., Lemire, D.: Scaling up Web service composition with the skyline operator. In: 2016 IEEE International Conference on Web Services (ICWS), pp. 147–154. IEEE (2016)
10. Mezura-Montes, E., Coello, C.A.C.: A simple multimembered evolution strategy to solve constrained optimization problems. IEEE Trans. Evol. Comput. **9**(1), 1–17 (2005)
11. Miyakawa, M., Sato, H., Sato, Y.: A study for parallelization of multi-objective evolutionary algorithm based on decomposition and directed mating. In: Proceedings of the 2019 3rd International Conference on Intelligent Systems, Metaheuristics & Swarm Intelligence, pp. 25–29. ACM (2019)

12. Mostafa, A., Zhang, M.: Multi-objective service composition in uncertain environments. IEEE Trans. Serv. Comput. (2015)
13. Rahi, K.H., Singh, H.K., Ray, T.: Investigating the use of sequencing and infeasibility driven strategies for constrained optimization. In: 2019 IEEE Congress on Evolutionary Computation (CEC), pp. 1642–1649. IEEE (2019)
14. Ramírez, A., Parejo, J.A., Romero, J.R., Segura, S., Ruiz-Cortés, A.: Evolutionary composition of QoS-aware Web services: a many-objective perspective. Expert Syst. Appl. **72**, 357–370 (2017)
15. Sadeghiram, S., Ma, H., Chen, G.: Cluster-guided genetic algorithm for distributed data-intensive Web service composition. In: 2018 IEEE Congress on Evolutionary Computation (CEC) (2018)
16. Sadeghiram, S., Ma, H., Chen, G.: Composing distributed data-intensive Web services using a flexible memetic algorithm. In: 2019 IEEE Congress on Evolutionary Computation (CEC), pp. 2832–2839 (2019)
17. Sadeghiram, S., Ma, H., Chen, G.: Composing distributed data-intensive Web services using distance-guided memetic algorithm. In: Hartmann, S., Küng, J., Chakravarthy, S., Anderst-Kotsis, G., Tjoa, A.M., Khalil, I. (eds.) DEXA 2019. LNCS, vol. 11707, pp. 411–422. Springer, Cham (2019). https://doi.org/10.1007/978-3-030-27618-8_30
18. Sadeghiram, S., Ma, H., Chen, G.: A memetic algorithm with distance-guided crossover: distributed data-intensive Web service composition. In: Proceedings of the Genetic and Evolutionary Computation Conference Companion, pp. 155–156 (2019)
19. Salcedo-Sanz, S.: A survey of repair methods used as constraint handling techniques in evolutionary algorithms. Comput. Sci. Rev. **3**(3), 175–192 (2009)
20. da Silva, A.S., Ma, H., Mei, Y., Zhang, M.: A hybrid memetic approach for fully-automated multi-objective Web service composition. In: 2018 IEEE International Conference on Web Services (ICWS), pp. 26–33. IEEE (2018)
21. da Silva, A.S., Mei, Y., Ma, H., Zhang, M.: Evolutionary computation for automatic Web service composition: an indirect representation approach. J. Heuristics **24**(3), 425–456 (2017). https://doi.org/10.1007/s10732-017-9330-4
22. Singh, H.K., Isaacs, A., Ray, T., Smith, W.: Infeasibility driven evolutionary algorithm (IDEA) for engineering design optimization. In: Wobcke, W., Zhang, M. (eds.) AI 2008. LNCS (LNAI), vol. 5360, pp. 104–115. Springer, Heidelberg (2008). https://doi.org/10.1007/978-3-540-89378-3_11
23. Tang, M., Ai, L.: A hybrid genetic algorithm for the optimal constrained Web service selection problem in Web service composition. In: IEEE Congress on Evolutionary Computation, pp. 1–8. IEEE (2010)
24. Xu, B., Zhang, H., Zhang, M., Liu, L.: Differential evolution using cooperative ranking-based mutation operators for constrained optimization. Swarm Evol. Comput. **49**, 206–219 (2019)

Heuristics Based Mosaic of Social-Sensor Services for Scene Reconstruction

Tooba Aamir[1], Hai Dong[1(✉)], and Athman Bouguettaya[2]

[1] School of Science, RMIT University, Melbourne, Australia
{tooba.aamir,hai.dong}@rmit.edu.au
[2] School of Computer Science, The University of Sydney, Sydney, Australia
athman.bouguettaya@sydney.edu.au

Abstract. We propose a heuristics-based social-sensor cloud service selection and composition model to reconstruct *mosaic* scenes. The proposed approach leverages *crowdsourced* social media images to create an image mosaic to reconstruct a scene at a designated location and an interval of time. The novel approach relies on the set of features defined on the bases of the image metadata to determine the *relevance* and *composability* of services. Novel heuristics are developed to filter out non-relevant services. Multiple machine learning strategies are employed to produce *smooth* service composition resulting in a mosaic of relevant images indexed by geolocation and time. The preliminary analytical results prove the feasibility of the proposed composition model.

1 Introduction

Advancements in smart devices, e.g., smartphones, and prevalence of social media, have established new means for information sensing and sharing [1,2]. Social media data, e.g., Twitter posts and Facebook statuses, have become a significant and accessible means for sharing the facts or opinions about any event. Crowdsourcing (i.e., sensing, gathering and sharing) of social media data is termed as *social sensing* [1,2]. Smart devices (i.e., *social-sensors* [1]) have the ability to embed sensed data directly into social media outlets (i.e., *social clouds* or *social-sensor clouds*) [4]. Monitoring the social media data (i.e., *social-sensor data*) provides multiple benefits in various domains. For example, traditional sensors like CCTVs are usually unable to provide complete coverage due to their limited field of view and sparsity. Social-sensor data, especially multimedia content (e.g., images) and related metadata may provide a multi-perspective coverage [1]. We focus on utilising offline social-sensor images, i.e., downloaded batches of social media images and their related metadata to assist in scene reconstruction.

The social-sensor images are featured with diverse formats, structures and sources. There are various inherent challenges for the efficient management and delivery of such multifaceted social-sensor images [3,4,7]. One of the significant challenges is *searching and analysing a massive amount of heterogeneous data*

© Springer Nature Switzerland AG 2020
Z. Huang et al. (Eds.): WISE 2020, LNCS 12342, pp. 503–515, 2020.
https://doi.org/10.1007/978-3-030-62005-9_36

in the inter-operable social cloud platforms [3]. *Service paradigm* is proposed to address this challenge. Service paradigm helps to convert social-sensor data into simplified and meaningful information, by abstracting away the data complexity [4,5]. Service paradigm abstracts a social-sensor image (i.e., an incident-related image posted on social media) as *an independent function*, namely *a social-sensor cloud service* (abbreviated as a SocSen service). A SocSen service has *functional* and *non-functional* properties. The functional attributes of the camera include taking images, videos and panoramas, etc. The spatio-temporal and contextual information of a social-sensor image is presented as the *non-functional properties* of a service. This abstraction allows a uniform and efficient delivery of social-sensor images as a SocSen service, without relying on *image processing* [4,5].

SocSen services provide efficient access to social media images for scene reconstruction and make them easy to reuse in multiple applications over diverse platforms. However, the challenge is designing an efficient selection and composition solution for context-relevant SocSen services based scene reconstruction. Context relevance includes covering the same incident or segment of an area in a given time. The existing techniques developed for SocSen service selection and composition are primarily dependent upon the prerequisite that objects in social media images can be explicitly identified and/or explicit definitions of relevance and composability are provided [4–6,12,13]. For example, two cars in two images can be automatically identified as identical by social media platforms. This type of object identification technologies is currently *unavailable* in most of the mainstream social media platforms. Moreover, the existing approaches analyse the spatio-temporal aspects of services and their relevance to queries as per the explicitly defined rules to achieve a successful service composition for scene reconstruction [4,5,12,13]. Defining these rules needs substantial human *intervention* and rely heavily on *domain experts' knowledge* in specific scenes. This dependency limits the generality of these approaches.

This paper proposes to employ *heuristics and machine learning to automatically select relevant and composable SocSen services based on the services' features and our designed heuristics*. The proposed approach provides a strategy to enable automatic SocSen service selection and composition for analysing scenes *without using image processing and objects identification technologies*. The proposed composition approach will form scenes by selecting spatially close and relevant images and placing them in a smooth mosaic-like structure.

The major contributions of our proposed work are:

- We design an approach enabling automatic and smooth mosaic scene construction based on social-sensor images, catering to most social media platforms that have not fully realised image processing and object identification.
- We explore various machine learning based classification strategies, including decision tree, support vector machine, and artificial neural network, to assess and select relevant and composable services for a composition.

2 Motivation Scenario

A typical scenario of area monitoring is used to illustrate the challenges in scene analysis. Let us assume, an incident occurred near the cross-section of *Road X* and *Road Y* during a *time period t*. The surveillance of the road segment through the conventional sensors, e.g., speed cameras and CCTV, is limited. Traffic command operations require the *maximum visual coverage* of the incident. The wide availability of smartphone users sharing images or posts on social networks might provide extra visual and contextual coverage. We assume that there are numerous social media images, i.e., *SocSen services* available over different social clouds, e.g., Facebook, Twitter, etc. These *SocSen services* can be used to fulfil the user's need for maximum coverage. Figure 1 shows a set of sample images taken around the queried location and time. This research proposes to *leverage these social media images to reconstruct the desired scene and provide users with extra visual coverage*. The query $q = < R, TF >$ includes an approximate region of interest and the approximate time frame of the queried incident.

- Query Region $R(P < x, y >, l, w)$, where $< x, y >$ is a geospatial co-ordinate set, i.e., decimal longitude-latitude position and l and w are the horizontal and vertical distances from the center $< x, y >$ to the edge of the region of interest, e.g., $(-37.8101008, 144.9634339, 10\,\text{m}, 10\,\text{m})$
- Query Time $TF(t_s, t_e)$, where t_s is the start time of the query, e.g., 13:20 AEST, 20/01/2017 and t_e is the end time of the query, e.g., 14:00 AEST, 20/01/17.

The process of scene reconstruction is considered as alleviating a service selection and composition problem. This research proposes to *leverage heuristics and machine learning based techniques for the selection and composition of SocSen services*. SocSen services possess multiple non-functional properties, adding value to unstructured social-sensors data. The non-functional properties of such services include (but not limited to) *spatial, temporal, and angular* information of the image. The main objective of this project is to *leverage heuristics and machine learning based techniques to select the services that are in the same information context, i.e., smoothly covering the area required by the user, by assessing their relevance and composability*. The composability assessment is based on *machine learning strategies with features defined upon non-functional attributes of the services*. Eventually, we aim to build a SocSen service composition to fulfil the user's requirement of visual coverage. The composition is a visual summary of the queried scene in the form of a mosaic-like structure by selecting the composable services covering the area. The composed visual summary is a set of spatio-temporally similar 2D images reflecting the queried incident. The quality of the composition relies on the smoothness and the spatial continuity of the mosaic. Therefore, adequately assessed composability between services is essential for a successful composition.

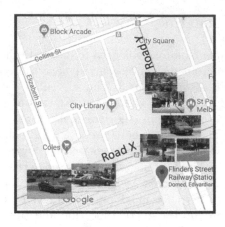

Fig. 1. Motivation scenario - a set of images from social networks

3   Model for Social-Sensor Cloud Service

In this section, we define concepts related to the social-sensor cloud service modelling.

An atomic SocSen service *Serv* is defined by:

- *Serv_id* is a unique service id of the service provider *SocSen*.
- *F* is a set of functional attributes of the service *Serv*.
- *nF* is a set of non-functional properties of service *Serv*.

The functional and non-functional attributes of an atomic SocSen service are defined as:

- The functional attributes of an atomic SocSen service capture the intended images, videos and/or panoramas.
- The non-functional attributes capture spatio-temporal and contextual behaviour depicted in an *atomic service*. The following are the minimal non-functional attributes associated with an atomic service:
  - Time, t, is the time of the service at which the image is taken. t can either be a single time-stamp or a short interval of time (t_s, t_e).
  - Location, $L(x,y)$, is the location of a service where x is the longitude position and y is the latitude position of the service.
  - Coverage Cov of the sensor. It defines the extent to which a service covers the scene. Coverage Cov is defined by VisD, dir, and α.
    * Angle *dir* is the orientation angle with respect to the North. It is obtained from the digital compass sensor.
    * Angle α is the angular extent of the scene covered. α is calculated from the camera lens property at the current zoom level. Angle α in degrees is calculated as:

$$\alpha = 2 * arctan\frac{d}{2f} \tag{1}$$

Where, f is the effective focal length of camera and d represents the size of the camera sensor in the direction of captured scene. For example, for 35 mm sensor, i.e., 36 mm wide and 24 mm high, $d = 36$ mm would be used to obtain the horizontal angle of view and $d = 24$ mm for the vertical angle.

* Visible distance $VisD$ is the maximum coverage distance.

4   Heuristics and Machine Learning for Social-Sensor Cloud Service Selection and Composition

We present a framework enabling heuristics and machine learning-based strategies to execute the selection and composition of the relevant and composable Soc-Sen services for reconstructing user queried scene. The query q can be defined as $q = (R, TF)$, giving the spatio-temporal region of interest of the required services. Query q includes: 1) $R = \{P < x, y >, l, w\}$, where P is a geospatial co-ordinate set, i.e., decimal longitude-latitude position and l and w are length and width distance from P to the edge of region of interest, and 2) $TF = \{t_s, t_e\}$, where t_s is the start time of the query and t_e is the end time of the query.

We propose a five-step heuristics and machine learning-based service selection and composition model to realise this goal. The five major steps are:

1. *Service indexing and spatio-temporal filtering.* We employ 3D R-Tree to spatio-temporally index the SocSen services [4]. The indexed services that are out of the spatio-temporal bounds defined by the user query are filtered.
2. *Features analysis and extraction.* We define a set of independent and dependent features based on the non-functional attributes of SocSen services. These features are believed to be related to the relevance and composability of Soc-Sen services for scene analysis.
3. *Heuristics based service filtering.* We introduce a heuristics-based algorithm to filter out non-relevant services based on angular and direction features.
4. *Machine learning classifier.* We explore several popular machine learning models, such as Decision tree, Support Vector Machine (SVM) and Neural Networks (NN) for assessing the relevance and composability between two services from the outputs of Step 3. The classifiers determine whether or not two services are relevant and composable based on the defined features (Table 1). The services that are non-composable with any other service are deemed as non-relevant. The details of the employed models and their configuration are discussed in Sect. 5.2.
5. *SocSen service composition.* We compose the composite services based on the one-to-one service composability to form an image mosaic.

4.1   Service Indexing and Spatio-Temporal Filtering

We need an efficient approach to select spatio-temporally relevant images (i.e. services), from a large number of social media images. Spatio-temporal indexing enables the efficient selection of services. We index services considering their

Table 1. Features of SocSen services

ID	Feature	Category	Source	Description
F1	$L(x,y)$	Independent	SocSen	Location of the service
F2	t	Independent	SocSen	Time of the service
F3	Dir	Independent	SocSen	Orientation angle with respect to the North
F4	α	Independent	SocSen	Angular extent of the scene covered
F5	$VisD$	Independent	SocSen	Maximum coverage distance
F6	$L_2(x, y)$	Independent	Calculated	Triangulated location of the service coverage
F7	$L_3(x, y)$	Independent	Calculated	Triangulated location of the service coverage
F8	α_O	Dependent	Calculated	Angular overlap between two services

spatio-temporal features using 3D R-tree [4]. 3D R-tree is used as an efficient spatio-temporal index to handle time and location-based queries. It is assumed that all available services are associated with a two-dimensional geo-tagged location $L(x, t)$ and time t. The leaf nodes of the 3D R-tree represent services. For the effective area of query q, we define a cube shape region BR using the user-defined rectangular query area R and start and end time of the scene, i.e., t_s and t_e. The region BR encloses a set of services relevant to q. Figure 2 illustrates the query region R and the bounded region BR across time t_s to t_e.

4.2   SocSen Service Features Analysis and Extraction

To create services based on images, we extract five basic image features from social media images' metadata. These features include services' geo-tagged locations, the time when photographs were captured, camera direction $\overrightarrow{dir}$, the maximum visible distance of the image $VisD$ and viewable angle α. These features are abstracted as the non-functional properties of a service. Utilising these non-functional properties, we define several features. The features include both independent and dependent features. The former consists of the triangulation features, i.e., geo-coordinates to form an approximate triangulated field of view (FOV). The latter includes angular overlap α_O between two images. These additional features are formulated as:

– The FOV of an image, i.e., the spatial coverage area of the service can be approximated as a triangle. We also use the trigonometry to calculate the field of view as linear measurement, i.e., $LFOV$. $LFOV$ is calculated by:

$$LFOV = 2 \times tan(\frac{\alpha}{2} \times VisD) \tag{2}$$

We calculate the triangular bounds of the service coverage using the triangulation rule (Eqs. 2 and 3) and the spherical law of cosines (Eqs. 4 and 5). In the spatial coverage area (Fig. 3), the service location $L(x,y)$ is also the location of the camera, $VisD$ is the maximum visible distance of the scene, and α is the viewable angle. The geometric parameters for the coverage triangle are calculated using:

$$| SFOV |= \sqrt{VisD^2 - 0.25(LFOV^2)} \tag{3}$$

$$\varphi_2 = asin(sin\varphi_1 \cdot cos\delta + cos\varphi_1 \cdot sin\delta \cdot cos\theta) \tag{4}$$

$$\lambda_2 = \lambda_1 + atan2(sin\theta \cdot sin\delta \cdot cos\varphi_1, cos\delta - sin\varphi_1 \cdot sin\varphi_2) \tag{5}$$

where, φ is latitude, λ is longitude, θ is the bearing (clockwise from north), and δ is the angular distance d/R. d is the distance travelled, and R is the earth's radius. Figure 4 shows an example of the service coverage estimated based on the FOV descriptors.

– Angular overlap α_O between services $Serv_i$ and $Serv_j$ is calculated as:

$$\alpha_O(Serv_i, Serv_j) = Serv_i.Dir - Serv_j.Dir \tag{6}$$

where, Dir is the orientation angle with respect to the North.

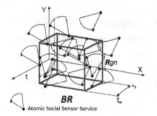

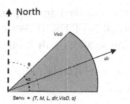

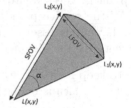

Fig. 2. SocSen service indexing illustration

Fig. 3. SocSen service coverage model

Fig. 4. SocSen service FOV illustration

4.3 Heuristics Based Service Filtering

We propose to filter out non-relevant services based on the direction related features in Table 1. The direction-related features rely on service coverage. The direction-related features help to assess and filter services that do not provide useful coverage of the user-required region. We propose features filtering based on a heuristic algorithm. The heuristic algorithm uses the directional and angular features of the social media images.

The classic SocSen service selection and composition techniques perform efficiently when the targeted problem completely satisfies spatial and temporal constraints. However, heuristics outperform traditional methods of composition in the noisy, sparse or discontinuous search space [9]. Heuristics have been extensively applied for feature selection, feature extraction and parameter fine-tuning. We employ hybridisation of heuristics and machine learning to improve the performance of machine learning techniques. We use 1) parameter control strategy (for SVM and decision tree) and 2) instance-specific parameter tuning strategy for fine-tuning of the parameters.

Parameter control strategy employs dynamic parameter fine-tuning by controlling the parameters during the training phase. We assume that the algorithms are robust enough to provide excellent results for a set of instances of the same problem with a fixed set of parameter values. Parameter control strategy focuses on a subset of the instances to analyse their response to the classifier.

Our proposed heuristic algorithm helps to select a subset of services for classifier implementation. The algorithm uses a rank-based selection strategy, which is defined in the following steps.

1. The distance and direction difference between any two services are measured.
2. The directional relevance value (Eq. 7) is evaluated for each pair of services, given distance and the directional difference value. The services are ranked from the highest to the lowest relevance.
3. All the services with relevance value above 0 are selected.
4. The distance of the services pair away the centre of the required region is measured, and the direction of the service pair with respect to the centre point is assessed (Eq. 8).
5. All the service pairs are then ranked upon position relevance. The position relevance is assessed as a ratio of the direction of a service pair from the centre to the distance of the services pair from the centre.

To keep it simple, we chose a relatively simple form of a 2-criteria relevance function, which is defined as follows:

$$
\begin{aligned}
Relevance(Serv_i, Serv_j) &= Serv_i \cdot Serv_j \\
&= \parallel Serv_i \parallel * \parallel Serv_j \parallel cos(\alpha_O(Serv_i, Serv_j))
\end{aligned}
\tag{7}
$$

where, $\parallel Serv \parallel$ is scalar magnitude of service coverage, and $Relevance$ is normalised between [-1,1]. If the services are pointing to the same direction, then relevance is positive. If the services are pointing to the opposite direction, then the relevance is negative.

$$
Dir(Serv_i, P) = \theta_{Serv_i} = cos^{-1}(\frac{(Serv_i \cdot P)}{\parallel Serv_i \parallel})
\tag{8}
$$

where, $Dir(Serv_i, P)$ is the direction of the service $Serv_i$ with respect to the centre point P of the user's query. P is a geospatial coordinate set, i.e., decimal longitude-latitude position (see Sect. 3).

4.4   SocSen Service Composition

Eventually, the composition is formed by placing composable services in a mosaic-like structure. 3D R-Tree indexing helps in the spatio-temporal formation of the queried scene. The one-on-one composability is extended to deduce the composability among multiple services. This is realised by allowing transfer among multiple composable service pairs. For example, if $Serv_a$ is composable with $Serv_b$, and $Serv_b$ is composable with $Serv_c$, then $Serv_a$ and $Serv_c$ are deemed composable. The composite services comprise of a set of selected atomic services to form a visual summary of the queried scene. The visual summary is an arrangement of the 2D images, forming a mosaic-like scene. Figure 1 shows a set of sample images taken around the same location and similar time. Figure 5 presents the eventually composed mosaic-like scene, i.e., the composite spatio-temporal SocSen services.

Fig. 5. Mosaic-like scene from the sample images

5   Experiment and Evaluation

5.1   Experiment Setup

Experiments are conducted to evaluate the impact of the heuristics and the performance of the classifiers on SocSen service selection and composition. To the best of our knowledge, there is no real-world spatio-temporal service dataset to evaluate our approach. Therefore, we focus on evaluating the proposed approach using a collected dataset. The dataset is a collection of 10,000 images from Jan 2016 to July 2019, in City of Melbourne, Australia. The images are downloaded from multiple social networks, e.g., flicker, twitter, google+. We retrieve and download the images related to different locations and time windows. To create services based on the images, we extract their geo-tagged locations, the time when an image was captured, and FOV information. Time, location, camera direction $\overrightarrow{dir}$, the maximum visible distance of the image $VisD$ and viewable angle α are abstracted as the non-functional property Cov of the service. We generate 70 different queries based on the locations in our dataset. Each query q is defined as $q = (R, t_s, t_e)$, giving the region of interest and time of the query (an example can be found in Sec. V). To evaluate the service composition performance using our proposed models (SVM, Decision Tree and ANN), we conduct several experiments with the features listed in Table 1. To obtain the ground truth, we conduct a user study to gauge the perceived composability of every pair of services for each user query. All the experiments are implemented in Java and MATLAB. All the experiments are conducted on a Windows 7 desktop with a 2.40 GHz Intel Core i5 processor and 8 GB RAM.

5.2   Machine Learning Based Composability Assessment and Composition

Systematic selection of the machine learning models is essential for the better performance of the proposed approach. We demonstrate the process of machine learning model selection by comparing multiple standard models with different parameter configurations. We have experimented with Decision tree, SVM and ANN for assessing the relevance and composability of services. The input data are divided into training data (70%), validation data (15%), and testing data (15%). We manually identify the total numbers of services, relevant services and relevant and selected services as the ground truth for each query. Precision, recall and F1-score are employed to assess the effectiveness of all the models.

We opt for the decision tree due to its simple structure and co-related nature of our features. We assess the decision tree along with different depth levels and features to find the parameters producing the optimal results. We eventually chose a 20 splits based decision tree that makes a coarse distinction between composability and non-composability between two services.

SVM is opted due to sparse and unpredictable nature of our services' parameters [8]. The success of SVM depends on the selection of the kernel and its parameters. In our case, we experimented Quadratic kernel, Cubic kernel and Gaussian RBF kernel. All the kernels are fine-tuned to achieve the best results in terms of precision, recall and f-score. Based on the experimental results, we have selected the Quadratic kernel and Gaussian RBF kernel for further comparisons.

ANNs are computing systems that learn to perform tasks by considering examples (i.e., training data) generally without being programmed with task-specific rules. We opt for feed-forward multilayer ANN due to their ability to learn and infer non-linear relations among training services and further detect similar relations in unseen services. We employ ANN comprised of an input layer, multiple hidden layers (sigmoid), and an output layer (softmax) [10]. The number of hidden layers is adjusted through parameter tuning. We use all the features (i.e. F1–F8 in Table 1) to generate a vector for each service. We attempt respectively two, three and four hidden layers with ten neurons to build a feed-forward neural network. We assess our composability approach for one-to-one and many-to-one composability assessment. One-to-one composability relationship depicts composability between two neighbouring services. Many-to-one composability relationship refers to composability of a cluster of services collaboratively depicting a scene. The results of the ANN with different layers and composability relations show that the 2-layered ANN for one-to-one composability detection.

5.3   Evaluation and Result Discussion

We conduct the following evaluations to assess the effectiveness of the proposed approach: 1) we assess the composition performance of the machine learning models to find out the most appropriate model for service composition; 2) we compare the composition performance of the machine learning models with and without the heuristics to testify the influence of the heuristics; 3) we compare the composition performance of the models with the existing tag and object identification based composition model (abbreviated as TaOIbM) [5] and an image processing based model - Scale-Invariant Feature Transform (SIFT) [11].

Performance of the Machine Learning Classifiers on Composition. In the first set of experiments, we analyse the performance of the tuned machine learning classifiers on service composition. We filter out services based on the dependent features. The services are then transferred to the classifiers for the relevance and composability assessment. Dependent features based filtration helps to eliminate the images with duplicated coverage and provides precise results. We assume 50% overlap as a general heuristic fitness threshold based on the existing studies [4,5]. We use precision, recall and F1-score to assess the performance of the classifiers on the service composition. The performance of the classifiers on service composition can be found in Fig. 6. It can be observed that the ANN shows slightly better overall performance than the SVM and the decision tree. The SVM Gaussian is somewhat behind the ANN with higher precision but lower recall and F1-score.

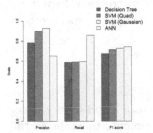 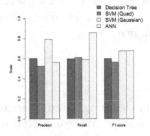 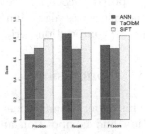

Fig. 6. Classifiers with heuristics

Fig. 7. Classifiers without heuristics

Fig. 8. Comparison with TaOIbM and SIFT

Impact of Heuristics on Composition. In the second set of experiments, we analyse the influence of the heuristics-based service filtering. We testify the performance of the machine learning classifiers with and without the heuristic algorithm. The results of the non-heuristics-based classifiers are shown in Fig. 7. It can clearly be observed that all the classifiers with heuristic outperform their counterparts. This observation validates that our heuristic algorithm is effective in filtering non-composable services.

Comparison with Tag and Object Identification based Model and Image Processing Model. The comparison result is shown in Fig. 8. First, the proposed model slightly outperforms TaOIbM. TaOIbM relies heavily on well defined composability models and tags as well as identified objects. The technologies of the latter are infeasible in most current social media platforms. Second, the proposed model has similar performance with SIFT on recall but worse performance on precision. Our model can only "guess" the image relevance and composability basing on the image metadata. In contrast, image processing models can precisely assess image relevance by processing image pixels. Therefore, it is reasonable that the precision of the former is lower than the latter. However, batch image processing is extremely time-consuming and infeasible in the enormous social media environment [2,3]. The major advantage of the proposed approach is that our proposed model does not rely on object recognition technologies and image processing.

Result Discussion. In our experiments, the heuristics are used to filter out non-composable services from the available SocSen services based on the features in Table 1. Decision tree, SVM and ANN are then used for the classification of the services passed by the heuristics into two groups: composable and non-composable services based on the features (i.e. F1–F8 in Table 1). In the experiments, the heuristic-based selection demonstrates its effectiveness in improving the performance of each classifier, by comparing the results of Fig. 6 and 7. Our model outperforms the tag and object recognition based model but has a reasonable distance with the image processing model on precision. In summary, the results of these experiments preliminarily validate the feasibility of our proposed SocSen service composition model for scene reconstruction. Especially the heuristics algorithm shows its effectiveness on improving the social media image composition performance.

6   Conclusion

This paper examines machine learning based on social-sensor cloud service selection and composition approaches. We conducted experiments to evaluate the proposed approach for accurate and smooth composition. In the future, we plan to focus on defining more effective features and exploring more machine learning models to further improve the selection and composition performance.

Acknowledgement. This research was partly made possible by DP160103595 and LE180100158 grants from the Australian Research Council. The statements made herein are solely the responsibility of the authors.

References

1. Rosi, A., Mamei, M., Zambonelli, F., Dobson, S., Stevenson, G., Ye, J.: Social sensors and pervasive services: approaches and perspectives. In: Proceedings of PERCOM (2011)
2. Castillo, C.: Big Crisis Data: Social Media in Disasters and Time-Critical Situations. University Press, New York (2016)
3. Stieglitz, S., Mirbabaie, M., Ross, B., Neuberger, C.: Social media analytics - challenges in topic discovery, data collection, and data preparation. Int. J. Inf. Manag. **39**, 156–168 (2018)

4. Aamir, T., Bouguettaya, A., Dong, H., Mistry, S., Erradi, A.: Social-sensor cloud service for scene reconstruction. In: Maximilien, M., Vallecillo, A., Wang, J., Oriol, M. (eds.) ICSOC 2017. LNCS, vol. 10601, pp. 37–52. Springer, Cham (2017). https://doi.org/10.1007/978-3-319-69035-3_3

5. Aamir, T., Dong, H., Bouguettaya, A.: Social-sensor composition for scene analysis. In: Pahl, C., Vukovic, M., Yin, J., Yu, Q. (eds.) ICSOC 2018. LNCS, vol. 11236, pp. 352–362. Springer, Cham (2018). https://doi.org/10.1007/978-3-030-03596-9_25

6. Neiat, A.G., Bouguettaya, A., Sellis, T., Dong, H.: Failure-proof spatio-temporal composition of sensor cloud services. In: Franch, X., Ghose, A.K., Lewis, G.A., Bhiri, S. (eds.) ICSOC 2014. LNCS, vol. 8831, pp. 368–377. Springer, Heidelberg (2014). https://doi.org/10.1007/978-3-662-45391-9_26

7. Bouguettaya, A., et al.: A service computing manifesto: the next 10 years. Commun. ACM 60(4), 64–72 (2017)

8. Durgesh, K.S., Lekha, B.: Data classification using support vector machine. J. Theor. Appl. Inf. Technol. 12(1), 1–7 (2010)

9. Calvet, L., de Armas, J., Masip, D., Juan, A.A.: Learnheuristics: hybridising meta-heuristics with machine learning for optimisation with dynamic inputs. Open Math. 15(1), 261–280 (2017)

10. Han, J., Kamber, M., Jian, P.: Classification: advanced methods. In: Data Mining Concepts and Techniques. Morgan Kaufmann Publishers (2012)

11. Lowe, D.G.: Distinctive image features from scale-invariant key points. Int. J. Comput. Vision 60(2), 91–110 (2004)

12. Aamir, T., Dong, H., Bouguettaya, A.: Social-sensor composition for tapestry scenes. In: IEEE Transactions on Services Computing (2020)

13. Aamir, T., Bouguettaya, A., Dong, H., Erradi, A., Hadjidj, R.: Social-sensor cloud service selection. In: Proceedings of ICWS (2017)

Fast Build Top-k Lightweight Service-Based Systems

Dandan Peng[1], Le Sun[1(✉)], and Rui Zhou[2]

[1] Jiangsu Collaborative Innovation Center of Atmospheric Environment and Equipment Technology (CICAEET), Nanjing University of Information Science and Technology, Nanjing 210044, China
sunle2009@gmail.com
[2] Department of Computer Science and Software Engineering, Swinburne University of Technology, Melbourne, VIC, Australia

Abstract. This work aims to solve the problem of automatically building top-k (the number of suggested results) lightweight service-based-systems (LitSBSs) based on user-given keywords and QoS constraints. Top-k LitSBSs are those service-based systems (SBSs) that optimally match the query keywords, contain the minimum number of component services and satisfy the user-given QoS constraints. A LitSBS is easy to be managed, executed, monitored, debugged, deployed and scaled. Its QoS performance (e.g. communication overhead, risk of failure and connection latency) is usually better than the non-lightweight SBSs in a similar network environment. We design a database-driven algorithm (called LitDB) to solve the problem. We conduct comprehensive experiments to demonstrate the time efficiency of LitDB. In addition, we use real mashup examples from programmableWeb to show its practicality.

Keywords: Service-based system · Keyword query · Database-driven · Top-k algorithm

1 Introduction

Service composition is a technology of developing service-based systems (SBSs) by composing existing services [4]. It can be applied in various fields such as app recommendation [10] and cloud computing [14,15]. Traditional service composition is too complicated for non-expert users. There have been works [6,16] focusing on simplifying the process of service composition. In particular, keyword-query-based service composition algorithms [3,4,12,19–21] support building SBSs automatically just based on user-given keywords that reflect users'

Supported by the National Natural Science Foundation of China (Grants No 61702274) and the Natural Science Foundation of Jiangsu Province (Grants No BK20170958), Australian Research Council Discovery Project DP180100212, the Postgraduate Research & Practice Innovation Program of Jiangsu Province (Grants No KYCX20_0980) and PAPD.

© Springer Nature Switzerland AG 2020
Z. Huang et al. (Eds.): WISE 2020, LNCS 12342, pp. 516–529, 2020.
https://doi.org/10.1007/978-3-030-62005-9_37

preferences [18]. That is, users only need to provide a few keywords to describe service textual properties, e.g. service functions [4,21], and input and output parameters [12]. Thus, non-professional users can build complex SBSs easily.

Some researches [2,12,17] develop keyword-based algorithms to query lightweight SBSs(LitSBSs). A LitSBS has the minimum number of component services among all the SBSs matching the query keywords. It is easy to be managed, executed, monitored, debugged, deployed and scaled. Its QoS performance (e.g. communication overhead, risk of failure and connection latency) is usually better than the non-lightweight SBSs in a similar network environment [11].

There are keyword-based SBS building algorithms [8,9] using relational databases (DB) to store services. Query techniques of DB are mature and robust, which guarantees the efficiency of basic query functions. The algorithm in [8] generates and stores all SBSs in a database in advance. It then traverses the database to find the best LitSBS. The algorithm in [9] precomputes all service combinations and stores them as paths in a database. It then uses two separate processes based on a planning graph and the database. However, both solutions [8,9] compute and store all possible SBSs in advance, which takes much time and storage space. To solve this problem, we propose a new database-driven algorithm for building LitSBSs based on user-given keywords (called LitDB). LitDB does not need to pre-store all possible SBSs.

Most existing works [2,4,8,9,12,20] approximately generate the optimal LitSBS that optimally matches user-given keywords, and rarely consider the matching degree between an SBS and the query keywords. In this paper, we use a keyword matching score to measure such matching degree, which adopts the term weighting technique of information retrieval. The proposed algorithm LitDB generates top-k LitSBSs that have the highest keyword matching scores and satisfy user-given QoS constraints.

The distinctive contributions of this work are as follows:

- We design an efficient LitSBS building algorithm (LitDB) utilizing the power and robustness of query and join techniques of relational databases. LitDB supports the building of k optimal lightweight SBSs (top-k).
- We use a keyword matching score: $kscore$ to measure the matching degree between an SBS and query keywords.
- We conduct extensive experiments to show the good time efficiency of the fast service composition algorithm. In addition, we use some practical mashup building examples from programmableWeb to demonstrate the rationality of our method.

The structure of this paper is as follows: Sect. 2 formally defines the problem we are going to solve; Sect. 3 describes the LitDB algorithm; Sect. 4 shows the experiment results and case studies; and Sect. 5 concludes the paper.

2   Problem Definition

2.1   Service Database

Services in a service domain are usually stored in a service library, which stores service functions, QoS properties and other service attributes. We build a database to represent a service library, called **service database**. A service database L is composed of four types of tables: the service table (e.g. Table 1), the parameter table (Table 4), the input (Table 2) and output (Table 3) tables. The service table stores the function description attributes and QoS properties of services.

Table 1. Service table TS

SID	Name	Service description	Response time (ms)	Throughput (rps)
s1	WeatherInfo	This service is used to get the information of weather	455	7000
s2	SportsByWeather	This service recommends sports based on the weather	185	5500
s3	GetHotelCost	This service is used to check hotel costs	280	9000
s4	Turkish-Airlines	This service queries the information of Turkish airlines based on the weather	190	2500
s5	FlightStats-Airlines	This service queries airline information based on flightstats	345	8000
s6	AirlineByWeather	This service is used to check whether an airline is scheduled based on the weather	220	1000
s7	GetDetailedInfo	This service recommends airlines and hotels based on tour information and weather	170	4500

Table 1 is part of a service table of L. The parameter table stores the input and output parameter collection used by all services in the service table. The input and output tables store the specific input and output parameters of each service. Because a service may have multiple input and output parameters, and a parameter may be the input or output parameter of multiple services, we store services and their input and output parameters in separate tables. This design follows the Boyce Codd Normal Form [1]. In the service table, a tuple represents a service. In the input (or output) table, a tuple represents one input (or output) parameter of a service. For example, service s4 has two input parameters: Weather and TourInfo (see Table 2 and Table 4).

For simplicity, we consider two services (s_i and s_j) can be composed if the output parameters of one service ($s_i.out$) has the same parameter with the input parameters of the other service ($s_j.in$), i.e. $s_i.out \cap s_j.in \neq \phi$ (More complicated or input-output matches were studied in [19]). For example, from Tables 1, 2, 3

Table 2. Input table TI **Table 3.** Output table TO **Table 4.** Parameter table TP

IID	SID	PID
i1	s1	p2
i2	s2	p1
i3	s3	p7
i4	s4	p1
i5	s4	p5
i6	s5	p5
i7	s6	p1
i8	s7	p1
i9	s7	p5

OID	SID	PID
o1	s1	p1
o2	s2	p3
o3	s3	p8
o4	s4	p6
o5	s5	p6
o6	s6	p4
o7	s7	p6
o8	s7	p7

PID	Name
p1	Weather
p2	Zipcode
p3	SportsOK
p4	AirlineOK
P5	TourInfo
p6	AirlineInfo
p7	Hotel
p8	Cost

and 4, we find a path $s1 - o1 - p1 - i7 - s6$, so $s1$ can be a precedent service of $s6$. We call the above path the service joining path (SJP).

A service-based system (SBS) is a composite system $S = \{s_1, ..., s_n\}$ that: (1) any service in S can be composed with at least one another service in S; (2) S comprises three types of services: a source service (s_{start}), a set of middle component services (s_{mid}) and a target service (s_{end}), where s_{start} does not have any ancestor service in S and s_{end} does not have any descendant service in S.

Given a set of keywords Q and a service database L containing a service table TS, the input and output tables (TI and TO), and a parameter table (TP). Assume services in TS have m function description attributes, and the attribute values of a service s is represented by E_s. If E_s contains a query keyword q_r ($q_r \in Q$), we say that s matches q_r with a certain matching degree. We apply the term weighting technique of information retrieval [13] to measure the matching degree between s and q_r, which is defined as the **keyword matching score** $kscore$ (see Formula 1).

$$kscore(s, E_s, q_r) = \sum_{\forall e_i \in E_s} kscore(s, e_i, q_r) \tag{1}$$

where $kscore(s, e_i, q_r)$ is defined by Formula 2.

$$kscore(s, e_i, q_r) = \frac{1 + \ln(1 + \ln(tf))}{(1 - \sigma) + \sigma \frac{dl_i}{avdl}} \cdot \ln(\frac{N + 1}{df}) \tag{2}$$

where, tf is the frequency of q_r in e_i; df is the number of tuples in E^i containing keyword q_r and E^i is the value union of the ith attribute of services in TS; dl_i is the character number in e_i; $avdl = (dl_1 + ... + dl_m)/m$; N is the number of services in TS; and σ is a constant (usually 0.2).

From Formula 2, $kscore$ is positively related to the frequency of a keyword (tf) in a function description attribute value of a service, and is inversely related to the character number of the attribute value (dl). That is, the more frequent the keyword occurred in the service attribute, the better the service matches with the keyword. In addition, $ln(\frac{N+1}{df})$ reduces the effect of the matching between the service and a stop word (e.g. terms *the*, *of* and *with*) on the matching degree.

An SBS S matches a set of Q means that the attribute values of component services in S contain Q. The matching degree between S and Q is calculated by Formula 3.

$$kscore(S,Q) = \frac{\sum_{s \in S} kscore(s, E_s, q_r)}{size(S)} \tag{3}$$

A lightweight SBS is defined in Definition 1.

Definition 1. *Given a set of query keywords Q, and a service table TS, a* **lightweight SBS** *(LitSBS, represented by $\tilde{S}$) matching Q is an SBS that (1) $\tilde{S} \subset TS$; (2) the attribute values of $\tilde{S}$ contains Q; and (3) for any non-lightweight SBS S with $kscore(S,Q) = kscore(\tilde{S},Q)$, the number of component services in S (represented by $|S|$) is larger than $|\tilde{S}|$.*

The top-k LitSBS is defined in Definition 2.

Definition 2. *Assume each service in a service database $L=\{TS, TI, TO, TP\}$ has l QoS parameters $\{p_1, ..., p_l\}$. Given a set of keywords Q and a set of QoS constrained conditions $C = \{c_1, ..., c_l\}$, the* **top-k LitSBSs** *are k SBSs with the highest kscores matching Q, and satisfy all QoS constraints C.*

After generating top-k LitSBSs, we can rank these LitSBSs based on their QoS performance. The QoS of an SBS is an aggregation of the QoS of its component services. For different QoS parameters, the aggregation methods are different [4]. For example, the throughout of an SBS equals to the minimum throughput of its component services. Assume we have got the aggregated values of QoS parameters $(q_1, q_2, ..., q_l)$ of an SBS S, the overall QoS performance of S is evaluated by a **quality score** *qscore* (see Formula 4)

$$qscore(S) = U(q_1, ..., q_l) \tag{4}$$

where U is a multi-objective value function. For example, a popular value function is the weighted average(see Formula 5)

$$qscore(S) = w_1 \bar{q}_1 + ... + w_n \bar{q}_l \tag{5}$$

where $w_i \in [0,1], \forall i \in \{1, ..., l\}, w_1 + ... + w_l = 1$; and $\bar{q}_i$ represents the normalized values of q_i.

The **problem** this work aims to solve is to design an algorithm that: based on a set of query keywords and user-given QoS constraints, it automatically queries the service database, and recommends the proposals of building top-k LitSBSs to the user.

3   LitDB: An Algorithm for Building Top-K LitSBSs

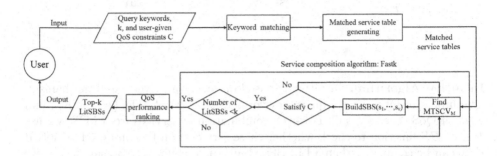

Fig. 1. Process of LitDB algorithm

We propose a database-driven algorithm (called LitDB) to efficiently build top-k LitSBSs based on user-query keywords. Figure 1 shows an overview of LitDB. It has four main steps: (1) *Keyword matching* finds the services containing the query keywords. (2) *Matched service table generating* builds a matched service table (MT) for each keyword, which contains the found services with respect to the keyword. Services in a service table are sorted in descending order in terms of their *kscores*. (3) Service composition algorithm finds the top-k LitSBSs. (4) After finding top-k LitSBSs, *QoS performance ranking* ranks them by their overall QoS performance. The QoS performance is evaluated by the *qscore*.

3.1   Matched Service Table Generating

Keyword matching finds all services that match query keywords. We use *kscore* (Formula 1) to measure the matching degree between a query keyword and a service. The higher the score, the better the service matches the query keywords. Based on the *kscores*, a matched keyword table (represented by MT) is generated for each query keyword. We use K_i to represent MT of the ith query keyword, $i \in \{1, ..., v\}$. For example, Table 5 shows the MTs of query keywords: Weather, AirLine and Hotel. Services in an MT are sorted in descending order in terms of their *kscores*.

3.2   Service Composition Algorithm

Based on the MTs of all query keywords, we design a very efficient service composition algorithm, called Fastk, to search the top-k LitSBSs matching Q. Before introducing Fastk, we present two basic algorithms: Intuitive and Enhanced. Intuitive algorithm is the most intuitive and time consuming. Enhanced improves the time efficiency of Intuitive by using a pruning strategy. Fastk is the most time efficient of the 3 algorithms.

Table 5. Matched keyword tables for keywords

K_1:Weather	
SID	kscore
s1	7
s2	5.3
s6	0.9

K_2:AirLine	
SID	kscore
s5	2
s4	1.5

K_3:Hotel	
SID	kscore
s3	5
s7	1.7

Intuitive Algorithm. Intuitive is based on an exhaustive search mechanism. The inputs of Intuitive are query keywords Q, k, a set of MTs ($K_1, ..., K_v$) and a set of QoS constraints C. For each combination of services from MTs, we use *BuildSBS* function to check whether these services can be composed or not. If they can be composed, then a LitSBS is built. If the built LitSBS satisfies C, we store it in R. Intuitive then calculates the *kscores* of the built LitSBSs. Finally it outputs the top-k LitSBSs.

BuildSBS (Algorithm 1) is used to check whether we can find an SBS contains the input services $\{s_1, ..., s_v\}$. The inputs of Algorithm 1 are a set of services $\{s_1, ..., s_v\}$, the size limit Γ of an SBS, a service table TS, an input table TI, an output table TO and a parameter table TP. We use Γ to limit the number of component services that can be composed by an SBS. It avoids taking too much time to search an SBS that has an arbitrary large size. The output of *BuildSBS* is a LitSBS that contains all service in $\{s_1, ..., s_v\}$. First, we move s_1 to a list (LS) which stores service compositions (line 2 in Algorithm 1). The expansion rule is that, if s in TS can be combined with a service in a service composition SC to form a SJP, then SC can be expanded (lines 9–11 in Algorithm 1). We use SQL query to search s_i: SELECT s_i FROM TS, TI, TO and TP, WHERE $s_i.IID.PID = s_j.OID.PID$. If a SC with the minimum size in LS contains $\{s_1, ..., s_v\}$, then we output the SC (lines 6–7 in Algorithm 1). The size of SC cannot exceed Γ (line 8 in Algorithm 1).

Enhanced Algorithm. Enhanced algorithm improves the efficiency of Intuitive algorithm by using an upper bound of *kscores*. The upper bound is called Most Top Service Composition Value ($MTSCV$), which is defined as follows: assume we have obtained the MTs $K = \{K_1, ...K_i, ..., K_v\}$ for query keywords $Q = \{q_1, ..., q_v\}$, and we have found k LitSBSs (may not be the top-k) by joining the tuples in $\{K_1, ...K_i, ..., K_v\}$. Services in each K_i are sorted in descending order of their *kscores*. The $MTSCV_i$ for an MT K_i ($\forall i \in \{1, ..., v\}$) is defined by Formula 6.

$$MTSCV_i = \frac{K_1.top.kscore + ... + f(K_i).kscore + ... + K_v.top.kscore}{|Q|} \quad (6)$$

where $K_i.top$ represents the service with the highest *kscore* in K_i. $f(K_i)$ represents the top unprocessed service in K_i. $|Q|$ is the number of query keywords. *Unprocessed* means the service has not been tested to be composed with the services in the other keyword matching tables. $f(K_i)$ is most likely to be composed

---

Algorithm 1. $BuildSBS(s_1, ..., s_v)$

Input: $\{s_1, ..., s_v\}$, Γ, TS, TI, TO, TP
Output: the most lightweight SBS (a LitSBS)
1: LS: a list stores service compositions, service compositions in LS are always sorted in ascending order of Γ
2: Add s_1 to LS
3: **while** $|LS| > 0$ **do**
4:     $SC = LS.\text{pop}(0)$
5:     **if** SC contains $\{s_1, ..., s_v\}$ **then**
6:         **return** SC
7:     **end if**
8:     **if** $|SC| < \Gamma$ **then**
9:         **for** each service s_i in TS can be combined with a service s_j in SC to form a SJP **do**
10:             $\widetilde{SC} = SC.\text{add}(SJP)$ and move $\widetilde{SC}$ to LS
11:         **end for**
12:     **end if**
13: **end while**
14: **return** $null$ //(There is no LitSBS contains services $\{s_1, ..., s_v\}$)

---

in the new top-k LitSBS among the unprocessed services in K_i. $MTSCV_i$ is the upper bound of the *kscore* of the next possible top-k LitSBS composing $f(K_i)$. Then the maximum $MTSCV_M$ is defined by Formula 7.

$$MTSCV_M = max_{i=1}^{v} MTSCV_i \tag{7}$$

The $MTSCV_M$ represents the overall upper bound of the *kscores* of all possible top-k LitSBSs composed by the remaining services (i.e. excluding the services in the found k LitSBSs).

Fastk Algorithm. We propose another service composition algorithm, called Fastk to further improve the time efficiency of Enhanced. Fastk follows a skyline idea. That is, in each iteration of building a new LitSBS, it always picks up the not-yet-processed service (s) that is most likely to be composed in the next optimal LitSBS (i.e. having the highest MTSCV). It then processes s by testing whether or not it can be composed with the previously processed services whose $MTSCV \geq s.MTSCV$. The previously processed services always have higher possibilities to be composed in a top-k LitSBS than the remaining unprocessed services (whose $MTSCV \leq s.MTSCV$).

Algorithm 2 shows the process of Fastk. We create a stack $N(K_i)$ for each K_i to keep the processed services of K_i. R keeps the possible top-k LitSBSs. P keeps the final top-k LitSBSs. And $f(K_i)$ keeps the top unprocessed service of K_i. We then check whether the service with the highest *kscore* of each K_i can be composed or not. If they can be composed, a LitSBS is built and put in R (line 4 in Algorithm 2). Then we calculate $MTSCV_i$ of each service in $f(K_i)$, $\forall i \in 1, ..., v$, and move the service with highest $MTSCV$ to $N(K_M)$ (lines 7–8

Algorithm 2. Fastk algorithm

Input: $Q, k, K_1, ..., K_v, C$
Output: top-k LitSBSs, P
1: $N(K_i) = \emptyset$: processed tuples of K_i
2: $f(K_i)$: top unprocessed tuple of K_i
3: $R = \emptyset$: possible top-k LitSBSs. LitSBSs in R are always sorted in descending order of $kscore$, $P = \emptyset$: final top-k LitSBSs
4: Add LitSBS $= BuildSBS\ (f(K_1), ..., f(K_v))$ to R
5: move $f(K_i)$ to $N(K_i)$ for $\forall i \in \{1, ..., v\}$
6: **while** $|P| \le k$ **do**
7:     Get tuple $s = f(K_M)$, where $MTSCV_M = max_{i=1}^{v} MTSCV_i$ (Calculate $MTSCV_i$ by using Formula 6)
8:     Add s to $N(K_M)$
9:     **for** each group $(s_1, ..., s_{M-1}, s_{M+1}, ..., s_v)$ **do**
10:         $\widetilde{S} = BuildSBS\ (s_1, ..., s_{M-1}, s, s_{M+1}, ..., s_v)$
11:         **if** $\widetilde{S}$ satisfies C **then**
12:             Add $\widetilde{S}$ to R
13:         **end if**
14:     **end for**
15:     Move the LitSBSs with $kscore \ge MTSCV_M$ in R to P
16: **end while**
17: **return** P

---

in Algorithm 2). For each composable service group, we build the corresponding LitSBS. If the LitSBS satifies C, we move it to R (lines 9–14 in Algorithm 2). Finally, The LitSBSs with $kscores \ge$ the current $MTSCV_M$ in R are moved to P (line 15 in Algorithm 2). The previous operations in lines 7–15 in Algorithm 2 are repeated until P contains k LitSBSs.

4   Experiment

Section 4.1 presents case studies to show the practicability of LitDB algorithm for building service-based system. Section 4.2 tests the time efficiency of Fastk based on the WSC-2009-web challenge datasets [7]. Our experiments are conducted on a Server with a Xeon CPU at 2.60 GHZ and 160 GB RAM, running Windows7 x64 Enterprise.

4.1   Case Studies

We use mashup examples from programmableweb.com to show the rationality of LitDB. LitDB recommends a proposal for building a set of optimally matched mashups given a set of keywords.

Assume we want to build a map typo. It has the heatmap function and users can search firebases from the map. We use three keywords to query the service database: $Q = \{map, heatmap, firebase\}$. For the query keyword map,

Table 6. Mapping with firebase

ServiceId	Service name	Input	Output	Response time (ms)	Throughput (rps)
1	Initmap	Interval, center position	Map	765	3400
2	Heatmap	Map	Heatmap	425	2300
3	InitFirebase	Heatmap	Firebase	100	1200
4	Google.maps.LatLng	Geometry.coordinates, geome-try.coordinates	LatLng	540	4700

Table 7. Create Google Map with markers and cluster markers

ServiceId	Service name	Input	Output	Response time (ms)	Throughput (rps)
1	CreateMap	Interval, center position	Map	455	5500
2	CreateMarker	Map	Map with markers	300	7800
3	Markercluste-ring	Map with markers	Markerclusters	250	2000

we calculate the *kscore* of all the attributes in service Initmap in Table 6:
$\sum_{e_i \in E} kscore(s, e_i, q_r) = \frac{1+\ln(1+\ln(3))}{(1-0.2)+0.2\frac{7}{8.5}} \cdot \ln(\frac{4+1}{3}) + \frac{1+\ln(1+\ln(2))}{(1-0.2)+0.2\frac{3}{8.5}} \cdot \ln(\frac{4+1}{2}) = 2.53$.
For the query keyword *heatmap*, we calculate the *kscore* of all the attributes in
service Heatmap in Table 6: $\sum_{e_i \in E} kscore(s, e_i, q_r) = \frac{1+\ln(1+\ln(1))}{(1-0.2)+0.2\frac{7}{4.5}} \cdot \ln(\frac{4+1}{1})$
$+ \frac{1+\ln(1+\ln(1))}{(1-0.2)+0.2\frac{7}{4.5}} \cdot \ln(\frac{4+1}{1}) = 2.9$. For the query keyword *firebase*, we calculate the *kscore* of all the attributes in service InitFirebase in Table 6:
$\sum_{e_i \in E} kscore(s, e_i, q_r) = \frac{1+\ln(1+\ln(1))}{(1-0.2)+0.2\frac{12}{7}} \cdot \ln(\frac{4+1}{1}) + \frac{1+\ln(1+\ln(1))}{(1-0.2)+0.2\frac{8}{7}} \cdot \ln(\frac{4+1}{1}) = 2.98$.
The output of Initmap is the same as the input of Heatmap, and the out-
put of Heatmap is the same as the input of InitFirebase, so they can be
composed. This service composition is a LitSBS S_1 (see Fig. 2 (a)) contains
$Q = \{map, heatmap, firebase\}$. The total *kscore* of S_1 by Formula 3 is:
$kscore(S, Q) = \frac{2.53+2.9+2.98}{3} = 2.8$.

(a) S_1 (b) S_2

Fig. 2. Mashups

After we find the top-k LitSBSs with the highest *kscores* and satisfying user-
required QoS constraints, we can also sort them descendingly by their *qscores*.
For example, from Fig. 2 (a), the throughput of S_1 is: $min(5500, 7800, 2000) =$

2000 and the response time is: $455 + 300 + 250 = 1050$. Assume the user sets the weights of throughput and response time as $w_1 = 0.5$ and $w_2 = 0.5$ respectively. Then $qscore(S_1) = 0.5 \times (1 - \frac{1050}{10000}) + 0.5 \times \frac{2000}{10000} = 0.5475$.

4.2   Time Efficiency of Fastk

We compare the time efficiency of Fastk with those of Enhanced and Intuitive. The time efficiency is evaluated with respect to the number of services (from 1000 to 9000 in five datasets), the value of k (from 1 to 10), and the number of query keywords from (2 to 5). Each test is performed five times to obtain the average performance of the algorithms. We use 5 WSC-2009-web challenge datasets [12]. and extract the service information. The service information includes service name, 4–5 input and output parameters, and two QoS properties (response time and throughput).

Table 8. Statistics of five datasets

Dataset	Numbers			Frequency of different $kscores$									
	Services	Inputs	Outputs	1	2	3	4	5	6	7	8	9	10
Dataset1	1000	2934	3006	77	112	95	97	134	110	114	100	95	66
Dataset2	3000	9119	9057	221	286	303	339	373	366	327	303	256	256
Dataset3	5000	15160	14997	343	455	523	570	577	601	565	522	474	370
Dataset4	7000	21081	21179	522	638	742	825	795	857	738	719	670	494
Dataset5	9000	26938	26938	618	812	973	995	1090	1048	1020	946	837	661

As our purpose is to test the time efficiency of Fastk, we do not practically calculate the service $kscores$ of query keywords. Instead, we simulate $kscores$ for each dataset by randomly setting $kscores$ in $\{1, 2, ..., 10\}$ [5]. Based on these $kscores$, we create matched service tables for each dataset. The statistics of the 5 datasets are shown in Table 8. The frequency of each $kscore$ at different values (from 1 to 10) in each dataset is also shown in Table 8.

Effect of Ns(number of Services). We fix $k = 5$ and the number of query keywords (Nq) $= 2$ to compare the time efficiency of Fastk, Intuitive and Enhanced. When Ns $= 1000, 3000, 5000, 7000$ and 9000 respectively. The average time is shown in Fig. 3(a). Figure 3(a) shows that the execution time of Fastk is much shorter than those of Enhanced and Intuitive with respect to different Ns values. When the number of services changes, the time consumed by Fastk does not change obviously. Intuitive has the highest execution time. Its execution time increases dramatically when Ns increases from 1000 to 9000. It consumes almost twice the execution time of Enhanced.

Effect of k. We fix Nq $= 5$ and Ns $= 1000$ to test Fastk, Enhanced and Intuitive when k increases from 1 to 10. The specific average time are shown in Fig. 3(b).

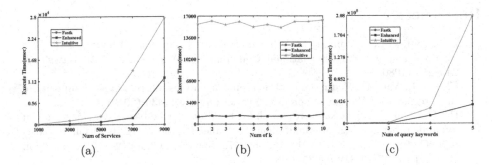

Fig. 3. Time for service composition algorithms

Intuitive takes the much more execution time than another two algorithms. Figure 3(b) also shows that the average execution time of the three algorithms all changes slightly when k varies from 1 to 10, especially the Intuitive algorithm. The reason is that, to get the top-k solutions, Intuitive needs to calculate the $kscores$ of all possible service compositions and then sort them, which takes a lot of time.

Effect of Nq (number of Query Keywords). We fix $k = 2$ and Ns = 3000 to test Fastk, Enhanced and Intuitive when Nq increases from 2 to 5. The average time are shown in Fig. 3(c). We can see that Fastk takes much shorter execution time than another two algorithms. In addition, when the Nq varies from 2 to 3, the execution time of the three algorithms are almost the same. When Nq increases from 3 to 5, the average time of Enhanced and Intuitive increases significantly. The reason is that the number of possible service compositions grows exponentially.

The above results shows that Fastk is more efficient than Intuitive and Enhanced. Figure 3(a) and (c) demonstrate the superiority of Fastk over Intuitive and Enhanced when the number of services and query keywords increase.

5 Conclusion

We design a database-driven algorithm (LitDB) to automatically build top-k lightweight SBSs based on user-given keywords. We use relational database as service repository, and develop an efficient top-k algorithm called Fastk. Case studies of building real mashups show the relationality of LitDB. Experiment results show its time efficiency. In the future, we plan to improve this work by considering the semantic matching between services and keywords. More efficient query techniques will also be explored.

References

1. Bernstein, P.A., Goodman, N.: What does Boyce-Codd normal form do? In: Proceedings of the Sixth International Conference on Very Large Data Bases, pp. 245–259 (1980)
2. Chen, M., Yan, Y.: Redundant service removal in Qos-aware service composition. In: International Conference on Web Services, pp. 431–439 (2012)
3. He, Q., et al.: Efficient keyword search for building service-based systems based on dynamic programming. In: Maximilien, M., Vallecillo, A., Wang, J., Oriol, M. (eds.) ICSOC 2017. LNCS, vol. 10601, pp. 462–470. Springer, Cham (2017). https://doi.org/10.1007/978-3-319-69035-3_33
4. He, Q., et al.: Keyword search for building service-based systems. IEEE Trans. Software Eng. **43**(7), 658–674 (2016)
5. Hristidis, V., Gravano, L., Papakonstantinou, Y.: Efficient IR-style keyword search over relational databases. In: Proceedings of the 29th International Conference on Very Large Data Bases, pp. 850–861. VLDB Endowment (2003)
6. Khanouche, M.E., Attal, F., Amirat, Y., Chibani, A., Kerkar, M.: Clustering-based and Qos-aware services composition algorithm for ambient intelligence. Inf. Sci. **482**, 419–439 (2019)
7. Kona, S., Bansal, A., Blake, M.B., Bleul, S., Weise, T.: WSC-2009: a quality of service-oriented web services challenge. In: 2009 IEEE Conference on Commerce and Enterprise Computing, pp. 487–490. IEEE (2009)
8. Lee, D., Kwon, J., Lee, S., Park, S., Hong, B.: Scalable and efficient web services composition based on a relational database. J. Syst. Softw. **84**(12), 2139–2155 (2011)
9. Li, J., Yan, Y., Lemire, D.: Scaling up web service composition with the skyline operator. In: International Conference on Web Services, pp. 147–154 (2016)
10. Peng, M., Zeng, G., Sun, Z., Huang, J., Wang, H., Tian, G.: Personalized app recommendation based on app permissions. World Wide Web **21**(1), 89–104 (2017). https://doi.org/10.1007/s11280-017-0456-y
11. Rodriguez-Mier, Pablo., Mucientes, Manuel, Lama, Manuel: A dynamic QoS-aware semantic web service composition algorithm. In: Liu, Chengfei, Ludwig, Heiko, Toumani, Farouk, Yu, Qi (eds.) ICSOC 2012. LNCS, vol. 7636, pp. 623–630. Springer, Heidelberg (2012). https://doi.org/10.1007/978-3-642-34321-6_48
12. Rodriguez-Mier, P., Mucientes, M., Lama, M.: Hybrid optimization algorithm for large-scale Qos-aware service composition. IEEE Trans. Serv. Comput. **10**(4), 547–559 (2015)
13. Singhal, A.: Modern information retrieval: a brief overview. IEEE Data Eng. Bull. Spec. Issue Text Databases **24**(4), 35–43 (2001)
14. Sun, L., Dong, H., Hussain, O.K., Hussain, F.K., Liu, A.X.: A framework of cloud service selection with criteria interactions. Futur. Gener. Comput. Syst. **94**, 749–764 (2019)
15. Sun, L., Ma, J., Wang, H., Zhang, Y., Yong, J.: Cloud service description model: An extension of USDL for cloud services. IEEE Trans. Serv. Comput. **11**(2), 354–368 (2018)
16. Wang, S., Zhou, A., Bao, R., Chou, W., Yau, S.S.: Towards green service composition approach in the cloud. IEEE Transactions on Services Computing, pp. 1–1 (2018)
17. Yan, Y., Xu, B., Gu, Z., Luo, S.: A Qos-driven approach for semantic service composition. In: 2009 IEEE Conference on Commerce and Enterprise Computing, pp. 523–526 (2009)

18. Yao, W., He, J., Wang, H., Zhang, Y., Cao, J.: Collaborative topic ranking: leveraging item meta-data for sparsity reduction. In: Twenty-Ninth AAAI Conference on Artificial Intelligence, pp. 374–380 (2015)
19. Yu, D., Zhang, L., Liu, C., Zhou, R., Xu, D.: Automatic web service composition driven by keyword query. In: World Wide Web, pp. 1–28 (2020)
20. Yu, Q., Bouguettaya, A.: Framework for web service query algebra and optimization. ACM Trans. Web **2**(1), 6 (2008)
21. Zhang, N., Wang, J., Ma, Y., He, K., Li, Z., Liu, X.: Web service discovery based on goal-oriented query expansion. J. Syst. Softw. **142**, 73–91 (2018)

Comparison of Text-Based and Feature-Based Semantic Similarity Between Android Apps

Md Kafil Uddin[✉], Qiang He, Jun Han, and Caslon Chua

Swinburne University University of Technology, Hawthorn 3122, Australia
{mdkafiluddin,qhe,jhan,cchua}@swin.edu.au

Abstract. Textual description of mobile apps comprised of apps' core functionality and describes the key features of any given app. Finding semantically similar apps help app users searching and looking for alternatives in the app stores as well as help the developers learn from their similar apps. A number of works have been done on apps' text-based semantic similarity. However, no attempts have been taken yet to find similar apps based on semantic similarity of their features. Moreover, varying length of textual descriptions affect semantic approach significantly. In this paper, we employ two different semantic similarity techniques to both the text-based and feature-based app descriptions. Together we investigate four different techniques and compare the similarities generated between the given android apps. We compare our results with a truth-set built on 50 similar and 50 non-similar apps in the same domain and across the other domains respectively. Experimental results demonstrate that feature-based approach generate better results in determining app similarity in the same domain and across the other domains irrespective of the length of apps' textual descriptions.

Keywords: Android app analysis · App improvement · App feature extraction · Semantic similarity measurement · App clustering

1 Introduction

Mining and categorization of software repositories is crucial for the developers to develop, improve and elicit requirements of their software application (a.k.a "app"). Application distribution platforms (also called app stores) provide a rich source of information about apps concerning their customers, businesses and technically focused attributes [21], which forms an ecosystem of users, developers and store owners [16]. In fact, app stores work as a communication channel between developers and users via the reviewing systems [24] and open a great opportunity for app developers to learn improvement requirements of their apps [27–29]. The two major stores, namely Google Play[1] and (Apple)

---

[1] https://play.google.com/.

© Springer Nature Switzerland AG 2020
Z. Huang et al. (Eds.): WISE 2020, LNCS 12342, pp. 530–545, 2020.
https://doi.org/10.1007/978-3-030-62005-9_38

App Store[2], contains 99% of the apps available. As of first quarter of 2020, Statistica[3] reported that about 2.56 million apps within 49 and 1.85 million apps within 27 categories were available for downloading Google Play and Apple app stores respectively.

Since these categories suffer from coarse-granularity [9], such broader categorisation of app stores help app developers keyword-based searching of apps only. App developers rather require searching of feature-based similar apps in order to find competitors and improve specific feature of their apps by learning from those competitors. Because, according to the developers, competitors are a great source of learning and improving an app under development (i.e., the *target app*) [31]. For example, mining reviews [18,19,29] of similar apps can improve the target app. Moreover, characterisation of similar app features help developers find out recent trends of popular and unpopular features in the store [30]. Consequently, allow app developers find out a lot of undiscovered requirements using prominent techniques like explicit content-rating interaction (ECRI) [15].

A recent report from top app developers recommends that other stack holders like businesses might choose developers of having experience of developing similar apps for businesses [5]. Others put more emphasis on checking similar apps before start developing an app from the scratch. They suggested - *"Before you do anything else, check the market for existing, similar mobile apps. Don't be put off if somebody has already done something similar-remember, there are a lot of apps out there, with many doing similar things to each other. As long as the app you create has qualities that do make it unique or slightly different, your users will love it! This is a faster and more cost-effective solution to create a mobile app than going down the custom app development route"* [1].

Previous literature on apps similarity measures [13,14] highly relies on apps textual descriptions only without considering its features. Moreover, these approaches require high volume of training dataset in order to find semantic similarity of the given pair of apps. A more closely related work [8] that compares the text-based app similarities based on the terms extracted from app descriptions. In this work, authors used their earlier works [9,21] to extract those terms called featurelets. However, these featurelets generates lower precision of features [22] and can not be always determined as functional features. In these works, authors employ WordNet [7] as a tool for measuring semantic similarities between the terms in the featurelets while other popular tool like Word2Vec [26] could also be considered in semantic measurements. A number of recent works [10,14,23,32] has shown Word2Vec as a popular tool for measuring semantic similarities between features.

Nonetheless, existing works have yet to exploit the power of both semantic tools on measuring similarities between android apps. In this work, we explore two different approaches, (1) *The Text-based approach:* use app the whole app description and (2) *The Feature-based approach:* use the features extracted from app description. Then, we employ semantic tools on both techniques and com-

---

[2] https://www.apple.com/au/ios/app-store/.
[3] https://www.statista.com/.

pare the results with varying conditions such as length of app descriptions, changing app domains etc. We also compare the efficiency of those techniques in terms of execution time.

In a nutshell, we answer the following research questions in this paper:

1. How are the similarities generated by the Word2Vec and WordNet in the Text-based approach?
2. How are the similarities generated by the Word2Vec and WordNet in the Feature-based approach?
3. How are the similarity scores vary with the length of app description?
4. Which technique is more effective in terms of measuring app similarities?
5. Which technique is more efficient in terms of execution time?

The rest of the paper is organised as follows. Section 2 presents the related work and the research gap that motivates this research. Section 3 presents the approaches for measuring semantic similarity. Section 4 describes the datasets and experiment settings. Section 5 presents the results and comparison of our experiments. Section 6 concludes the paper and outlines some future work.

2   Related Work

In this section, we review the works pertaining to app feature extraction (Sect. 2.1) where app's functional features are extracted from app's textual descriptions. We also review the two popular and widely used semantic tools used in natural language (Sect. 2.2) and a number of techniques in app similarity measurement (Sect. 2.3).

2.1   App Feature Extraction

Extraction of app features has been widely studied for various reasons such as app classification, malware detection, identification similar apps, feature recommendations, software evaluation and maintenance. Harman et al. [21] proposed a greedy method to extract app features from app descriptions by considering specific 'patterns' in writing app descriptions by the developers. Others extract functional keywords from app descriptions in order to facilitate app users faster searching. However, these keywords are rather un-comprehensible by human [22]. The SAFE [22] feature extraction approach overcome these problems by proposing a rule-based approach which provide better accuracy in identification of features from both reviews and descriptions. More importantly, this approach does not require large training set and configuration data. However, this method is useful for only small datasets where the number of apps are very low. Because, the app-feature matrix generated in this method is computationally expensive when the number of apps is large. Moreover, there is a loss of information while processing rules for feature extraction from the descriptions. In addition to that, after extraction of features, a further processing is required in order to identify similarity between them based on their descriptions. For example, finding similarity between two apps, appA and appB, is not possible without further calculation of semantic similarity between them.

2.2 Text Semantic Tools

A possible solution of the problem mentioned above might be using WordNet [7] for obtaining semantic similarity between apps. WordNet is a popular tool which has been widely used in a number of researches [9,33]. However, these methods use word-by-word semantics that does not take the whole text (i.e., app description) into account, might potentially increase the cost of computation. Another tool called Word2Vec [17] calculates the semantic similarity scores between the text descriptions using a technique called skip-grams. Recent studies show that Word2Vec is the state-of-the-art tool for measuring semantic similarity scores of word tokens represented by vectors [11,25].

2.3 App Similarity Measures

A recent work of Al-Subaihini et al. [8] empirically compare the text-based app similarities based on topic modelling and keyword terms. However, this work is built on their earlier works [9,21] where unique terms are extracted from app descriptions. These terms called featurelets which generates lower precision of features [22] and can not be always determined as functional features. Chen et al. [13] proposed a framework called SimApp that extracts 10 modalities from each app such as name, category, developer, description, size, update, permissions, content rating, reviews and image. Later, they developed a similarity model using the Online Kernel Weight Learning (OKWL) algorithm that learn kernel values from a set of apps and assign those values as weight of kernel in order to identify similar apps for any given app. Moreover, a recent work [14] of the same author effectively uses Word2Vec as a tool for building training corpus on app descriptions in order to compute semantic relevance between apps. In our Text-based approach, we follow Chen's work [14] for calculating semantic similarity between android apps. In addition, we exploit another popular semantic tool, called WordNet, while building the Text-based approach.

3 Techniques of Measuring Semantic Similarity

In order to measure similarity between given pair (e.g., appA and appB) of android apps (our crawled app dataset is from Google Play), it is necessary to process textual description first. Since app functionality lies in its description, the given pair of apps are considered similar when semantic similarity is found between the descriptions or between the features extracted from the descriptions. From each given pair, the first one, i.e., appA, is considered as *the target app* (developer's app under improvement) and the other one, i.e., appB, is considered as *the candidate app* from the store. Moreover, app descriptions are written in natural language where different text (such as voice calling and audio calling) might have the same meaning. Therefore, similarity measure between the apps must also comply with their semantic meanings.

In the following sections (Sect. 3.1, Sect. 3.2), we discuss the two approaches to identify semantic similarity between the target app and the candidate app.

1. **Text-based approach:** Process textual descriptions and employ semantic tools to calculate semantic similarity.
2. **Feature-based approach:** Extract features from app descriptions and employ semantic tools to calculate semantic similarity.

3.1   Text-Based Semantic Similarity

Before subjecting app descriptions to similarity analysis, we applied standard techniques of natural language processing (NLP) such as filtering stop words, filtering domain related words like 1-letter, 2-letter, app names, connecting words and removal of auxiliary verbs. As mentioned earlier (in Sect. 3), in the Text-based approach, following an existing work [14] we also exploit WordNet as an additional tool for measuring semantic similarity between android apps. We apply both the semantic tools (Word2Vec and WordNet) on the pre-processed app descriptions. The two approaches for measuring text-based semantic similarity are listed below.

A) **Text Processing+Word2Vec (TPW2V):** Use Word2Vec after text processing.
B) **Text Processing+WordNET (TPWN):** Use WordNET after text processing.

To achieve this, at first, we cleanup app's text descriptions by removing all the unnecessary symbols, numbers, hypen, characters, tags and spaces. Then, we discard the URLs, emails and app names from the app descriptions and remove all the English STOPWORDS, like "to", "for", "the", etc., from the app descriptions. STOPWORDS usually have little lexical content, and their presence in a text fails to distinguish it from other texts. We remove them by using the Python libraries NLTK [2]. Finally, we manually built our own "Custom Dictionary" relating to app domain for the removal of unnecessary words such as *1-letter, 2-letters, app names, connecting words, auxiliary verbs, adverbs, praising words,* etc. A portion of the dictionary has been shown below:

['http", "https", "inc", "etc", "able", "typical", "yet","otherwise", "welcome", "none", "done", "ago", "recently", "still", "wait", "today", "soon", "always", "app", "also", "even", "ever", "available", "please", "much", "almost", "many", "2g", "x", "facebook", "viber", ...]

In this way, app's functional features are preserved without major loss of information. Then, we tokenize the app description using word-tokenizer and generate Parts-of-Speech (POS) for each word-token using the Stanford POS-taggers library [4], the world's richest POS-taggers library for Natural Language Processing (NLP). This is the final step before we feed tokenized app descriptions into the WordNet and Word2Vec model in order to calculate the functional similarity score of the target app relevant to the candidate app. The whole implementation of Text-based semantic similarity calculation (TPW2V and TPWN) has been shown in the Algorithm 1. *Code:* https://github.com/mdkafil/semantic_similarity.

Algorithm 1: Text-based Semantic Similarity Calculation

   Input: $Desc_t$, $Desc_c$ //Description of target and candidate app respectively
   Result: TP_{w2v}, TP_{wn}, Text-based Similarity Score
1 $wordnet \leftarrow$ wn // initialize the WordNet from NLTK
2 $wordmodel \leftarrow$ wiki.en.vec // initialize the model
3 $M \leftarrow$ word2vec.Keyedvectors($wordmodel$) //load trained wiki vectors
4 **if** $Desc_t$ *is Empty* **then**
5 | return "Error"
6 **else**
7 | //Text-Processing of Target App
8 | $Desc_t \leftarrow$ removeStopWords($Desc_t$)
9 | $Desc_t \leftarrow$ processText($Desc_t$, $regEx$)
10 | $Desc_t \leftarrow$ tokenizeDesc($Desc_t$){StanfordTagger}
11 **end**
12 **foreach** $Desc_c \in \mathbb{C}$ **do**
13 | **if** $Desc_c$ *is Empty* **then**
14 | | return "Error"
15 | **else**
16 | | //Text-Processing of Candidate App
17 | | $Desc_c \leftarrow$ removeStopWords($Desc_c$)
18 | | $Desc_c \leftarrow$ processText($Desc_c$, $regEx$)
19 | | $Desc_c \leftarrow$ tokenizeDesc($Desc_c$){StanfordTagger}
20 | **end**
21 | $TP_{w2v} \leftarrow$ w2v($Desc_t$,$Desc_c$,M)
22 | $TP_{wn} \leftarrow$ wn($Desc_t$,$Desc_c$) //calculate Wu-Palmer similarity
23 **end**
24 Return TP_{w2v},TP_{wn}

The algorithm follows 3-steps. In the first step (line 1–4), loads the semantic models (word2vec or WordNet). In the second step (line 5–21), pre-process the app description of both the target app and the candidate app into tokenized form. In the final step (line 22–23), feed the tokenized descriptions in to the models. We use Wu-Palmer (WUP) path similarity for our WordNet Model Since it provides better semantic relatedness [33]. We explain the detail implementation of the semantic tool, Word2Vec, in the next section.

3.2 Feature-Based Semantic Similarity

Functional features of an app lies in the description. In this step, developer's claimed features are extracted from its description. Then, semantic tools such as Word2Vec and WordNet are employed to measure similarities between the features of the target app and the candidate app. The two feature-based semantic methods are-

A) **Feature Extraction + Word2Vec (FXW2V):** Use Word2Vec after extracting Features.

B) **Feature Extraction + WordNET (FXWN)**: Use WordNET after extracting Features.

Algorithm 2: Feature-based Semantic Similarity Calculation

Input: $Desc_t$, $Desc_c$ //Description of target and candidate app respectively
Result: FX_{w2v}, FX_{wn}, Feature-based Similarity Score
1 $wordnet \leftarrow$ wn // initialize the WordNet from NLTK
2 $wordmodel \leftarrow$ wiki.en.vec // initialize the model
3 $M \leftarrow$ word2vec.Keyedvectors($wordmodel$) //load trained wiki vectors
4 **if** $Desc_t$ *is Empty* **then**
5 | return "Error"
6 **else**
7 | //Feature-Processing of Target App
8 | $Feature_t\{...\} \leftarrow \emptyset$
9 | $Feature_t\{...\} \leftarrow$ SAFE($Desc_t$)
10 **end**
11 **foreach** $Desc_c \in \mathbb{C}$ **do**
12 | **if** $Desc_c$ *is Empty* **then**
13 | | return "Error"
14 | **else**
15 | | //Feature-Processing of Candidate App
16 | | $Feature_c\{...\} \leftarrow \emptyset$
17 | | $Feature_c\{...\} \leftarrow$ SAFE($Desc_c$)
18 | **end**
19 | $FX_{w2v} \leftarrow$ w2v($Feature_t$,$Feature_c$,M)
20 | $FX_{wn} \leftarrow$ wn($Feature_t$,$Feature_c$) //calculate Wu-Palmer similarity
21 **end**
22 Return FX_{w2v},FX_{wn}

Functional features are extracted from app descriptions. In this phase, our purpose is to find functional similarity between the target app and the candidate apps. A recent work from Johann et al. proposes a method called Safe [22] extracts features from app descriptions. SAFE extracts features by using 18 parts-of-speech (POS) and 5 sentence patterns that are frequently followed by the developers in the app description. This method claims the higher feature extraction results among the existing works. Therefore, we choose SAFE as our feature extraction tool. However, SAFE has few drawbacks since it considers app description as formal text and apply rules and patterns on it. During our SAFE implementation, we improve the following drawbacks: *(1) Feature mentioned in double-quotation, (2) Feature starts with the word "Feature" and (3) Capitalized words, mentioning specific feature.*

After the feature extraction, we feed extracted features into the WordNet and Word2Vec model in order to calculate the functional similarity score of the target app relevant to the candidate app. However, there is a loss of information while processing rules for feature extraction from the descriptions. The whole implementation of Feature-based semantic similarity calculation (FXW2V and FXWN) has been shown in the Algorithm 2. The algorithm follows similar steps like the previous one where extracted app features (feature extraction shown in line 4–18) are taken as input parameters in the semantic models (line 19–20).

In order to simplify our approach, we use Gensim [6], a popular implementation of Word2Vec. As there is no Word2Vec model trained on an app store corpus, we use the one trained on Wikipedia [3], which contains over 2 millions word-vectors with more than 600 billions of word-tokens. According to Chan et al. [12], if two arguments co-occur in the same Wikipedia article, the content of the article can be used to check whether the two entities are related. Since all our app descriptions are written in English, to calculate functional similarity scores accurately, we use the Word2Vec model trained on English Wikipedia pages (line 2,3 and 19). *Code:* https://github.com/mdkafil/semantic_similarity

4   Experiment Setup

We compare Text-based and Feature-based semantic approaches through a series of experiments. Section 4.1 introduces datasets used throughout the experiments. Section 4.2 describes validation methods for effectiveness comparison. Section 4.3 presents evaluation metrics for validating effectiveness of both the approaches.

4.1   Datasets

To compare and evaluate efficiency and effectiveness of both Text-based and Feature-based semantic similarity methods, we conducted a range of experiments on our app datasets. For our experimental purpose, we developed our own crawler that crawls a recent dataset for a period of 3 months (Nov, 2019 to Jan, 2020) from Google Play store. Our dataset consists of 5000 apps of all categories that includes app descriptions necessary for measuring app similarities. We removed those categories and sub categories that has less 30 apps in it. Finally, our curated dataset contains 2000 apps of 10 categories excluding the Games category because of its exceptional nature [19]. These 10 categories are: Communication, Education, Book and Ref., Health and Fitness, Shopping, Finance, Lifestyle, Productivity, Transport and Business. Each of these categories has apps between 195 to 205.

Analysis of our dataset shows that we have app descriptions of varying length starting from 59 words to 390. This ensures that our dataset has a great diversity of app length. However, only 15% of apps found as short length (length < 75 words) whereas 85% found as longer than 75 words. 23% of the dataset found longer than 300 words. On average, app descriptions are of length 200 to 230 words. We randomly choose 10 apps as the *target app* from each category, resulting in a total of 100 *target* apps and 2000 *candidate* apps from all 10 categories.

4.2   Validation Methods

Both the approaches are compared in terms of (1) *effectiveness*, effectiveness in generating app similarities and (2) *efficiencies*, time required to execute both the approaches. In order to validate effectiveness, we compare the results against a truth-set of 100 apps. Comparison of effectiveness and efficiency of both the approaches has been discussed in Sect. 5.2 and Sect. 5.3.

4.3   Evaluation Metrics

To evaluate our classification and ranking, we used 3 popular metrics: Precision (Eq. 1), Recall (Eq. 2) and F-score (Eq. 3), defined below [20].

$$Precision = \frac{TP}{TP + FP} \tag{1}$$

$$Recall = \frac{TP}{TP + FN} \tag{2}$$

$$F - score = 2 * \frac{Precision * Recall}{Precision + Recall} \tag{3}$$

where TP is the number of apps correctly found as relevant to the similar apps, FP is the number of apps not correctly found as relevant to the similar apps and FN is the number of apps incorrectly found as relevant to the similar apps. The F-score is the harmonic mean of precision and recall. We calculate them against the truth-set of 100 apps.

5   Results Analysis and Discussion

In this section, we discuss the results of app similarity scores generated through both the approaches and compare them in terms of effectiveness and efficiencies. Finally, we discuss the limitation of our evaluation in the threats to validity.

5.1   Approach Application and Example Results

We apply both the Text-based (TPW2V, TPWN) and Feature-based (FXW2V, FXWN) approaches for each of the target apps against all the candidate apps in our dataset. In order to compare effectiveness of semantic approaches, we made sure to keep track of at least one of the 4 criteria, namely: *(1) similar apps:* target and the candidate apps are similar, *(2) non-similar:* candidate app is non-similar to the target app, *(3) short length:* length of description of candidate app is <75 words and *(4) good length:* length of description of candidate app is >= 75 words. Due to the space limitation, we report in Table 1, an example of similarity scores generated through Text-based Approach (TPW2V, TPWN) and Feature-based

Table 1. Example Results: Text-based Approach and Feature-based Approach (where, *WhatsApp* as target app)

App Types	Candidate apps	Text-Based Approach		Feature-Based Approach	
		TPW2V	TPWN	FXW2V	FXWN
Similar Apps	WhatsApp	**1.0**	**0.99**	**1.0**	**0.99**
	Facebook Messenger	0.95	0.73	0.94	0.80
	Skype	0.97	0.72	0.94	0.74
Non-Similar Apps	Career Service	**0.82**	**0.54**	**0.0**	**0.0**
	Adblocker Browser	0.92	0.59	0.81	0.49
	Xfinity WiFi Hotspots	0.93	0.63	0.0	**0.53**
length <= 75 words	Ring App	0.84	0.51	0.0	0.41
	Me@Walmart	**0.87**	**0.69**	**0.0**	**0.0**
	UBC Mobile	0.73	0.53	0.0	0.0
length > 75 words	Kik	0.92	0.60	0.87	0.55
	Telegram	0.94	0.72	0.91	0.68
	Viber	0.97	0.80	0.95	0.73

approach (FXW2V, FXWN). Table 1 shows the similarity scores are generated for the target app (i.e, *WhatsApp*) against a number of candidate apps.

As can be seen in Table 1, all the similarity scores are ranging from 0 to 1. A similarity score 1 means, the target app and the candidate app are fully similar where as a similarity score 0 means, the target app and the candidate app are not similar at all. Lets take an example of WhatsApp as a candidate app. Since all the scores in Table 1 are calculated against WhatsApp as the target app, hence, the similarity scores for WhatsApp to WhatsApp is nearly 1 which means that apps are fully similar. In case of other similar apps like Skype and Facebook Messenger, scores are likely similar for both the approaches. Likewise, no significant score differences found between the approaches in case of good length candidate apps (such as Kik, Telegram and Viber) except the variation of scores between Word2Vec and WordNet semantic tools.

In general, Word2Vec produces higher similarity scores compared to the WordNet among both the text-based and feature-based approaches. In our experiment, we found TPW2V always generates very high similarity scores irrespective of description length. TPW2V even produces unexpected results like high similarity scores in case of non-similar apps. Since Word2Vec word model works on the intersection of descriptions between two comparable apps, most often, the similarity scores are higher than expected.

For example, Text-based approach produces a high similarity scores of 0.82 by TPW2V and 0.54 by TPWN, for a non-similar candidate app called Career Service. A similar trend is shown in short length app as well. For example, the candidate app, Me@Walmart has no functional feature matching with WhatsApp. The Feature-bases approach produce similarity score 0 where as a similarity score is found by Text-based approach which is as high as 0.87. Therefore, we compare the effectiveness of both the approaches in details in the section below. *Sample data available on:* https://github.com/mdkafil/semantic_similarity.

5.2 Comparison of Effectiveness Between Semantic Approaches

In order to validate effectiveness, we compare the results against a truth set of 100 apps. We run a survey among 15 participants where 8 of them are PhD students in software engineering field, 2 of them are app developers and 5 of them are software professionals having experience of 3 to 12 years. We build a truth set of 100 apps having 50 similar and 50 non-similar apps where 11 of them are of short length (< 75 words) and the rest of them are of good length i.e., descriptions $>= 75$ words. Table 2 reports the overall effectiveness comparison between the Text-based approach and the Feature-based approach. We compare the result in terms of 4-types of candidate apps.

(1) Similar apps: As can be seen in Table 2, when candidate apps are of similar types, both Text-based approach (TPW2V,TPWN) and Feature-based (FXW2V,FXWN) approach produces almost similar scores, therefore, the number of true positive scores always dominate the result. The precision and recall value is always higher than 0.8 and F-score is as high as 0.9.

(2) Good length: In the case of good length app descriptions (length $= >75$ words), it becomes difficult to find out the significant differences between similar apps. Both the approaches have higher similarity scores, hence, the number of true positive is high. Therefore, the precision and recall goes above 0.85. In case of Text-based approach false positive score sore is higher than feature-based approach. Therefore, precision is slightly lower for Text-based approach than Feature-based approach.

(3) Short length: For short length descriptions (length < 75 words), more cases of functionally non-similar apps are found as similar by Text-based approach. Consequently, a higher false negative score lowers the recall value to 0.53. However, a similar trend also shown in Feature-based approach. Hence, the recall value always lies in a range of 0.6 to 0.7 in Feature-based approach.

(4) Non-similar: Though both the approaches find similarities between apps, the purpose of developers is to find functional feature similarities between the apps rather than non-functional features or actual text similarities. Since the app description contains functional feature information along with non-functional features (i.e., feature benefits) and additional aspects of the feature, therefore, word2vec produces higher scores due to matching with functional features and non-functional feature words in the description. In our experiment, we found that app descriptions contain >60% non-functional words. Therefore, the higher the non-functional features or aspect words contribute to the similarity scores, the higher the false positive score is. Hence, Text-based approach produces lowest precision (0.34 for TPW2V, 0.41 for TPWN) and lowest F-score (0.45 for TPW2V, 0.53 for TPWN) compared to the Feature-based approach.

Overall, Feature-based approaches (FXW2V and FXWN) perform better irrespective of app types (i.e., varying length and varying similarity). However, this approach depends on the extraction of functional features, hence, the effectiveness of these two methods solely relies on the extraction of features. Experiment shows that FXWN works better for both low and high length app descriptions. However, FXW2V produces poor results when one of the app has very

Table 2. Effectiveness comparison between Text-based and Feature-based Approach

App Types	Evaluation Metric	Text-Based Approach		Feature-Based Approach	
		TPW2V	TPWN	FXW2V	FXWN
Similar Apps	Precision	0.91	0.87	0.91	0.92
	Recall	0.97	0.95	0.93	0.91
	F-score	**0.93**	**0.90**	**0.91**	**0.91**
Non-Similar Apps	Precision	**0.34**	**0.41**	0.86	0.91
	Recall	0.69	0.75	0.95	0.92
	F-score	**0.45**	**0.53**	0.90	0.91
length <= 75 words	Precision	0.83	0.81	**0.71**	0.79
	Recall	**0.53**	0.67	0.68	0.64
	F-score	0.64	0.73	0.69	0.70
length > 75 words	Precision	0.89	0.87	0.95	0.92
	Recall	0.92	0.95	0.89	0.96
	F-score	**0.90**	**0.90**	**0.92**	**0.93**

low app features. This happens because of word2vec works on word embedding which consider the context. The lower, the number of functional features generates poor context situation which results in poor similarity scores in Word2Vec.

5.3   Comparison of Efficiency Between Semantic Approaches

Comparison of efficiency between semantic approaches shown in Fig. 1 and Fig. 2. All our experiments run on 3 cloud instances of same configuration (*Ubuntu Xenial, 16 Core and 32GB RAM*).

As can be seen in Fig. 1 and Fig. 2, Feature-based approach requires higher execution time (MS for milliseconds) compared to the Text-based approach. Since SAFE feature extraction works on rule based approach, Feature-based approach requires additional cost for processing the rules and the patterns. In case of Text-based approach in Fig. 1 and Fig. 2, TPWN requires higher cost than TPW2V because of its complexity since WordNet works on lexical path similarity between the terms. However, Word2Vec works faster than Wordnet since Word2Vec works based on word vectors which associates the initial cost of loading the training vectors (in our experiment, we use wiki model).

Following the above principles, in Fig. 2, there is a significant cost difference found among all approaches for short length and good length apps except the TPW2V. This is because, the execution time becomes almost insignificant for both short and good length apps with the cost of initial loading time for word model. However, for FXW2V, the cost of feature extraction overrides the efficiency of Word2Vec model. Overall, though effective in finding functional similarity between apps, Feature-based approach poses a greater bottleneck in terms of efficiency because of the cost of extracting features.

5.4   Discussion

While building the truth-set, we solely relied human judgment. As the relevance of similarity is subjective, we reduced the bias running the survey to a focused

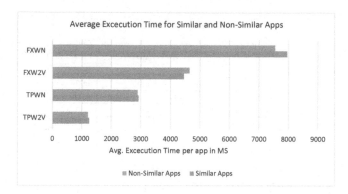

Fig. 1. Approach Efficiency: Similar Vs. Non-similar Candidate Apps

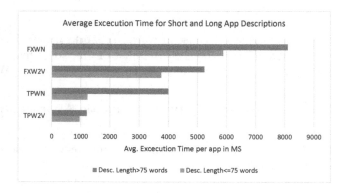

Fig. 2. Approach Efficiency: Short Length Vs. Good Length Candidate App

group of researchers and developers. An external validity threat is the representativeness of the target apps that we choose in the experiments. Our experimental results are applicable only to those target apps. To minimize this threat, we randomly chose 100 target apps from 10 different app categories.

6   Conclusion and Future Work

App descriptions are the source of app's functional features. Since developers focus on improvement of specific features by learning from their competitors, a closer look at functional features from their competitors help improve the app. In this paper, we empirically analyse semantic similarities between android apps in order to help app developers identify their competitors. We compare two approaches, *(1) Text-based approach*: employ semantic tools on tokenized text descriptions and *(2) Feature-based approach*: employ semantic tools features extracted from descriptions. We use two popular semantic tools namely, Word2Vec and WordNet on each of the approaches.

Experiments are run on varying app length and type for each of the approaches and results are validated in terms of effectiveness and efficiency. Effectiveness of the approaches are compared against a truth-set of 100 apps built on human judgement whereas efficiency is compared against time required for semantic calculation. Experimental results reveal that, Feature-based approaches perform better irrespective of app types (i.e., length and similarity). However, though effective, Feature-based approach poses a greater bottleneck in terms of efficiency because of the cost of extracting features.

In our future work, we aim to identify the impact of low-level app application package (APK) on android app similarity and compare the low-level similarity with high-level functional similarity.

References

1. Diy app builders. https://www.mobiloud.com/blog/build-mobile-app-without-experience/. Accessed 13 July 2020
2. Nltk, natural language toolkit. http://www.nltk.org. Accessed 30 Jan 2020
3. Pre-trained vectors. https://fasttext.cc/docs/en/pretrained-vectors.html. Accessed 01 Feb 2019
4. Stanford pos-tagger library: The richest parts-of-speech tagger library online. https://nlp.stanford.edu/software/tagger.shtml/. Accessed 01 Feb 2019
5. Top app developers (2019). https://www.businessofapps.com/app-developers/. Accessed: 2020–01-30
6. Word2vec model. https://radimrehurek.com/gensim/models/word2vec.html. Accessed 01 Feb 2019
7. Wordnet:lexical dictionary. https://wordnet.princeton.edu. Accessed 30 Jan 2019
8. Al-Subaihin, A., Sarro, F., Black, S., Capra, L.: Empirical comparison of text-based mobile apps similarity measurement techniques. Empirical Softw. Eng. **24**(6), 3290–3315 (2019)
9. Al-Subaihin, A.A., Sarro, F., Black, S., Capra, L., Harman, M., Jia, Y., Zhang, Y.: Clustering mobile apps based on mined textual features. In: Proceedings of the 10th ACM/IEEE International Symposium on Empirical Software Engineering and Measurement, ACM, p. 38. ACM (2016)
10. Baeza-Yates, R., Jiang, D., Silvestri, F., Harrison, B.: Predicting the next app that you are going to use. In: Proceedings of the Eighth ACM International Conference on Web Search and Data Mining, pp. 285–294 (2015)
11. Bojanowski, P., Grave, E., Joulin, A., Mikolov, T.: Enriching word vectors with subword information. Trans. Assoc. Comput. Linguistics **5**, 135–146 (2017)
12. Chan, Y.S., Roth, D.: Exploiting background knowledge for relation extraction. In: Proceedings of the 23rd International Conference on Computational Linguistics, Association for Computational Linguistics, pp. 152–160 (2010)
13. Chen, N., Hoi, S. C., Li, S., Xiao, X. Simapp: a framework for detecting similar mobile applications by online kernel learning. In: Proceedings of the 8th ACM International Conference on Web Search and Data Mining, pp. 305–314. ACM (2015)
14. Chen, N., Hoi, S. C., Li, S., and Xiao, X. Mobile app tagging. In: Proceedings of the 9th ACM International Conference on Web Search and Data Mining, pp. 63–72 (2016)

15. Du, J., Rong, J., Wang, H., Zhang, Y.: Helpfulness prediction for online reviews with explicit content-rating interaction. In: Cheng, R., Mamoulis, N., Sun, Y., Huang, X. (eds.) WISE 2020. LNCS, vol. 11881, pp. 795–809. Springer, Cham (2019). https://doi.org/10.1007/978-3-030-34223-4_50

16. Genc-Nayebi, N., Abran, A.: A systematic literature review: opinion mining studies from mobile app store user reviews. J. Syst. Softw. **125**, 207–219 (2017)

17. Goldberg, Y., Levy, O.: word2vec explained: deriving mikolov et al'.s negative-sampling word-embedding method. arXiv :1402.3722 (2014)

18. Gu, X., Kim, S.: What parts of your apps are loved by users?"(t). In: 2015 30th IEEE/ACM International Conference on Automated Software Engineering (ASE), pp. 760–770. IEEE (2015)

19. Guzman, E., Maalej, W.: How do users like this feature? a fine grained sentiment analysis of app reviews. In: 2014 IEEE 22nd International Requirements Engineering Conference (RE), pp. 153–162. IEEE (2014)

20. Han, J., Pei, J., Kamber, M.: Data Mining: Concepts and Techniques. Elsevier (2011)

21. Harman, M., Jia, Y., Zhang, Y.: App store mining and analysis: MSR for app stores. In: Proceedings of the 9th IEEE Working Conference on Mining Software Repositories, pp. 108–111. IEEE Press (2012)

22. Johann, T., Stanik, C., Maalej, W., et al.: Safe: a simple approach for feature extraction from app descriptions and app reviews. In: 2017 IEEE 25th International Requirements Engineering Conference (RE), pp. 21–30. IEEE (2017)

23. Lu, M., Liang, P.: Automatic classification of non-functional requirements from augmented app user reviews. In: Proceedings of the 21st International Conference on Evaluation and Assessment in Software Engineering, pp. 344–353 (2017)

24. Martin, W., Sarro, F., Jia, Y., Zhang, Y., Harman, M.: A survey of app store analysis for software engineering. IEEE Trans. softw. Eng. **43**(9), 817–847 (2017)

25. Mikolov, T., Grave, E., Bojanowski, P., Puhrsch, C., Joulin, A.: Advances in pre-training distributed word representations. In: Proceedings of the International Conference on Language Resources and Evaluation (LREC 2018) (2018)

26. Mikolov, T., Sutskever, I., Chen, K., Corrado, G.S., Dean, J.: Distributed representations of words and phrases and their compositionality. Adv. Neural Inf. Process. Syst. **113**, 3111–3119 (2013)

27. Palomba, F., Linares-Vasquez, M., Bavota, G., Oliveto, R., Di Penta, M., Poshyvanyk, D., De Lucia, A. User reviews matter! tracking crowdsourced reviews to support evolution of successful apps. In: 2015 IEEE International Conference on Software Maintenance and Evolution (ICSME), pp. 291–300. IEEE (2015)

28. Palomba, F., et al.: Crowdsourcing user reviews to support the evolution of mobile apps. J. Syst. Softw. **137**, 143–162 (2018)

29. Panichella, S., Di Sorbo, A., Guzman, E., Visaggio, C.A., Canfora, G., Gall, H.C.: How can i improve my app? classifying user reviews for software maintenance and evolution. In: 2015 IEEE International Conference on Software Maintenance and Evolution (ICSME), pp. 281–290. IEEE (2015)

30. Sarro, F., Al-Subaihin, A.A., Harman, M., Jia, Y., Martin, W., Zhang, Y.: Feature lifecycles as they spread, migrate, remain, and die in app stores. In: 2015 IEEE 23rd International Requirements Engineering Conference (RE), pp. 76–85. IEEE (2015)

31. Villarroel, L., Bavota, G., Russo, B., Oliveto, R., Di Penta, M.: Release planning of mobile apps based on user reviews. In: Proceedings of the 38th International Conference on Software Engineering, pp. 14–24. ACM (2016)

32. Vu, P.M., Nguyen, T.T., Pham, H.V., Nguyen, T.T.: Mining user opinions in mobile app reviews: a keyword-based approach (t). In: 2015 30th IEEE/ACM International Conference on Automated Software Engineering (ASE), pp. 749–759. IEEE (2015)

33. Wu, Z., Palmer, M.: Verbs semantics and lexical selection. In: Proceedings of the 32nd Annual Meeting on Association for Computational Linguistics, Association for Computational Linguistics, pp. 133–138 (1994)

Subjective Metrics-Based Cloud Market Performance Prediction

Ahmed Alharbi and Hai Dong[✉]

School of Science, RMIT University, Melbourne, Australia
{ahmed.alharbi,hai.dong}@rmit.edu.au

Abstract. This paper explores an effective machine learning approach to predict cloud market performance for cloud consumers, providers and investors based on social media. We identified a set of comprehensive subjective metrics that may affect cloud market performance via literature survey. We used a popular sentiment analysis technique to process customer reviews collected from social media. Cloud market revenue growth was selected as an indicator of cloud market performance. We considered the revenue growth of Amazon Web Services as the stakeholder of our experiments. Three machine learning models were selected: linear regression, artificial neural network, and support vector machine. These models were compared with a time series prediction model. We found that the set of subjective metrics is able to improve the prediction performance for all the models. The support vector machine showed the best prediction results compared to the other models.

Keywords: Cloud market performance prediction · Subjective metrics · Social media · Sentiment analysis

1 Introduction

Cloud computing is increasingly becoming the technology of choice as the next-generation platform for most organization [16,23]. Cloud computing services are provided by vendors under the following categories: Software-as-a-Service (SaaS), Platform-as-a-Service (PaaS), and Infrastructure-as-a-Service (IaaS) are available in the cloud market. Different organizations such as banks, universities, etc., are the main consumers of cloud services.

The cloud market is divided into two types of participants: those who use cloud services - *cloud customer*, and those who need to invest in the cloud market - *cloud investors/providers*. *Both* the parties are interested in cloud service performance in the market. Sufficient knowledge of the cloud market allows the market participants to be aware of opportunities, advantages, and risks associated with it. This can include knowledge of how the market is performing in terms of market competition, market growth, customer satisfaction, etc. [15], which can all help in making informed decisions regarding the market. In fact, a comprehensive analysis of the market is of value to the cloud customer due to

© Springer Nature Switzerland AG 2020
Z. Huang et al. (Eds.): WISE 2020, LNCS 12342, pp. 546–558, 2020.
https://doi.org/10.1007/978-3-030-62005-9_39

the long-term commitment involved. Such customers could be those who make IT decisions after determining the influence of emerging technologies. They must understand how the market performs in terms of customer satisfaction, service quality, and many other factors before deciding to participate in the market. Investors also require information on market performance before making any decision to invest in stocks. In fact, many industry reports have been prepared by different organisations in relation to all the participants [3,5]. Most of these reports are concerned with the evaluation of cloud market size and growth as well as numerous other concerns. These reports tend to be produced after thorough monitoring of the market or after accessing to cloud providers' performance/sales.

The performance information provided by cloud market reports are based on two types of metrics: *objective metrics* and *subjective metrics*. The objective metrics are based on actual facts without any individual bias. Examples of such metrics include market size and growth. According to 451Research [5], the collection of such data is frequently implemented through extensive market monitoring. Meanwhile, the subjective metrics are based on personal perceptions, opinions or feelings, an example being the information regarding the barriers to cloud adoption [4]. Subjective measures are frequently obtained using industry surveys [2,3]. Industries utilise either objective **or** subjective measures, or a combination of both to produce industry reports. However, it is a challenge to apply objective metrics [25] due to the information sensitivity, while it can be *expensive, time consuming* and *less timely* to obtain the information required for objective performance. On the other hand, the gathering of subjective market information can be more practical, since the sharing of subjective evaluation information is preferred over objective evaluation information sharing [25]. Market participants often tend to share similar information on *social media.*

Numerous individuals are becoming increasingly involved in sharing their opinions on a variety of subjects in social media platforms [6,7]. It has become increasingly popular in recent years [19]. Consumers express their complaints or their satisfaction with certain brands and services. It has been reported that one in three social media users prefer to use this outlet rather than contact the company by phone. Moreover, 50% of social media users share their complaints/concerns about a service/brand at least once a month [4]. Hence, social media can be viewed as a *free and up-to-date* source of personal wisdom [26].

Purchase decisions can be affected by consumer's perceptions on services/goods. Thus, subjective measures can be used to predict the future behaviour of the participants. With this subjective measure evaluation of market performance, a better comprehension of objectively measured results can be gained. Any given customer review can include perceptions related to various areas of a product or service experience [10]. In turn, these perceptions will influence the readers' perceptions and, subsequently, their purchase decisions [18]. Here, perceptions related to risks, barriers and benefits can be particularly influential on a consumer's purchasing intention [20]. Future behaviour can be predicted based on this intention to purchase [21].

According to Jayaratna et al. [12], subjective measures are applicable in evaluating cloud market performance. They identified unique subjective metrics that can assess cloud market performance. However, the subjective information is limited to *product/service quality*. In addition, attention has not been given to *use machine learning techniques* for predicting economical/financial performance of cloud services in the cloud market.

We aim to *explore a set of more comprehensive subjective metrics* that may affect cloud market performance and *analyse the effectiveness of machine learning techniques* in predicting cloud market performance. A comprehensive set of subjective cloud service evaluation metrics were collected via literature survey. The values of subjective metrics were collected from social media and quantified using sentiment analysis. We applied three mainstream machine learning prediction techniques: *linear regression, artificial neural network*, and *support vector machine* to ascertain which model predicts cloud market performance most accurately. We considered *cloud market revenue growth* as the indicator of cloud market performance. This paper makes the following contributions:

1. We identified a set of more comprehensive subjective metrics that may influence the cloud market performance.
2. We explored an effective machine learning model to predict the cloud market performance based on the identified subjective metrics.

The paper is structured as follows: Sect. 2 describes our solution. Section 3 depicts the results of our experiment. Section 4 concludes the paper.

2   Methodology

We used two sets of metrics (subjective and objective) to evaluate cloud market performance. Cloud market revenue was considered as the objective metrics because it is an important factor in market performance evaluation [15]. Cloud user's perception on different cloud aspects was used as the subjective metrics. We explored an effective machine learning approach that predicts cloud market revenue growth based on the cloud user's perceptions on different cloud aspects. The main objective of the proposed approach is to predict the cloud market performance based on a comprehensive set of subjective metrics.

2.1   Cloud Service Evaluation Subjective Metrics Identification

Subjective metrics of customer perception have shown effectiveness in predicting cloud market performance over the objective metrics [12]. Therefore, in this research, we aim to explore a comprehensive set of subjective metrics that may influence cloud market performance.

Certain drivers and boundaries drive consumers to/from cloud services according to the surveys conducted by RightScale [2] and the Australian Bureau of Statistics [1]. We can thus assume that drivers and boundaries are important for cloud consumers. The knowledge of the benefits and challenges are likely to influence a customer's decision regarding the purchase of cloud services. Therefore, the benefits and challenges of cloud services will act as drivers of cloud market revenue. Jayaratna et al. [12] introduced 13 drivers and boundaries related to product/service quality.

Based on the above, we combined Jayaratna et al. [12]'s 13 drivers and boundaries with three more drivers: after-sales experience, market responsiveness and marketing execution from Gartner [9], as a result of our literature survey. Table 1 presents the combined drivers and boundaries used in this paper. We focused on consumer's perception on same cloud aspects. In this paper, we aim to predict cloud market performance based on the 16 cloud aspects.

Table 1. Subjective aspects affecting cloud purchase decisions

Cloud aspects	Description
Greater scalability	Flexible to either up-scale or down-scale
Faster access to infrastructure	Easy access to infrastructure without having to purchase them
Managing multiple services	Overheads and difficulty in managing multiple services
Security concerns	Concerns over data breaches, privacy, access control, etc.
Cost savings	Cost saved by transfer to cloud infrastructure
Higher availability	High availability of cloud services
Lack of control	Uncertainty of data location. Uncertainty regarding legal issues, and dispute resolution
Higher performance	Higher performance of cloud compared to on-premise infrastructure
Lack of expertise/resources	Lack of specialised people or sufficient resources for managing cloud services
IT staff efficiency	Increase of productivity
Provider lock-in	Difficulties with changing cloud computing service provider
Business continuity	Ability to continually operate even through disasters
Move from CapEx to OpEx	Changing from capital expenditure to operating expense
After-sales experience	Ability to provide acceptable customer services
Market responsiveness	Ability to enhance the products' after sales
Marketing execution	Ability to deliver the product that the customer expected

2.2   Data Collection

We collected two types of data: objective data and subjective data. The Amazon web service (AWS) quarterly revenue was considered as objective. The subjective data was collected in terms of AWS customer reviews.

(1) Objective Data:
IaaS is defined as the capability to provide computer resources, storage, and other fundamental computing resources to the customer. Accordingly, we consider the revenue generated by offering these types of cloud services. Amazon is considered as the leading cloud provider because it holds the largest market share in the IaaS segment. Therefore, we consider Amazon revenue as a cloud market performance. AWS revenue is reported for the whole segment (IaaS, PaaS, etc.). In this paper, the AWS quarterly revenue growth [22] from the 4th quarter of 2015 to the

4th quarter of 2018 was considered as an indicator of AWS performance in the market. Revenue growth was the increase in a company's revenue from one period to the next. Revenue growth identifies over time trends in business. The AWS was calculated based on the following Eq. [12]:

$$AWSQRG_y = \frac{Revenue_q - Revenue_{q-1}}{Revenue_{q-1}} \tag{1}$$

where $AWSQRG_y$ is the AWS quarterly revenue growth of quarter q, $Revenue_q$ is the revenue made in q, and $Revenue_{q-1}$ is the revenue made in the quarter before q.

(2) Subjective Data:
Social media is a rich source of customer opinion. Obvious differences can be seen among the customer perceptions collected from social media websites such as Twitter and Facebook and from customer review websites. However, social media websites also contain references to cloud services that are not based on experiences, such as references related to discussions or news on topics about newly-released features. In contrast, cloud customers' review websites are exclusively focused on the customer experience. These review sites contain comprehensive information on users' experiences. Moreover, any user can post comments about a cloud service/provider on social media, and the platforms do not verify whether a post is from a genuine cloud service customer. However, most of the popular review sites do go through this process before posting the feedback. Therefore, we can conclude that customer review websites provide more valuable information on cloud services than popular social media websites.

We examined six cloud social media, which are the most popular of those currently active. These are 'G2 Crowd'[1], 'Trust Radius'[2], 'Clutch'[3], 'Gartner Peer Insights'[4], 'WhoIsHostingThis'[5] and 'Spiceworks'[6], all of which guarantee the genuineness of the reviews are based on the genuine customer experience. These sources contain reviews of a variety of cloud providers, including Amazon, Google, and Microsoft. The reviews are written by customers from different domains (i.e. education, health, insurance) and organisations of different scale (small, mid-sized and enterprise). We assumed that the customers' reviews can be considered as a representative of the overall market customers because the number of customer reviews on these websites exceeds 1000.

2.3 Perception Analysis on Cloud Aspects

VADER sentiment analysis technique [11] is a well-acknowledged sentiment detection tool. VADER takes into account punctuation, capitalization, degree

---

[1] https://www.g2crowd.com.
[2] https://www.trustradius.com.
[3] https://clutch.co/.
[4] https://www.gartner.com/reviews/home.
[5] https://www.whoishostingthis.com.
[6] https://www.spiceworks.com.

modifiers, and the use of contrastive conjunction. In this paper, we used VADER sentiment analysis to process the user reviews. Sentiment analysis was performed to find customers' perception on each cloud aspect separately. First, we built a vocabulary that contains the most frequent words to refer to each specific aspect. We used NVivo software[7], which has been designed for qualitative and mixed-methods research, to build the vocabulary. We fed the customer reviews of each quarter to the NVivo software. Then, we found the most frequent words that related to the cloud aspects. For example, secure, secured, securely, security were repeated more than 40 times in 2016 4th quarter reviews. Therefore, we assumed that those words will fit into the security cloud aspect.

Next, the sentiments of the AWS user reviews of each quarter were analysed using VADER. VADER is based on lexicons of sentiment related-words. Every word in the lexicon is rated as to whether it is positive or negative. VADER hired many domain experts to manually rate the words. Each word in the lexicon is rated between -4 (extremely negative) and $+4$ (extremely positive). VADER checks a piece of text to see if any of the words in the text are present in the lexicon. VADER produces four sentiment scores: positive, neutral, negative, and compound. The positive, neutral, and negative scores are ratios for the proportions of text that falls into those categories. The compound score is a metric that calculates the sum of all the lexicon ratings[8] which have been normalized between -1 (most extreme negative) and $+1$ (extremely positive). Finally, we considered the compound sentiment results as the overall perception of a particular cloud aspect. The customers' perception of cloud aspect A in quarter Q is defined as follows:

$$Perception_{A,Q} = \frac{Compound_{A,Q}}{Reviews_{A,Q}} \tag{2}$$

where $Compound_{A,Q}$ is the compound sentiment value, and $Reviews_{A,Q}$ is the number of reviews referring to aspect A in quarter Q.

2.4   Sensitivity Analysis

We designed the following machine learning models to evaluate cloud market performance based on the subjective metrics.

(1) Linear Regression Analysis (LR):
Linear regression is a linear approach used to ascertain the relationship between a dependent variable and one or more independent variables. In our approach, the independent variables are sets of cloud aspects, while the dependent variable is the AWS revenue growth. The linear regression has the following Equation:

$$Y = a + bX_1 + cX_2 + ... + mX_n \tag{3}$$

where Y is used to represent the AWS revenue growth, and X_1, X_2, ... X_n represent the user perception of each cloud aspect.

---

[7] https://www.qsrinternational.com/nvivo/home.
[8] http://comp.social.gatech.edu/papers/.

(2) Artificial Neural Network (ANN):
ANN models are mathematical models inspired by the functioning of the nervous system. ANNs are based on learning which is a characteristic of adaptive systems which are capable of improving their performance on a problem as a function of previous experience [14]. In this paper, we used a Multlayer Perceptron. The Levenberg-Marquardt backpropagation algorithm the most widely-used training algorithm for time series prediction [17] was used to train the network. The input activation function was obtained using the following equation:

$$s = \sum w_{ij}x_j + b \tag{4}$$

where x_i is the user perception of each cloud aspect, w_j is the weight, and b is the bias.

The output neuron uses the sigmoid activation function obtained with the following equation:

$$f(s) = \frac{1}{1 + e^{-s}} \tag{5}$$

where s is the sum of products (sop) between each user perception on each cloud aspect and its corresponding weight. s is an input activation function.

The error function ascertains how close the predicted output is from the target output. The error function was obtained using the following equation:

$$E = \frac{1}{2} \sum_{i=1}^{N} (O_i - T_i) \tag{6}$$

where N is the total number of output data, O_i is the actual AWS revenue growth of ith data and T_i is the predicted AWS revenue growth of ith data.

(3) Support Vector Machine (SVM):
The nu-SVR model, which is applicable for modelling continuous time series, was chosen. The nu-SVR model was found to be reliable and robust, even for models based on small training samples or data burdened by noise based on experience from applications in a different area [27]. In our approach, we used the nu-SVR model with a radial basis function kernel (RBF) and which takes the user perception of cloud services and AWS revenue growth as the target. The RBF kernel can be defined as follows [27]:

$$\exp(\gamma \times \sum | u - v |^2) \tag{7}$$

where γ is the parameter, u is the actual AWS revenue growth, and v is the predicted AWS revenue growth.

3   Experiments

A set of experiments was conducted to evaluate the efficiency of the proposed prediction models. First, we describe the dataset and how is it obtained.

We then compare the parameter of each model. Next, we compare the prediction performance between four predictions models. We ran two experiments in order to analyse the impact of our new subjective metrics. First, we used the user perceptions on the 13 cloud aspects identified in Jayaratna et al.'s [12]. Then, we used the user perceptions on the 16 cloud aspects in order to find the impact of the new subjective metrics. We evaluated all the proposed models in terms of Mean Squared Error (MSE), Root Mean Squared Error (RMSE) and Theil's U-statistics.

3.1 Datasets

We divided the data set into the sets of training data and testing data according to an approximate ratio of 2:1. AWS quarterly revenue growth and user perceptions of each cloud aspect from the 4th quarter of 2015 to the 4th quarter of 2017 are the training data. The data from the 2018 1st quarter to 2018 4th quarter are the testing data. Figure 1 presents the AWS quarterly revenue growth data. We collected approximately 1100 user reviews for analysis.

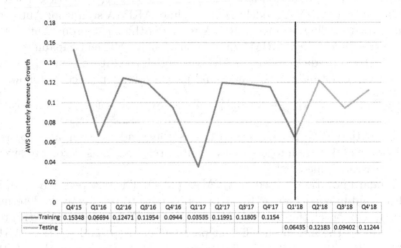

Fig. 1. AWS quarterly revenue from 2015 4th quarter to 2018 4th quarter

3.2 Sentiment Analysis

First, we identified the most frequent words from the customer reviews of each quarter. For example, After-sales aspect has ten words which are customer service, satisfaction, good service, after-sales, client service, product service, troubleshooting, assistance, customer care and support. Next, we categorized the customer reviews based on the 16 cloud aspects that were introduced in the previous section. Next, the sentiment of the AWS user reviews were analysed.

We processed each review to get the sentiment results. We then calculated the user perception using Eq. 2. Table 2 shows the sentiment results and the user perceptions for After-sales for 2016 4th quarter as an example.

Table 2. Sentiment and user perceptions results for After-sales for 2016 Q4

Review	1	2	3	4	5	6	7	8	9	10	11	12
Compound	0.4215	0.3612	0.3182	0.7092	0.9196	0.8876	0.9681	0.4091	0.9702	0.1027	0.9715	0.9829
User perception	0.66848334											

3.3 Tuning Process for the Proposed Models

We tested four models in the experiment, including an objective model (ARIMA) based on previous quarterly revenue growth and three subjective models based on previous quarterly revenue growth and customer perceptions. Here we introduced the optimal tuning result for each model.

(1) ARIMA model:
ARIMA model is a prediction model based on the objective metrics. Therefore, we considered the ARIMA model as our baseline. ARIMA stands for Autoregressive integrated moving average. ARIMA model is the most prominent methods in time series forecasting. ARIMA models have shown efficient capability to generate short-tram prediction [24]. In ARIMA model, the future value of a variable is a linear combination of past values and past errors, expressed as follows:

$$Y_t = \phi + \phi_1 Y_{t-1} + \phi_2 Y_{t-2} + ... + \phi_p Y_{t-p} + \varepsilon - \phi_1 \varepsilon_{1t-1} - \phi_2 \varepsilon_{2t-2} - ... - \phi_q \varepsilon_{qt-q} \tag{8}$$

where Y_t is the AWS revenue growth, ε is the random error at t, ϕ_i is the coefficients, p and q are integers that are often referred to as autoregressive and moving average, respectively.

A standard notation is used of ARIMA (p,d,q). The autoregressive (p), integrated (d) and moving average (q) parameters have to be effectively determined in order to construct the best ARIMA model for AWS revenue growth prediction. We tested the model on different parameters of (p) and (q) (i.e., ARIMA (1,0,0), (1,1,0), (2,1,0), (0,1,1), (2,0,0) and (3,0,0)). We found that ARIMA (1,0,0), (2,1,0) and (2,0,0) achieve the best performance. All the experiments were conducted using Python.

(2) Linear Regression Analysis (LR):
At first, we tested the 13 cloud aspects as the independent variables to identify the prediction model. The prediction model based on the 13 cloud aspects can be represented in the Eq. 9. Then, we tested the 16 cloud aspects as the independent variables to identify the prediction model. The prediction model can be represented in the Eq. 10

$$RG = -0.151 * Per_{IT} - 0.069 * Per_{Perf} + 0.075 * Per_{Cost}$$
$$- 0.138 * Per_{control} - 0.113 * Per_{Ex} - 0.156 * Per_{Provider} + 0.042 * Per_{Security} + 0.519 \tag{9}$$

$$RG = -0.017 * Per_{BC} + 0.054 * Per_{Cost} - 0.065 * Per_{control} - 0.138 * Per_{Ex}$$
$$- 0.132 * Per_{Provider} - 0.092 * Per_{execution} - 0.115 * Per_{Security} + 0.515 \quad (10)$$

where RG is the revenue growth, Per_{BC} is the perception of business continuity, Per_{Cost} is the perception of cost savings, $Per_{control}$ is the perception of lack of control, Per_{Ex} is the perception of lack of expertise/resources, $Per_{Provider}$ is the perception of provider lock-in, $Per_{execution}$ is the perception of marketing execution, $Per_{Security}$ is the perception of security concerns, Per_{IT} is the perception of IT staff efficiency and Per_{Perf} is the perception of higher performance.

(3) Artificial Neural Network (ANN):
We needed to establish the network parameters to build the neural network. The input data was randomly divided into training data (70%), validation data (15%) and testing data (15%). The best validation performance for the 13 cloud aspects is obtained at epoch 2 and for the 16 cloud aspects is obtained at epoch 3. All experiments were conducted using the neural networks toolbox of MAT-LAB.

(4) Support Vector Machine (SVM):
Several levels of gamma constant of radial function was tested in the parameter settings experiments. We chose four different values for gamma 0.01, 0.1, 1, and 5 [13]. We found that the best performance is obtained when the value of gamma is 5 for the both cloud aspect combinations. All experiments were performed on windows using Chang and Lin's library developed for SVM implementations [8].

3.4 Experimental Result and Discussion

We predicted the AWS revenue growth form the 1st quarter of 2018 to the 4th quarter of 2018 using four different prediction models: ARIMA, LR, ANN, and SVM. Figure 2 depicts the results of predicted revenue growth compared to the actual revenue growth. It can be seen that there is a significant gap between the actual and predicted revenue growth for LR with 13 and 16 cloud aspects prediction models. However, the other models were able to predict AWS revenue growth for all quarters.

All the experimental results prove the positive impact of the new set of subjective matrices and the prediction performance of the proposed models. All the models based on the 16 cloud factors outperform their counterparts based on the 13 cloud factors as described in Table 3. We can confidently conclude that *After-seals, Market responsiveness* and *Market execution* improve the prediction performance. Additionally, the SVM prediction model with 16 cloud aspects produced the best results with the smallest error. The SVM prediction models outperformed ARIMA, ANN and LR models. The reason is that SVM implements the structural risk minimization principle, resulting in better generalization than the other techniques. In contrast, ANN needs a larger training set to perform more accurate predication, while ARIMA purely bases on the previous revenue data and disregards the impact of the subjective metrics on the avenue growth. LR preformed the worst due to the existence of nonlinear relationship between revenue growth and the subjective metrics.

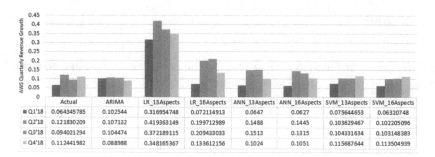

	Actual	ARIMA	LR_13Aspects	LR_16Aspects	ANN_13Aspects	ANN_16Aspects	SVM_13Aspects	SVM_16Aspects
Q1'18	0.064345785	0.102544	0.316954748	0.072114913	0.0647	0.0627	0.075644653	0.06320748
Q2'18	0.121830209	0.107132	0.419363149	0.199712989	0.1488	0.1445	0.103629467	0.102205096
Q3'18	0.094021294	0.104474	0.372189115	0.209433033	0.1513	0.1315	0.104331634	0.103148383
Q4'18	0.112441982	0.088988	0.348165367	0.133612156	0.1024	0.1051	0.115687644	0.113504939

Fig. 2. Actual vs. predicted

Table 3. Prediction performance of ARIMA, LR, ANN, and SVM

Prediction model	MSE	RMSE	Theil's U-statistics
ARIMA	0.000583622	0.024	0.394573683
LR - 13 cloud aspects	0.071319997	0.267058041	1.455854282
LR - 16 cloud aspects	0.004973533	0.03526164	0.504616664
ANN - 13 cloud aspects	0.001027297	0.0321	0.444061429
ANN - 16 cloud aspects	0.000493797	0.0222	0.404017853
SVM - 13 cloud aspects	0.000143942	0.011997583	0.354241995
SVM - 16 cloud aspects	**0.000117719**	**0.010849839**	**0.343687709**

4   Conclusion

We explored a machine learning based approach to predict the cloud market performance using social media subjective metrics. We identified a set of comprehensive subjective metrics (including After-seals, Market responsiveness, and Market execution). We processed the user reviews collected from social media using VADER sentiment analysis, a popular sentiment analysis technique on each subjective metric. We applied three mainstream machine learning: linear regression, artificial neural network, and support vector machine. Experimental results show that the new set of subjective metrics have a positive impact on predicting cloud market performance. The proposed model is able to predict cloud market performance for both cloud consumers and cloud investors/providers. In the future work, we plan to apply deep learning models to asses cloud market performance.

References

1. Australian Bureau of Statistics. Tech. rep., Paid Cloud Computing in Australian Business (2015). http://www.abs.gov.au/ausstats/abs@.nsf/mf/8129.0
2. State of the Cloud Report. Tech. rep., RightScale (2016)

3. Top Markets Report Cloud Computing. Tech. rep., U.S. Department of Commerce (2017)
4. State of the Media - the Social Media Report 2012. Tech. rep., Nielsen (2017)
5. Research: Market size, share and forecast data for emerging technology markets. Tech. rep., 451 Research.com (2014)
6. Aamir, T., Dong, H., Bouguettaya, A.: Social-sensor composition for tapestry scenes. IEEE Trans. Serv. Comput. 1 (2020)
7. Ali, K., Dong, H., Bouguettaya, A., Erradi, A., Hadjidj, R.: Sentiment analysis as a service: a social media based sentiment analysis framework. In: ICWS, pp. 660–667 (2017)
8. Chang, C.C., Lin, C.J.: LIBSVM: a library for support vector machines. ACM Trans. Intel. Syst. Technol. 2, 271–2727 (2011)
9. Gartner: Magic quadrant for cloud infrastructure as a service, worldwide. Tech. rep., gartner.com (2018)
10. Goh, K.Y., Heng, C.S., Lin, Z.: Social media brand community and consumer behavior: quantifying the relative impact of user-and marketer-generated content. Inf. Syst. Res. 24(1), 88–107 (2013)
11. Hutto, C.J., Gilbert, E.: Vader: a parsimonious rule-based model for sentiment analysis of social media text. In: ICWSM (2014)
12. Jayaratna, M.S.H., Bouguettaya, A., Dong, H., Qin, K., Erradi, A.: Subjective evaluation of market-driven cloud services. In: ICWS, pp. 516–523 (2017)
13. Lai, L.K.C., Liu, J.N.K.: Stock forecasting using support vector machine. In: Proceedings of Management Science and Engineering Management, pp. 1607–1614 (2010)
14. Luk, K., Ball, J.E., Sharma, A.: A study of optimal model lag and spatial inputs to artificial neural network for rainfall forecasting. J. Hydrol. 227(1–4), 56–65 (2000)
15. Mauboussin, M.J.: The true measures of success. Harvard Bus. Rev. 90(10), 46–56 (2012)
16. Mistry, S., Bouguettaya, A., Dong, H.: Economic Models for Managing Cloud Services. Springer, Cham (2018). https://doi.org/10.1007/978-3-319-73876-5
17. Moré, J.J.: The Levenberg-Marquardt algorithm: implementation and theory. In: Watson, G.A. (ed.) Numerical Analysis. LNM, vol. 630, pp. 105–116. Springer, Heidelberg (1978). https://doi.org/10.1007/BFb0067700
18. Mudambi, S.M., Schuff, D.: Research note: what makes a helpful online review? A study of customer reviews on Amazon.com. MIS Q. 34, 185–200 (2010)
19. Nepal, S., Paris, C., Bouguettaya, A.: Trusting the social web: issues and challenges. World Wide Web 18(1), 1–7 (2015)
20. Ruiz-Mafe, C., Sanz-Blas, S., Aldás-Manzano, J.: Drivers and barriers to online airline ticket purchasing. J. Air Transp. Manag. 15, 294–298 (2009)
21. Shahabuddin, S.: Forecasting automobile sales. Manag. Res. News 32(7), 670–682 (2009)
22. Statista: Quarterly revenue of Amazon Web Services from 1st quarter 2014 to 2nd quarter 2018 (in million U.S. dollars) (2018)
23. Sun, L., Dong, H., Hussain, O.K., Hussain, F.K., Liu, A.X.: A framework of cloud service selection with criteria interactions. Future Gener. Comput. Syst. 94, 749–764 (2019)
24. Tascikaraoglu, A., Uzunoglu, M.: A review of combined approaches for prediction of short-term wind speed and power. Renew. Sustain. Energy Rev. 34, 243–254 (2014)
25. Vij, S., Bedi, H.: Are subjective business performance measures justified? IJPPM 65, 603–621 (2016)

26. Yu, S., Kak, S.: A survey of prediction using social media. arXiv preprint arXiv:1203.1647 (2012)
27. Zhu, D., Ji, B., Meng, C., Shi, B., Tu, Z., Qing, Z.: The performance of ν-support vector regression on determination of soluble solids content of apple by acousto-optic tunable filter near-infrared spectroscopy. Anal. Chimica Acta **598**, 227–234 (2007)

Towards Web-Scale and Energy-Efficient Hybrid SDNs: Deployment Optimization and Fine-Grained Link State Management

Yufei Liu[1], Shang Cheng[1], and Xinli Huang[1,2](✉)

[1] School of Computer Science and Technology, East China Normal University, Shanghai 200062, China
xlhuang@cs.ecnu.edu.cn
[2] Shanghai Key Laboratory of Multidimensional Information Processing, Shanghai 200062, China

Abstract. As one of the most promising solutions for the next generation Web-scale datacenter networks, hybrid software-defined networks (SDN) have received extensive attention and in-depth research in recent years. Specifically, the issues of deployment optimization and fine-grained link state management in hybrid SDNs are extremely crucial for building Web-scale and energy-efficient SDN-driven datacenters. In this paper, we introduce an energy-efficient deployment optimization scheme for hybrid SDNs, in order to reduce the deployment cost and energy consumption, so that the hybrid networks can approach or even achieve the centralized control ability of pure SDN networks. In addition, we propose fine-grained link state management algorithms with energy consumption sensing ability. With the algorithms, the link state of the hybrid SDNs can be controlled in a fine-grained way, and can be managed by dynamically clearing or temporarily closing redundant links, so as to further reduce link energy consumption. We setup experiments to demonstrate the functionality effectiveness and performance gains of the proposed scheme and algorithms, and the extensive simulation results show that, the proposed solution of this paper outperforms its competitors, producing an arresting increase of energy efficiency. Specifically, under the premise of only 20% of SDN switch deployment budget, the proposed scheme in this paper can reduce the total energy consumption by about 40%.

Keywords: Web-scale datacenter · Software-defined networks · Hybrid deployment optimization · Link state management · Energy efficiency

1 Introduction

Hybrid Software-Defined Network (SDN) refers to a hybrid networking architecture where both SDN centralized paradigm and legacy decentralized paradigm coexist to different degrees and interact with each other, to configure, control,

© Springer Nature Switzerland AG 2020
Z. Huang et al. (Eds.): WISE 2020, LNCS 12342, pp. 559–573, 2020.
https://doi.org/10.1007/978-3-030-62005-9_40

change, and manage network behaviors, for the purpose of optimizing network performance, energy consumption and resource utilization. With such a hybrid deployment, network operators can guide the traffic, balance the load, and manage the network more easily. The controller only needs to pay attention to the specific traffic or flow, and the forwarding and processing of most packets are done by the traditional routing protocols.

Despite these advantages, hybrid SDNs also suffer the problems of heterogeneous paradigms management and interaction consistency, which are critical for successful network deployment, operation and application [1]. For example, SDN switches in hybrid SDNs may be improperly deployed or an inefficient optimization solution may be adopted, resulting in functionality mis-operation, performance degradation or energy inefficiency. Forwarding decisions made by the controller may conflict with traditional routing protocols due to the isolated control domains, which may lead to forwarding loops or black-holes.

In this paper, we focus on the problems described above, and manage to introduce new ideas and solutions for deployment optimization and fine-grained link state management in hybrid SDNs, with the main purpose of building performance- and energy-efficient SDN-driven datacenter networks. The work and main contributions of the paper are summarized as follows.

- Firstly, we introduce an energy-efficient deployment optimization scheme for hybrid SDNs, in order to reduce the deployment cost and energy consumption, so that the hybrid networks can approach or even achieve the centralized control of traffic in pure SDN networks, and realize green and efficient hybrid network deployment.
- Secondly, we propose fine-grained link state management algorithms with energy consumption sensing ability. With the algorithms, the SDN link state can be controlled in a fine-grained way, and can be managed by dynamically clearing or temporarily closing redundant links, in order to further reduce link energy consumption.
- And finally, we setup experiments to demonstrate functionality effectiveness and performance gains of the proposed scheme and algorithms, and the extensive simulation results show that, our solution outperforms its competitors, producing an arresting increase of energy efficiency. Specifically, under the premise of only 20% of SDN switch deployment budget, our proposed scheme can reduce the total link energy consumption by about 40%.

The rest of the paper is organized as follows. We review related work in Sect. 2, and then present in detail the problem formulation and design rationale in Sect. 3, the methodology and implementation in Sect. 4, respectively. We describe experimental setup and report the results and analysis in Sect. 5, followed by the conclusion and future work of this paper in the last section.

2   Related Work

Focusing on how to build performance- and energy-efficient hybrid SDN networks, a lot of efforts have been made to investigate the issues of hybrid deployment optimization and behavior management.

Casado et al. [2] proposed the Fabric framework to optimize the network traffic, which can only be applied to pure SDN networks. Based on Fabric, [3,4] developed methods to divide the OSPF domain into several independent subdomains connected by SDN switches. These two schemes initially consider node selection strategies, but lack of actual deployment verification, and the proposed independent sub-domain partitioning method will produce additional power consumption in the network and may lead to overloading or even congestion of some links.

In response to the above considerations, some researchers seek to propose solutions to optimize the node selection and control the SDN switch behavior. [5–7] proposed to minimize the maximum link utilization in the whole network by optimizing the traffic separation ratio of SDN switches in hybrid SDN networks, thus reducing the overall network load. [8,9] focused on how to guide more links that cannot participate in forwarding in traditional routing protocols to participate in forwarding by SDN switches. [7,10] concentrated on the problem of how to maximize the proportion of controllable traffic in a given hybrid network, so as to control the network, in the centralized way, as much as possible.

Although the aforementioned solutions try to optimize the behavior of SDN switches, they did not consider the optimization of switch deployment location and took few into account of efforts on power consumption saving. [11,12] proposed a heuristic algorithm to select nodes with lower cost and higher transit traffic during the deployment process, without considering the operation cost and energy consumption of links in the network. To reduce the resources occupancy and energy consumption of the network, [13,15] presented that the links connected to the SDN switches could be opened and closed freely and the redundant links without traffic could be closed temporally to saving network resources. However, these proposals can only work on a given hybrid network model, without considering the hybrid deployment optimization. It also may affect the network performance due to link closure, such as path unreachable, link congestion, etc.

Based on the above analysis, in this paper, we focus on the problems of deployment optimization and fine-grained link state management in hybrid SDNs, and manage to introduce new ideas, methods and solutions, with the main purpose of building performance- and energy-efficient hybrid SDN networks.

3   Problem Formulation

We formulate a graph $G = (V, E)$ to represent the topology of a network, where V is the set of the switches, and E is the set of the links. The rest of the notations used in this paper are listed in Table 1.

In this paper, we concentrate on deployment optimization and fine-grained link state management in hybrid SDNs. The optimization objective is to reduce the total link utilization and energy consumption of the network after SDN switch deployment to the lowest level.

Therefore, the first optimization problem we need to formalize is how to reduce the total load in the network as much as possible by adjusting the SDN switch deployment scheme, so as to minimize the maximum link utilization to reduce the network resource occupation. Formula 1 and Formula 2 used to represent the network load before and after the deployment of SDN switches:

$$\sum_{e_{i,j} \in E} Cap(e_{i,j}) \tag{1}$$

$$\sum_{e_{i,j} \in E_{SDN}} Cap(e_{i,j}) \times Pri(Turn(e_{i,j})) + \sum_{e_{i,j} \in E_{leg}} Cap(e_{i,j}) \tag{2}$$

Table 1. Notations used in this paper.

Parameter	Description
v_i	The switch i
v_i^p	The number of ports for the switch i
V_{SDN}	The set of SDN switches in the network
V_{leg}	The set of traditional switches in the network
E_{SDN}	The set of links controlled by SDN switches
E_{leg}	The set of links controlled by traditional switches
$e_{i,j}$	The link from switch i to switch j
$f(e_{i,j})$	The total traffic on link $e_{i,j}$
$f(e_{i,j}^{s,d})$	The traffic on link $e_{i,j}$ for $TF_{s,d}$
$In_j^{s,d}$	The set of the traffic $TF_{s,d}$ that enters switch v_j
$Out_j^{s,d}$	The set of the traffic $TF_{s,d}$ that leaves switch v_j
$Cap(e_{i,j})$	Maximum allowable load on link $e_{i,j}$
TM	The matrix of network traffic
$TF_{s,d}$	The traffic between switch s and switch d, $TF_{s,d} \in TM$
$Cost(v_i)$	The cost of deploying an SDN switch at v_i
$flow(v_i)$	The traffic through v_i based on a traditional routing protocol
$pass(v_i)$	The logarithm of the traffic through v_i based on a traditional routing protocol
W	The matrix of link weight
$Turn(e_{i,j})$	The state of link $e_{i,j}$
$Budget$	Total budget for deployment of SDN switches

The second optimization problem we need to formalize is how to control and manage the operation state of the link and its ports, in a fine-grained way, to reduce the energy consumption of the link in the network as much as possible. We use $Pri(Turn(e_{i,j}))$ to represent the energy consumption coefficient of the link $e_{i,j}$. The traditional switch can only run at full load, with the value of $Turn(e_{i,j})$ that connected to the traditional switch is 4. And the link $e_{i,j}$ connected to the SDN switch could have four states: high, medium, low and off, and the corresponding value of $Turn(e_{i,j})$ are 0, 1, 2 and 4. The default energy consumption coefficient in this paper is as follows:

$$Pri(Turn(e_{i,j})) = \begin{cases} 0 & \text{if } Turn(e_{i,j}) = 0 \\ \sqrt{\frac{Turn(e_{i,j})}{4}} & \text{if } Turn(e_{i,j}) = 1, 2, 4 \end{cases} \tag{3}$$

Besides, Eq. 3 should satisfy the following constraints:

- For any node v_j in the network, the total traffic into v_j plus the traffic began from v_j is equal to the total traffic out of v_j plus the traffic ended in v_j:

$$\sum_{e_{i,j} \in In_j^{s,d}} f(e_{i,j}^{s,d}) + \sum TF_{j,d} = \sum_{e_{j,m} \in Out_j^{s,d}} f(e_{j,m}^{s,d}) \quad s, d, j \in V \tag{4}$$

- For link $e_{i,j}$, the total traffic is equal to the sum of each pair of traffic on the link and should be less than the current allowable load of the link:

$$f(e_{i,j}) = \sum_{s,d \in V} f(e_{i,j}^{s,d}) \leq \frac{1}{4} Cap(e_{i,j}) \times Turn(e_{i,j}) \quad i, j \in V \tag{5}$$

- If node j is a traditional switch, it could select at most one link to forward:

$$card(Out_j^{s,d}) = 0, 1 \quad \text{if } j \in V_{leg} \tag{6}$$

- The deployment cost of SDN switches cannot be more than the budget. And the deployment cost of node v_i is proportional to the number of its ports:

$$\sum_{v_i \in V_{SDN}} Cost(v_i) \leq Budget \quad \text{where } Cost(v_i) \propto v_i^p \tag{7}$$

- The traffic of each link $e_{i,j}$, the deployment cost of each node v_i and the weight of each traffic $e_{i,j}^{s,d}$ are non-negative:

$$f(e_{i,j}), \ Cost(v_i), \ f(e_{i,j}^{s,d}), \ Budget, \ W(e_{i,j}) \geq 0 \tag{8}$$

4   Methodology and Implementation

In this section, we describe the methodology and implementation of the proposed deployment optimization scheme and fine-grained link state management algorithms for hybrid SDNs.

4.1   Multi-path Load Balancing

This part of work mainly includes the establishment of hybrid SDN network optimization model and the control of multi-path forwarding behavior of SDN switch.

In order to solve the problem of route loop in hybrid SDN, we propose to control the forwarding behavior of the ports of SDN switch, guarantee the correctness of forwarding behavior of SDN switch when choosing non-shortest path, and hereby realize correct and reliable hybrid SDN network routing. Considering the definition of upstream and downstream nodes in the network topology, only the SDN switch node can forward packets to the upstream nodes. Once there is a loop in the calculation process, it will go back to the previous multi-path forwarding location for re-selection. In addition, we implement a flow traction control method for the forwarding behavior of SDN switch to allow multi-path forwarding. And finally, we realize multi-path load balancing and traffic distribution optimization in hybrid SDN by changing the separation ratio of SDN switch to traffic.

On the basis of multi-path load balancing, we further investigate how to improve the controllability of the whole network traffic, so that the hybrid networks can approach or even achieve the centralized control ability of the pure SDN networks. In hybrid SDNs, the traffic that flows through at least one SDN switch during the forwarding process is considered as controllable traffic. Therefore, we also implement a multi-path load balancing algorithm with mandatory traffic (as is shown in Algorithm 1). We consider that traditional switches only need to forward all the traffic to the nearest SDN switch, and the controller is able to obtain all the information of the traffic passing through the SDN switch, which plays the key role of traffic monitoring and guidance.

Algorithm 1. Multi-path Load Balancing with Mandatory Traffic

---

Input:
   $G = (V, E)$, V_{SDN}, W, TM, $Budget$
Output:
   Maximum link utilization U
1: $Topology$ = Routing (W, G);
2: $Tradtopology$ = Dijkstra (W, G);
3: **for** each $TF_{s,d} \in TM$ **do**
4:    **if** $v_s \in V_{SDN}$ **then**
5:      continue;
6:    **else**
7:      Select the SDN switch v_i closest to v_s;
8:      $e_{i,j}$=shortest_link(v_i, v_s);
9:      Connect $e_{i,j}$ with the links between v_i and v_d in G;
10:     $TF_{s,d} = TF_{s,i} + TF_{i,d}$;
11:     Update TM;
12:    **end if**
13: **end for**
14: U=Multi-path Load Balancing(G, V_{SDN}, W, TM);
15: **return** U;

4.2 Fine-Grained Link State Management

In hybrid SDN networks, not all links participate in forwarding, and some links may only participate in a very small amount of forwarding. The SDN switch can query and define the running state directly by the controller, and can configure and manage the link directly connected with it by controlling the port. This link is defined as the potential optimized link. By implementing fine-grained management on the state of such potentially to-be-optimized links, the links connected with SDN switches can be opened and closed freely, or some idle or redundant links can be temporarily closed based on the dynamic demand of the service, so as to reduce the energy consumption and resource occupation of the links, ultimately reduce the network operation cost, and realize green and efficient SDN hybrid deployment.

Algorithm 2. Fine-grained Link State Management (Greedy-algorithm-based)

Input:
$G = (V, E)$, V_{SDN}, W, TM
Output:
  $Turn(e_{i,j})$, Energy saving rate
  $Energy\_saved$
1: Determine E_{SDN} according to V_{SDN};
2: Initialize Energy_saved $= 0$;
3: $validity, utilize(E_{SDN})$ = Multi-path
  Load Balancing(G, V_{SDN}, W, TM);
4: **while** $validity==1$ **do**
5:   List[] $idlelist$=null;
6:   **for** each $e_{i,j} \in E_{SDN}$ **do**
7:     $idlevalue(e_{i,j}) = utilize(e_{i,j})$-$\frac{1}{4}$
      ($Turn(e_{i,j})$-1);
8:     $idlelist$.add(idle($e_{i,j}$));
9:   **end for**
10:   **if** $idlelist$.length ==0 **then**

11:       break;
12:   **end if**
13:   sort($idlelist$);
14:   Select the biggest of
      $idle(e_{i,j}), e_{m,n}$;
15:   $Turn(e_{m,n}) = Turn(e_{m,n}) - 1$;
16:   $validity, utilize(E_{SDN})$ =
      Load_Banlancing(G, V_{SDN}, W, TM);
17:   **if** $validity==1$ **then**
18:       update($Energy\_saved$);
19:   **else**
20:       $Turn(e_{m,n}) = Turn(e_{m,n}) + 1$;
21:       $idlelist$.delete($e_{m,n}$) and goto
        line 12;
22:   **end if**
23: **end while**
24: **return** $Turn(e_{i,j})$, $Energy\_saved$;

However, the optimization and adjustment of the link state and the evaluation process of the adjustment result under the background of this problem have been proved to be the NP-hard problems. A single adjustment scheme cannot meet the complex and changeable service requirements in the hybrid deployment application scenario. Therefore, we propose a greedy-algorithm-based link state optimization and adjustment algorithm (as is shown in Algorithm 2) based on load level and restricted constraints. As a fast or online dynamic link state management method, it can be applied to application scenarios with high real-time requirements. In addition, we also propose a genetic-algorithm-based heuristic algorithm (as is shown in Algorithm 3) for link state optimization and adjustment. As a slow or offline link state management method, it can be applied to large scale application scenarios with multi-service hybrid deployment.

4.3   SDN Switch Deployment Optimization

Limited by the budget and cost, it is, in practice, unrealistic to deploy a large number of SDN switches in the actual network. Therefore, how to optimize the deployment of SDN switch, reduce the number of equipment deployment, improve the deployment efficiency, reduce the deployment cost, and realize energy efficiency from the perspective of deployment innovation and cost efficiency are the main problems to be solved here. We design and implement two

---

Algorithm 3. Fine-grained Link State Management (Genetic-algorithm-based)

Input:
$\quad$ $G = (V, E)$, V_{SDN}, W, TM
Output:
$\quad$ $Turn(e_{i,j})$, Energy saving rate
$\quad$ $Energy\_saved$
1: Determine E_{SDN} according to V_{SDN};

2: Initialize $Energy\_Saved$=0;
3: Randomly initialize $Turn(e_{i,j})$ of E_{SDN};
4: Iteration and form the next generation through selection, crossover and mutation;
5: For each generated sequence, perform Multi-path load balancing to determine whether it can be forwarded correctly in the link state, and if not, it requires a punitive reduction in fitness;
6: Calculate the benefit of each set of link state on network resources, and utilize it as fitness for next generation selection;
7: Update $Energy\_saved$ and the corresponding link state;
8: Repeat Line 3-7 until the set generation is reached;
9: **return** $Turn(e_{i,j})$, $Energy\_saved$;

SDN switch deployment optimization algorithms (one is based on the greedy algorithm and is shown in Algorithm 4, the other is based on the genetic algorithm and is shown in Algorithm 5), from the switch selection, switch deployment location, switch deployment sequence selection, and other aspects to achieve the optimization and innovation of hybrid deployment scheme.

5   Experimentation and Evaluation

In this section, we setup experiments to demonstrate the functionality effectiveness and performance gains of the proposed scheme and algorithms. The factors that affect the energy efficiency of hybrid SDN networks are very complex. Here, we mainly focus on the energy efficiency improvement brought by the SDN switch deployment optimization and fine-grained link state management in hybrid SDN networks.

Table 2 shows the topology configuration information and parameter settings of the hybrid SDN networks that are adopted for the experiments and simulations. In particular, Aarnet, Abvt, and Sprint are selected for experimental topologies from The Internet Topology Zoo, and Random is selected as a random deployment scheme. The experiments randomly generate 10 groups of traffic matrices for each topology, where the network utilization rate of 4 groups of traffic matrices is less than 25%, the utilization rate of 4 groups is between 25% and 50%, and the remaining 2 groups are higher than 50%.

5.1   Maximum Link Utilization

In this experiment, we choose Abvt as the network topology, investigate and compare the maximum link utilization of different algorithms under different

---

Algorithm 4. SDN Switch Deployment Optimization (Greedy-algorithm-based)

Input:
   $G = (V, E)$, V_{SDN}, W, TM, $Budget$
Output:
   V_{SDN}

1: **for** each $v_i \in V$ **do**
2:     $topology$ = Routing(W, G);
3:     Initialize $Cost(v_i)$;
4: **end for**
5: **for** each $TF_{s,d} \in TM$ **do**
6:     **for** each $v_i \in V$ **do**
7:       Initialize $In_j^{s,d}$ according to $topology$;
8:       $pass(v_i) += |In_j^{s,d}|$;
9:       $flow(v_i) += |In_j^{s,d}| \times TF_{s,d}$;
10:     **end for**
11: **end for**
12: **if** consider flow **then**
13:     sort($pass(v_i) \div Cost(v_i)$);
14:     Select the nodes V_{SDN}^1 with the largest ratio of traffic to cost, with the total price less than $Budget$;
15: **end if**
16: **if** consider pass **then**
17:     sort($flow(v_i) \div Cost(v_i)$);
18:     Select the nodes V_{SDN}^2 with the largest ratio of logarithm of traffic to cost, with the total price less than $Budget$;
19: **end if**
20: **if** consider degree **then**
21:     sort($degree(v_i) \div Cost(v_i)$);
22:     Select the nodes V_{SDN}^3 with the largest ratio of node degree to cost, with the total price less than $Budget$;
23: **end if**
24: $V_{SDN} = Energy\_saving(G = (V, E), V_{SDN}, W, TM)$;
25: **return** bestof($V_{SDN}^1, V_{SDN}^2, V_{SDN}^3$);

Table 2. Topologies used for the experiments and evaluations.

Topology	Number of nodes	Number of links
Aarnet	19	24
Abvt	23	31
Sprint	11	18
Random	9	13

given deployment budget constraints of the SDN switches, including the multi-path load-balancing algorithm, the SOTE algorithm [14], the multi-path load balancing algorithm combined with the deployment optimization, and the OSPF protocol. For a traditional switch in a hybrid SDN network, all nodes in the network are traditional OSPF devices, so we can use an OSPF environment to represent a pure traditional switch network.

The experimental results are shown in Fig. 1. Here, the abscissa represents the deployment budget of the SDN switch and the ordinate represents the ratio of maximum link utilization changes before and after the deployment of the corresponding optimization algorithm. When the SDN switches are not deployed, our scheme do not separate the traffic, so the maximum utilization rate of the link is very high. With the increase of SDN switch deployment budget, the maximum link utilization of the network begins to decrease. When the budget is increased to 50% or more, most of the links in the network are controlled by

---

Algorithm 5. SDN Switch Deployment Optimization (Genetic-algorithm-based)

Input:
$\quad G = (V, E),\ V_{SDN},\ W,\ TM,\ Budget$
Output:
$\quad V_{SDN}$
1: Initialize $Cost$, set generation $Times$, and population size $Length$;
2: int $m, n, x, y = 0$;
3: **for** $m < Length$ **do**
4:   **for** each $e_{i,j} \in E$ **do**
5:     $e_{i,j} =$ Random(0,100);
6:   **end for**
7:   Select several nodes V_{SDN} with the maximum weight and the sum of costs not exceeding the total budget;
8:   $Energy\_saved_m =$ $Energy\_saving(G =$ $(V, E), V_{SDN}, W, TM)$;
9:   Update($Best\_saved, V_{SDN}$);
10:   $m + +$;
11: **end for**

12: **for** $n < Times$ **do**
13:   Set $(Topsaved, Midsaved, Lowsaved)$ by sort($Energy\_saved_m$);
14:   **for** $x < Length$ **do**
15:     Randomly select the weight set, generate the next generation of weight set by crossover and mutation;
16:     Select the new set of V_{SDN} according to the weight;
17:     $Energy\_saved_x =$ $Energy\_saving(G =$ $(V, E), V_{SDN}, W, TM)$;
18:     Update($Best\_saved, V_{SDN}$);
19:     $x + +$;
20:   **end for**
21:   $n + +$;
22: **end for**
23: **return** V_{SDN};

---

the SDN switches, which can participate in forwarding flexibly. Experimental results show that when the ratio of SDN switches in the network is more than a half, the scheme proposed in this paper outperforms the SOTE algorithm.

Next, we force all the traffic to pass through the SDN switches and observe the changes of the maximum link utilization. Figure 2 shows the results of the experiments. Here, FRS [7] is a routing algorithm requiring every flow traverses an appropriate SDN switch. When there is no SDN switch in the network, all traffic is uncontrollable, so this part is omitted. When there is just a few SDN switches in hybrid SDN, all traffic will flow to the switches, causing extremely high link load, the ratio of maximum link utilization changes were even more than 100%. Due to the multi-path load balancing algorithm does not force traffic flowing through the SDN switches, so the maximum link utilization will be always lower. FRS will select an appropriate SDN switch as the transit node to control all traffic in the network, so when the deployment budget of the SDN switches is less than 50%, performance gains of our method is better than that of FRS. When most of the nodes in the network are SDN switches, the maximum link utilization of the multi-path load balancing algorithm with mandatory traffic is quite similar to that of the simple multi-path load balancing algorithm.

5.2   Link Energy Consumption

We compare the link energy consumption of the algorithms proposed in this paper with the HEATE algorithm [15]. According to the conversion relationship

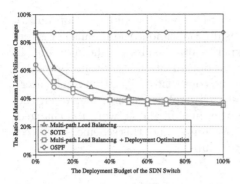
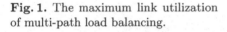
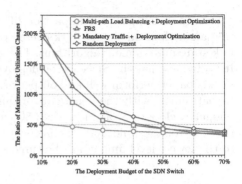

Fig. 1. The maximum link utilization of multi-path load balancing.

Fig. 2. The maximum link utilization of multi-path load balancing with mandatory traffic.

between link energy consumption and the maximum load set by the link, we use a function representing the energy consumption coefficient in Eq. 3, which is used to describe the percentage of current network energy consumption as a percentage of original energy consumption.

The results are shown in Fig. 3, and the proportion of the SDN switches in the network is shown in parentheses. We can see that HEATE performs well in the case of just a few SDN switches being deployed in the network, but this advantage is gradually lost with the increase of the number of the deployed SDN switches. Specifically, when the deployment budget reached 40%, the energy consumption of our scheme proposed in this paper will be reduced smoothly. When the SDN switches in the network are more than half, most links are controlled by SDN, and the network energy consumption will be maintained at a stable and relatively low level. When the proportion of the SDN switch exceeds 60% in hybrid SDNs, the performance gain of our approach, with regard to the link energy consumption savings, is close to or better than that of the HEATE algorithm.

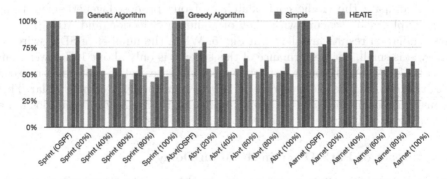

Fig. 3. Comparison of link state management algorithms on link energy consumption.

Next, we compare the impact of different SDN switch deployment optimization algorithms on the network energy consumption under the premise of given link-state management algorithms. The experimental results are shown in Fig. 4. It is obvious that the SDN switch deployment optimization algorithms proposed in this paper perform well and outperform the competitor algorithm in all cases. In particular, with the genetic-algorithm-based deployment scheme, when the budget is close to 30%, more than 50% of the links in the network are changed into controllable links, and the link energy consumption of this scheme is reduced to the lowest level among all the three algorithms.

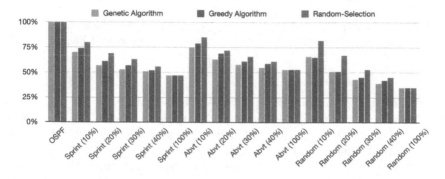

Fig. 4. Comparison of SDN switch deployment optimization on energy consumption.

5.3 Traffic Controllability

In this experiment, we investigate the impact of the SDN switch deployment schemes on the traffic controllability. Here, traffic controllability is measured by the controllable traffic proportion, which refers to the proportion of controllable traffic in the total traffic of traffic matrix. If the traffic between the source node and the destination node passes through at least one SDN switch, the traffic is considered as controllable traffic, otherwise, it is uncontrollable traffic.

We compare the traffic controllability of our proposed genetic-algorithm-based deployment scheme with that of Random Selection and MUcPF [16]. The simulation results are shown in Fig. 5. When the number of SDN switches deployed is small, the performance of this scheme is only about 10% lower than that of MUcPF. As the budget of deployment increases, when the budget exceeds 20%, the performance of our scheme and MUcPF gradually becomes similar. This means that our method can be applied to real-world application scenarios more accurately and achieve near ideal performance results.

5.4 Impact of Path Diversification on Delay

In this experiment, we test and investigate the impact of path diversification on delay, by calculating the ratio of the average path length of all traffic in

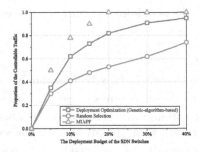

Fig. 5. The proportion of controllable network traffic.

the hybrid network (Hybrid) to the average path length of all traffic in the traditional network (Origin). The experimental results are shown in Table 3. The results indicate that multi-path forwarding in the hybrid network will increase the length of some links and increase the path delay by 15%–43%, depending on the topology and the number of the SDN switches deployed.

Table 3. The path extension ratio after multi-path forwarding for different topologies.

Topology	Hybrid/origin	Topology	Hybrid/origin
Abvt (20%)	1.25	Sprint (20%)	1.15
Abvt (40%)	1.34	Sprint (40%)	1.23
Abvt (60%)	1.42	Sprint (60%)	1.33
Abvt (100%)	1.43	Sprint (100%)	1.41

5.5 Running Time of Optimization Process

In this experiment, we test running time of deployment optimization process related to the scheme proposed in this paper. The experiment is conducted in the case where 20% of SDN switches are deployed in the topology of Abvt. The results are shown in Table 4.

The results further demonstrate, from the perspective of empirical research, that the genetic-algorithm-based approach needs to go through a large number of iterative processes, and the running time is relatively longer, so it is not suitable for the application scenarios with high real-time performance requirements. However, because it can avoid the local optimal solution and is easy to implement, it is suitable for offline and independent calculation before deployment, and can be used in scenarios with multi-service hybrid deployment, to achieve fine-grained link state management. On the contrary, the operation speed of the greedy-algorithm-based approach has almost an order of magnitude improvement, which means that it can be used as a fast or online solution for the application scenarios with high real-time requirements.

Table 4. Running time of different optimization processes in this paper.

Experimental content	Running time
Multi-path load balancing	0.13 s
Fine-grained link state management (greedy-algorithm-based)	12.16 s
Fine-grained link state management (greedy-algorithm-based)	1.33 s
SDN switch deployment optimization (genetic-algorithm-based) Fine-grained link state management (greedy-algorithm-based)	54 s
SDN switch deployment optimization (greedy-algorithm-based) Fine-grained link state management (greedy-algorithm-based)	3.71 s
SDN switch deployment optimization (genetic-algorithm-based) Multi-path load balancing with mandatory traffic	21.11 s

6   Conclusions and Future Work

In this paper, we investigate the issues of deployment optimization and fine-grained link state management in hybrid SDNs. In order to reduce the deployment cost and energy consumption, we introduce an energy-efficient deployment optimization scheme for hybrid SDNs, so that the hybrid networks can approach or even achieve the centralized control ability of the pure SDN networks. In order to further reduce link energy consumption, we then propose fine-grained link state management algorithms with energy consumption sensing ability. With the algorithms, the SDN link state can be controlled in a fine-grained way, and can be managed by dynamically clearing or temporarily closing redundant links. We setup experiments to demonstrate the functionality effectiveness and performance gains of the proposed scheme and algorithms, and the extensive simulation results show that, our proposed solution outperforms its competitors, producing an arresting increase of energy efficiency. Specifically, under the premise of only 20% of SDN switch deployment budget, the proposed scheme in this paper can reduce the total link energy consumption by about 40%.

The issues of deployment optimization and behavior management of hybrid SDNs are extremely crucial for building performance- and energy-efficient SDN-driven datacenters. Meanwhile, the factors that affect the energy efficiency of hybrid SDN networks are very complex. In future work, we plan to carry out our research in two directions: (i) establish a more accurate and practical hybrid SDN network optimization model to better depict the behavior of the network, and (ii) consider the experiments based on the actual application and real traffic in the real-world deployment scenarios of large-scale hybrid SDN networks.

Acknowledgements. This work was supported in part by the National Key Research and Development Plan of China under Grant 2019YFB2012803, in part by the Key Project of Shanghai Science and Technology Innovation Action Plan under Grant 19DZ1100400 and Grant 18511103302, in part by the Key Program of Shanghai Artificial Intelligence Innovation Development Plan under Grant 2018-RGZN-02060, and in part by the Key Project of the "Intelligence plus" Advanced Research Fund of East China Normal University.

References

1. Huang, X., Cheng, S., Cao, K., Cong, P., Wei, T., Hu, S.: A survey of deployment solutions and optimization strategies for hybrid SDN networks. IEEE Commun. Surv. Tutor. **21**(2), 1483–1507 (2019)
2. Casado, M., Koponen, T., Shenker, S., et al.: Fabric: a retrospective on evolving SDN. In: Proceedings of the First Workshop on Hot Topics in Software Defined Networks, Helsinki, Finland (2012)
3. Caria, M., Das, T., Jukan, A.: Divide and conquer: partitioning OSPF networks with SDN. In Proceedings of the International Conference on Integrated Network Management (IM 2015), Ottawa, ON, Canada (2015)
4. Caria, M., Jukan, A.: The perfect match: optical bypass and SDN partitioning. In: Proceedings of HPSR, Turin, Italy (2015)
5. Agarwal, S., Kodialam, M., Lakshman, T.: Traffic engineering in software defined networks. In: Proceedings of INFOCOM, Turin, Italy (2013)
6. Guo, Y., Wang, Z., Yin, X., et al.: Traffic engineering in SDN/OSPF hybrid network. In: Proceedings of ICNP, Raleigh, NC, USA (2014)
7. Ren, C., Wang, S., Ren, J., et al.: Enhancing traffic engineering performance and flow manageability in hybrid SDN. In: Proceedings of GLOBECOM, Washington, DC, USA (2016)
8. Das, T., Caria, M., Jukan, A., et al.: A techno-economic analysis of network migration to software-defined networking. preprint (2013)
9. Das, T., Caria, M., Jukan, A., et al.: Insights on SDN migration trajectory. In: Proceedings of ICC, Washington, DC, USA (2015)
10. Hu, Y., Wang, W., Gong, X., et al.: Maximizing network utilization in hybrid software-defined networks. In: Proceedings of GLOBECOM, CA, USA (2015)
11. Lukovszki, T., Rost, M., Schmid, S.: It's a match!: near-optimal and incremental middlebox deployment. ACM SIGCOMM Comput. Commun. Rev. **46**(1), 30–36 (2016)
12. Jia, X., Jiang, Y., Guo, Z.: Incremental switch deployment for hybrid software-defined networks. In: Proceedings of LCN, Dubai, United Arab Emirates (2016)
13. Wang, H., Li, Y., Jin, D., et al.: Saving energy in partially deployed software defined networks. IEEE Trans. Comput. **65**(5), 1578–1592 (2016)
14. Nascimento, M.R., Rothenberg, C.E., Salvador, M.R., Corrêa, C.N., de Lucena, S.C., Magalhães, M.F.: Virtual routers as a service: the routeflow approach leveraging software-defined networks. In: Proceedings of the 6th International Conference on Future Internet Technologies, pp. 34–37 (2011)
15. Wei, Y., Zhang, X., Xie, L., et al.: Energy-aware traffic engineering in hybrid SDN/IP backbone networks. J. Commun. Netw. **18**(4), 559–66 (2016)
16. Kar, B., Wu, H.K., Lin, Y.D.: The budgeted maximum coverage problem in partially deployed software defined networks. IEEE Trans. Netw. Serv. Manag. **13**(3), 394–406 (2017)

Correction to: Web Information Systems Engineering – WISE 2020

Zhisheng Huang, Wouter Beek⬤, Hua Wang⬤, Rui Zhou⬤,
and Yanchun Zhang⬤

Correction to:
Z. Huang et al. (Eds.): *Web Information Systems*
***Engineering – WISE 2020*, LNCS 12342,**
https://doi.org/10.1007/978-3-030-62005-9

In Chapter 12, the author's name was corrected to Deepak P.

In Chapter 24, a co-author listed on the Consent to Publish form was inadvertently forgotten. This mistake has been corrected and the forgotten co-author has been added.

The updated version of these chapters can be found at
https://doi.org/10.1007/978-3-030-62005-9_12
https://doi.org/10.1007/978-3-030-62005-9_24

Author Index

Printed in the United States
by Baker & Taylor Publisher Services